Lecture Notes in Mathematics

Volume 2388

This series reports on new developments in all areas of mathematics and their applications - quickly, informally and at a high level. Mathematical texts analysing new developments in modelling and numerical simulation are welcome. The type of material considered for publication includes:

1. Research monographs
2. Lectures on a new field or presentations of a new angle in a classical field
3. Summer schools and intensive courses on topics of current research.

Texts which are out of print but still in demand may also be considered if they fall within these categories. The timeliness of a manuscript is sometimes more important than its form, which may be preliminary or tentative. Please visit the LNM Editorial Policy (https://drive.google.com/file/d/1MOg4TbwOSokRnFJ3Z R3ciEeKs9hOnNX_/view?usp=sharing).

Titles from this series are indexed by Scopus, Web of Science, Mathematical Reviews, and zbMATH.

Mustapha Mokhtar-Kharroubi

Peripheral Spectra of Perturbed Positive Semigroups

Applications to Transport Theory and Related Fields

 Springer

Mustapha Mokhtar-Kharroubi
Laboratoire de Mathématiques de Besançon
Université Marie et Louis Pasteur
Besançon, France

ISSN 0075-8434   ISSN 1617-9692 (electronic)
Lecture Notes in Mathematics
ISBN 978-3-032-11172-2   ISBN 978-3-032-11173-9 (eBook)
https://doi.org/10.1007/978-3-032-11173-9

Mathematics Subject Classification: 47A75, 47D06, 47G20, 47A55

This Springer imprint is published by the registered company Springer Nature Switzerland AG
The registered company address is: Gewerbestrasse 11, 6330 Cham, Switzerland

*La science va sans cesse se raturant
elle-même. Ratures fécondes.*

—Victor Hugo

*It should not be forgotten that the foundations
of statistical mechanics originated by
Boltzmann and Gibbs abound with
asymptotic problems of great significance
and great difficulty.*

—Kurt Otto Friedrichs

*Comme il faut une différence des
températures des sources pour une machine,
ainsi une différence d'ordre-désordre pour le
travail de l'esprit. Tout ordre ou tout
désordre et rien ne va.*

—Paul Valéry

To Latifa

Preface

This monograph focuses, in particular though not exclusively, on the spectral theory of integro-differential equations arising in neutron transport theory. This field dates back to the 1950s with the advent of nuclear programs around the world, although its roots run deeper, particularly in astrophysics, through the study of radiative transfer, and in the kinetic theory of gases. The operators involved are notoriously non-self-adjoint, and their spectral analysis, especially in physical L^1-spaces, is significantly more complex than that of classical self-adjoint or skew-adjoint equations in mathematical physics. Fortunately, their connection to positive operators (i.e., operators that leave the cone of positive functions invariant) allows for a fruitful approach to the analysis of their peripheral spectrum, the only part relevant for addressing long-time asymptotic behavior, a key issue in neutron transport theory. Spectral problems in this field can be traced back to a seminal 1955 paper by J. Lehner and M. G. Wing, in which the analysis of a highly simplified neutron transport model in slab geometry revealed unexpected complexity and laid the groundwork for subsequent studies of more general models and theoretical developments. Owing to the contributions of numerous physicists, engineers, and mathematicians, spectral problems in neutron transport theory, and in its various extensions and offshoots, for instance in population dynamics, have evolved over the years into a rich and well-defined mathematical field. Essentially concerned with jump perturbations of flows, this field, which also has a probabilistic counterpart, lies at the intersection of the spectral theory of positive semigroups, first-order differential equations, and integral equations and is notably driven by concrete problems arising in kinetic theory and related areas. Despite the significant progress achieved in various directions since the 1960s, neutron transport theory remains fertile with complex spectral problems of great interest that are either unexplored or only partially understood. Drawing on powerful tools developed over the past two decades, this monograph examines many of these problems, both within transport theory and in related areas. It also brings to light a number of open questions, thereby offering new perspectives for future research.

The way we view those who came before us is likely both subjective and selective. For my part, I hold a special thought for S. Ukaï, whose work left a lasting mark on all branches of kinetic theory. Following his passing in 2012, he was widely recognized for his decisive contributions to nonlinear kinetic theory and fluid mechanics, beginning in the 1970s. Yet his earlier work, dating back to the 1960s, was primarily devoted to the spectral theory of neutron transport, a body of work that profoundly influenced subsequent literature. I have also a thought for I. Vidav, I. Marek, and J. Voigt, who contributed remarkably to establishing the spectral properties of positive operators as a central theme in transport theory.

The author shares with L. Arlotti and J. Banasiak a functional-analytic perspective on transport theory. Their 2006 book has been a constant source of inspiration, and the present monograph may be viewed as its continuation, particularly in the direction of spectral analysis.

As we all know, research is never a solitary endeavor, and this work has certainly been shaped by many enriching exchanges with colleagues over the years. I am particularly grateful L. Arlotti, J. Banasiak, J. Glück, F. A. Khodja, M. Pierre, A. Rhandi, E. Ricard, R. Rudnicki, D. Seifert, and M. Tyran-Kamińska. A special thank you goes to B. Lods, my long-time accomplice, to whom this monograph owes so much. Thanks also to J. Banasiak (again!), J. Goldstein, R. Nagel, and B. Perthame for their encouragement regarding this book. Finally, my warm thanks go to the reviewers for their insightful comments, which greatly helped shape the final version, and to the editorial team for giving this monograph a home in their collection.

Besançon, France
October 2025

M. Mokhtar-Kharroubi

Competing Interests The author has no competing interests to declare that are relevant to the content of this manuscript.

Contents

Chapter 1
Introduction

1.1 History of the Subject and Perspective

Spectral problems for Neutron Transport trace back to the 1950s, within the framework of Nuclear Reactor Theory. The operators involved are first-order differential with respect to the spatial variable and integral with respect to the velocity variable. This combination, along with their non-normal nature, makes their analysis particularly intricate. Fortunately, the connection between neutron transport operators and the theory of positive operators opened the way for analyzing their peripheral spectra and for gaining insight into the dynamics governed by such operators.

Linear Transport Theory has reached a high level of maturity, as evidenced by the publication of several monographs over the years [33, 79, 144, 178, 245] . In particular, topics such as spectral theory, scattering theory and inverse problems in Neutron Transport are addressed in [245] while more recent developments, up to 2013, were presented during a CIMPA School held in Muizenberg (Cape Town) [268]. This field remains fertile with open problems and new perspectives, many of which are explored in the present monograph. Building on powerful tools developed over the past two decades, this work offers new functional-analytic insights into the peripheral spectrum, time asymptotics, and scattering theory of perturbed positive C_0-semigroups on L^1-spaces. These developments find applications in a variety of areas, including Neutron Transport Theory, Population Dynamics, Kolmogorov differential equations, weighted graphs, linearized nonlocal Allen-Cahn equations and perturbed convolution semigroups. Despite the variety of fields considered, this monograph maintains a strong unity, grounded in shared underlying formal structure and the common mathematical tools, chiefly the use of weak compactness arguments and the spectral theory of positive operators. Notably, many of the phenomena arising in Neutron Transport Theory and Population Dynamics are modeled by semiflows with jumps, leading to a broad class of linear integro-differential equations with

---

The original version of the chapter has been revised. A correction to this chapter can be found at
https://doi.org/10.1007/978-3-032-11173-9_15

M. Mokhtar-Kharroubi, *Peripheral Spectra of Perturbed Positive Semigroups*, Lecture Notes in Mathematics 2388, https://doi.org/10.1007/978-3-032-11173-9_1

various types of boundary conditions. These models, presented in the next section, serve as the primary motivation for the unified approach developed throughout the monograph. In fact, this variety reflects the different types of collision (or jump) phenomena modeled across these fields. The focus of the monograph is primarily on the well-posedness of the resulting equations in suitable L^1-spaces and on their long-time behavior, analyzed through the peripheral spectrum of the associated operators. The choice of L^1-spaces is naturally dictated by the physical context of the problems and their underlying probabilistic structure. The Cauchy problems considered in this monograph are fundamentally of perturbative nature and can be abstractly formulated as

$$\frac{df}{dt} = Tf + Bf, \quad f(0) = f_0$$

where T is a generator of a positive C_0-semigroup on $L^1(\nu)$ and

$$B : D(T) \subset L^1(\nu) \to L^1(\nu)$$

is a positive (possibly) unbounded perturbation. Due to the potential unboundedness of the perturbation in physically relevant problems, the well-posedness of such Cauchy problems is far from elementary and requires specialized tools specific to L^1-spaces. Indeed, two generation theorems play a central role in this monograph: the Kato–Voigt perturbation theorem for substochastic (i.e., positive and contractive) C_0-semigroups, and the Desch perturbation theorem for positive C_0-semigroups, which notably does not rely on a dissipativity condition. With respect to the time asymptotic problems addressed in this monograph, these two generation theorems apply to different settings.

The idea for this monograph gradually took shape two years ago, when the author envisioned extending the (abstract) peripheral L^1-spectral results for neutron transport semigroups, originally developed for bounded collision operators [251, 266, 281], to the more challenging case of unbounded ones. This endeavor also aimed to build upon and generalize classical works [146, 343, 344, 368, 375], with the specific goal of addressing general space-inhomogeneous linear Boltzmann equations with power-like hard sphere potentials. This project proved to be feasible following the extension of the concept of regular collision operators, originally introduced in [207] for bounded operators, and a deeper understanding of space-homogeneous linear Boltzmann equations in various weighted L^1-spaces [211]. In the course of this work, several classical results from neutron spectral theory were also revisited and significantly improved. Various known results in the kinetic theory of particle swarms in weakly ionized gases [18, 19, 28, 29, 127–129, 308] are also revisited and improved in several respects. This endeavor also provided the opportunity to address additional spectral problems. Among these, particular attention is given to more complex collision operators than the standard inelastic ones, originally introduced in the 1970s by Larsen and Zweifel [192] and later revisited in [324]. These involve additional "downshift" and "Bragg" (elastic) collision operators, analogous to those encountered in semiconductor theory [22, 219, 220]. Another important class of new spectral problems arises in the context of neutron transport equations with mass-preserving boundary operators, connecting incoming and outgoing fluxes, in multidimensional spatial domains; these problems, inspired to some extent by the kinetic theory of gases, are substantially more complex than the classical ones with vacuum boundary conditions. Diffusive neutron equations, of significant interest in nuclear reactor calculations where the advection operator is replaced by a diffusion operator, constitute yet another class of novel problems. Curiously, aside from some multigroup formulations, spectral problems associated with diffusive neutron equations have never been thoroughly investigated in full generality. Yet, while their analysis is by no means elementary, it is considerably less complex than that of the standard neutron transport equations. Classical works from the 1960s on spatially homogeneous and isotropic kinetic models for neutron thermalization in L^2-spaces [92, 289, 333, 351, 352] are also revisited here, but now within the framework of L^1-spaces, as jump semigroups on abstract spaces $L^1(\Omega, \nu)$. Specializing to the case $\Omega = \mathbb{N}$ with a (weighted) counting measure or more generally to any countable measure space (Ω, ν) enables us to deal with the peripheral spectral theory for Kolmogorov differential equations and weighted graphs. Furthermore, the presence of a detailed balance principle motivates the construction of a general axiomatic Hilbertian form-perturbation theory; this development goes beyond the traditional needs of nuclear reactor theory and is expected to be of independent mathematical interest, with potential applications extending beyond the scope of Kinetic Theory. On the other hand, the literature on Population Dynamics (see e.g. [14, 15, 51, 68, 218]) features many kinetic-type models that bear strong mathematical similarities to those in Transport Theory. It is therefore quite natural that certain recent developments in Population Dynamics, particularly in the study of growth-fragmentation equations, have been integrated with those from Neutron Transport. This convergence of ideas forms a central core from which the present monograph originated. Besides Transport Theory, two functional analysis areas have undergone significant development over the past fifteen years. The first concerns the theory of honesty for perturbed substochastic C_0-semigroups in L^1-spaces [261], which traces back to a classical work by Kato in the 1950s on Kolmogorov differential equations [180]. This theory was revisited in the early 2000s by different authors, particularly to address problems arising in Transport Theory and Population Dynamics [33]. It is worth noting that this theme also admits a non-commutative counterpart in the Banach space of trace-class operators on a Hilbert space, which is relevant in the study of quantum dynamical semigroups; this direction dates back to the 1970s with the pioneering work of Davies [96] and has inspired more recent developments [25, 258, 379] that are not considered in the present monograph. The second theme concerns the use of Desch's perturbation theorem in L^1-spaces [106, 371], particularly through the application of weak compactness arguments [260]. It also includes the spectral analysis of Schrö dinger-like semigroups (or, more generally, absorption semigroups) in L^1-spaces with Kato-class potentials, using local weak compactness arguments [271]; these functional analytic developments are likewise revisited and significantly improved in this monograph. These developments are further complemented by results on partially integral stochastic C_0-semigroups [135, 319] which are particularly relevant when dealing with semigroups arising from semiflows with jumps; such semigroups may not be integral but can still exhibit partial integrality. The fruitful interplay between developments in Functional Analysis and Transport Theory, together with several related fields, forms the foundation of this multifaceted monograph which presents a synthetic treatment of diverse areas.

This monograph is not centered on a single objective or the resolution of specific problems. Rather, it is motivated by a collection of theoretical questions and ideas arising from various fields, which we seek to organize into a coherent and productive framework. A much more detailed description of the content of this monograph is provided in the next section which offers an overview of the various integro-differential equations arising in Transport Theory and Population Dynamics that motivate the study, situating them within the broader context of both classical and contemporary literature. It also outlines the central mathematical ideas that connect the different areas addressed throughout the monograph. In addition to setting the stage, it serves as a guide to navigating this multifaceted work, even though the individual chapters each have their own introductions and are largely self-contained. Each chapter focuses on a specific class of problems. Numerous open questions, of varying scope and significance, are scattered throughout the monograph, particularly in the "Comments" sections at the end of the chapters, with the aim of encouraging further research. While some known results are recalled without proofs (but with appropriate references), the core of the monograph consists of either substantial improvements of existing results or entirely new contributions. In addition, some recent developments from the literature are also included for completeness and context. This monograph is primarily intended for applied mathematicians working in Kinetic Theory, Probability Theory, Mathematical Biology, and more broadly from PDE with an interest in peripheral spectral theory. It is also designed to attract pure mathematicians from Operator Semigroup Theory towards very rich and inspiring applied contexts. While the monograph is chiefly aimed at professional researchers, it may also serve as a valuable resource for Ph.D. students and postdoctoral researchers working in the aforementioned fields. The reader is expected to be familiar with basic concepts from semigroup theory and spectral theory, which are covered in many standard references, for instance, Chap. 9 of [46] provides a smooth and pedagogical introduction. A working knowledge of the fundamental properties of positive operators and positive C_0-semigroups is also assumed; these are briefly reviewed in Sect. 2.3 of Chap. 2. This chapter serves as a Mathematical Toolbox, where a number of more specialized results, used repeatedly throughout the monograph, are collected. Apart from Sect. 2.3, the Toolbox can largely be skipped on a first reading and is intended primarily for reference when specific results are invoked later in the text. To a large extent, the individual chapters are self-contained, allowing the monograph to be approached flexibly according to the reader's interests.

1.2 What the Monograph Is About?

In this section, we provide a more precise overview of the structure and content of the monograph. The theoretical topics explored are strongly motivated by Transport Theory and several related areas, forming an interconnected whole. To appropriately introduce and contextualize these topics, a brief detour through these applied areas is therefore essential. Transport Theory is a multifaceted field at the intersection of Physics, Engineering, and Mathematics. It is concerned with the mathematical description of the diffusion of a very large number of "particles"[1] through a given host medium. Since tracking each individual particle is neither feasible nor useful, a statistical description is adopted. This is typically formulated in terms of a particle density function $f(x, \zeta, t)$ representing, at time t, the average number of particles located at position x with velocity ζ.[2] One of the central problems in Transport Theory is the understanding of the long-time asymptotic behavior of the particle density. Two emblematic examples of this theory are the *diffusion of neutrons* through the uranium fuel elements of a nuclear reactor (or in neutron *scattering experiments*), and the *diffusion of light photons* in planetary and stellar atmospheres. In such cases, the governing equations are naturally linear due to the negligible interactions between particles; for instance, neutron-neutron collisions are rare events, as the ratio of neutrons to atoms in the host medium is extremely small, on the order of 10^{-11}. In contrast, the motion of rarefied gas molecules, streaming and interacting through binary collisions, leads to the well-known nonlinear Boltzmann equation for dilute gases. Nonlinearity also arises in various collisionless contexts involving self-consistent fields.[3] The early development of transport theory was strongly influenced by astrophysical studies of radiative energy transfer in stellar and planetary atmospheres, as well as by the modeling of neutron and gamma transport in nuclear systems. For a more detailed exposition of the relevant physical problems in Transport Theory, we refer the reader to the monograph by Duderstadt and Martin [114].[4]

There also exists a rich body of work in Population Dynamics that features kinetic-type models, some of which will be presented below. It is worth noting that many of the phenomena studied in both Transport Theory and Population Dynamics are naturally modeled by *semiflows with jumps*, and therefore, a priori fall within the probabilistic framework of piecewise deterministic processes [95, 163, 320].

---

[1] The term "particles" is used generically and may refer to various physical entities such as neutrons, photons, molecules, ions, or even stars and galaxies (!), depending on the context.

[2] This statistical perspective emerged in the context of the kinetic theory of gases in the late 19th century, most notably through the work of Boltzmann [81].

[3] In addition to Plasma Dynamics, Vlasov–Poisson systems are widely used in Stellar Dynamics to describe the distribution of stars in a galaxy or even the distribution of galaxies in the universe. See, for instance, the course [112] and the references therein. The present monograph does not delve into the nonlinear kinetic theory literature; we simply refer the reader to the survey by Villani [360] for a comprehensive overview of the Boltzmann equation and related topics.

[4] Neutron transport theory evolved somewhat independently from other areas of Kinetic Theory due to the highly specialized nature of neutron-related problems in nuclear systems and the intense focus this field received after World War II within atomic energy programs.

This monograph is concerned exclusively with linear kinetic-type equations. Besides linearity, these equations share a fundamental feature: *positivity* lies at the core of their analysis, i.e., solutions remain positive whenever the input data are positive, a property that is not only natural from the modeling perspective but also central to the mathematical theory. The importance of positivity in neutron transport theory was recognized very early by Garrett Birkhoff, one of the pioneers of the theory of ordered vector spaces and positive operators [58–62].[5] Since its origins in the early 20th century within the context of positive matrices, the theory of Banach lattices and positive operators has developed significantly, both as a mathematical discipline in its own right and under the influence of various applied fields. Among the standard modern references on the subject are the monographs by Schaefer [327] and by Aliprantis and Burkinshaw [3]. It has long been recognized that positive operators possess distinctive peripheral spectral properties, which are collectively referred to as Perron–Frobenius theory (see e.g. [1, 110, 133, 137, 179, 323, 326]). These properties are particularly well-suited to the study of long-time asymptotics of positive C_0-semigroups (see e.g. [13, 46, 90, 223, 286, 288]). The field of linear Transport Theory gives rise to a wide class of non-normal operators, whose spectral analysis, at least for the simplest physical models in slab geometry, can be traced back to the 1950s, with the seminal work of Lehner and Wing [204].[6] The peripheral spectral analysis of such kinetic operators relies in a deep and nontrivial way on the theory of positive operators on Banach lattices.[7] As noted more than sixty years ago by Wigner [377], linear transport theory is a mathematically rich and intricate domain. Despite the publication of several monographs [33, 79, 144, 178, 245], many aspects of the theory remain far from fully understood.

It should be emphasized that the aim of this monograph is not to provide a comprehensive survey of the existing literature on Transport Theory or Operator Semigroups. Rather, its objective is to present a broad and coherent set of new functional analytic results that offer fresh perspectives across several fields of applied interest.

---

[5] Garrett Birkhoff concluded his article [61] with the following remark:
I think the problem of constructing rigorous mathematical theories of criticality, in neutron chain reactors, will supply useful and interesting problems to mathematicians, for many years to come ! I hope reactor physicists will regard the technical solution of such problems with due respect; in recent years, the phrase "by physical intuition" has been too freely used as a substitute for "I do not know why"!.

[6] This paper, which addressed the simplest Neutron Transport model in slab geometry, marked the beginning of spectral theory in Neutron Transport. The authors attempted to expand the solution over a complete set of eigenvectors, only to discover the unexpected presence of a half-space of essential spectrum alongside finitely many real eigenvalues. The commonly held assumption among physicists, that a complete eigenvector expansion should exist, was shown to be unfounded. It became clear early on that one could not expect more than a partial spectral understanding, centered on peripheral spectra and their implications for long-time behavior of solutions [174, 359].

[7] Neutron transport theory has benefited significantly from the abstract theory of positive operators; conversely, this applied field has inspired numerous developments in the theory of positive semigroups. This exemplifies a well-known phenomenon of cross-fertilization between Mathematical Physics and pure Mathematics (see, e.g., [202]), although one must remain cautious when evaluating the actual impact of mathematical progress on physical theory [332].

Because of the lack of positivity, *linearized* kinetic equations are not considered in this monograph. For a discussion of the physical and mathematical differences between linear and linearized Boltzmann equations, we refer the reader to Chap. 5 of [114] and Chap. IV of [80]. Before detailing the precise objectives of this monograph, it is useful to illustrate the variety of linear evolution equations that arise in Transport Theory. This diversity reflects the wide range of collision phenomena encountered in the field. In some cases, even differential (Landau-type) operators in the velocity variable may appear (see, e.g., [102, 103, 107]).

Non-autonomous linear kinetic equations have received relatively little attention in the literature. Aside from general existence and uniqueness results for linear (non-autonomous) initial and boundary value problems (see [144] Chap. XI, and [295]), we refer to [262] for the existence of time-periodic (resp. time-almost-periodic) solutions to linear transport equations with time-periodic (resp. time-almost-periodic) cross sections. Spectral results for perturbed propagators have also been studied in the context of linearized Vlasov-Poisson-Boltzmann systems [136]. In this monograph, however, we restrict attention to autonomous equations. We begin with the "simplest" such case

$$\frac{\partial f(t, \zeta)}{\partial t} + \left(\Sigma_a(\zeta) + \Sigma_s(\zeta)\right) f(t, \zeta) = \int_{\mathbb{R}^3} c(\zeta, \zeta') f(t, \zeta') d\zeta'$$

(ζ is a velocity) where $\Sigma_a(\zeta)$ is the absorption cross section and

$$\Sigma_s(\zeta) = \int_{\mathbb{R}^3} c(\zeta', \zeta) d\zeta'$$

is the scattering cross section. This equation describes the evolution of a space homogeneous density of "particles" $f(t, \zeta)$ at time $t \geq 0$ with velocity $\zeta \in \mathbb{R}^3$ [92, 289, 333, 351, 352]. This is a jump equation where the jump operator

$$\varphi \to C\varphi = \int_{\mathbb{R}^3} c(\zeta, \zeta') \varphi(\zeta') d\zeta',$$

called a collision (or scattering) operator in the context of Transport Theory, appears naturally as a perturbation of a multiplication operator. This is not a pure jump because of possible absorption cross sections Σ_a. Moreover, in the context of nuclear reactor theory, we cannot ignore the possibility of the occurence of a fission operator

$$\varphi \to F\varphi = \int_{\mathbb{R}^3} c_f(\zeta, \zeta') \varphi(\zeta') d\zeta',$$

typically a rank one operator [114], to be added to the collision operator C; this explains why positive L^1 semigroups arising in Transport Theory may be substochastic (i.e., contractive), stochastic (i.e., mass-preserving) or non contractive. Similar equations occur in the description of *non local dispersal* in Mathematical Biology, (see e.g. [93, 167]) and in the context of *Schrödinger* equations in momentum spaces (i.e., via Fourier transform) [351]. We mention also *Chapman-Kolmogorov* equations (linear jump equations in sequence space $l^1(\mathbb{N})$) for neutron population in a multiplying assembly [48, 49]; similar models also appear in Mathematical Biology [224].

In Kinetic Theory of *particle swarms* in a weakly ionized gas and in Semiconductor Theory, the transport of charged particles involves jump perturbations of flows induced by an external force

$$\frac{\partial f}{\partial t} + a(\zeta).\nabla_\zeta f + \Sigma(\zeta)f(t,\zeta) = \int_{\mathbb{R}^3} c(\zeta,\zeta')\Sigma(\zeta')f(t,\zeta')d\zeta' \qquad (1.2.1)$$

where $\int_{\mathbb{R}^3} c(\zeta',\zeta)d\zeta' = 1$ and $a : \mathbb{R}^3 \to \mathbb{R}^3$ is a smooth divergence free vector field [18, 19, 28, 29, 127–129, 308]; in this case, the collision operator appears as a natural perturbation of a first order differential operator.

In nuclear *reactor theory* [114, 190], neutron transport equations involve an additional space variable x living in a smooth open set $\Omega \subset \mathbb{R}^3$ and a transport operator $\zeta.\nabla_x$

$$\frac{\partial f(t,x,\zeta)}{\partial t} + \zeta.\nabla_x f(t,x,\zeta) + (\Sigma_a(x,\zeta) + \Sigma_s(x,\zeta)) f(t,x,\zeta) \quad (1.2.2)$$

$$= \int_{\mathbb{R}^3} c(x,\zeta,\zeta')f(t,x,\zeta')\nu(d\zeta') \quad (x,\zeta) \in \Omega \times \mathbb{R}^3$$

with

$$\Sigma_s(x,\zeta) = \int_{\mathbb{R}^3} c(x,\zeta',\zeta)\nu(d\zeta'),$$

where the velocity measure $\nu(d\zeta)$ is a Borel measure on $\mathbb{R}^3$ which can be absolutely continuous with respect to the volumic Lebesgue measure on $\mathbb{R}^3$ or with respect to surface Lebesgue measure on finitely many spheres about the origin (multigroup models) or even a combination of the two.[8] Note that the collision operator is an integral operator in velocity variable only and appears naturally as a perturbation of a first order operator in space variable only. Even if we do not consider them here, we mention the interest of more general vector fields

$$\zeta.\nabla_x + F(x,\zeta).\nabla_\zeta$$

considered also in the literature [17, 113, 122, 123, 152, 281, 381]. The natural boundary condition in nuclear reactor theory is

---

[8] We will see that some results for neutron transport hold for *arbitrary* Borel velocity measure $\nu(d\zeta)$ exceeding thus the physical examples. Moreover, the spectral analysis of *collisionless* neutron transport semigroups and *collisional* ones relies on *different* assumptions on $\nu(d\zeta)$. Thus, it is wiser not to make too precise assumptions on $\nu(d\zeta)$ early on. See the comment 20 in Sect. 10.18.

$$f_- := f_{|\Gamma_-} = 0,$$

where

$$\Gamma_- = \left\{(x, \zeta) \in \partial\Omega \times \mathbb{R}^3, \ \zeta.n(x) < 0\right\}$$

and $n(x)$ is the outward normal at $x \in \partial\Omega$; this is the so-called vacuum (or non-incoming) boundary condition. The mathematical literature on Neutron Transport dates back at least to the 1950s and is, of course, far too vast to be summarized here. It reflects the contributions of numerous physicists, engineers, and mathematicians over several decades, across a wide range of directions.[9] As in many other fields, asymptotic phenomena play a fundamental role in neutron transport theory, which explains the central importance of peripheral spectral analysis, (see Chap. 5 of [114]). A representative selection from the spectral literature, focusing on certain theoretical aspects, can be found in [245], with further developments up to 2013 presented in [268].

Much more complex boundary conditions, motivated by the kinetic theory of gases, are of the form

$$f_- = H(f_+)$$

where H is a (local in $x \in \partial\Omega$) linear jump *boundary* operator relating the outgoing flux f_+ and the incoming flux f_- of "particles" [80]. The most relevant case is a stochastic (i.e., mass-preserving) boundary operator H between suitable (boundary) functional spaces. Typically H is a convex combination of a deterministic (e.g. specular reflection) part and a diffuse (e.g. Maxwell-type) part; see [80] Chap. III for more information.

In Transport Theory, jump (or collision) operators possess the distinctive feature of being local with respect to the spatial variable. This spatial locality introduces significant mathematical challenges, which are, to some extent, balanced by the non-locality of these operators in the velocity variable. The spatial domain $\Omega \subset \mathbb{R}^3$ can be bounded, the whole space $\mathbb{R}^3$ (or the 3-dimensional torus) or, in some cases, an exterior domain.

We mention collisionless transport equations with stochastic boundary operators, which form an interesting class of models

$$\begin{cases} \frac{\partial f(t,x,\zeta)}{\partial t} + \zeta.\nabla_x f(t, x, \zeta) = 0 \\ f_- = H(f_+) \end{cases} \tag{1.2.3}$$

which have also their own interest in Kinetic Theory (see e.g. [212, 213, 273, 365] and references therein) and in the theory of stochastic billiards [91]. This model may be considered as a *boundary* perturbation of the simpler model with $H = 0$. Due to the complexity of the trace theory for transport equations [82, 83], a priori this class

---

[9] See, e.g., the massive bibliography in [335], on the historical development of the theory of radiative transfer and its connection to nuclear reactor theory.

of problems is not covered by G. Greiner's abstract boundary perturbation theory [148] nor by its later developments [67, 290].

In nuclear reactor theory, in addition to the prompt neutrons that are emitted instantaneously during a fission event, a fraction of neutrons, known as delayed neutrons, are emitted after a short time delay as decay products of certain radioactive fission fragments. These delayed neutrons introduce an additional source term in the standard neutron transport equation. Due to such delayed neutrons, the collision operator

$$C : \varphi \rightarrow \int_{\mathbb{R}^3} c(x, \zeta, \zeta')\varphi(x, \zeta')d\zeta' \tag{1.2.4}$$

is complemented by a fission operator

$$F : \varphi \rightarrow \int_{\mathbb{R}^3} c_f(x, \zeta, \zeta')\varphi(x, \zeta')d\zeta'$$

and a source term

$$S = \Sigma_{j=1}^{J}\lambda_j f_j$$

depending on $f(t, x, \zeta)$ through J equations

$$\frac{df_j(t, x)}{\partial t} + \lambda_j f_j(t, x) = \int_{\mathbb{R}^3} f(t, x, \zeta')d\zeta' \ (1 \le j \le J)$$

where $\lambda_j > 0 \ (1 \le j \le J)$ are the radioactive decay constants, (see e.g. [114] and Chap. 4 of [245]). We end up with a *system* kinetic model with a matrix collision operator.

Much more complex jump (collision) operators were introduced in [192]. Indeed, besides the usual inelastic collision operator (1.2.4), a *"downshift"* collision operator

$$K_d\varphi(x, \zeta) = \sum_{m=1}^{\overline{m}} \int_{\{|\zeta'|=\omega_n(|\zeta|)\}} k_d^{(m)}(x, \zeta')\varphi(x, \zeta')dS_{\omega_n(|\zeta|)}(\zeta') \tag{1.2.5}$$

occurs ($dS_{\omega_n(|\zeta|)}$ is the surface Lebesgue measure on the sphere with radius

$$\omega_n(|\zeta|) = \sqrt{|\zeta|^2 + \frac{2E_m}{N}}$$

and N is the mass of the neutron) with a *"Bragg"* (elastic) collision operator

$$K_b\varphi(x, \zeta) = \int_{\{|\zeta'|=|\zeta|\}} k_b(x, \zeta, \zeta')\varphi(x, \zeta')dS_{|\zeta|}(\zeta') \tag{1.2.6}$$

where $dS_{|\zeta|}(\zeta')$ the surface Lebesgue measure on the sphere with radius $|\zeta|$. These additional collision operators introduce a loss of compactness in the velocity variable, which gives rise to *"spectral curves"* in the essential spectrum, as observed in [192]. Similar types of jump operators also appear in semiconductor transport theory (see [22, 219, 220]), another vibrant area of Kinetic Theory, see Jüngel's monograph [175]. In such models, the collision operator takes the form

$$\widehat{C} := K_d + K_b + C.$$

This structure significantly complicates the spectral analysis. However, it proves useful to regard these models as neutron transport equations with collision operator $K_d + K_b$ perturbed by inelastic collision operator C, thereby allowing a more tractable functional analytic treatment.

In nuclear reactor theory, mainly for numerical calculation reasons, we find also the following neutron *diffusion* equation for the "flux" of neutrons where v is the velocity, $\omega = \frac{v}{|v|} \in S^2$ and $E = \frac{1}{2}m\,|v|^2$ is the kinetic energy of the neutron with mass m

$$\frac{d\varphi}{dt}(x, \omega, E) - |v|\,\mathrm{div}_x\,(D(x)\nabla_x\varphi(x, \omega, E)) + \Sigma(x, \omega, E)\varphi(t, x, \omega, E)$$

$$= \int_\alpha^\beta \int_{S^2} c(x, \omega, E, \omega', E')\varphi(t, x, \omega', E')dE'd\omega'$$

$$+ \int_\alpha^\beta \int_{S^2} c_f(x, \omega, E, \omega', E')\varphi(t, x, \omega', E')dE'd\omega',$$

where $\nabla_x\varphi$ is gradient with respect to the space variable $x \in \Omega \subset \mathbb{R}^3$, $D(x)$ is the matrix of diffusion, div_x is a divergence in space variable, $E \in (\alpha, \beta)$ with $0 < \alpha < \beta \leq +\infty$ and $d\omega'$ is the Lebesgue surface measure on S^{d-1}, (see [94, Chap. I, Part A, Sect. 5]); see also [150, 342] for multigroup versions. Thus the first order operator $\zeta.\nabla_x$ is replaced by an elliptic operator (in space variable) depending on the velocity variable. The collision operator, an integral operator in velocity, appears here as a perturbation of an elliptic operator in space variable. Several kinds of boundary conditions with respect to the space variable are possible e.g. Neumann, Dirichlet, Robin or periodic boundary condition. Actually, the Robin boundary condition

$$\varphi(t, x, \omega, E) + \lambda(x)\frac{\partial\varphi}{\partial\nu}(t, x, \omega, E) = 0 \ (x \in \partial\Omega)$$

seems to be the most relevant one where $\lambda(.) \geq 0$ is the so-called extrapolation length. Similar equations can also be derived as singular limits of the standard neutron transport equations via appropriate scalings of time and space, typically in terms of a small parameter representing the mean free path tending to zero. This asymptotic regime leads naturally to diffusion-type equations. A classical approach to such diffusion limits relies on Hilbert expansions, as developed in [42] and originally rooted in the methodology of [193]. A probabilistic counterpart, expressing the convergence of transport processes to diffusion processes, is presented in [297]. An important technical condition is that the collision frequency must remain bounded away from zero; otherwise, the limiting diffusion equation will exhibit more complex boundary behavior, as discussed in [43]. From a spectral point of view, the diffusion limit can be understood in a strikingly intuitive manner, especially for neutron transport equations posed on the torus. As shown in [246], diagonalization of the transport operator in Fourier space reveals that the point spectrum of the Laplacian gradually "emerges" from the spectrum of the transport operator as the mean free path vanishes, while the essential spectrum recedes to infinity; this gives a near-"visual" insight into the transition from kinetic to diffusive behavior. We also note that alternative scalings can lead to fractional diffusion limits, as explored in [228]. In a similar spirit, we mention the much deeper and more complex problems of rigorously deriving the Boltzmann equation from many-particle systems, as well as the transition from Kinetic Theory to Hydrodynamics [132, 321] directly related to the celebrated Sixth Hilbert Problem on the axiomatization of physical theories [143].

Although this monograph does not address it, we note the existence of a growing body of literature on transport theory on graphs: see e.g. [36] and the numerous works cited therein.

On the other hand, a wide range of linear and nonlinear mathematical models arises in the literature on Population Dynamics (see e.g. [15, 51, 218, 302, 320]). Many of these are linear integro-differential equations that exhibit strong mathematical similarities with linear kinetic equations; (see e.g. [14, 68, 195]). In this monograph, we focus on three such models that have attracted considerable attention in recent years. The first model concerns *diffusive* structured populations with generalized Wentzell–Robin boundary conditions

$$
\begin{cases}
\quad\quad u_t(s, t) + (\gamma(s)u(s, t))_s \\
= (d(s)u_s(s, t))_s - \mu(s)u(s, t) + \int_0^m \beta(s, y)u(y, t)dy \\
[(d(s)u_s(s, t))_s]_{s=0} - b_0 u_s(0, t) + c_0 u(0, t) = 0, \\
[(d(s)u_s(s, t))_s]_{s=m} + b_m u_s(m, t) + c_m u(m, t) = 0
\end{cases}
\tag{1.2.7}
$$

with $m < +\infty$ and $b_0 - \gamma(0) > 0,\quad b_m + \gamma(m) > 0$ (or with $m = +\infty$ where the last boundary condition is omitted). The (possibly) unbounded reproduction operator

$$
u \rightarrow \int_0^m \beta(., y)u(y)dy
$$

plays a role analogous to that of a collision operator in Neutron Transport. It naturally arises as a perturbation of an elliptic operator subject to generalized Wentzell–Robin boundary conditions. For a detailed discussion of the biological interpretation and modeling aspects of this framework, we refer to [124]. For further context, we also mention the existence of a related body of literature on the spectral analysis of growth equations involving nonlocal boundary conditions and spatial diffusion with respect to an additional spatial variable; see [66] and the references therein.

The second model is *growth-fragmentation* equations

$$\begin{cases} \frac{\partial}{\partial t}u(x,t) + \frac{\partial}{\partial x}\left(r(x)u(x,t)\right) + a(x)u(x,t) \\ = \int_x^{+\infty} a(y)b(x,y)u(y,t)dy, \quad (x,t>0) \end{cases} \qquad (1.2.8)$$

with a nonnegative fragmentation kernel $b(.,.)$ such that

$$\int_0^y xb(x,y)dx = y, \quad (y>0)$$

a positive continuous growth rate $r(.)$ such that

$$\int_0^\infty \frac{1}{r(\tau)}d\tau = +\infty$$

and with various boundary conditions. The fragmentation operator

$$\varphi \to \int_x^{+\infty} a(y)b(x,y)\varphi(y)dy$$

appears here as a perturbation of a first order operator. Such equations arise in the modeling of various physical and biological phenomena involving aggregates that undergo both growth and fragmentation. Typical biological applications include models of phytoplankton dynamics [16, 32], and prion proliferation [76, 121]. The growth-fragmentation equation also appears as the linear part of the more general growth-coagulation-fragmentation equation, in which the coagulation mechanism is described by an additional quadratic integral term; see [37]. For a broader overview of contexts where such equations naturally arise, ranging from biology to materials science, we refer to [131] and the references therein; see [39, 40, 54, 55, 126, 235, 280] for more recent developments and also [56] for the probabilistic context.

The third model concerns growth equations without fragmentation, subject to a *nonlocal* McKendrick-von Foerster-type boundary condition

$$\begin{cases} \frac{\partial}{\partial t}u(x,t) + \frac{\partial}{\partial x}\left(r(x)u(x,t)\right) + a(x)u(x,t) = 0 \\ \lim_{x\to 0_+}\left(r(x)u(x)\right) = \int_0^{+\infty}\beta(y)u(y)dy. \end{cases} \qquad (1.2.9)$$

This model bears strong structural resemblance to the collisionless transport equations (1.2.3), particularly in the presence of a nonlocal boundary operator, and can be interpreted as a perturbation of the underlying transport dynamics with $\beta = 0$. A notable special case of this framework is given by the age-dependent population equation, corresponding to the choice $r(x) = 1$, which is significantly simpler and has been extensively studied in the literature; see e.g. [147] or ([120, Chap. VI, p. 216]).

From a physical standpoint, the L^1-norm is very natural in Transport Theory and Population Dynamics, as it represents the mean number of "particles". The natural character of the L^1-framework also stems from the probabilistic foundation underlying both fields [95, 163, 320]. Similarly to elliptic operators [98], the L^1-theory carries a probabilistic flavor. This explains why L^1-theory plays a central role in this monograph. Fortunately, it often happens that equations well-posed in all L^p-spaces share the same peripheral spectral (gap) theory across all p, so that the L^1-analysis provides spectral results applicable to L^p-spaces as well. This is the case, for example, for certain L^2-operators defined by symmetric Dirichlet forms, which can be handled via the L^1-framework, and also for some non-symmetric operators such as neutron transport operators (see Chap. 14). Contrary to common belief, in many respects the L^1-treatment of these equations turns out to be simpler and more powerful than their L^2 counterparts. In particular, the weak compactness arguments frequently employed in this monograph are easier to apply than compactness techniques and allow the recovery of compactness results thanks to the Dunford-Pettis property of L^1-spaces.

We have excluded from our study linearized kinetic models because they lack positivity.[10] This exclusion naturally applies to the linearized Boltzmann equation, which arises in a highly specific nonlinear context [355] and must be analyzed accordingly; for a rich spectral literature on linearized Boltzmann equations, see [382] and the extensive references therein. On the other hand, some nonlinear equations, both within Kinetic Theory and beyond, lead to linearized kinetic-type equations that do possess full positivity properties. A notable example is found in nuclear reactor theory: to describe *fluctuations* beyond mean value theory, a probabilistic framework for neutron chain fissions is developed in terms of nonlinear transport equations with nonlocal polynomial nonlinearities [50, 171, 293, 298, 299, 356, 357] ([245, Chap. 10]) [163, 265]. Linearizing these nonlinear equations about the trivial solution recovers the usual neutron transport equations with dual boundary conditions

$$f_+ := f_{|\Gamma_+} = 0.$$

Another example stems from the search of traveling waves for nonlocal *Allen-Cahn* equations

$$u_t = J * u - u + f(u)$$

where $J * u$ is the convolution of u and $J \in C^1(\mathbb{R}) \cap L^1_+(\mathbb{R})$ with $\int_{\mathbb{R}} J(x)dx = 1$. The function f is smooth with three zeros ± 1 and $a \in (-1, 1)$ satisfying $f'(\pm 1) < 0$ and $f'(a) > 0$. A typical example is given by $f(u) = (u - a)(1 - u^2)$. There exist traveling wave solutions of the form $u(t, x) = \phi(x - ct)$ where

$$c = \frac{\int_{-1}^1 f(u)du}{\int_{-\infty}^\infty (\phi'(x))^2 \, dx},$$

---

[10] Actually, any bounded operator O in L^1 space admits a modulus (i.e. a minimal positive dominating operator) $|O|$ [86] which is weakly compact if and only if O is [3]. Therefore, by using domination arguments, some results can be obtained without relying explicitly on positivity; however, this monograph does not explore this aspect in depth.

ϕ is continuous, nondecreasing and

$$\begin{cases} c\phi' + J * \phi - \phi + f(\phi) = 0 \\ \phi(-\infty) = -1, \ \phi(+\infty) = 1. \end{cases} \tag{1.2.10}$$

The linearized operator of (1.2.10) about the traveling wave ϕ

$$L : \psi \in D(L) \to c\psi' + J * \psi - \psi + f'(\phi)\psi, \tag{1.2.11}$$

investigated in [44], shares many positivity properties with kinetic equations. This operator is a convolution perturbation of a first order operator.

The structure of all the above equations suggests naturally a mathematical treatment of *perturbative* type.[11] Their first common feature is the positivity of the C_0-semigroups that govern them; we will note below another key common feature.

In summary, this extensive sample of equations and the diversity of their mathematical properties provided the initial motivation for the unified and systematic treatment developed in this monograph. The approach is fundamentally perturbative in nature and has been made possible by powerful tools developed over the last two decades. Our objective, however, is primarily functional analytic and extends well beyond the concrete applications mentioned above. Indeed, we aim to delve as deeply as possible into the peripheral spectral properties and time asymptotics of various classes of abstract perturbed C_0-semigroups in L^1-spaces, which encompass many applications of practical interest, such as Transport Theory and related fields. From a mathematical standpoint, much of the construction relies on weak compactness techniques and spectral properties of positive operators. Moreover, for several C_0-semigroups that act consistently on all L^p-spaces, the peripheral spectral properties established in L^1 extend naturally to the entire scale of L^p-spaces, thus conferring a special status to the L^1-framework. Finally, alongside numerous developments in various directions, open questions of differing significance are scattered throughout this monograph, with the hope of inspiring further research and exploration.

This monograph primarily addresses two key issues related to perturbed Cauchy problems in abstract Lebesgue spaces $L^p(\nu)$ (mostly with $p = 1$) of the form

$$\frac{df}{dt} = Tf + Bf, \ \ f(0) = f_0$$

where T is a generator of a positive C_0-semigroup and

$$B : D(T) \to L^p(\nu)$$

---

[11] We point out that differential (Landau) collision operators are *not* amenable to the perturbative construction of this monograph; see e.g. [102] or ([144, p. 362]) for the Fokker-Planck model of electron scattering where the collision operator is a *Legendre* operator $-\frac{\partial}{\partial\mu}\left((1 - \mu^2)\frac{\partial}{\partial\mu}\right)$ and the "velocity" $\mu \in [-1, 1]$ is actually the direction cosine of propagation.

is a positive perturbation:

(1) Is $T + B$ (or some extension $G \supset T + B$) the generator of a positive C_0-semigroup $(S(t))_{t \geq 0}$? In general, well-posedness is established through perturbation arguments. For example, in Transport Theory, the perturbation typically takes the form of a collision operator. However, well-posedness in this setting is far from trivial, since the collision operator is often unbounded due to the unboundedness of the cross sections appearing in various kinetic equations. Similarly, in Population Dynamics, the perturbation usually corresponds to a reproduction operator.

(2) The second issue concerns the understanding of the time asymptotics ($t \to +\infty$) of the semigroup $(S(t))_{t \geq 0}$. We show that this problem encompasses a rich variety of scenarios. In particular, some cases depend on the peripheral spectral analysis of $(S(t))_{t \geq 0}$. For these, when properly scaled, $S(t)$ converges either exponentially in operator norm or just strongly, depending on the existence of a spectral gap, to a spectral projection as $t \to +\infty$.[12] Partly linked to point spectral theory, some stochastic (i.e., mass-preserving) models may exhibit a *sweeping* phenomenon, where the total mass escapes to infinity or concentrates near a null set as $t \to +\infty$. Finally, in contrast to these behaviors, another possible asymptotic regime arises, more closely related to scattering theory.

Semigroup theory, which already holds a central role in Markov process theory [6, 27, 63, 141, 288, 320], also provides the natural framework for spectral analysis. In this monograph, for a given perturbed stochastic (i.e., mass-preserving) C_0-semigroup, peripheral spectral analysis refers to either establishing the existence of a spectral gap or, at minimum, proving the existence of an invariant density and the asymptotic stability of the semigroup, that is, its strong convergence to its ergodic projection. Checking the existence of a spectral gap amounts to estimating the essential type of the semigroup. To accomplish this, a careful analysis of the remainders in the Dyson-Phillips expansion is required, particularly to verify that some of these remainder terms are weakly compact [245]. In more complex cases, such as partly elastic neutron transport operators, no remainder term is weakly compact, necessitating a double perturbation analysis to reach the desired conclusion. When a spectral gap is absent, weak compactness arguments remain crucial for proving the existence of an invariant density; asymptotic stability in this scenario is established via a 0–2 law for semigroups, which involves showing that certain Dyson-Phillips remainder terms depend continuously on time t in the operator norm topology [269]. This approach forms the core of our strategy in this monograph, supplemented by additional analytical tools.

In the beautiful book by Lasota and Mackey [194], the existence of an invariant density for a given stochastic operator P is established via an analysis of weak limits of Cesàro means of P using the concept of *lower bound functions* (see [194], Chap. 5). At present, it remains unclear whether this approach can be successfully adapted to

---

[12] There also exists a robust approach to proving strong convergence using relative entropy techniques, which have been applied even in nonlinear contexts [176, 216, 231, 304, 361, 366].

the neutron transport problems addressed in this monograph. Another noteworthy result is the *Foguel alternative* (also discussed in [194], Chap. 5), which provides a useful means for analyzing the occurrence of sweeping phenomena when invariant densities are absent. For more recent developments in this direction, we refer to the monograph [320]. This perspective has led to the identification of interesting sweeping behaviors in different contexts of Transport Theory [212, 273].

We emphasize that our discussion has focused primarily on the existence of spectral gaps for stochastic C_0-semigroups, or on asymptotic stability in the absence of such gaps. This is because these are the only two asymptotic regimes that arise in the various fields considered in this monograph. However, we note that more intricate asymptotic behaviors, such as time-asymptotic periodicity, can occur in other contexts. For instance, such phenomena arise in the setting of transport on networks; see [46, Chap. 18]. We also note that if a semigroup is partially integral, i.e., it dominates a kernel operator, then the peripheral point spectrum of its generator is either empty or reduced to the spectral bound [134]. As a consequence, asymptotic periodicity cannot occur in such settings. This observation explains why the various perturbed semigroups studied in this monograph, each admitting partially integral Dyson–Phillips expansions, do not exhibit asymptotic periodic behavior. Conversely, it is not surprising that asymptotic periodicity arises in the semigroups encountered in transport on networks, which typically consist in shift operators; such operators cannot be partially integral. In contrast, the semigroups studied in this monograph arise as jump perturbations of semiflows, this constitutes their second key common feature. It is precisely these jump mechanisms that underlie the partially integral structure of the corresponding Dyson–Phillips expansions.

In the case of asymptotic stability without spectral gaps, this monograph does not address the considerably more intricate issue of convergence rates. Nonetheless, we note that general results in this direction, based on quantified Tauberian approaches, are available in the context of Transport Theory [211, 213, 215, 274].

From the panoramic view provided by this introduction, well-posedness and time asymptotics represent only the visible part of a broader construction grounded in a rich body of functional analytic results that are of independent interest. We begin by investigating several classes of abstract perturbed C_0-semigroups in Lebesgue spaces, before turning to a range of more applied contexts, including Kolmogorov differential equations, weighted graphs, Transport Theory, structured population models, linearized nonlocal Allen–Cahn equations and perturbed convolution semigroups. Despite the diversity of fields considered, this monograph maintains a strategic unity through common underlying mathematical tools, primarily based on weak compactness methods and the spectral theory of positive operators. Naturally, the specific problems and technical details vary across areas, which justifies the full autonomy of the different chapters. We are confident, however, that the mathematical framework developed here can be adapted to a much broader class of problems sharing a similar structure.

We make no claim to exhaustiveness in the bibliography. Our aim has been to highlight references with significant relevance to the themes of this monograph; we apologize in advance for any inadvertent omissions. This monograph is organized as follows:

Chapter 2 serves as a *Mathematical Toolbox*, gathering a collection of both classical and more specialized results that are used repeatedly throughout this monograph. In particular, we recall important spectral properties of (perturbed) positive operators and positive C_0-semigroups on Banach lattices. While many of the results are well known and included without proofs (but carefully referenced), we also present several new results along with complete proofs. We provide variational characterizations of the leading eigenvalue, in terms of Max-inf and Inf-sup formulations, for a suitable class of positively perturbed operators with positive resolvent. Two key perturbation theorems in L^1-spaces are highlighted and play a foundational role in our analysis: Desch's perturbation theorem, which deals with positive C_0-semigroups (see below), and Kato-Voigt's perturbation theorem, concerning substochastic (i.e., positive contraction) C_0-semigroups (see below). We also present sufficient conditions involving weak compactness or quasi-compactness[13] under which Desch's theorem can be applied. Furthermore, we explain how these results extend to general Banach lattices when the unperturbed semigroup is holomorphic. Finally, we address the subtle and technically involved question of determining the domain of Kato-Voigt perturbed generators, a subject that has given rise to a rich *honesty* theory.

A C_0-semigroup $(V(t))_{t \geq 0}$ on a Banach space (with generator T) is said to have a *spectral gap* if

$$\omega_{ess}(V) < \omega(V) \tag{1.2.12}$$

where $\omega(V)$ (resp. $\omega_{ess}(V)$) is the type (resp. the essential type) of $(V(t))_{t \geq 0}$; in this case, $e^{-\omega(V)t} V(t)$ converges exponentially in operator norm ($t \to +\infty$) to a spectral projection of T. We point out that (1.2.12) implies that

$$\sigma(T) \cap \{\lambda \in \mathbb{C}; \ \mathrm{Re}\,\lambda > \omega_{ess}(V)\} \tag{1.2.13}$$

is a non empty set of isolated eigenvalues with finite algebraic multiplicities but, a priori, this consequence alone, which expresses that T has a "spectral gap", is not sufficient to deal with time asymptotics $\left(e^{-\omega(V)t} V(t)\right)_{t \geq 0}$ [359]. We also recall several important results concerning the stability of the essential type under perturbations of C_0-semigroups. These results are instrumental in the spectral analysis of perturbed semigroups, particularly when investigating the existence of spectral gaps. Finally, we include a discussion of more recent developments on the asymptotic stability of positive, *partially integral*, bounded C_0-semigroups in Lebesgue spaces; these results apply even in the absence of spectral gaps, and they provide a robust framework for understanding long-time behavior in a wide variety of applications.

Given the wide range of problems addressed in this monograph, it is not possible to delve into all the significant results in this general introduction. Instead, we provide only a broad overview of some key themes and refer the reader to the introductions of the individual chapters for more detailed information.

---

[13] Note that quasi-compactness has long played an important role in probability theory [74].

Let $(\Omega, \mathcal{E}, \nu)$ be a sigma-finite measure space and $L^1(\Omega, \mathcal{E}, \nu)$ with norm

$$\|\varphi\|_{L^1(\Omega,\mathcal{E},\ \nu)} = \int_\Omega |\varphi(x)|\, \nu(dx).$$

For the simplicity of notations we write simply $L^1(\nu)$ instead of $L^1(\Omega, \mathcal{E}, \nu)$ and denote by $L^1_+(\nu)$ the positive cone of $L^1(\nu)$. The following Kato-Voigt perturbation theorem plays an ubiquitous role in this monograph.

Theorem (Kato [180], Voigt [368]) Let $(U(t))_{t\geq 0}$ be a substochastic C_0-semigroup on $L^1(\nu)$ with generator T and let $B : D(B) \subset L^1(\nu) \to L^1(\nu)$ be positive (i.e., $B\varphi \in L^1_+(\nu)$ if $\varphi \in D(B) \cap L^1_+(\nu)$) and such that $D(B) \supset D(T)$ and

$$\int_\Omega (T\varphi + B\varphi)\, d\nu \leq 0 \ \ (\varphi \in D_+(T) := D(T) \cap L^1_+(\nu)). \tag{1.2.14}$$

Then there exists a unique extension $G \supset T + B : D(T) \to L^1(\nu)$ which generates a minimal substochastic C_0-semigroup $(V(t))_{t\geq 0}$ in $L^1(\nu)$.

In this monograph, we refer to such perturbed C_0-semigroups $(V(t))_{t\geq 0}$ (resp. their generators) as the Kato-Voigt semigroups (resp. Kato-Voigt generators). If the inequality (1.2.14) is an equality, then $(V(t))_{t\geq 0}$ is stochastic (i.e., mass-preserving) if and only if $G = \overline{T + B}$. In this case, we say that $(V(t))_{t\geq 0}$ is a honest stochastic Kato-Voigt semigroup. If $G = T + B$ then we say that $(V(t))_{t\geq 0}$ is genuinely honest.

Chapter 3 is devoted to time asymptotics of Kato-Voigt semigroups. We highlight and analyze three very different types of asymptotic behavior:

(1) Existence of *invariant density* with asymptotic stability
(2) *Sweeping* behavior (i.e., the total mass escapes to infinity or concentrates around a null-set as $t \to +\infty$)
(3) *Scattering theory,* i.e., existence of a bounded operator Ω^- such that

$$\left\| V(t)\varphi - U(t)\Omega^-\varphi \right\| \to 0 \ (t \to +\infty) \ (\varphi \in L^1(\nu)).$$

The analysis relies on whether $(U(t))_{t\geq 0}$ is strongly stable or not and, in the case that $(U(t))_{t\geq 0}$ is strongly stable, on whether $r(B(0 - T)^{-1})$ is an eigenvalue of $B(0 - T)^{-1}$ or not. These regimes will appear in more applied contexts in Chaps. 8 and 10.

To introduce the following chapter, we recall another key perturbation theorem which does not rely on dissipativity. Its significance lies in the fact that many naturally occurring non-contractive L^1-semigroups appear in Transport Theory and Population Dynamics. Desch's perturbation theorem, which addresses such cases, plays a ubiquitous role throughout this monograph:

Theorem (Desch [106]) Let $(U(t))_{t\geq 0}$ be a positive C_0-semigroup on $L^1(\nu)$ with generator T and let $B : D(T) \subset L^1(\nu) \to L^1(\nu)$ be positive. Then

$$\lim_{\lambda \to +\infty} r\left(B\left(\lambda - T\right)^{-1}\right) < 1$$

is a necessary and sufficient condition for

$$G := T + B : D(T) \subset L^1\left(\nu\right) \to L^1\left(\nu\right)$$

to generate a positive C_0-semigroup $(V(t))_{t \geq 0}$.

We refer to such C_0-semigroups $(V(t))_{t \geq 0}$ as the *Desch* semigroups.

Chapter 4 is devoted to the peripheral spectral analysis of Desch semigroups using weak compactness arguments. We demonstrate how Desch's generation theorem can be established under weak compactness (or quasi-compactness) conditions. Several scenarios where Desch semigroups exhibit a spectral gap are explored, including the spectral impact of absorption terms $a \in L_+^\infty\left(\nu\right)$ with small oscillations. We also examine the extension of this theory to Lebesgue spaces $L^p\left(\nu\right)$. Two distinct approaches to such L^p-extensions are presented: a predual construction and a more classical interpolation method. Similarly to Chap. 3, this chapter lays the groundwork for natural applications discussed later in the monograph.

Chapter 5 is devoted to spectral analysis of *absorption semigroups*. Let $(U(t))_{t \geq 0}$ be a positive C_0-semigroup on $L^p\left(\nu\right)$ with generator T and let

$$V : \Omega \to [-c, +\infty] (c > 0)$$

be measurable and finite a.e. A general theory [370] allows the construction of positive C_0-semigroups $(U_V(t))_{t \geq 0}$ generated by the formal operator

$$T_V = \text{``} T - V \text{''}$$

for singular (i.e., unbounded) potentials V; actually more general indefinite potentials V are considered in [370]. In general,

$$T_V \supset T - V.$$

The occurrence of absorption semigroups is now very common in Physics as well as in Geometry and Probability (see e.g. [27, 78, 105, 149, 285, 338]). We recall first several known results from [370], in particular a result peculiar to L^1-theory of absorption (substochastic) C_0-semigroups, namely the key contraction property

$$D(T_V) \subset D(V) \text{ and } \left\| V(\lambda - T_V)^{-1} \right\|_{\mathcal{L}(L^1(\nu))} \leq 1 (\lambda > 0). \tag{1.2.15}$$

This contraction property was fully exploited in [271] to develop a general theory of resolvent compactness for absorption generators T_V (when $(\lambda - T)^{-1}$ is "locally" weakly compact) or compactness of absorption C_0-semigroups $(U_V(t))_{t \geq 0}$ (when $(U(t))_{t \geq 0}$ is "locally" weakly compact). Spectral gap results are also provided in [271], although their formulations, relying on kernel estimates, are quite abstract and do not appear to have direct practical applications, for instance, in the study of Schrödinger equations on unbounded domains. We say that T_V has a spectral gap if

$$\sigma_{ess}(T_V) \subset \{\operatorname{Re}\lambda \leq (s(T_V) - \varepsilon)\} \text{ for some } \varepsilon > 0$$

and $(U_V(t))_{t \geq 0}$ has a spectral gap if $\omega_{ess}(U_V) < \omega(U_V)$.

One of the main objects of Chap. 5 is to improve [271] in different directions. In particular, if $(\lambda - T)^{-1}$ is "locally" weakly compact, we show that (1.2.15) alone is enough to imply a *local* spectral gap for T_V in the sense that there exists a function $c \to \varepsilon(c) > 0$ such that

$$\sigma_{ess}(T_V) \cap \{|\operatorname{Im}\lambda| \leq c\} \subset \{\operatorname{Re}\lambda \leq (s(T_V) - \varepsilon(c))\}$$

where $\varepsilon(.)$ is linked to $\liminf_{x \to \infty} V(x)$. We give also another approach when the potential V is locally bounded; (see Remark 5.10.1 for a more general assumption). This approach is significantly more powerful than the previous one, as it does not rely on the contraction property (1.2.15) or on the substochasticity of the semigroup $(U(t))_{t \geq 0}$. It applies uniformly across all L^p-spaces and, crucially, yields a global spectral gap, not merely a local one. Furthermore, it provides spectral gap results directly at the semigroup level, not just for the generators. Finally, we consider also Kato-Voigt perturbations of absorption semigroups, i.e., perturbation of absorption semigroups $(U_V(t))_{t \geq 0}$ where the perturbation P of the generator is of the form $P\varphi = K(V\varphi)$, $\varphi \in D(V)$ where $K : L^1(\nu) \to L^1(\nu)$ is subtochastic, and study the spectral properties and time asymptotics of such perturbed C_0-semigroups. The results of this chapter will be partially applied to the study of growth equations with nonlocal boundary conditions (see Sect. 11.6), as well as to the spectral properties of perturbed convolution semigroups, weighted Laplacians on $\mathbb{R}^d$, and Witten Laplacians on 1-forms (see Chap. 13). Moreover, these general results on absorption semigroups offer new insights into the essential spectra of general non-compact weighted Riemannian manifolds (see Sect. 14.4).

Chapter 6 is devoted to jump equations in $L^1(\nu)$ of the form

$$\begin{cases} \frac{d\varphi}{dt} = -\sigma(x)\varphi(x,t) + \int_\Omega p(x,y)\varphi(y,t)\nu(dy) \\ \qquad\qquad \varphi(x,0) = \varphi_0(x) \end{cases} \tag{1.2.16}$$

where $\sigma = \sigma_a + \sigma_s$ and

$$0 < \sigma_s(x) := \int_\Omega p(y,x)\nu(dy) < +\infty \ \ \nu\text{-a.e.}$$

The integral part in (1.2.16) can also be written as

$$\int_\Omega p_s(x,y)\sigma_s(y)\varphi(y,t)\nu(dy)$$

where $p_s(x, y) := \frac{p(x,y)}{\sigma_s(y)}$ is the kernel of a stochastic (i.e., mass-preserving) operator on $L^1(\nu)$

$$P_s : \psi \to \int_\Omega p_s(x, y)\psi(y)\nu(dx).$$

More complex jump equations arising in Kinetic Theory will be addressed in Sect. 10.6. The jump equations discussed above are motivated by classical space-homogeneous models encountered in nuclear reactor theory [92, 289, 333, 351, 352], which were originally studied in L^2-spaces. Similar types of equations are also used to describe nonlocal dispersal in Mathematical Biology (see, e.g., [93, 167]). We demonstrate how the abstract peripheral spectral theory developed in previous chapters applies to these jump equations in L^1-spaces. In the concrete settings considered in Chaps. 9 and 10, the weak compactness arguments crucially depend on the appropriate choice of the measure $\nu(dx)$. We also show how the Kato-Voigt perturbation theory developed in Chap. 3 applies effectively when a (generalized) detailed balance condition

$$\omega(x)p(x, y) = \omega(y)p(y, x) \qquad\qquad (1.2.17)$$

is satisfied where $\omega : \Omega \to (0, +\infty)$ is a priori an arbitrary measurable function. Finally, we show how the theory can be extended to $L^p(\nu)$ spaces, either via a predual approach or through interpolation techniques. Besides space-homogeneous kinetic equations, the methods developed in this chapter find natural applications in Kolmogorov differential equations and in the analysis of weighted graphs where Ω is a countable set.

Chapter 7 continues the analysis of jump semigroups in two directions:

1. We assume the (generalized) detailed balance condition (1.2.17) and build a systematic theory of jump semigroups in $L^2(\mu)$ where

$$\mu(dx) = \omega(x)\nu(dx). \qquad\qquad (1.2.18)$$

We point out that while the jump operator $\psi \to P\psi$ is always σ-bounded in $L^1(\nu)$, this operator need not be σ-bounded in $L^2(\mu)$. The good news is that $\psi \to P\psi$ is always σ-form-bounded in $L^2(\mu)$, i.e., in the sense of symmetric quadratic forms. This observation opens the door to a general *form-perturbation* theory in $L^2(\mu)$. In full generality, the corresponding nonnegative quadratic form need not be closed and the associated symmetric operator has a distinguished positive self-adjoint extension, the Friedrichs extension, which is the hilbertian counterpart of the (minus) Kato-Voigt generator. We give a thorough spectral analysis with time asymptotics in this Hilbert space framework. The distinction between "genuine honesty" and "honesty" in $L^1(\nu)$-Kato-Voigt perturbation theory seems to have its hilbertian counterpart in the distinction between self-adjointness and essential self-adjointness. More generally, we clarify the link between the previous $L^1(\nu)$-theory and the $L^2(\mu)$-theory so that the two theories illuminate each other. This dual viewpoint enriches the overall understanding of jump semigroups.

2. While a systematic theory in $L^2(\mu)$ is always feasible in full generality under (1.2.17), developing an analogous theory in $L^1(\mu)$-spaces requires an additional condition. It turns out that this is essentially a replica, via a suitable similarity transformation, of the predual theory previously developed in $L^1(\nu)$ to extend the construction to $L^p(\nu)$ ($1 \leq p < +\infty$). As a result, the spectral results in $L^1(\mu)$ mirror those already established in Chap. 6. Finally, the extension to $L^p(\mu)$ is much simpler than $L^p(\nu)$-case because of the symmetry of the jump operator with respect to the new measure (1.2.18).

Chapter 8 is concerned with jump perturbations of *Frobenius-Perron* semigroups.

We revisit the Kato-Voigt perturbation theory of absorption semigroups, as discussed in Chaps. 3 and 5, focusing on the case where $(U(t))_{t\geq 0}$ is a C_0-semigroup generated by an autonomous flow. More precisely, let (Ξ, ν) be a sigma-finite measure space and let

$$\pi : (-\infty, +\infty) \times \Xi \to \Xi$$

be a flow on Ξ, i.e., π is measurable and the transformations

$$\pi_t : \Xi \to \Xi \ \ (t \in \mathbb{R})$$

are bijective, *measure-preserving* and satisfy the group property

$$\pi_0(x) = x, \ \ \pi_{t+s}(x) = \pi_t(\pi_s(x)), \ \ t, s \in \mathbb{R}, \ x \in \Xi.$$

Let $U(t) : L^1(\Xi, \nu) \to L^1(\Xi, \nu)$ be the Frobenius-Perron operator associated to the flow π_t,[14] i.e.,

$$(U(t)f)(x) = f(\pi_{-t}(x)).$$

It is well known that the C_0-group $(U(t))_{t\in\mathbb{R}}$ is stochastic (see e.g. [194, Theorem 3.2.1]). We refer to the stochastic C_0-group $(U(t))_{t\in\mathbb{R}}$ as the *Frobenius-Perron* C_0-group. We denote by T its generator. Let

$$\Sigma : (\Xi, \nu) \to [0, +\infty]$$

be measurable and finite a.e. We consider the weighted Frobenius-Perron semigroup $(U_\Sigma(t))_{t\geq 0}$, defined by

$$(U_\Sigma(t)f)(x) = e^{-\int_0^t \Sigma(\pi_{-s}(x))ds} f(\pi_{-t}(x)),$$

---

[14] In the literature on ergodic theory, the interest of the spectral properties of this operator is to capture the *statistical* properties (ergodicity, mixing...) of the deterministic map π_t. We are not concerned with these aspects in this monograph. We note that ergodic theory goes back to Maxwell's and Boltzmann's kinetic theory of gases and was born as a mathematical theory around 1930 by the groundbreaking works of John von Neumann and George David Birkhoff (the father of Garrett Birkhoff; we refer e.g. to [111, 117, 194].

(see e.g. [320, 367]) which is nothing but the absorption semigroup (defined in Chap. 5) for the absorption Σ. We denote by T_Σ the generator of $(U_\Sigma(t))_{t\geq 0}$. Let

$$K : L^1(\Xi, \nu) \to L^1(\Xi, \nu)$$

be a substochastic operator and $P : D(\Sigma) \to L^1(\Xi, \nu)$ be defined by

$$P\varphi = K(\Sigma\varphi), \quad \varphi \in D(\Sigma).$$

According to the general Kato-Voigt perturbation theory, there exists a unique extension $J \supset T_\Sigma + P$ which generates a minimal substochastic C_0-semigroup $(\mathcal{J}(t))_{t\geq 0}$ on $L^1(\Xi, \nu)$. What we gain here is an explicit "computation" of the type of $(U_\Sigma(t))_{t\geq 0}$ or, equivalently, the spectral bound of its generator

$$s(T_\Sigma) = -\lim_{t\to+\infty} \inf_{y\in\Xi} t^{-1} \int_0^t \Sigma(\pi_s(y)))ds.$$

Various characterizations are provided to determine whether $s(T_\Sigma) = 0$ or $s(T_\Sigma) < 0$. This allows for a more precise analysis of the peripheral spectrum and long-time asymptotics of $(\mathcal{J}(t))_{t\geq 0}$ in the case where K is stochastic and weakly compact. We also obtain a more detailed understanding of the opposite time-asymptotic regime, which is closely related to scattering theory. This chapter is particularly relevant in Transport Theory, for example, in the kinetic theory of particle swarms (see [18, 19, 28, 29, 127–129, 308]) we consider in Sect. 10.13.

Chapter 9 is devoted to jump equations on countable measure spaces (E, m). We demonstrate how the abstract results of Chap. 6 apply in this contex; in particular, we highlight the consequences that follow from the countability of E. We deal first with Kolmogorov's differential equations

$$\frac{df}{dt} = -\sigma f + Bf \tag{1.2.19}$$

on the standard sequence space $l^1(\mathbb{N})$ with the counting measure where

$$B := Q^t - diag\, Q$$

(Q is the so-called Q-matrix) and

$$\sigma_j := -q_{jj} = \sum_i b_{ij} \ (\forall i \in \mathbb{N}).$$

We show that for general birth-and-death models, the (generalized) detailed balance condition (1.2.17) is always satisfied. As a consequence, the dichotomy between asymptotic stability and sweeping always holds for Kolmogorov C_0-semigroups governing (1.2.19); this dichotomy is known to hold in more general settings (see, e.g., [320, Theorem 6.1]), where it is established using different methods. We introduce also a class of Kolmogorov's differential equations on $l^1(\mathbb{N})$ that exhibit quasi-compactness properties; in this case, the Kolmogorov C_0-semigroup has a spectral gap or is asymptotically stable depending on whether $\inf \sigma_j > 0$ or $\inf \sigma_j = 0$.

We study also Kolmogorov's differential equations with detailed balance condition in weighted spaces $l_p^1(\mathbb{N})$ where p is a weight with respect to counting measure; (we do not try to rewrite here the weighted hilbertian theory given in Chap. 7). Note that in $l_p^1(\mathbb{N})$ the stochastic character of the Kolmogorov C_0-semigroup is lost. This construction is inspired by Transport Theory (see Sect. 10.3) and allows, via a *reverse* construction (given $\Psi \in l_+^1(\mathbb{N})$ and a positive sequence $\omega := \left(\omega_j\right)_j$ such that $\left(\omega_j^{-\frac{1}{2}}\right)_j \in l^1(\mathbb{N})$), to build a whole class of Kolmogorov's differential equations, indexed by the choice of Ψ and ω, with detailed balance and such that $B := Q^t - diag\, Q$, given by

$$\begin{cases} b_{ij} = \frac{\sqrt{\omega_j}}{\sqrt{\omega_i}} \Psi\left(\left|i^2 - j^2\right|\right) \; (i \neq j) \\ \qquad\qquad b_{ii} = 0, \end{cases}$$

is a compact operator on the weighted space $l_{\sqrt{\omega}}^1(\mathbb{N})$. In particular, the corresponding Kolmogorov C_0-semigroup has a spectral gap in the weighted space $l_{\sqrt{\omega}}^1(\mathbb{N})$ if $\inf \sigma_j > 0$ where

$$\sigma_j := \sum_i b_{ij} = \sum_i \frac{\sqrt{\omega_j}}{\sqrt{\omega_i}} \Psi\left(\left|i^2 - j^2\right|\right).$$

If $\inf \sigma_j = 0$ then the corresponding Kolmogorov C_0-semigroup extends uniquely to $l_{\frac{1}{\sqrt{\omega}}}^1(\mathbb{N})$ as a stochastic asymptoticaly stable C_0-semigroup in $l_{\frac{1}{\sqrt{\omega}}}^1(\mathbb{N})$. In particular, the choice

$$\sqrt{\omega_j} = 1 + j^s \; (s > 1)$$

provides us with a peripheral spectral theory of Kolmogorov's differential equations in higher moment spaces. The construction is not able to capture the first moment space, i.e. the case $s = 1$.

Finally, we consider spectral gaps for symmetric *weighted graphs*, i.e., for l^2-operators L_2 defined by Dirichlet forms on $l^2(E, m)$

$$Q(u) = \frac{1}{2} \sum_{x \in E} \sum_{y \in E} b(x, y)\, (u(x) - u(y))^2 + \sum_{x \in E} c(x) u(x)^2;$$

where (E, m) is a countable measure space and $c(.) \geq 0$ is the so-called killing term. There exists a rich literature on weighted graphs (see e.g. [184, 185] and references therein) which we complement here with a spectral gap result. Indeed, as a consequence of the abstract spectral results of Chap. 5, we obtain a lower bound of $\sigma_{ess}(L_2)$

$$\sigma_{ess}(L_2) \subset \left[\lim_{\mathcal{P}_c(E)} \inf \frac{c}{m},\ +\infty \right)$$

where $\lim \inf$ is relative to the directed sets $\mathcal{P}_c(E)$ of subsets of E with finite complements.

Chapter 10, devoted to Transport Theory, is the longest one and contains numerous developments. A comprehensive overview of the state of the art on spectral theory for neutron transport up to 1997 can be found in [245] and in the references therein. More recent advances, covering developments up to 2013, are presented in the proceedings of the CIMPA School held in Muizenberg (Cape Town) [268]. In this chapter, we revisit and enhance several key topics in Neutron Transport, advancing them in multiple directions.[15] It is not possible to delve into the mathematical details here. Each section is dedicated to a significant topic. In Sect. 10.3, we revisit some classical works on space-homogeneous and isotropic equations in neutron thermalization for monoatomic gas models [92, 289, 333, 351, 352]. Isotropy allows for angular averaging, reducing the problems to one-dimensional formulations. These classical works were originally formulated in weighted L^2-spaces; here, we show how to treat them in more physically relevant weighted L^1-spaces. In Sect. 10.4, we revisit the multidimensional physical free gas model considered by Suhadolc [343] in weighted L^1-space focusing exclusively on the space-homogeneous case. We show how compactness results can be established within the natural space $L^1(\mathbb{R}^3; dx)$, that is, the weight is no longer required, provided the absorption cross section is sufficiently strong. In Sect. 10.5, we deal with general space-homogeneous linear Boltzmann equations with power-like hard sphere potentials in various weighted L^1-spaces. In Sect. 10.6, we continue the investigation of conservative space-homogeneous equations, incorporating more complex downshift and Bragg scattering (jump) phenomena that arise in nuclear reactor theory [192]; we build a general spectral gap theory for these models. In Sect. 10.7, we show how generation, mean ergodicity and sweeping results for space-homogeneous equations (with arbitrary velocity measures) are inherited by space-nonhomogeneous equations. Section 10.8 is devoted to a systematic analysis of the peripheral spectrum of space-nonhomogeneous neutron transport semigroups (with suitable abstract velocity measures) and to its implications for their long-time asymptotic behavior. This program was initially developed in [251, 266] for bounded collision operators on spatial domains of finite volume. Here, we extend it to unbounded (but Σ-bounded) collision operators where Σ denotes the collision frequency. In Sect. 10.9, we demonstrate how the abstract framework developed in Sect. 10.8 applies to general space-nonhomogeneous linear Boltzmann equations with power-like hard sphere potentials, thereby extending known results on the subject [343, 375] in various directions. In Sect. 10.10, we show how the abstract (Max-inf and Inf-sup) variational characterizations of the leading eigenvalue given in Sect. 2.5 can be applied to neutron transport generators. Section 10.11 is devoted

---

[15] Some unpublished results are also included such as those in Sect. 10.14 which reports on some results given in a conference held in Bedlewo (Poland) in 2016 [270] and Sect. 10.16 which contains some results from an unpublished mansucript [247].

to the criticality eigenvalue problem and the time asymptotics of the associated C_0-semigroup. In Sect. 10.12, we study stochastic neutron transport semigroups on the whole space $\mathbb{R}^d$. Although several key compactness results extend to this setting (for cross sections vanishing at infinity), the existence of an invariant density remains an open problem. Conversely, in the opposite direction, we develop a general scattering theory that applies to cross sections which are not necessarily compactly supported in space. In Sect. 10.13, we present general spectral and scattering theories of particle swarms in weakly ionized gas. In Sect. 10.14, we analyze the spectral properties and time asymptotics of neutron transport semigroups with partly elastic collision operators on the d-dimensional torus. Unlike the fully inelastic case, the presence of an elastic component induces a loss of compactness in velocity. The key aspect of the analysis hinges on a subtle interplay between the elastic and inelastic parts. In contrast to the previous sections which focus on vacuum or periodic boundary conditions, Sect. 10.15 is devoted to the significantly more complex subject of neutron transport semigroups with nonlocal, mass-preserving boundary operators that relate incoming and outgoing fluxes; most of the existing literature on this topic is restricted to the one-dimensional spatial case. The previous sections focus primarily on the so-called asymptotic (discrete) spectrum of neutron transport semigroups; in Sect. 10.16, we turn to the analysis of their L^1 *essential* spectrum. Finally, in Sect. 10.17, we explore the spectral theory of a class of diffusion models used in nuclear reactor theory, where the spatial transport operator is replaced by a diffusion operator that depends on the velocity variable as a parameter.

Chapter 11 is devoted to the spectral analysis of three biological models arising in Population Dynamics. Section 11.3 addresses the spectral gap theory of *diffusive* structured populations models (1.2.7) with generalized Wentzell-Robin boundary conditions.

Other sections are devoted to *growth–fragmentation* equations (1.2.8) studied in various functional settings under different structural assumptions. In particular, three natural L^1 functional spaces emerge as especially relevant to the analysis

$$\begin{cases} X_1 := L^1(\mathbb{R}_+; \, xdx), \\ X_0 := L^1(\mathbb{R}_+; \, dx), \\ X_{0,1} := L^1(\mathbb{R}_+; \, (1+x)\,dx), \end{cases}$$

where X_1 is the "finite mass" space, X_0 is the "finite agregates number" space and $X_{0,1}$ is the "finite mass and agregates number" space. In addition, we may consider conservative models in the sense that

$$\int_0^y xb(x, y)dx = y, \quad (y \geq 0) \tag{1.2.20}$$

or models with mass loss

$$\int_0^y xb(x, y)dx = y\,(1 - \eta(y)), \quad 0 \leq \eta(y) \leq 1 \, (y \geq 0). \tag{1.2.21}$$

Finally, we have two physically motivated assumptions on the growth rate

$$\int_0^1 \frac{1}{r(\tau)} d\tau < +\infty, \quad \int_1^\infty \frac{1}{r(\tau)} d\tau = +\infty, \tag{1.2.22}$$

or

$$\int_0^1 \frac{1}{r(\tau)} d\tau = +\infty, \quad \int_1^\infty \frac{1}{r(\tau)} d\tau = +\infty. \tag{1.2.23}$$

We note that (1.2.22) needs a boundary condition, e.g.

$$\lim_{x\to 0} r(x)u(x) = 0 \tag{1.2.24}$$

while (1.2.23) needs no boundary condition. The mathematical analysis of such equations proves to be remarkably rich. Indeed, the interplay between different assumptions and functional settings gives rise to several distinct theoretical frameworks. For instance, one section is devoted to the spectral gap theory for growth-fragmentation equations (1.2.8) in the space $X_{0,1}$ under the assumption (1.2.22) with mass loss assumption (1.2.21) at infinity when the sublevel sets of $a(.)$ are "thin at infinity" in a suitable sense, e.g. when $\lim_{x\to+\infty} a(x) = +\infty$. In another section, we explore an opposite regime. Specifically, we show that conservative growth-fragmentation equations (1.2.8) in the space X_1 under Assumption (1.2.23) and

$$\int_0^{+\infty} \frac{a(\tau)}{r(\tau)} d\tau < +\infty$$

exhibit a *runaway* phenomenon. In fact, we restrict ourselves to the specific case $r(x) = \alpha x$ ($\alpha > 0$) and

$$\int_0^{+\infty} \frac{a(\tau)}{\tau} d\tau < +\infty$$

in order to carry out explicit and optimal calculations. This scattering regime is fundamentally different from the asynchronous exponential growth associated with the presence of a spectral gap. To the best of our knowledge, such a runaway phenomenon has not previously been identified in the context of growth–fragmentation equations.

In the sections above, the fragmentation operator plays a central role. In another section, we consider the growth equations without fragmentation equipped with McKendrick-von Foerster boundary condition (1.2.9) under (1.2.22). We identify a key $X_{0,1}$-vector measure that ensures the weak compactness of the difference between the C_0-semigroups corresponding to $\beta = 0$ and $\beta \neq 0$. In particular, this boundary perturbation preserves the essential type. This framework enables us to establish *two* general spectral gap theories in the space $X_{0,1}$. The first, inspired by growth–fragmentation theory, relies on the thinness at infinity of the sublevel sets of $a(.)$. The second theory, which does not rely on this condition (e.g. $a(.)$ may be bounded at infinity), utilizes spectral gap results for absorption semigroups presented in Chap. 5.

Chapter 12 is devoted to the spectral theory of linearized nonlocal Allen-Cahn equations (1.2.11). A spectral gap result for the generators in an L^2-space framework is established in [44]. In this chapter, besides presenting various preliminary results of independent interest, we develop a systematic L^1 spectral gap theory for the nonlocal Allen-Cahn generators and their associated C_0-semigroups. This approach relies on L^1-compactness tools and further demonstrates how the theory extends naturally to all L^p-spaces.

Chapter 13 is devoted to the spectral properties of absorption semigroups $(U_V(t))_{t\geq 0}$ with generator T_V corresponding to indefinite potentials

$$V = V^{(+)} - V^{(-)}$$

where $V^{(+)}$ and $V^{(-)}$ are not necessarily the standard positive and negative parts of V. The unperturbed C_0-semigroup $(U(t))_{t\geq 0}$, with generator T, is a symmetric convolution semigroup acting on $L^p(\mathbb{R}^d)$

$$U(t) : f \in L^p(\mathbb{R}^d) \to \int_{\mathbb{R}^d} f(x - y)m_t(dy) \in L^p(\mathbb{R}^d)$$

where $\{m_t\}_{t\geq 0}$ is a family of symmetric (with respect to the origin) Borel sub-probability measures on $\mathbb{R}^d$ such that $m_0 = \delta_0$ (Dirac measure at zero), $m_t * m_s = m_{t+s}$ and $m_t \to m_0$ vaguely as $t \to 0_+$. Such convolution semigroups are closely related to Lévy processes and encompass many examples of practical interest, including Gaussian semigroups, α-stable semigroups, relativistic Schrödinger semigroups, and others. Notably, $(U(t))_{t\geq 0}$ is a positive contraction C_0-semigroup on $L^p(\mathbb{R}^d)$ ($1 \leq p < +\infty$). This class of semigroups provides a natural and rich setting for the abstract spectral theory of absorption semigroups developed in Chap. 5. The main questions addressed are:

(1) Is T_V a generator of a C_0-semigroup?
(2) Is T_V resolvent compact? Is $(U_V(t))_{t\geq 0}$ compact?
(3) Does T_V have a (local) spectral gap? Does $(U_V(t))_{t\geq 0}$ exhibit a spectral gap?

As noted previously, absorption semigroups arise naturally in various fields such as Physics, Probability and Geometry [27, 78, 105, 149, 285, 338].

The questions outlined above were previously studied in [271]. Our goal here is to revisit and extend the results of [271] in several directions, most notably concerning the spectral gap theory developed therein. The original approach, based on abstract kernel estimates, lacks practical applicability in unbounded geometric settings. We therefore aim to provide improvements that address these limitations. We demonstrate how the theoretical results from Chap. 5 come into play through the positive part $V^{(+)}$, while the negative part $V^{(-)}$ serves to extend the class of Kato potentials for convolution semigroups. Besides general spectral results for convolution semigroups perturbed by potentials, we improve several spectral results from [271] concerning *weighted* Laplacians on $\mathbb{R}^d$ with applications to Poincaré inequalities for probability measures.[16] We also enhance spectral results for *Witten* Laplacians on 1-forms in $\mathbb{R}^d$ with applications to Helffer–Sjöstrand's covariance formula, notably without requiring any convexity assumptions on the weight function.

Chapter 14 which ends this monograph, contains several unrelated comments and results that complement earlier chapters. We provide a proof of Desch's theorem in AL-spaces and show how Miyadera perturbations are connected to a suitable bounded variation (BV) condition. A general spectral gap result is established for Laplacians on noncompact weighted Riemannian manifolds. We also discuss the possibility of extending the L^1-based results for neutron transport to L^p-spaces via a suitable predual construction. Finally, for a specific class of neutron transport operators, we briefly explain how the theory of symmetrizable operators in Hilbert spaces enables analysis beyond the peripheral spectrum, specifically by capturing parts of the real point spectrum.

---

[16] In fact, in Chap. 14 (Sect. 14.4), we provide more general statements on weighted Laplacians on noncompact Riemannian manifolds without boundary.

Chapter 2
Mathematical Toolbox

2.1 Chapter Aims

In this chapter, we recall several classical results on positive operator semigroups on AL-spaces, with a particular emphasis on perturbation settings. Special attention is given to peripheral spectral properties, which are central to the asymptotic analysis of such semigroups. We also present two fundamental generation theorems: Desch's perturbation theorem and the Kato–Voigt perturbation theorem, the latter accompanied by its associated honesty theory.

2.2 Chapter Overview

For the reader's convenience, this chapter recalls (without proofs) various mathematical results and tools required in the sequel. Some of these are classical, while others are less well known. The aim is to make the monograph as self-contained as possible. Certain additional results, which are new, are presented with full proofs. Throughout this monograph, we assume the reader is familiar with the basic notions of C_0-semigroup theory and spectral theory, as can be found, e.g., in Chap. 9 of [46], which provides a clear and pedagogical introduction to the subject. The reader is also encouraged to consult Sect. 2.3, where we briefly recall some basic properties of positive operators and positive C_0-semigroups that are used frequently throughout the monograph, often without explicit mention. By contrast, the other sections of this Mathematical Toolbox, which cover more specialized topics, may be skipped on a first reading and referred to as needed. The goal is not to provide a comprehensive survey of the extensive literature on positive C_0-semigroups, but rather to recall, as clearly and coherently as possible, certain specific results that are essential for the developments in this monograph. Section 2.3 recalls some basic definitions and spectral properties of positive operators on Banach lattices. In Sect. 2.4, we present

© The Author(s), under exclusive license to Springer Nature Switzerland AG 2026

M. Mokhtar-Kharroubi, *Peripheral Spectra of Perturbed Positive Semigroups*, Lecture Notes in Mathematics 2388, https://doi.org/10.1007/978-3-032-11173-9_2

several key results: a strict comparison of spectral radii for positive operators under domination, a criterion for the strict positivity of the spectral radius of positive irreducible power-compact operators, and a result on the monotonicity of the essential spectral radius in L^1-spaces. Section 2.5 is devoted to various characterizations of the spectral bound for perturbed resolvent positive operators on Banach lattices. In Sect. 2.6, we recall Desch's perturbation theorem for positive C_0-semigroups in L^1-spaces, which plays a central role in this monograph. For the reader's convenience, and because some technical aspects of the proof are used elsewhere, we provide a complete proof in Chap. 14. This proof differs slightly from the original given in [371]. One of the key advantages of Desch's theorem is that it does not rely on a dissipativity assumption, making it applicable beyond the setting of contraction C_0-semigroups. We also show how to apply the theorem using weak compactness or quasi-compactness arguments. Although this result is specific to L^1-spaces, it admits an extension to general Banach lattices when the unperturbed semigroup is holomorphic. We likewise demonstrate how to exploit this more general version using compactness or quasi-compactness techniques. In Sect. 2.7, we recall the notions of essential spectrum, essential spectral radius, and essential type. Section 2.8 is devoted to several spectral stability results for perturbed generators and perturbed C_0-semigroups, based on (weak) compactness arguments. In Sect. 2.9, we recall a result concerning strong integrals of bounded, strongly measurable $\mathcal{W}(X, Y)$-valued mappings, where X, Y are Banach spaces, and $\mathcal{W}(X, Y)$ denotes the space of weakly compact operators from X to Y. In Sect. 2.10, in addition to classical results on mean ergodicity in L^1-spaces, we present recent results on the strong convergence to the ergodic projection for positive partially integral C_0-semigroups on Lebesgue spaces. Finally, Sect. 2.11 is devoted to the *Kato–Voigt* perturbation theorem for substochastic C_0-semigroups in L^1-spaces, which also plays a key role in this monograph. We recall the core aspects of the *honesty* theory that has developed around this result. Additional remarks and clarifications are provided in the Comments section at the end of the chapter.

2.3 Positive Operators and Spectra

In this section, we recall the basic concepts of Banach lattices, positive operators, and positive C_0-semigroups, along with their key spectral properties, which will be used tacitly, without explicit mention, throughout this monograph. These notions are drawn primarily from [46, Chaps. 10 and 12] and [90, Chaps. 7 and 8)]. A set E with an order relation $\leq$ is said to be an ordered set if

(i) $x \leq x$ for all $x \in E$
(ii) $x \leq y$ and $y \leq x$ imply $x = y$
(iii) $x \leq y$ and $y \leq z$ imply $x \leq z$.

Definition 2.3.1 A real vector space E ordered by an order relation $\leq$ is called a vector lattice if any two elements $x, y \in E$ have a least upper bound denoted by $x \vee y = \sup(x, y)$ and a greatest lower bound denoted by $x \wedge y = \inf(x, y)$ and the following properties are satisfied:

(i) If $x \leq y$ then $x + z \leq y + z$ $(x, y, z \in E)$

(ii) If $0 \leq x$ then $0 \leq tx$ for $x \in E$ and $0 \leq t \in \mathbb{R}$.

If $(E, \leq)$ is a vector lattice then we define its positive cone

$$E_+ := \{x \in E, \ x \geq 0\}$$

and by

$$x_+ := x \vee 0, \ x_- := (-x) \vee 0 \text{ and } |x| := x \vee (-x)$$

the positive part, the negative part and the absolute value of $x \in E$. Two elements $x, y \in E$ are called orthogonal or lattice disjoint (denoted by $x \perp y$) if $|x| \wedge |y| = 0$. Then we have

Proposition 2.3.1 ([46, Proposition 10.4]) *Let $(E, \leq)$ be a vector lattice and $x, y \in E$. Then*

(i) $x = x_+ - x_-$

(ii) $|x| = x_+ + x_-$

(iii) $x_+ \perp x_-$ *and the decomposition of x into a difference of two positive orthogonal elements is unique.*

Definition 2.3.2 (i) A norm $\|\,\|$ on a vector lattice E is called a lattice (or Riesz) norm if $|x| \leq |y|$ implies $\|x\| \leq \|y\|$. A Banach lattice is a real Banach space endowed with an order relation $\leq$ such that $(E, \leq)$ is a vector lattice and the norm is a lattice norm.

Note that $y = |y|$ for $y \in E_+$ so $|(|x|)| = |x|$ $(x \in E)$ and then the definition of a lattice norm shows that

$$\|x\| = \||x|\|, \ (x \in E).$$

Basic examples are provided by real Lebesgue spaces $L^p(\Omega, \nu)$ with σ-finite measure ν $(1 \leq p \leq +\infty)$ where the ordering $f \leq g$ is defined by $f(s) \leq g(s)$ ν-a.e. In particular

$$L^p_+(\Omega, \nu) = \{f \in L^p(\Omega, \nu), \ f(s) \geq 0 \ \nu\text{-a.e.}\}.$$

Definition 2.3.3 A subspace I of a Banach lattice E is called an ideal if $x \in I$ implies $|x| \in I$ and $0 \leq x \leq y \in I$ implies $x \in I$.

Proposition 2.3.2 *Let $1 \leq p < +\infty$. The closed ideals I of $L^p(\Omega, \nu)$ are of the form*

$$I = \{f \in L^p(\Omega, \nu), \ f(s) = 0 \ a.e. \ s \in Y\}$$

where Y is a measurable subset of Ω.

To deal with spectral theory of an operator T acting on a *real* Banach space E, it is necessary to extend T to a complexified version of E. Similarly, a complex Banach lattice E_c is a complexification of a real Banach lattice E defined as the complex vector space of couples (x, y) $(x, y \in E)$ with natural addition and with scalar multiplication given by

$$(a + ib)(x, y) = (ax - by, ay + bx) \quad (a, b \in \mathbb{R})$$

endowed with the norm
$$\|(x, y)\| = \|\,|(x, y)|\,\|$$

where $|(x, y)| \in E$, given by

$$|(x, y)| := \sup_{0 \le \theta \le 2\pi} (x \sin \theta + y \cos \theta),$$

is the natural extension to E_c of the modulus $|\,|$ in E. It can be shown that the supremum in E above exists. By identifying $(x, 0) \in E_c$ with $x \in E$, the space E is isometrically isomorphic to a real linear subspace $E_\mathbb{R}$ of E_c. We can thus use a more convenient notation

$$(x, y) = (x, 0) + (0, y) = (x, 0) + i\,(y, 0) = x + iy.$$

Definition 2.3.4 Let E, F be complex Banach lattices. A linear operator $T : E \to F$ is positive if $T(E_+) \subset F_+$. We denote this by $T \ge 0$ or $T \in \mathcal{L}_+(E, F)$.

Here is a characterization of positive operators.

Proposition 2.3.3 ([46, Lemma 10.18]) *Let* E, F *be complex Banach lattices and a linear operator* $T : E \to F$. *The following assertions are equivalent.*
   (i) T is positive
   (ii) For all $x \in E_\mathbb{R}$, $(Tx)_+ \le Tx_+$ and $(Tx)_- \le Tx_-$
   (iii) For all $x \in E$ $|Tx| \le T(|x|)$

Note that every positive linear operator must be bounded, (see [46, Theorem 10.20]) and we have

Proposition 2.3.4 ([46, Proposition 10.22]) *Let* E, F *be complex Banach lattices and a positive linear operator* $T : E \to F$. *Then* $\|T\|_{\mathcal{L}(E,F)} = \sup\{\|Tx\|,\ x \in E_+,\ \|x\| \le 1\}$. *If S and T are two positive operators such that $S \le T$ (in the sense $Sx \le Tx$ for all $x \in E_+$) then* $\|S\|_{\mathcal{L}(E,F)} \le \|T\|_{\mathcal{L}(E,F)}$.

Here is a key spectral result.

Proposition 2.3.5 ([46, Theorem 10.24 and Lemma 10.25]) *Let E be a complex Banach lattice and a positive linear operator* $T : E \to E$ *with spectral radius* $r(T)$. *Then* $r(T) \in \sigma(T)$ *(the spectral radius belongs to the spectrum) and the resolvent* $(\lambda - T)^{-1} \ge 0$ *if and only if $r(T) < \lambda \in \mathbb{R}$.*

Definition 2.3.5 Let E be a complex Banach space and $G : D(G) \subset E \to E$ be a densely defined closed linear operator whose spectrum $\sigma(G)$ is included in a left half plane. We define its spectral bound by

$$s(G) := \sup\{\operatorname{Re}\lambda; \ \lambda \in \sigma(G)\}$$

with the convention that $s(G) = -\infty$ if $\sigma(G)$ is empty.

We say that a C_0-semigroup $\mathcal{U} = (U(t))_{t \geqslant 0}$ on a Banach lattice E is positive if $U(t) \in \mathcal{L}_+(X)$ $(t \geq 0)$. It can be shown that $(U(t))_{t \geqslant 0}$ is positive if and only if $(\lambda - G)^{-1} \geq 0$ for λ large enough. We give now the semigroup versions of the above spectral results.

Proposition 2.3.6 ([46, Corollary 12.9]) *Let* $(U(t))_{t \geqslant 0}$ *be a positive* C_0-*semigroup with generator* G *on a Banach lattice* E *and let* $s(G)$ *be the spectral bound of* G. *If* $\sigma(G)$ *is not empty then* $s(G) \in \sigma(G)$.

Proposition 2.3.7 ([46, Corollary 12.10]) *Let* $(U(t))_{t \geqslant 0}$ *be a positive* C_0-*semigroup with generator* G *on a Banach lattice* E *and let* $s(G)$ *be the spectral bound of* G. *Then* $(\lambda - G)^{-1} \geq 0$ *if and only if* $s(G) < \lambda \in \mathbb{R}$. *In this case,* $r((\lambda - G)^{-1}) = \frac{1}{\lambda - s(G)}$ $(s(G) < \lambda \in \mathbb{R})$.

It can be shown that the dual E' of a real Banach lattice E can be ordered by

$$p \leq q \ (p, q \in E') \Leftrightarrow \langle p, x \rangle \leq \langle q, x \rangle \ (x \in E_+)$$

($\langle ., . \rangle$ denotes the duality pairing) and is also a Banach lattice, see the details in [90, p. 275]. We use both notations $\langle p, x \rangle$ and $\langle x, p \rangle$ interchangeably for $p \in E'$ and $x \in E$.

We point out that in a Banach lattice E, the notation $x > 0$ $(x \in E)$ means $x \in E_+$ and $x \neq 0$. In particular, in a Lebesgue space $L^p(\Omega, \nu)$, $f > 0$ means that $f \in L^p_+(\Omega, \nu)$ does not vanish a.e. Here are useful definitions.

Definition 2.3.6 Let E be a Banach lattice. We say that $x \in E_+$ is quasi-interior if $\langle p, x \rangle > 0$ for all $p \in E'$ such that $p > 0$. A functional $p \in E'_+$ is strictly positive if $\langle p, x \rangle > 0$ for all $x > 0$.

Note that in Lebesgue spaces $L^p(\Omega, \nu)$ $(1 \leq p \leq +\infty)$, $f \in L^p_+(\Omega, \nu)$ is quasi-interior means $f(s) > 0$ ν-a.e.

Definition 2.3.7 A positive linear operator T on a Banach lattice E is said to be positivity improving if Tx is quasi-interior for all $x > 0$.

The concept of irreducibility is fundamental.

Definition 2.3.8 Let E be a Banach lattice. A positive linear operator T on E is irreducible if $\{0\}$ and E are the only closed ideals invariant under T. A positive C_0-semigroup $(U(t))_{t \geqslant 0}$ on E is irreducible if and only $\{0\}$ and E are the only closed ideals invariant under $(U(t))_{t \geqslant 0}$.

Note that the irreducibility of $(U(t))_{t \geqslant 0}$ is more general than the irreducibility of *each* operator $U(t)$ $(t \geq 0)$. Here is another formulation of the irreducibility.

Proposition 2.3.8 *Let E be a Banach lattice and T be a positive operator. Then T is irreducible if and only if for any $x \in E$, $x > 0$ and $p \in E'$, $p > 0$ there exists $n \in \mathbb{N}$ such that $\langle T^n x, p \rangle > 0$.*

In particular, any positivity improving operator is irreducible. We give also alternative formulations of the irreducibility for C_0-semigroups.

Proposition 2.3.9 ([90, Proposition 7.6]) *Let E be a Banach lattice and let $(U(t))_{t \geqslant 0}$ be a positive C_0-semigroup with generator G on E. Then the following statements are equivalent.*
(i) $(U(t))_{t \geqslant 0}$ is irreducible
(ii) For each $x \in E$, $x > 0$ and $p \in E'$, $p > 0$ there exists $t \geq 0$ such that $\langle U(t)x, p \rangle > 0$.
(iii) The resolvent $(\lambda - G)^{-1}$ is positivity improving for all (for some) $\lambda > s(G)$.
(iv) The resolvent $(\lambda - G)^{-1}$ is irreducible for all (for some) $\lambda > s(G)$.

We end this section by

Proposition 2.3.10 (See e.g. [376]) *Let $(U(t))_{t \geqslant 0}$ be a positive C_0-semigroup on a Lebesgue space $L^p(\Omega, \nu)$. Then the type (or growth bound) of $(U(t))_{t \geqslant 0}$ coincides with the spectral bound of its generator.*

2.4 Spectral Comparisons

It is elementary to see that if O_i $(i = 1, 2)$ are two positive operators on a Banach lattice X such that $O_1 \leq O_2$ (i.e., $O_2 - O_1 \in \mathcal{L}_+(X)$) then $r(O_1) \leq r(O_2)$. The following statement provides us with useful strict comparisons.

Proposition 2.4.1 ([133, Theorem 3.9 (1)]) *Let X be a complex Banach lattice and O_i $(i = 1, 2)$ be two positive operators on X such that $O_1 \leq O_2$ and $O_1 \neq O_2$. Suppose that O_2 is irreducible. If $r(O_2)$ is a pole of the resolvent of O_2 then $r(O_1) < r(O_2)$.*

In the same spirit, albeit much more involved, we have

Proposition 2.4.2 (See e.g. [328, Theorem A]) *Let X be a complex Banach lattice and let O be a positive irreducible operator on X. If some power of O is compact then $r(O) > 0$.*

Unlike the (trivial) monotonicity to the spectral radius for positive operators, the monotonicity of the essential spectral radius (see (2.7.1) below the definition) is not true in general. Fortunately, this is the case in *AL*-space (see Sect. 2.6) and therefore in L^1 spaces.

Theorem 2.4.1 ([225, Corollary 2.10]) *Let X be an AL-space and let O_i ($i = 1, 2$) be two positive linear operators on X such that $O_1 \leq O_2$ (i.e., $O_2 - O_1 \in \mathcal{L}_+(X)$) then $r_{ess}(O_1) \leq r_{ess}(O_2)$.*

Remark 2.4.1 If X is an L^1-space and if O_i ($i = 1, 2$) are two positive linear operators on X such that $O_1 \leq O_2$ then O_1 is weakly compact if O_2 is; this fact, used repeatedly in this monograph, follows simply from the equi-integrability criterion of weak compactness. On the other hand, if X is an L^p-space ($1 < p < \infty$) then a more complex result (see [4]), occasionally used in this monograph, asserts that O_1 is compact if O_2 is.

2.5 Spectral Bound of Perturbed Resolvent Positive Operators

We start with

Definition 2.5.1 Let X be a complex Banach lattice and $T : D(T) \subset X \to X$ be an unbounded linear operator. We say that T is resolvent positive if the resolvent $(\lambda - T)^{-1}$ exists for λ large enough and is positive.

Definition 2.5.2 Let X be a complex Banach lattice and $T : D(T) \subset X \to X$ be a closed operator. A linear operator $B : D(T) \to X$ is said to be T-bounded if it is continuous on $D(T)$ endowed with its graph norm, i.e., there exist two positive constants α, β such that $\|Bx\| \leq \alpha \|Tx\| + \beta \|x\|$ ($x \in D(T)$).

We recall first a key characterization of the spectral bound of positively perturbed resolvent positive operators.

Proposition 2.5.1 ([371]) *Let X be a complex Banach lattice and $T : D(T) \subset X \to X$ be resolvent positive with spectral bound $s(T)$. Let $B : D(T) \to X$ be T-bounded and positive, i.e., $B : D(T) \cap X_+ \to X_+$. Then*

$$A := T + B : D(T) \to X$$

is resolvent positive if and only if $r\left(B\left(\lambda - T\right)^{-1}\right) < 1$ for real λ large enough. In this case

$$s(A) = \inf\left\{\lambda > s(T),\ r\left(B\left(\lambda - T\right)^{-1}\right) < 1\right\}$$

and

$$(\lambda - A)^{-1} = (\lambda - T)^{-1} \sum_{j=0}^{\infty} \left(B(\lambda - T)^{-1}\right)^j \ (\lambda > s(A)).$$

Corollary 2.5.1 *Let X be a complex Banach lattice and $T : D(T) \subset X \to X$ be resolvent positive with spectral bound $s(T)$. Let $B : D(T) \to X$ be T-bounded and positive and let*

$$A := T + B : D(T) \to X$$

be resolvent positive. If $B(\lambda - T)^{-1}$ is irreducible then $(\lambda - A)^{-1}$ is positivity improving. In particular, if $T + B$ is a generator then the corresponding semigroup is irreducible.

Proof Let $x \in X_+$ and $x' \in X'_+$ be non trivial. Since $(\lambda - T')^{-1}x'$ is not trivial then the irreducibility of $B(\lambda - T)^{-1}$ and

$$\langle (\lambda - A)^{-1}x, x' \rangle = \langle \sum_{j=0}^{\infty} \left(B(\lambda - T)^{-1} \right)^j x, (\lambda - T')^{-1}x' \rangle$$

show that $\langle (\lambda - A)^{-1}x, x' \rangle > 0$. This ends the proof. $\qquad\qquad\square$

If $B (\lambda - T)^{-1}$ is irreducible and some power of $B (\lambda - T)^{-1}$ is compact then variational (Max-inf and Inf-sup) characterizations of the perturbed spectral bound $s(T + B)$ are given in [256][1] where the results are stated under the assumption that B is bounded. In this section, we weaken these assumptions by dropping the boundedness of B and by replacing the power compactness of $B (\lambda - T)^{-1}$ just by its essential compactness, i.e.,

$$r_{ess}(B (\lambda - T)^{-1}) < r(B (\lambda - T)^{-1}), \quad (\lambda > s(T)). \tag{2.5.1}$$

These new results, of interest for neutron transport (see Sect. 10.10), are given in the next subsections.

2.5.1 Max-Inf Characterization

Let $D_+(T) := D(T) \cap X_+$. We start with

Lemma 2.5.1 *Let X be a complex Banach lattice and $T : D(T) \subset X \to X$ be resolvent positive with spectral bound $s(T)$. Let $B : D(T) \to X$ be T-bounded and positive. Suppose that (2.5.1) is satisfied and that*

$$A := T + B : D(T) \to X$$

is resolvent positive. Let

$$E := \{\lambda > s(T); \ \exists \varphi \in D_+(T), \ \varphi \neq 0, \ A\varphi - \lambda\varphi \in X_+\}. \tag{2.5.2}$$

Then

$$E = \widehat{E} := \left\{\lambda > s(T); \ r(B (\lambda - T)^{-1}) \geq 1\right\}.$$

---

[1] A longer version of [256] appeared previously as a prepublication [248].

Proof Let $\lambda \in E$. Then $\lambda > s(T)$ and there exists $\varphi \in D_+(T)$, $\varphi \neq 0$, such that $A\varphi - \lambda\varphi \in X_+$, i.e., $T\varphi + B\varphi \geq \lambda\varphi$. Since $(\lambda - T)^{-1}$ is positive then $\varphi \leq (\lambda - T)^{-1}B\varphi$. Then $\psi := B\varphi \neq 0$ and

$$\psi \leq B(\lambda - T)^{-1}\psi$$

so $\psi \leq \left(B(\lambda - T)^{-1}\right)^k \psi$ $(\forall k \in \mathbb{N})$. Hence $\left\|\left(B(\lambda - T)^{-1}\right)^k\right\| \geq 1$ $(\forall k \in \mathbb{N})$ and $r(B(\lambda - T)^{-1}) \geq 1$, i.e., $\lambda \in \widehat{E}$. Conversely, let $\lambda > s(T)$ and $r(B(\lambda - T)^{-1}) \geq 1$. It follows from (2.5.1) that

$$\alpha := r(B(\lambda - T)^{-1}) \geq 1$$

is an eigenvalue of $B(\lambda - T)^{-1}$ associated with a nonnegative eigenvector ψ, i.e.,

$$B(\lambda - T)^{-1}\psi = \alpha\psi.$$

Then $\varphi := (\lambda - T)^{-1}\psi \in D_+(T)$ is not trivial and $\frac{1}{\alpha}B\varphi = \psi$ so

$$\lambda\varphi - T\varphi = \frac{1}{\alpha}B\varphi \leq B\varphi$$

and $T\varphi + B\varphi - \lambda\varphi \geq 0$. $\qquad\square$

Corollary 2.5.2 *Let X be a complex Banach lattice and $T : D(T) \subset X \to X$ be resolvent positive with spectral bound $s(T)$. Let $B : D(T) \to X$ be T-bounded and positive. Suppose that (2.5.1) is satisfied and that*

$$A := T + B : D(T) \to X$$

is resolvent positive. Then either $s(A) = s(T)$ or $E \neq \emptyset$ and

$$s(A) = \sup\{\lambda; \lambda \in E\}$$

where E is the set (2.5.2).

Proof Note first that we always have $s(A) \geq s(T)$. Let $s(A) > s(T)$. Note that

$$\lambda \in (s(T), +\infty) \to r(B(\lambda - T)^{-1})$$

is nonincreasing. According to Lemma 2.5.1,

$$(s(T), s(A)] = \widehat{E}$$

and we are done since $E = \widehat{E}$. $\qquad\square$

We have

Lemma 2.5.2 *Let X be a complex Banach lattice and $T : D(T) \subset X \to X$ be resolvent positive with spectral bound $s(T)$. Let $B : D(T) \to X$ be T-bounded and positive. Suppose that $B\,(\lambda - T)^{-1}$ satisfies (2.5.1) and is irreducible. Then*

$$\lambda \in (s(T), +\infty) \to r(B\,(\lambda - T)^{-1}) \tag{2.5.3}$$

is continuous and decreasing.

Proof Note first that $r(B\,(\lambda - T)^{-1}) > 0$ and (2.5.3) is non increasing. Since $r(B\,(\lambda - T)^{-1})$ is an isolated eigenvalue of $B\,(\lambda - T)^{-1}$ (with simple algebraic multiplicity) then, by Proposition 2.4.1, Eq. (2.5.3) is decreasing. Finally, since $\lambda \in (s(T), +\infty) \to B\,(\lambda - T)^{-1}$ is continuous in operator norm then the continuity of (2.5.3) is consequence of classical Kato's results on the continuity with respect to a parameter λ of finite sets of eigenvalues of a bounded operator depending continuously on the parameter λ (see [183, Chap. 4]). $\square$

We are ready to show

Theorem 2.5.1 *Let X be a complex Banach lattice and $T : D(T) \subset X \to X$ be resolvent positive with spectral bound $s(T)$. Let $B : D(T) \to X$ be T-bounded and positive. Suppose that $B\,(\lambda - T)^{-1}$ satisfies (2.5.1) and is irreducible. Let*

$$A := T + B : D(T) \to X$$

be resolvent positive. For each $\varphi \in D_+(T)$, $\varphi \neq 0$ let

$$\tau(\varphi) := \sup \{\lambda > s(T); \ A\varphi - \lambda\varphi \in X_+\}$$

with the convention that $\tau(\varphi) = s(T)$ if $\{...\} = \emptyset$. Then

$$s(T) < s(A) \text{ if and only if } \sup_{\varphi \subset D_+(T), \ \varphi \neq 0} \tau(\varphi) > s(T);$$

in such a case,

$$s(A) = \sup_{\varphi \in D_+(T), \ \varphi \neq 0} \tau(\varphi) = \max_{\varphi \in D_+(T), \ \varphi \neq 0} \tau(\varphi); \tag{2.5.4}$$

moreover, if $Ker(B) \cap X_+ = \{0\}$ then $\tau(\varphi) = s(A)$ if and only if $A\varphi = s(A)\varphi$.

Proof Let $s(A) > s(T)$. According to Proposition 2.5.1 and to Lemma 2.5.2, $s(A)$ is characterized as the unique $\lambda > s(T)$ such that

$$r(B\,(\lambda - T)^{-1}) = 1. \tag{2.5.5}$$

Hence there exists $h \in X_+$ such that

$$B \left(s(A) - T \right)^{-1} h = h$$

so $\psi := \left(s(A) - T \right)^{-1} h \in D_+(T)$ is not trivial and

$$s(A)\psi - T\psi = B\psi$$

i.e., $A\psi - s(A)\psi = 0$. Thus

$$\sup_{\varphi \in D_+(T),\ \varphi \neq 0} \tau(\varphi) > s(T). \tag{2.5.6}$$

Note that for all $\varphi \in D_+(T)$, $\varphi \neq 0$ such that $\tau(\varphi) > s(T)$, we have

$$A\varphi - \tau(\varphi)\varphi \in X_+$$

and, by Corollary 2.5.2, $\tau(\varphi) \leq s(A)$ whence

$$\sup_{\varphi \in D_+(T),\ \varphi \neq 0} \tau(\varphi) = \max_{\varphi \in D_+(T),\ \varphi \neq 0} \tau(\varphi) = s(A).$$

Finally, let $\varphi \in D_+(T)$, $\varphi \neq 0$ such that $\tau(\varphi) = s(A)$, i.e.,

$$T\varphi + B\varphi - s(A)\varphi \geq 0.$$

Then

$$\varphi \leq \left(s(A) - T \right)^{-1} B\varphi \tag{2.5.7}$$

and $\psi := B\varphi$ is such that

$$\psi \leq B \left(s(A) - T \right)^{-1} \psi. \tag{2.5.8}$$

We claim that (2.5.7) must be an equality. Indeed, if not then $Ker(B) \cap X_+ = \{0\}$ implies that (2.5.8) is not an equality. According to (2.5.1), 1 which is the spectral radius of the $\left(B \left(s(A) - T \right)^{-1} \right)'$ (see (2.5.5)) is an algebraically simple eigenvalue associated to a strictly positive eigenfunctional ψ'. If

$$B \left(s(A) - T \right)^{-1} \psi - \psi \neq 0$$

then (2.5.8) implies

$$\langle B \left(s(A) - T \right)^{-1} \psi, \psi' \rangle > \langle \psi, \psi' \rangle$$

which leads to a contradiction since

$$\langle B \left(s(A) - T \right)^{-1} \psi, \psi' \rangle = \langle \psi, \left(B \left(s(A) - T \right)^{-1} \right)' \psi' \rangle = \langle \psi, \psi' \rangle.$$

Hence $B \left(s(A) - T \right)^{-1} \psi = \psi$ and $\varphi = \left(s(A) - T \right)^{-1} B\varphi$, i.e., $A\varphi = s(A)\varphi$. $\square$

According to Corollary 2.5.1, if $s(T) < s(A)$ then the eigenvector of A corresponding to $s(A)$ is quasi-interior and then (2.5.4) becomes

$$s(A) = \sup_{D_{++}(T)} \tau(\varphi) = \max_{D_{++}(T)} \tau(\varphi).$$

where

$$D_{++}(T) = \{\varphi \in D(T); \ \varphi \ \text{quasi-interior}\}.$$

If $X = L^p(\nu)\,(1 \le p < +\infty)$, we can express Theorem 2.5.1 slightly differently.

Theorem 2.5.2 *Let $X = L^p(\nu)\,(1 \le p < +\infty)$ and $T : D(T) \subset X \subset X \to X$ be resolvent positive with spectral bound $s(T)$. Let $B : D(T) \to X$ be T-bounded and positive. Suppose that $B\,(\lambda - T)^{-1}$ satisfies (2.5.1) and is irreducible. Let*

$$A := T + B : D(T) \to X$$

be resolvent positive. Then $s(A) > s(T)$ if and only if there exists some $\psi \in D_{++}(T)$ such that $\inf \frac{A\psi}{\psi} > s(T)$. In this case,

$$s(A) = \max_{\varphi \in D_{++}(T)} \left(\inf \frac{A\varphi}{\varphi} \right),$$

where $\inf \frac{A\varphi}{\varphi}$ is the essential infimum of $\frac{A\varphi}{\varphi}$.

Proof We know that

$$s(A) = \max_{\varphi \in D_{++}(T)} \tau(\varphi)$$

and

$$\tau(\varphi) = \sup\{\lambda > s(T); \ A\varphi - \lambda\varphi \in X_+\}$$
$$= \begin{cases} s(T) \ \text{if} \ \inf \frac{A\varphi}{\varphi} \le s(T) \\ \inf \frac{A\varphi}{\varphi} > s(T) \end{cases}$$

ends the proof. $\qquad\qquad\square$

2.5.2 Inf-Sup Characterization

We give now *dual* results. We start with

Lemma 2.5.3 *Let X be a complex Banach lattice and $T : D(T) \subset X \to X$ be resolvent positive with spectral bound $s(T)$. Let $B : D(T) \to X$ be T-bounded and positive. Suppose that $B\,(\lambda - T)^{-1}$ satisfies (2.5.1) and is irreducible. Let*

$$A := T + B : D(T) \to X$$

be resolvent positive. Let

$$F := \left\{ \lambda > s(T); \ \exists \varphi \in D(T) \cap X_{++}, \ \lambda \varphi - A\varphi \in X_+^* \right\}$$

where $X_+^* := X_+ \setminus \{0\}$. If $Ker(B) \cap X_+ = \{0\}$ then

$$F = \widehat{F} := \left\{ \lambda > s(T); \ r(B \, (\lambda - T)^{-1}) < 1 \right\}.$$

Proof Let $\lambda \in \widehat{F}$. According to Proposition 2.5.1, $\lambda > s(A)$ and

$$(\lambda - A)^{-1} = (\lambda - T)^{-1} \sum_{j=0}^{\infty} \left(B(\lambda - T)^{-1} \right)^j \geq 0.$$

Let $\psi \in X_+^*$ and $\varphi = (\lambda - A)^{-1}\psi$. According to Corollary 2.5.1, $(\lambda - A)^{-1}$ is positivity improving so $\varphi \in D_{++}(T)$ and $\lambda \varphi - A\varphi \in X_+^*$, i.e., $\lambda \in F$. Conversely, let $\lambda \in F$ then $\lambda \varphi - T\varphi - B\varphi \in X_+^*$ so

$$\varphi - (\lambda - T)^{-1} B\varphi \in X_+^*.$$

Let $\psi = B\varphi$. Since $Ker(B) \cap X_+ = \{0\}$ then

$$\psi - B(\lambda - T)^{-1}\psi \in X_+^*.$$

We note that the eigenfunctional ψ' of $(B \, (\lambda - T)^{-1})'$ relative to

$$r\left[(B \, (\lambda - T)^{-1})' \right] = r\left(B \, (\lambda - T)^{-1} \right)$$

is strictly positive, i.e., $\langle \psi', h \rangle > 0$ for any $h \in X_+^*$. Indeed, there exists an integer n (depending on ψ' and h) such that

$$\langle \psi', (B \, (\lambda - T)^{-1})^n h \rangle > 0$$

so $\langle \left((B \, (\lambda - T)^{-1})' \right)^n \psi', h \rangle > 0$ and

$$\left(r((B \, (\lambda - T)^{-1})') \right)^n \langle \psi', h \rangle > 0.$$

Thus

$$\langle \psi - B(\lambda - T)^{-1}\psi, \psi' \rangle > 0$$

i.e.,

$$\langle \psi, \psi' \rangle - \langle \psi, (B \, (\lambda - T)^{-1})'\psi' \rangle > 0$$

or

$$\langle \psi, \psi' \rangle - r \left[(B (\lambda - T)^{-1})' \right] \langle \psi, \psi' \rangle > 0.$$

Hence $r \left[(B (\lambda - T)^{-1})' \right] < 1$ and $\lambda \in \widehat{F}$. $\qquad\square$

We are ready to state

Theorem 2.5.3 *Let X be a complex Banach lattice and $T : D(T) \subset X \to X$ be resolvent positive with spectral bound $s(T)$. Suppose that $B (\lambda - T)^{-1}$ satisfies (2.5.1) and is irreducible. Let*

$$A := T + B : D(T) \to X$$

be resolvent positive. For each $\varphi \in D_{++}(T)$ let

$$\widehat{\tau}(\varphi) := \inf \left\{ \lambda > s(T); \ \ \lambda \varphi - A\varphi \in X^*_+ \right\}$$

with the convention that $\widehat{\tau}(\varphi) = +\infty$ if $\{...\} = \emptyset$. Then $s(A) = \inf_{\varphi \in D_{++}(T)} \widehat{\tau}(\varphi)$.

Proof Note that according to Lemma 2.5.3, $F = (s(A), +\infty)$ and $s(A) = \inf \{\lambda \in F\}$. For all $\varphi \in D_{++}(T)$ such that $\widehat{\tau}(\varphi) < +\infty$ and all $\lambda > \widehat{\tau}(\varphi)$, we have $\lambda \varphi - A\varphi \in X^*_+$ so $s(A) \leq \widehat{\tau}(\varphi)$ and consequently $s(A) \leq \inf_{\varphi \in D_{++}(T)} \widehat{\tau}(\varphi)$. Conversely, let $\lambda > s(A)$ be arbitrary. According to Lemma 2.5.3, there exists $\varphi \in D_{++}(T)$ such that $\lambda \varphi - A\varphi \in X^*_+$ so $\widehat{\tau}(\varphi) \leq \lambda$. Hence $\inf_{\varphi \in D(T) \cap X_{++}} \widehat{\tau}(\varphi) \leq s(A)$. Finally $s(A) = \inf_{\varphi \in D_{++}(T)} \widehat{\tau}(\varphi)$. $\qquad\square$

In $L^p(\nu)$ $(1 \leq p < +\infty)$, Theorem 2.5.3 can expressed as

Theorem 2.5.4 *Let $X = L^p(\nu)$ $(1 \leq p < +\infty)$ and $T : D(T) \subset X \to X$ be resolvent positive with spectral bound $s(T)$. Suppose that $B (\lambda - T)^{-1}$ satisfies (2.5.1) and is irreducible. Let*

$$A := T + B : D(T) \to X$$

be resolvent positive. Then

$$s(A) = \inf_{\varphi \in D_{++}(T)} \left(\sup \frac{A\varphi}{\varphi} \right)$$

where $\sup \frac{A\varphi}{\varphi}$ is the essential supremum of $\frac{A\varphi}{\varphi}$.

2.6 Desch's Theorem in AL-Spaces and Related Results

This theorem first appeared in an unpublished manuscript by Desch [106]. A simpler and more practical proof, based on Miyadera perturbations in Banach spaces [362], was later provided in [371]. Desch's perturbation theorem plays a central and ubiquitous role in this monograph. For the reader's convenience, we present a complete proof, slightly different from the one in [371], in Chap. 14. Some technical details from this proof are also used elsewhere in the monograph. Although we do not make use of Miyadera perturbations in general Banach spaces within this work, we recall the relevant theory (without proof) in Chap. 14 for completeness. We also explain how these perturbations relate to a suitable bounded variation (BV) property, which sheds light on why Miyadera perturbations are particularly well suited to positive C_0-semigroups in L^1-spaces.

An *AL*-space is a Banach lattice X with additive norm on the positive cone, i.e.,

$$\|x + y\| = \|x\| + \|y\|, \quad x, y \in X_+.$$

We point out that any *AL*-space is isomorphic to some $L^1(\nu)$-space, (see [327, Chap. II, Theorem 8.5]). We need first

Lemma 2.6.1 *Let $(U(t))_{t\geq 0}$ be a positive C_0-semigroup on an AL-space X with generator T and let*

$$B : D(T) \subset X \to X$$

be positive. Let $U_0(t) = U(t)$ and inductively

$$U_{j+1}(t)x = \int_0^t U_j(t - s)BU(s)x\,ds, \quad (x \in D(T)).$$

Then $U_j(t)$ extends uniquely as a bounded operator on X and $t \in [0, +\infty) \to U_j(t) \in \mathcal{L}(X)$ is strongly continuous.

Proof See the proof of Theorem 14.2.1. $\qquad\qquad\qquad\qquad\qquad\qquad\square$

We are ready to state

Theorem 2.6.1 *Let $(U(t))_{t\geq 0}$ be a positive C_0-semigroup on an AL-space X with generator T and let $B : D(T) \subset X \to X$ be positive. Then*

$$G := T + B : D(T) \subset X \to X$$

generates a positive semigroup $(V(t))_{t\geq 0}$ if and only if

$$\lim_{\lambda \to +\infty} r\left(B(\lambda - T)^{-1}\right) < 1.$$

In addition, $(V(t))_{t\geq 0}$ is given by the Dyson-Phillips series

$$V(t) = \sum_{j=0}^{\infty} U_j(t)$$

(converging in operator norm uniformly in $t \in [0, c]$, $c < +\infty$) and we have the Duhamel equations

$$V(t)\varphi = U(t)\varphi + \int_0^t V(t-s)BU(s)\varphi ds, \quad (\varphi \in D(T)) \tag{2.6.1}$$

and

$$V(t)\varphi = U(t)\varphi + \int_0^t U(t-s)BV(s)\varphi ds, \quad (\varphi \in D(T)). \tag{2.6.2}$$

Proof See the proof of Theorem 14.2.1.   □

We mention a useful technical result (see also Remark 14.2.1).

Theorem 2.6.2 ([226]) *Let the assumptions in Theorem 2.6.1 be satisfied. If $U(.)$ is continuous on $(0, +\infty)$ in operator norm then so are the terms $U_j(.)$ of the Dyson-Phillips expansion of $(V(t))_{t\geq 0}$ and therefore $V(.)$ is also continuous on $(0, +\infty)$ in operator norm.*

Here is a way to use Desch's theorem in $L^1(\nu)$.

Theorem 2.6.3 ([260]) *Let $(U(t))_{t\geq 0}$ be a positive C_0-semigroup on $L^1(\nu)$ with generator T and let*

$$B : D(T) \subset L^1(\nu) \to L^1(\nu)$$

be positive.
(i) If some power $\left(B(\lambda - T)^{-1}\right)^n$ is (weakly) compact in $L^1(\nu)$ then

$$T + B : D(T) \to L^1(\nu) \tag{2.6.3}$$

generates a C_0-semigroup $(V(t))_{t\geq 0}$ in $L^1(\nu)$.
(ii) If $B = B^{(1)} + B^{(2)}$ ($B^{(1)}, B^{(2)} \geq 0$) with $\lim_{\lambda \to +\infty} r\left(B^{(1)}(\lambda - T)^{-1}\right) < 1$ and $B^{(2)}(\lambda - T)^{-1}$ weakly compact in $L^1(\nu)$ then (2.6.3) generates a C_0-semigroup $(V(t))_{t\geq 0}$ in $L^1(\nu)$.

We give now a more general result in terms of *quasi-compactness* assumptions. As far as we know, this result appears here for the first time.

Definition 2.6.1 Let X be a Banach space. We say that $B \in \mathcal{L}(X)$ is quasi-compact if there exists $n \in \mathbb{N}$ and a compact operator C such that $\|B^n - C\| < 1$.

It is known (see [225, Proposition 2.2]) that $B \in \mathcal{L}(X)$ is quasi-compact if and only if

$$r_{ess}(B) < 1$$

where $r_{ess}(.)$ refers to the essential spectral radius (see (2.7.1) below for the definition of $r_{ess}(.)$). Note that if $B \in \mathcal{L}(X)$ is a positive quasi-compact operator on a Banach lattice X, the operator C above and $B^n - C$ need not be positive. We have

Theorem 2.6.4 *Let $(U(t))_{t\geq 0}$ be a positive C_0-semigroup with generator T on an AL-space X and let B be a positive T-bounded operator. If there exist $\lambda_0 > s(T)$ and an integer j such that $\left(B(\lambda_0 - T)^{-1}\right)^j$ is a quasi-compact contraction then $T + B : D(T) \subset X \to X$ generates a C_0-semigroup $(V(t))_{t\geq 0}$ in X.*

Proof Since $\left(B(\lambda_0 - T)^{-1}\right)^j$ is quasi-compact then there exist $n \in \mathbb{N}$ and a compact operator C such that

$$\left\| \left(B(\lambda_0 - T)^{-1}\right)^{nj} - C \right\| < 1.$$

Let

$$D := \left(B(\lambda_0 - T)^{-1}\right)^{nj} - C.$$

It follows from the positivity of B that

$$
\begin{aligned}
\left(B(\lambda - T)^{-1}\right)^{(n+1)j} &= \left(B(\lambda - T)^{-1}\right)^j \left(B(\lambda - T)^{-1}\right)^{nj} \\
&\leq \left(B(\lambda - T)^{-1}\right)^j \left(B(\lambda_0 - T)^{-1}\right)^{nj} \quad (\lambda \geq \lambda_0) \\
&= \left(B(\lambda - T)^{-1}\right)^j D + \left(B(\lambda - T)^{-1}\right)^j C
\end{aligned}
$$

where C is compact and $\|D\| < 1$. Since $\left(B(\lambda_0 - T)^{-1}\right)^j$ is a contraction then

$$\left\| \left(B(\lambda - T)^{-1}\right)^j \right\| \leq \left\| \left(B(\lambda_0 - T)^{-1}\right)^j \right\| \leq 1 \quad (\lambda \geq \lambda_0)$$

and

$$\left\| \left(B(\lambda - T)^{-1}\right)^j D \right\| \leq \left\| \left(B(\lambda_0 - T)^{-1}\right)^j \right\| \|D\| < 1 \quad (\lambda \geq \lambda_0).$$

Note that

$$
\begin{aligned}
T(\lambda - T)^{-1} &= (T - \lambda)(\lambda - T)^{-1} + \lambda(\lambda - T)^{-1} \\
&= -I + \lambda(\lambda - T)^{-1}
\end{aligned}
$$

is uniformly bounded as $\lambda \to +\infty$. For $x \in D(T)$

$$T(\lambda - T)^{-1}x = (\lambda - T)^{-1}Tx \to 0 \ (\lambda \to +\infty)$$

so the density of $D(T)$ in X gives

$$T(\lambda - T)^{-1}x = (\lambda - T)^{-1}Tx \to 0 \ (\lambda \to +\infty) \ (x \in X),$$

i.e., $(\lambda - T)^{-1}x \to 0 \ (\lambda \to +\infty)$ in $D(T)$ endowed with its graph norm. Since B is T-bounded than

$$B(\lambda - T)^{-1}x \to 0 \ (\lambda \to +\infty) \ (x \in X)$$

and consequently $B(\lambda - T)^{-1}x \to 0$ $(\lambda \to +\infty)$ uniformly on compact sets of X. Hence

$$\left\| B(\lambda - T)^{-1}C \right\| \to 0 \ (\lambda \to +\infty)$$

since C is compact. Similarly

$$\left\| \left(B(\lambda - T)^{-1}\right)^{j} C \right\| \to 0 \ (\lambda \to +\infty).$$

Finally, for λ large enough

$$\left\| \left(B(\lambda - T)^{-1}\right)^{(n+1)j} \right\| \leq \left\| \left(B(\lambda_0 - T)^{-1}\right)^{j} \right\| \|D\| + \left\| \left(B(\lambda - T)^{-1}\right)^{j} C \right\| < 1$$

and consequently $r(B(\lambda - T)^{-1}) < 1$. This ends the proof by virtue of Theorem 2.6.1. $\qquad\qquad\qquad\qquad\qquad\qquad\qquad\qquad\qquad\qquad\qquad\qquad\qquad\qquad\qquad\Box$

We point out that W. Desch's theorem is peculiar to AL-spaces, (see [7, Example 3.3] for a counter example in L^p-spaces for $p > 1$); however, for holomophic C_0-semigroups, the result is true in all Banach lattices.

Lemma 2.6.2 ([8]) *Let X be a complex Banach lattice and T be a generator of a holomorphic positive C_0-semigroup $(U(t))_{t\geq 0}$ on X and let $B : D(T) \to X$ be T-bounded and positive. Then $\lim_{\lambda\to+\infty} r\left(B\left(\lambda - T\right)^{-1}\right) < 1$ is a necessary and sufficient condition for*

$$G := T + B : D(T) \to X$$

to be a generator of a positive holomorphic C_0-semigroup $(V(t))_{t\geq 0}$.

We can then extend Theorems 2.6.3 and 2.6.4 to holomorphic C_0-semigroups in Banach lattices. As far as we know, this result appears here also for the first time.

Theorem 2.6.5 *Let X be a complex Banach lattice and T be a generator of a holomorphic positive C_0-semigroup $(U(t))_{t\geq 0}$ on X and let $B : D(T) \to X$ be T-bounded and positive.*

(i) If some power $\left(B(\lambda - T)^{-1}\right)^{n}$ is compact in X then $T + B : D(T) \to X$ generates a holomorphic positive C_0-semigroup $(V(t))_{t\geq 0}$ in X.

(ii) Let $B = B^{(1)} + B^{(2)}$ $(B^{(1)}, B^{(2)} \geq 0)$ with $\lim_{\lambda\to+\infty} r\left(B^{(1)}(\lambda - T)^{-1}\right) < 1$ and $B^{(2)}(\lambda - T)^{-1}$ compact in X then $T + B : D(T) \to X$ generates a holomorphic positive C_0-semigroup $(V(t))_{t\geq 0}$ in X.

Proof (i) The same proof as that of Theorem 2.6.3 (i) shows that there exists an iterate $\left(B(\lambda - T)^{-1}\right)^{m}$ such that

$$\left\| \left(B(\lambda - T)^{-1}\right)^{m} \right\|_{\mathcal{L}(X)} \to 0 \ (\lambda \to +\infty)$$

and therefore, by the spectral mapping theorem

$$\lim_{\lambda \to +\infty} r\left(B\left(\lambda - T\right)^{-1}\right) = 0.$$

We conclude with Lemma 2.6.2.

(ii) Lemma 2.6.2 insures that $T + B^{(1)} : D(T) \to X$ generates a positive holomorphic semigroup and

$$(\lambda - T - B^{(1)})^{-1} = (\lambda - T)^{-1} \sum_{j=0}^{\infty} \left(B^{(1)}(\lambda - T)^{-1}\right)^j .$$

Since $B^{(2)}(\lambda - T)^{-1}$ is compact then so is $B^{(2)}(\lambda - T - B^{(1)})^{-1}$ and consequently the part (i) of the proof insures that $T + B^{(1)} + B^{(2)} : D(T) \to X$ generates a positive holomorphic semigroup. $\qquad\square$

Finally, by using Lemma 2.6.2 and arguing as in Theorem 2.6.4, we can also use quasi-compactness arguments in the context of holomorphic C_0-semigroups on Banach lattices.

Theorem 2.6.6 *Let X be a Banach lattice and $(T(t))_{t\geq 0}$ be a positive holomorphic C_0-semigroup with generator T. Let B be a positive T-bounded operator. If there exist $\lambda_0 > s(T)$ and an integer j such that $\left(B(\lambda_0 - T)^{-1}\right)^j$ is a quasi-compact contraction then $T + B : D(T) \subset X \to X$ generates a positive holomorphic C_0-semigroup in X.*

There exist generation results using quasi-compactness arguments [349] in the more general context of Banach spaces but under dissipativity conditions which we do not assume here.

2.7 Essentially Compact Semigroups

For reader's convenience we recall

Definition 2.7.1 Let G be a closed operator on a complex Banach space X. We say that $\lambda_0 \in \mathbb{C}$ is an isolated eigenvalue of G with finite algebraic multiplicity if λ_0 is isolated in $\sigma(G)$ and is a pole of the resolvent $(\lambda - G)^{-1}$ with degenerate associated spectral projection. The algebraic multiplicity of λ_0 is the dimension of the range of the spectral projection. Even if it is not standard in the literature, we call discrete spectrum of G (and denote it by $\sigma_{discr}(G)$) the set of isolated eigenvalues of G with finite algebraic multiplicities.

We recall that a C_0-semigroup $\mathcal{U} = (U(t))_{t\geq 0}$ on a Banach space X has a type (or growth bound) $\omega(\mathcal{U}) = \lim_{t\to\infty} t^{-1} \ln \|U(t)\|$ and

$$r(U(t)) = e^{\omega(\mathcal{U})t}.$$

We denote by $\mathcal{K}(X)$ the ideal of compact operators on X and by

$$C(X) := \frac{\mathcal{L}(X)}{\mathcal{K}(X)}$$

the Calkin algebra endowed with the quotient norm

$$\left\| \widehat{T} \right\|_{C(X):} = \inf_{K \in \mathcal{K}(X)} \| T + K \| ; \quad \widehat{T} = T + \mathcal{K}(X).$$

This norm is multiplicative, i.e.,

$$\left\| \widehat{T_1}\widehat{T_2} \right\|_{C(X):} \leq \left\| \widehat{T_1} \right\|_{C(X):} \left\| \widehat{T_2} \right\|_{C(X):} .$$

The essential spectrum of $T \in \mathcal{L}(X)$, is defined as the spectrum of (the representative of) T in the Calkin algebra, i.e.,

$$\sigma_{ess}(T) = \sigma(T + \mathcal{K}(X))$$

(where $\sigma(T + \mathcal{K}(X))$ is the spectrum of $\widehat{T} = T + \mathcal{K}(X)$ in the Calkin algebra) and its essential spectral radius is

$$r_{ess}(T) = r(T + \mathcal{K}(X)) = \sup \{ |\lambda| ; \lambda \in \sigma_{ess}(T) \} \qquad (2.7.1)$$

where $r(T + \mathcal{K}(X))$ is the spectral radius of $\widehat{T} = T + \mathcal{K}(X)$ in the Calkin algebra. Note that $\sigma_{ess}(T)$ is also the complement of the Fredholm domain of T, (see [116, p. 40]). There exist several non equivalent definitions of the essential spectrum but the essential radius is the same for all of them (see [116, Corollary 4.11, p. 44]). We note that $r_{ess}(T)$ is the radius of the smallest disc outside which the spectrum of T consists at most of isolated eigenvalues with finite algebraic multiplicities.

Similarly, if $\mathcal{U} = (U(t))_{t \geq 0}$ is a C_0-semigroup on a Banach space X then there exists an essential type $\omega_{ess}(\mathcal{U}) \in [-\infty, \omega(\mathcal{U})]$ such that

$$r_{ess}(U(t)) = e^{\omega_{ess}(\mathcal{U})t} \quad (t \geq 0).$$

Definition 2.7.2 We say that a C_0-semigroup $\mathcal{U} = (U(t))_{t \geq 0}$ is essentially compact (or has a spectral gap) if $\omega_{ess}(\mathcal{U}) < \omega(\mathcal{U})$.

More generally, a bounded operator O is essentially compact (or has a spectral gap) if $r_{ess}(O) < r(O)$. This notion should not be confused with quasi-compactness which refers to $r_{ess}(O) < 1$. The interest of the existence of a spectral gap lies in the following very classical result.

Theorem 2.7.1 (See e.g. [359, 367], [46, Theorem 14.4]) *Let* $\mathcal{U} = (U(t))_{t \geq 0}$ *be a C_0-semigroup on a complex Banach space X with generator G. If* $\mathcal{U} = (U(t))_{t \geq 0}$ *has a spectral gap then for any* $\eta \in \mathbb{R}$ *such that*

$$\omega_{ess}(\mathcal{U}) < \eta \leq \omega(\mathcal{U})$$

$\sigma(G) \cap \{\mathrm{Re}\,\lambda \geq \eta\}$ *consists of a finite and non empty set of eigenvalues* $\{\lambda_1, \ldots, \lambda_m\}$ *with finite algebraic multiplicities. If* Q *is the finite rank spectral projection associated with this set of eigenvalues then there exists*

$$\gamma < \inf \left\{ \mathrm{Re}\,\lambda_j,\ 1 \leq j \leq m \right\}$$

such that

$$\|U(t)(I - Q)\| = O(e^{\gamma t}).$$

In other terms, the time asymptotic behavior $(t \to +\infty)$ *of* $\mathcal{U} = (U(t))_{t \geq 0}$ *is determined by its finite rank part* $U(t)Q$.

Remark 2.7.1 (Important) We point out that if the semigroup $(U(t))_{t \geq 0}$ is not continuous in operator norm on $(0, +\infty)$, then, a priori, the existence of a spectral gap of its generator does not imply the existence of a spectral gap for the semigroup itself. This classical observation goes back to [359]. In the literature, the term "spectral gap" is sometimes ambiguous, as it often refers to a spectral gap of the generator rather than of the semigroup. Unless explicitly stated otherwise, throughout this monograph we always refer to spectral gaps of semigroups.

In this monograph, once we establish that a positive C_0-semigroup $(U(t))_{t \geq 0}$ possesses a spectral gap, we do not elaborate further on its time asymptotic behavior, as this is fully covered by Theorem 2.7.1. On the other hand, if the semigroup C_0-semigroup $(U(t))_{t \geq 0}$ does not have a spectral gap, we (when possible) provide an explicit statement on the asymptotic stability of the rescaled semigroup $\left(e^{-s(G)t} U(t)\right)_{t \geq 0}$, under the assumption that the spectral bound $s(G)$ of its generator G is an eigenvalue. In the context of positive semigroups on Banach lattices, Theorem 2.7.1 admits a much simpler formulation, which further facilitates its application.

Theorem 2.7.2 (See e.g. [90, Theorem 9.11]) *Let* $(U(t))_{t \geq 0}$ *be a positive* C_0-*semigroup on a complex Banach lattice* X *with generator* G. *Suppose that* $(U(t))_{t \geq 0}$ *has a spectral gap. Then the peripheral spectrum of* G *reduces to* $s(G)$ *which is an isolated eigenvalue with finite algebraic multiplicity. If* $(U(t))_{t \geq 0}$ *is irreducible then the dominant eigenvalue of* G *is algebraically simple; the corresponding eigenfunction* e *is quasi-interior and the one dimensional projection associated with* $s(G)$ *is given by* $Qx = \langle e^*, x \rangle_{X^*, X}\, e$ *where* e^* *is a dual positive eigenfunctional of* G^* *associated with* $s(G^*) = s(G)$. *Finally*

$$U(t) = e^{ts(G)} Q + U(t)(I - Q) \text{ with } \|U(t)(I - Q)\| = O(e^{\gamma t}) \text{ and } \gamma < s(G).$$

In the context of stochastic C_0-semigroups in $L^1(\nu)$, we can get rid of the irreducibility assumption in Theorem 2.7.2. Indeed, 0 is a semi-simple eigenvalue of G (i.e., a simple pole of the resolvent $(\lambda - G)^{-1}$) and we have

$$U(t) = Q + U(t)(I - Q) \text{ with } \|U(t)(I - Q)\| = O(e^{\gamma t}) \text{ and } \gamma < 0$$

where Q is the (a priori multi-dimensional) spectral projection associated with the 0-eigenvalue of G, see the proof of [214, Corollary 4.13].

2.8   Spectra of Perturbed Operators

Spectral stability results play an important role in this monograph. We recall some of them.

Definition 2.8.1 Let $T : D(T) \subset X \to X$ be a closed operator on a Banach space X. The essentiel resolvent set of T (denoted by $\rho_e(T)$) is the union of its resolvent set $\rho(T)$ and the isolated eigenvalues of T with finite algebraic multiplicities.

The analytic Fredholm alternative [315] gives the following spectral result of interest for perturbed generators.

Theorem 2.8.1 ([364]) *Let $T : D(T) \subset X \to X$ be a closed operator on a Banach space X and let $B : D(T) \to X$ be T-bounded. Let there exist some $m \in \mathbb{N}$ such that $\left(B(\lambda - T)^{-1}\right)^m$ is compact then $\rho_e(T)$ and $\rho_e(T + B)$ have the same unbounded component. (If $X = L^1(\nu)$ then we can replace "compact" by "weakly compact").*

The analogous result for C_0-semigroups is provided by:

Theorem 2.8.2 ([245, Theorems 2.6 and 2.10]) *Let $(U(t))_{t \geq 0}$ be a C_0-semigroup on a Banach space X with generator T and let $B \in \mathcal{L}(X)$. Let $V(t) = \sum_{j=0}^{\infty} U_j(t)$ be the Dyson-Phillps expansion of the C_0-semigroup $(V(t))_{t \geq 0}$ with generator $T + B$. Then:*

(i) Let $m \in \mathbb{N}$. Then $U_m(t)$ is compact for all $t \geq 0$ if and only if the m-th remainder term $R_m(t) := \sum_{j=m}^{\infty} U_j(t)$ is compact for all $t \geq 0$. (If $X = L^1(\nu)$, then we can replace "compact" by "weakly compact").

(ii) If some remainder $R_m(t)$ is compact for all $t \geq 0$ then $(U(t))_{t \geq 0}$ and $(V(t))_{t \geq 0}$ have the same essential type, i.e., $\omega_{ess}(V) = \omega_{ess}(U)$. (If $X = L^1(\nu)$ then we can replace "compact" by "weakly compact").

We give also some useful technical results.

Theorem 2.8.3 ([245, Theorem 2.7 and Corollary 2.2]) *Let $(U(t))_{t \geq 0}$ be a C_0-semigroup on a Banach space X with generator T and let $B \in \mathcal{L}(X)$. Let $m \in \mathbb{N}$. Then*

(i) $t \to U_m(t)$ is continuous in operator norm if and only if $t \to R_m(t)$ is continuous in operator norm.

(ii) If $U_m(t)$ is compact for all $t \geq 0$ then $t \to U_{m+1}(t)$ is continuous in operator norm.

For *unbounded* perturbations B, we have the following result in the context of Desch's perturbation theory in $L^1(\nu)$.

Theorem 2.8.4 ([373], see also [245, Chap. 7]) *Let* $(U(t))_{t\geq 0}$ *be a positive C_0-semigroup on $L^1(\nu)$ with generator T and let*

$$B : D(T) \subset L^1(\nu) \to L^1(\nu)$$

be positive such that $\lim_{\lambda\to+\infty} r\left(B\left(\lambda - T\right)^{-1}\right) < 1$. *Let* $(V(t))_{t\geq 0}$ *be the semigroup with generator* $T + B : D(T) \subset X \to X$. *If some remainder* $R_m(t) = \sum_{j=m}^{\infty} U_j(t)$ *(of the Dyson-Phillips expansion) is weakly compact for t large enough, then* $\omega_{ess}(V) \leq \omega_{ess}(U)$.

2.9 Strong Integral of $\mathcal{W}(X, Y)$-valued Mappings

Here is a useful tool for *strongly* measurable mappings.

Theorem 2.9.1 ([330]) *Let* (Ω, μ) *be a finite measure space and let X, Y be two Banach spaces. Let $\mathcal{W}(X, Y) \subset \mathcal{L}(X, Y)$ be the closed subspace of weakly compact operators. If*

$$U : (\Omega, \nu) \to \mathcal{W}(X, Y)$$

is bounded and strongly measurable, i.e.,

$$\omega \in (\Omega, \nu) \to U(\omega)x \in Y \ \ (x \in X)$$

is measurable, then

$$x \in X \to \int_{\Omega} U(\omega)x \, \mu(d\omega) \in Y$$

(defined by a strong integral) belongs to $\mathcal{W}(X, Y)$.

Remark 2.9.1 Theorem 2.9.1 is also true if we replace $\mathcal{W}(X, Y)$ by $\mathcal{K}(X, Y)$, the closed subspace of compact operators [189, 372].

2.10 On Strong Convergence to Ergodic Projection

We recall first some basic notions around ergodic C_0-semigroups. Let X be a Banach space and let $(Z(t))_{t\geq 0}$ be a bounded C_0-semigroup on X with generator G. We say that $(Z(t))_{t\geq 0}$ is mean ergodic if

$$\lim_{t\to+\infty} \frac{1}{t} \int_0^t Z(s)x\,ds \text{ exists } \forall x \in X.$$

For short, we will say "ergodic" instead of "mean ergodic". The following classical result provides various characterizations of ergodic C_0-semigroups.

Theorem 2.10.1 (See e.g. [97, Theorem 5.1, p. 123]) *Let $(Z(t))_{t \geqslant 0}$ be a contraction C_0-semigroup on a Banach space X and let G be its generator. Then the following assertions are equivalent:*

(i) $\forall x \in X$, $\lim_{t \to +\infty} \frac{1}{t} \int_0^t Z(s)x ds$ exists.
(ii) $\forall x \in X$, $\lim_{\lambda \to +0_+} \lambda(\lambda - G)^{-1}x$ exists.
(iii) $\forall x \in X$, $\{\lambda(\lambda - G)^{-1}x; \ 0 < \lambda \leq 1\}$ is relatively weakly compact.
In this case,

$$\lim_{t \to +\infty} \frac{1}{t} \int_0^t Z(s)x ds = \lim_{\lambda \to +0_+} \lambda(\lambda - G)^{-1}x$$

and

$$P : x \in X \to \lim_{t \to +\infty} \frac{1}{t} \int_0^t Z(s)x ds$$

is a projection on $Ker(G)$, the kernel of G, along $\overline{R(G)}$, the closure of the range of G; in particular $X = Ker(G) \oplus \overline{R(G)}$.

Besides reflexive Banach spaces where all uniformly bounded C_0-semigroups are ergodic (see e.g. [140, Theorem 8.20, p. 58]), we recall an ergodicity theorem in L^1 spaces.

Theorem 2.10.2 (See e.g. [97, Theorem 7.3, p. 174]) *Let $(Z(t))_{t \geqslant 0}$ be a positive uniformly bounded C_0-semigroup on $L^1(\nu)$ with generator G. Suppose that $(Z(t))_{t \geqslant 0}$ is irreducible and that $Ker(G) \neq \{0\}$. Then*

(i) $Ker(G)$ is one-dimensional and is spanned by some $e > 0$ a.e.
(ii) $Ker(G')$ is one-dimensional and is spanned by some $\varphi > 0$ a.e.
(ii) $(Z(t))_{t \geqslant 0}$ is ergodic and the ergodic projection is given by

$$Pf = (\int_\Omega \varphi f d\nu)e \ \ \forall f \subset L^1(\nu), \tag{2.10.1}$$

where φ and e are chosen so that $\int_\Omega \varphi e d\nu = 1$.

Here is a useful sufficient condition of mean ergodicity in $L^1(\nu)$ spaces.

Proposition 2.10.1 ([12, Proposition 4.3.14, p. 268]) *Let $(Z(t))_{t \geq 0}$ be a bounded positive C_0-semigroup with generator G on $L^1(\nu)$. If there exists $f \in L^1(\nu)$, $f > 0$ ν-a.e. such that $Z(t)f \leq f$ $(t \geq 0)$ then $(Z(t))_{t \geq 0}$ is mean ergodic.*

In this monograph, our focus extends beyond mere ergodic convergence; we are interested in the strong convergence of positive C_0-semigroups on $L^1(\nu)$. A natural and general strategy for studying such convergence is through a spectral approach, relying on Ingham's Tauberian theorem (see, e.g., [88, Theorem 1.1]). This method can be adapted to perturbation settings, as illustrated in [215, Remark 6.4]. See also

[12] for a comprehensive treatment of Tauberian techniques. However, in this work, we adopt a less general but more flexible route, more directly suited to the settings we consider. To prepare for this, we begin by introducing a few foundational definitions.

Definition 2.10.1 We say that $O \in L^1(\nu)$ is substochastic if O is a positive contraction. We say that $O \in L^1(\nu)$ is stochastic if $\|Of\|_{L^1(\nu)} = \|f\|_{L^1(\nu)}$ for all $f \in L^1_+(\nu)$. Similarly, a C_0-semigroup $(Z(t))_{t \geq 0}$ on $L^1(\nu)$ is said to be substochastic (resp stochastic) if $Z(t)$ is substochastic (resp stochastic) for all $t \geq 0$.

Definition 2.10.2 Any $f \in L^1_+(\nu)$ with $\|f\|_{L^1(\nu)} = 1$ is called a density. An invariant density of a stochastic semigroup $(Z(t))_{t \geq 0}$ on $L^1(\nu)$ (with generator G) is a density f such that $Z(t)f = f$ $(t \geq 0)$ or equivalently if $f \in D(G)$ and $Gf = 0$. The invariant density f is said to be asymptotically stable if $\lim_{t \to +\infty} \|Z(t)\varphi - f\| = 0$ for any density φ. We say also that $(Z(t))_{t \geq 0}$ is asymptotically stable.

Definition 2.10.3 A positive C_0-semigroup $(Z(t))_{t \geq 0}$ on $L^1(\nu)$ is strongly stable if $\lim_{t \to +\infty} \|Z(t)\varphi\| = 0$ for all $\varphi \in L^1(\nu)$.

The asymptotic stability of stochastic C_0-semigroups having an invariant density is a key issue in the literature on piecewise deterministic processes [320]. Two approaches of the strong convergence to ergodic projection are given here.

2.10.1 Strong Convergence Versus Partially Integral Semigroups

Several results on the asymptotic stability of stochastic C_0-semigroups exist (see [320] and references therein). Here, we restrict ourselves to one such result. We start with the following

Definition 2.10.4 A stochastic C_0-semigroup $(Z(t))_{t \geq 0}$ on $L^1(\nu)$ is partially integral for some $t > 0$ if there exists a non trivial positive integral operator K_t with kernel $k_t(x, y)$ such that

$$Z(t)f(x) \geq \int_\Omega k_t(x, y) f(y) \nu(dy) \quad (f \in L^1_+(\nu)).$$

We are ready to state

Theorem 2.10.3 ([305]) *Let $(Z(t))_{t \geq 0}$ be a stochastic C_0-semigroup on $L^1(\nu)$. Suppose that $(Z(t))_{t \geq 0}$ is partially integral for some $t > 0$ and has a unique invariant density $f > 0$ a.e. Then $(Z(t))_{t \geq 0}$ is asymptotically stable.*

This result turns out to be true in $L^p(\nu)$ spaces or even in general Banach lattices with order continuous norm with an appropriate (abstract) concept of kernel operators which extends the usual notion of integral operator in $L^p(\nu)$. We mention that there exist quite general results for integral C_0-semigroups, (see [11, 145] and references therein); however, most semigroups occuring in the applied contexts of this monograph are partially integral only.

Theorem 2.10.4 ([134]) *Let $(Z(t))_{t\geq 0}$ be a bounded irreducible C_0-semigroup with generator G on a Banach lattice X with order continuous norm. Suppose that $(Z(t))_{t\geq 0}$ has a non trivial fixed point e and, for some $t > 0$, $Z(t)$ is partially integral. Then*

$$\lim_{t\to +\infty} Z(t)x = \langle e', x\rangle e \ \ (x \in X)$$

for some strictly positive fixed point e' of the dual semigroup $\left(Z'(t)\right)_{t\geq 0}$.

The proof of Theorem 2.10.4 is quite involved and is based in particular on an extension to Banach lattices of the 0–2 law of C_0-semigroups in Lebesgue spaces [145] (see also [286, Theorem 2.6, p. 346]) and on the following key result on triviality of the peripheral point spectrum

Theorem 2.10.5 ([134]) *Let $(Z(t))_{t\geq 0}$ be a bounded irreducible C_0-semigroup with generator G on a Banach lattice X with order continuous norm. Suppose that for some $t > 0$, $Z(t)$ is partially integral. Then the peripheral point spectrum of G is trivial, i.e., $\sigma_p(G) \cap \{i\mathbb{R}\} \subset \{0\}$.*

We refer to [135] and references therein for further developments on the subject. See also [13] for a well-documented survey on the long-time behavior of positive irreducible C_0-semigroups, including several applications. A useful consequence of Theorem 2.10.3 in a perturbation context is

Theorem 2.10.6 *Let $(U(t))_{t\geq 0}$ be a substochastic C_0-semigroup on $L^1(\nu)$ with generator T and let*

$$B : D(T) \to L^1(\nu)$$

be positive and such that $\lim_{\lambda\to +\infty} r\left(B\left(\lambda - T\right)^{-1}\right) < 1$. Let the conservativity assumption

$$\int_\Omega (T\varphi + B\varphi)\, d\nu = 0 \ \ (\varphi \in D(T)) \tag{2.10.2}$$

be satisfied. Then $G := T + B : D(T) \to L^1(\nu)$ generates a stochastic C_0-semigroup $(V(t))_{t\geq 0}$. Suppose that $(V(t))_{t\geq 0}$ is irreducible with $Ker(G) \neq \{0\}$ and denote by P its ergodic projection (2.10.1) on $Ker(G)$. If there exists some positive integer m such that, for some $t > 0$, $R_m(t) = \sum_{j=m}^{+\infty} U_j(t)$ is a non-trivial (i.e., non-zero) compact operator, then $\lim_{t\to +\infty} V(t)f = Pf \ \ \forall f \in L^1(\nu)$.

Proof The generation of $(V(t))_{t\geq 0}$ follows from Theorem 2.6.1 and the stochasticity follows from the conservativity assumption (2.10.2). The compactness of $R_m(t)$ implies that $R_m(t)$ is an integral operator (see e.g. [115, p. 508]). Hence the stochastic C_0-semigroup $(V(t))_{t\geq 0}$ is partially integral since $R_m(t)$ is a non zero integral operator and

$$V(t)f = W_m(t)f + R_m(t)f \geqslant R_m(t)f \ \forall f \in L^1_+(\nu).$$

We conclude with Theorem 2.10.3. $\square$

We give now a characterization of the existence of an invariant density (i.e., $Ker(G) \neq \{0\}$) in the perturbation context above.

Proposition 2.10.2 ([269, Theorem 5]) *Let* $(U(t))_{t \geqslant 0}$ *be a substochastic strongly stable* C_0-*semigroup on* $L^1(\nu)$ *with generator* T *and let* $B : D(T) \to L^1(\nu)$ *be positive and such that* $\lim_{\lambda \to +\infty} r\left(B\left(\lambda - T\right)^{-1}\right) < 1$. *Let the conservativity assumption* $\int_{\Omega} (T\varphi + B\varphi)\, d\nu = 0$ $(\varphi \in D(T))$ *be satisfied. Then* $G := T + B$ *has a non trivial kernel if and only if* 1 *is an eigenvalue of* $B(0_+ - T)^{-1}$ *associated to an eigenvector belonging to the range of* T.

In case the stochastic C_0-semigroup $(V(t))_{t \geq 0}$ has no invariant density then a *sweeping* phenomenon may occur. We give here just one result in this direction; see [320] for more information. We recall first that any positive operator O on $L^1(\nu)$ can be extended to nonnegative measurable functions h by monotonicity as a pointwise limit

$$Oh := \lim_{n \to +\infty} Oh_n$$

where $h_n \in L^1_+(\nu)$ and $h_n \to h$ monotonically and pointwisely and this limit does not depend of the choice of the approximating sequence $(h_n)_n \subset L^1_+(\nu)$ [125, Chap. I].

Theorem 2.10.7 ([318]) *Let* $(Z(t))_{t \geq 0}$ *be a stochastic* C_0-*semigroup on* $L^1(\nu)$. *Suppose that* $(Z(t))_{t \geq 0}$ *is partially integral and has no invariant density. Let there exists a measurable function* $h > 0$ *a.e. such that* $Z(t)h \leq h$. *Then, for any measurable subset* $\Xi \subset \Omega$ *such that* $\int_{\Xi} h(y)\nu(dy) < +\infty$,

$$\int_{\Xi} V(t)f\, d\nu \to 0 \ (t \to \infty), \quad (f \in L^1_+(\nu)).$$

2.10.2 The 0–2 Law of Semigroups in a Perturbation Context

In a perturbation context, the 0–2 law for C_0-semigroups in Lebesgue spaces [145] provides a proof of strong L^1-convergence that does not rely a priori on Theorems 2.10.4 or 2.10.5.

Theorem 2.10.8 ([269]) *Let* $(U(t))_{t \geqslant 0}$ *be a substochastic* C_0-*semigroup on* $L^1(\nu)$ *with generator* T. *Let* $B : D(T) \to L^1(\nu)$ *be positive, non-trivial (i.e., non-zero) such that* $r\left(B\left(\lambda - T\right)^{-1}\right) < 1$ *for real* λ *large enough and*

$$\int_{\Omega} (T\varphi + B\varphi)\, d\nu \leq 0 \ (\varphi \in D_+(T)).$$

Let $(V(t))_{t \geqslant 0}$ be the substochastic C_0-semigroup generated by

$$A = T + B : D(T) \to L^1(\nu).$$

Suppose that $(V(t))_{t \geqslant 0}$ is irreducible with $Ker(A) \neq \{0\}$ and denote by P the ergodic projection (2.10.1) on $Ker(A)$. If there exists some positive integer m such that the mapping

$$(0, +\infty) \ni t \to R_m(t) = \sum_{j=m}^{+\infty} U_j(t) \in \mathcal{L}(L^1(\nu))$$

is continuous in operator norm, then $(V(t))_{t \geq 0}$ is asymptotically stable.

Remark 2.10.1 (*Open question*) If B is a bounded operator then the weak compactness of some $U_n(t)$ $(t \geq 0)$ implies the compactness of $U_{2n+1}(t)$ $(t \geq 0)$ and the continuity in operator norm of the mapping $t \to U_{2n+2}(t)$ (or equivalently the continuity in operator norm of the mapping $t \to R_{2n+2}(t)$) see [245, Chap. 2]. Thus, if B is bounded, weak compactness arguments provide us with a way to check the continuity in operator norm of some remainder term of the Dyson-Phillips expansion. The case of general T-bounded $B's$ is an open question even if we strongly suspect this result to be true.

Remark 2.10.2 In this monograph, we have chosen to impose irreducibility assumptions frequently, even when it is sometimes possible to proceed without them, though doing so would come at the expense of much more cumbersome and less transparent statements.

2.11 Honesty Theory of Kato-Voigt Semigroups

2.11.1 Kato-Voigt Perturbation Theorem

An important perturbation result for substochastic C_0-semigroups used repeatedly in this monograph is the following.

Theorem 2.11.1 *Let $(U(t))_{t \geq 0}$ be a substochastic C_0-semigroup on $L^1(\nu)$ with generator $T : D(T) \subset L^1(\nu) \to L^1(\nu)$ and let*

$$B : D(B) \subset L^1(\nu) \to L^1(\nu)$$

be positive (i.e., $B\varphi \in L^1_+(\nu)$ if $\varphi \in D(B) \cap L^1_+(\nu)$) and such that $D(B) \supset D(T)$ and

$$\int_\Omega (T\varphi + B\varphi)\, d\nu \leq 0 \quad (\varphi \in D_+(T)). \tag{2.11.1}$$

Then there exists a unique extension $G \supset T + B : D(T) \to L^1(\nu)$ which generates a minimal substochastic C_0-semigroup $(V(t))_{t \geq 0}$ in $L^1(\nu)$.

By minimal we mean that $\widehat{V}(t) \geq V(t)$ for any substochatic C_0-semigroup $\left(\widehat{V}(t)\right)_{t \geq 0}$ generated by an extension of $T + B$. Theorem 2.11.1 originates from the classical paper by Kato on Kolmogorov's differential equations [180]. An alternative proof was later provided by Voigt [369] as a consequence of Miyadera perturbation theory [362]. In this monograph, we refer to the substochastic C_0-semigroup $(V(t))_{t \geq 0}$, constructed in Theorem 2.11.1, as the Kato-Voigt semigroup, and its generator $G \supset T + B$ as the Kato-Voigt generator. This theorem has been revisited by several authors to address problems arising in Mathematical Biology, Kinetic Theory, and related fields, and has inspired a variety of developments, some of them extending even beyond the L^1-setting (see [20, 30, 34, 130, 347–350]); in particular, it plays an important role in the monograph [33].

We point out that in general

$$T + B : D(T) \to L^1(\nu)$$

need not be closed (but is closable) and G may be even a proper extension of $\overline{T + B}$. A priori, three cases may occur:

- $G = T + B$. This is the genuinely honest case which occurs if and only if $r_\sigma\left(B(\lambda - T)^{-1}\right) < 1 \ (\lambda > 0)$.[2]
- $T + B$ is *not* closed and $G = \overline{T + B}$.
- $G \supsetneq \overline{T + B}$. We name this the dishonest case.

We point out that the dishonest case may arise in certain (very) singular situations, see [33].

Definition 2.11.1 We say that the Kato-Voigt semigroup $(V(t))_{t \geq 0}$ is honest if $G = T + B$ (*genuine honesty*) or $G \neq T + B$ and $G = \overline{T + B}$.

A general framework built around the Kato-Voigt perturbation theorem, known as honesty theory, is developed in the monograph [33]. This theory was further revisited and extended in [261], where several new developments were introduced. The aim of this section is to present, without proofs, a selection of results from [261] that will be relevant for our purposes. We note first that the dissipativity condition (2.11.1) implies

$$\left\| B(\lambda - T)^{-1} \right\|_{\mathcal{L}(L^1(\nu))} \leq 1 \ (\lambda > 0) \tag{2.11.2}$$

and consequently

$$T + rB : D(T) \to L^1(\nu) \ (r \in [0, 1)$$

---

[2] To avoid an ambiguity, in this section, we denote a spectral radius by r_σ instead of r since r is used for a different purpose.

generates a contraction C_0-semigroup $\left(V^{(r)}(t)\right)_{t\geq 0}$ in $L^1(\nu)$ and its resolvent is given by

$$(\lambda - T - rB)^{-1} = (\lambda - T)^{-1} \sum_{j=0}^{\infty} r^j \left(B(\lambda - T)^{-1}\right)^j \quad (\lambda > 0)$$

where the convergence holds in operator norm. This can be proved for instance by means of Theorem 2.6.1.

Note that $(0, +\infty) \ni \lambda \to B(\lambda - T)^{-1}\varphi$ is non increasing so (2.11.2) implies that the limit $\lim_{\lambda \to 0_+} B(\lambda - T)^{-1}\varphi$ exists in $L^1(\nu)$ (for all $\varphi \in L^1_+(\nu)$ and then for all $\varphi \in L^1(\nu) = L^1_+(\nu) - L^1_+(\nu)$) and defines a contraction on $L^1(\nu)$. We will denote symbolically this contraction by $B(0_+ - T)^{-1}$ or simply $B(0 - T)^{-1}$.

Since $V^{(r)}(t)$ is a contraction for any $r \in [0, 1)$ then it follows, by the monotone convergence theorem, that the strong limit of $\left(V^{(r)}(t)\right)_{t\geq 0}$ as $r \to 1_-$ exists

$$V(t) = s \lim_{r \to 1_-} V^{(r)}(t)$$

providing us with a substochastic C_0-semigroup $(V(t))_{t\geq 0}$ in $L^1(\nu)$ with generator G. Moreover, by the monotone convergence theorem, the resolvent of G is given by

$$(\lambda - G)^{-1} f = s \lim_{r \to 1_-} (\lambda - T - rB)^{-1} f \quad (\lambda > 0) \ (f \in L^1_+(\nu))$$

and consequently, since $L^1(\nu) = L^1_+(\nu) - L^1_+(\nu)$, for all $f \in L^1(\nu)$ and $\lambda > 0$

$$(\lambda - G)^{-1} f = (\lambda - T)^{-1} \sum_{j=0}^{\infty} \left(B(\lambda - T)^{-1}\right)^j f \qquad (2.11.3)$$

where, a priori, the series converges strongly only (not in operator norm). Similarly, the semigroup is given by the Dyson-Phillips expansion

$$V(t) = \sum_{j=0}^{\infty} U_j(t) \qquad (2.11.4)$$

which converges strongly only (not in operator norm) uniformly in $t \in [0, c]$ $(0 < c < +\infty)$. Moreover (see [314]) we have the Duhamel equation on $D(T)$

$$V(t)\varphi = U(t)\varphi + \int_0^t V(t-s)BU(s)\varphi ds, \quad (\varphi \in D(T)) \qquad (2.11.5)$$

but a priori not the second one (2.6.2) because $D(T) \ (\subsetneq D(G))$ need not be invariant under $(V(t))_{t\geq 0}$. We note however that under additional conditions, Eq. (2.6.2) is true for $\varphi \in D(G)$, see Corollary 3.4.1 below.

2.11.2 Outline of Honesty Analysis

The analysis of whether $G = \overline{T + B}$ is quite involved and follows the following construction. Let the nonnegative functional

$$\widehat{c} : f \in D(G) \to - \int_{\Omega} Gf \, d\nu$$

and let

$$c : f \in D(T) \to - \int_{\Omega} (T + B) f \, d\nu$$

be the restriction of $\widehat{c}$ to $D(T)$. It turns out that c extends "continuously" (in a suitable sense) to $D(G)$ as a nonnegative functional in the following way. Let

$$c_\lambda((\lambda - G)^{-1}f) := \lim_{r \to 1_-} c\left(\lambda - T - rB\right)^{-1}f\right) \quad (\lambda > 0, \ f \in L^1_+(\nu)).$$

By construction

$$c_\lambda((\lambda - G)^{-1}f) \leq \widehat{c}((\lambda - A)^{-1}f) \quad (\lambda > 0, \ f \in L^1_+(\nu)).$$

We extend c_λ to $D(G)$ by

$$c_\lambda((\lambda - G)^{-1}f) = c_\lambda((\lambda - G)^{-1}f_+) - c_\lambda((\lambda - G)^{-1}f_-) \ (\lambda > 0, \ f \in L^1(\nu)).$$

Proposition 2.11.1 ([261, Proposition 1.1])
  (i) The restriction of c_λ to $D(T)$ is equal to c.
  (ii) c_λ is λ-independent.
  (iii) The extension of c to $D(G)$ by $c := c_\lambda$ is continuous for the graph norm of G, c and $\widehat{c}$ coincide on $D(\overline{T + B})$ and

$$c(f) \leq \widehat{c}(f), \quad f \in D(G) \cap L^1_+(\nu).$$

We introduce a continuous nonnegative functional β_λ on $L^1(\nu)$ by

$$\beta_\lambda(f) = \widehat{c}((\lambda - G)^{-1}f) - c((\lambda - G)^{-1}f).$$

Then

Proposition 2.11.2 ([261, Lemma 1.4 and Corollary 1.5])
  (i) $\beta_\lambda(f) = \lim_{j \to \infty} \int \left(B(\lambda - T)^{-1}\right)^j f \, d\nu$
  (ii) If $\beta_\lambda \neq 0$ then β_λ is an eigenvector of the dual operator $\left(B(\lambda - T)^{-1}\right)'$ in $L^\infty(\nu)$ associated to the eigenvalue 1; it is the maximal eigenvector ≤ 1.

We have a characterization of $D(\overline{T + B})$.

Proposition 2.11.3 ([261, Proposition 1.6]) *Let* $f \in D(G) \cap L^1_+(\nu))$. *Then* $f \in D(\overline{T + B})$ *if and only if* $c(f) = \widehat{c}(f)$.

Note that for any $f \in L^1(\nu)$

$$\int_s^t V(r) f \, dr \in D(G) \text{ and } V(t) f - V(s) f = G \int_s^t V(r) f \, dr$$

and by integration

$$\|V(t) f\| = \|V(s) f\| - \widehat{c}(\int_s^t V(r) f \, dr) \ (f \in L^1_+(\nu)).$$

We are ready to introduce the concept of honesty.

Definition 2.11.2 Let $f \in L^1_+(\nu)$. The trajectory $\{V(t) f; \ t \geq 0\}$ is said to be honest if

$$\|V(t) f\| = \|V(s) f\| - c(\int_s^t V(r) f \, dr) \ (f \in L^1_+(\nu)) \ (0 \leq s \leq t < +\infty)$$

or equivalently if
$$\widehat{c}(\int_s^t V(r) f \, dr) = c(\int_s^t V(r) f \, dr).$$

The semigroup $(V(t))_{t \geq 0}$ is said to be honest if any trajectory emanating from $f \in L^1_+(\nu)$ is honest or equivalently (see Proposition 2.11.3) if $G = \overline{T + B}$.

Remark 2.11.1 In the conservative case

$$\int_\Omega (T\varphi + B\varphi) \, d\nu = 0 \ (\varphi \in D(T)),$$

the functional c is equal to zero; in this case, if a trajectory $\{V(t) f; \ t \geq 0\}$ is not honest then $\|V(t) f\| < \|f\|$, i.e., a mass loss occurs and the C_0-semigroup $(V(t))_{t \geq 0}$ is not stochastic despite the conservativity assumption.

We give now a characterization of honest trajectories.

Theorem 2.11.2 ([261, Theorem 2.2]) *Let* $f \in L^1_+(\nu)$. *Then the trajectory* $\{V(t) f; \ t \geq 0\}$ *is honest if and only if*

$$\beta_\lambda(f) = \lim_{j \to \infty} \int \left(B(\lambda - T)^{-1} \right)^j f \, d\nu = 0$$

for some (or equivalently all) $\lambda > 0$.

Some sufficient conditions of honesty are given in the following proposition.

Proposition 2.11.4 ([261, Proposition 2.6 and Remark 2.3]) *Let $\lambda > 0$.*

(i) Let $f \in L^1_+(\nu)$. If $B(\lambda - T)^{-1} f \le f$ then the trajectory $\{V(t)f;\ t \ge 0\}$ is honest.

(ii) Let $f \in D(T) \cap L^1_+(\nu)$. If $Tf + Bf \le \lambda f$ then the trajectory $\{V(t)f;\ t \ge 0\}$ is honest.

(iii) If a trajectory $\{V(t)f;\ t \ge 0\}$ with $f > 0$ a.e. is honest then all trajectories emanating from $L^1_+(\nu)$ are honest.

Remark 2.11.2 In this section, we have presented a *honesty theory* for "additive" perturbations. In the (concrete) context of Transport Theory with stochastic boundary operators, it is possible to build in the same spirit (albeit technically quite different) a honesty theory for "boundary" perturbations [259], (see also [21]).

2.12 Comments

Much more information on the spectral properties of positive operators and semi-groups on Banach lattices, as well as their time asymptotics, can be found in [46, 90] and, of course, in the comprehensive treatise [286]. We also particularly mention [101] which, although less systematic in its presentation, offers a broader and more varied perspective on the subject, enriched with numerous examples. We refer to [13] for a broad and insightful overview of positive irreducible C_0-semigroups and their long-time behavior. For results on asymptotic stability of partially integral C_0-semigroups, we mention [135, 319], whose approaches are particularly well suited to semigroups arising from semiflows with jumps which can be treated within the partially integral framework. The spectral radius of positive operators exhibits many deep and useful properties, as discussed in [1]; see also [133] for results on the strict monotonicity of the spectral radius in domination settings.

Variational characterizations of the leading eigenvalue, similar to those presented in Sect. 2.5, also exist in the context of second-order elliptic operators; see [52].

We also mention the existence of a probabilistic approach to the existence of invariant densities for semiflows with jumps, developed in [57]. While this approach does not appear to have been applied to neutron transport problems, it offers promising potential and certainly merits further investigation in that context.

For completeness, we mention a robust approach to strong convergence based on relative entropy techniques [176, 216, 231, 304, 361, 366]. These methods have been successfully employed even in nonlinear contexts.

In Sect. 2.8, we focused on compactness results of remainder terms of Dyson-Phillips expansions for perturbed C_0-semigroups; we refer also to [72] for further results in this direction. Some developments in other directions are given in [69, 346]. It was conjectured in [245, p. 25], in the more general context of perturbed C_0-semigroups in Banach spaces (at least when B is a bounded operator), that a partial spectral mapping theorem holds

$$\sigma\left(V(t)\right) \cap \left\{|\mu| > e^{t\omega_{ess}(V)}\right\} = e^{t\{\sigma(A)\cap\{\operatorname{Re}\lambda > \omega_{ess}(V)\}\}}$$

provided there exists $n \in \mathbb{N}$ such that $t \to R_n(t)$ is continuous in operator norm. This conjecture was confirmed in [71], where the critical type $\omega_{crit}(V)$ replaces the essential type $\omega_{ess}(V)$ after the introduction of the notions of *critical* spectrum σ_{crit} and *critical* type $\omega_{crit}(V)$ [287]. These concepts prove more suitable than the essential spectrum and essential type when dealing with problems exhibiting lack of compactness; see also [253, 254] for applications to neutron transport in unbounded geometries.

See the open question in Remark 2.10.1. Theorem 2.9.1 with $Y = L^1(\nu)$ has a quite simple proof relying on the Dunford-Pettis criterion of weak compactness (see [249]). Theorem 2.9.1 is also true for other subspaces of $\mathcal{L}(W, Y)$, see [372].

In this monograph, we use quite often the fact that in $L^1(\nu)$ spaces the essential spectrum of a bounded operator is stable under weakly compact perturbations; this classical result can be found in [181]. Actually, if we denote by $\mathcal{W}(L^1(\nu))$ the closed ideal of weakly compact operators on $L^1(\nu)$ then the spectrum of (the class of) T in the algebra $\frac{\mathcal{L}(L^1(\nu))}{\mathcal{W}(L^1(\nu))}$ is the same as its spectrum in the Calkin algebra $\frac{\mathcal{L}(L^1(\nu))}{\mathcal{K}(L^1(\nu))}$ [139]; see also [196]. Behind this result lies the fact that L^1 spaces possess the Dunford-Pettis property, whereby any weakly compact operator maps a weakly compact set into a relatively compact one; in particular, the square of a weakly compact operator is compact (see [115], Corollary 13, p. 510). This fact is frequently used throughout this monograph.

We have commented extensively on asymptotically stable C_0–semigroups and, conversely, on sweeping C_0–semigroups; this suffices for the problems considered in this monograph. However, more complex behaviors may arise. For such richer asymptotic $(t \to \infty)$ decompositions of certain partially integral stochastic C_0-semigroups into asymptotically stable and sweeping components, we refer to [306].

In the same spirit, this monograph focuses on C_0-semigroups with spectral gaps and their "finite-dimensional" time asymptotics, motivated by the physical problems under consideration. In these cases, the spectral bound of the generator is strictly dominant, placing us in the so-called regime of balanced exponential growth. However, more complex regimes can arise. For instance, if a bounded irreducible semigroup on Lebesgue spaces has zero as the spectral bound of its generator and this point is a pole of the resolvent, and if the boundary point spectrum of the generator is not reduced to a single point, then the semigroup is asymptotically periodic (see [46, Theorem 14.19]). In the context of perturbed semigroups, it is interesting to investigate potential connections among three time-asymptotic regimes: asymptotic periodicity, decomposition into asymptotically stable and sweeping components, and the scattering theory regime discussed in Sect. 3.4.

Note that, according to Theorem 2.10.5, the regime of asymptotic periodicity cannot occur if the semigroup is partially integral, that is, if it dominates a kernel operator. This explains why the various perturbed C_0-semigroups studied in this monograph, for which the Dyson-Phillips expansion exhibits a weakly compact (or at least partially integral) remainder, do not display asymptotic periodicity.

Kato's classical paper [180] has a notable non-commutative counterpart by Davies [96] in the framework of quantum dynamical semigroups. Similarly, Sect. 2.11 admits a non-commutative extension to the Banach space of trace class operators on Hilbert spaces, as developed in [258]. A broader generalization within *state spaces* is provided in [25], with a non-autonomous version explored in [26]. For further developments in the realm of quantum dynamical semigroups, we also refer to [379].

Chapter 3
Time Asymptotics of Kato-Voigt Semigroups

3.1 Chapter Aims

In this chapter, we consider Kato-Voigt stochastic semigroups $(\mathcal{J}(t))_{t\geq 0}$ in L^1 spaces which govern abstract Cauchy problems of the form

$$\frac{df}{dt} = Mf + Pf, \quad f(0) = f_0$$

where M is a generator of a substochastic C_0-semigroup $(\mathcal{M}(t))_{t\geq 0}$ on $L^1(\nu)$ and $P : D(M) \subset L^1(\nu) \to L^1(\nu)$ is a positive operator subject to a suitable conservativity assumption. In the case where $(\mathcal{M}(t))_{t\geq 0}$ is strongly stable and the spectral radius of the stochastic operator $P(0_+ - M)^{-1}$ is an eigenvalue, we establish a dichotomy for $(\mathcal{J}(t))_{t\geq 0}$ between asymptotic stability and sweeping. We also show that a scattering-type regime may arise when $(\mathcal{M}(t))_{t\geq 0}$ fails to be strongly stable.

3.2 Chapter Overview

In this chapter, we systematically explore various time asymptotic regimes associated with the Kato-Voigt semigroups constructed in Theorem 2.11.1. As mentioned earlier, this theorem originates from Kato's classical work on Kolmogorov's differential equations [180] and was subsequently revisited by several authors, notably Voigt [369]. This abstract class of C_0-semigroups encompasses numerous concrete applications in Transport Theory, Mathematical Biology, and related fields [33, 320].

To our knowledge, the analysis of time asymptotics for such perturbed C_0-semigroups, at this level of generality, has not been carried out before; see however ([320, Chaps. 5 and 6]) for numerous related developments. This chapter aims to provide several functional analytic results addressing this subject. We recall the

© The Author(s), under exclusive license to Springer Nature Switzerland AG 2026
M. Mokhtar-Kharroubi, *Peripheral Spectra of Perturbed Positive Semigroups*, Lecture Notes in Mathematics 2388, https://doi.org/10.1007/978-3-032-11173-9_3

general framework: Let $(\mathcal{M}_1(t))_{t\geq 0}$ be a substochastic C_0-semigroup on $L^1(\nu)$ with generator M_1 and let

$$P : D(P) \subset L^1(\nu) \to L^1(\nu)$$

be positive (i.e., $P\varphi \in L^1_+(\nu)$ if $\varphi \in D(P) \cap L^1_+(\nu)$) such that $D(P) \supset D(M_1)$ and

$$\int M_1\varphi + P\varphi \leq 0 \ (\varphi \in D_+(M_1)). \tag{3.2.1}$$

According to Theorem 2.11.1, there exists a (minimal) substochastic Kato-Voigt C_0-semigroup $(\mathcal{J}_1(t))_{t\geq 0}$ generated by an extension

$$J_1 \supset M_1 + P.$$

This chapter is devoted to time asymptotics of $(\mathcal{J}_1(t))_{t\geq 0}$. Several preliminary results of independent interest are presented, and three mutually exclusive types of time behavior are analyzed. The analysis relies on whether $(\mathcal{M}_1(t))_{t\geq 0}$ is strongly stable or not and, in the case of strong stability of $(\mathcal{M}_1(t))_{t\geq 0}$, on whether $r(P(0-M_1)^{-1})$ is an eigenvalue of $P(0-M_1)^{-1}$ or not. More precisely, the first section is devoted to a class of honest Kato-Voigt C_0-semigroups $(\mathcal{J}_1(t))_{t\geq 0}$ such that $(\mathcal{M}_1(t))_{t\geq 0}$ is strongly stable and $r(P(0-M_1)^{-1})$ is an eigenvalue of $P(0-M_1)^{-1}$. We show that this class exhibits a general dichotomy between a sweeping behavior of $(\mathcal{J}_1(t))_{t\geq 0}$, characterized by the escape of total mass from the integrability sets of a suitable non-integrable function, and the presence of an asymptotically stable invariant density, (see Theorem 3.3.1). This result will manifest in various forms in the more concrete problems addressed in the subsequent chapters. In the second section, we examine the case where $(\mathcal{M}_1(t))_{t\geq 0}$ is not strongly stable, more precisely when it extends to a positive bounded C_0-group, and show that a different time asymptotic regime emerges, related to scattering theory; namely, the existence of wave operators, (see Theorems 3.4.1 and 3.4.2). This scattering regime will occur later in concrete contexts. Some remarks, comments or open questions are relegated to the end of this chapter in the section "Comments".

3.3 Alternative for Stochastic Kato-Voigt Semigroups

We recall (see (2.11.5)) the Duhamel identity

$$\mathcal{J}_1(t)\varphi = \mathcal{M}_1(t)\varphi + \int_0^t \mathcal{J}_1(t-s)P\mathcal{M}_1(s)\varphi ds \ \ \varphi \in D(M_1). \tag{3.3.1}$$

This section explores a dichotomy property for stochastic Kato-Voigt C_0-semigroups. We start first with a key observation.

Lemma 3.3.1 *Let $(\mathcal{J}_1(t))_{t \geq 0}$ be stochastic. Then $P\,(0 - M_1)^{-1}$ is stochastic if and only if $(\mathcal{M}_1(t))_{t \geq 0}$ is strongly stable.*

Proof Let $\varphi \in D_+(M_1)$. According to the Duhamel identity and the stochasticity of $(\mathcal{J}_1(t))_{t \geq 0}$

$$\|\varphi\| = \|\mathcal{J}_1(t)\varphi\| = \|\mathcal{M}_1(t)\varphi\| + \int_0^t \|\mathcal{J}_1(t - s)P\mathcal{M}_1(s)\varphi\|\,ds$$

$$= \|\mathcal{M}_1(t)\varphi\| + \int_0^t \|P\mathcal{M}_1(s)\varphi\|\,ds$$

$$= \|\mathcal{M}_1(t)\varphi\| + \left\|\int_0^t P\mathcal{M}_1(s)\varphi ds\right\|.$$

Since P is M_1-bounded and $s \to \mathcal{M}_1(s)\varphi \in D(M_1)$ is continous ($D(M_1)$ endowed with the graph norm) then $\int_0^t P\mathcal{M}_1(s)\varphi ds = P\int_0^t \mathcal{M}_1(s)\varphi ds$ and

$$\|\varphi\| = \lim_{t \to +\infty} \|\mathcal{M}_1(t)\varphi\| + \left\|P\,(0 - M_1)^{-1}\,\varphi\right\| \tag{3.3.2}$$

so $(\mathcal{M}_1(t))_{t \geq 0}$ is strongly stable if and only if $P\,(0 - M_1)^{-1}$ is stochastic. $\qquad\square$

As a useful consequence of Lemma 3.3.1, we have

Corollary 3.3.1 *Suppose that $(\mathcal{J}_1(t))_{t \geq 0}$ is stochastic and $D(J_1) = D(M_1)$. If M_1 is injective and if*

$$\lim_{t \to +\infty} \|\mathcal{M}_1(t)\varphi\| \neq 0 \ (\varphi \neq 0) \tag{3.3.3}$$

then $(\mathcal{J}_1(t))_{t \geq 0}$ has no invariant density.

Proof The proof is based on the following fact: If $(\mathcal{J}_1(t))_{t \geq 0}$ has an invariant density then 1 is an eigenvalue of $P\,(0 - M_1)^{-1}$ associated to an eigenfunction ψ. Indeed, if we assume this fact then (3.3.2) implies the contradiction

$$\left\|P\,(0 - M_1)^{-1}\,\psi\right\| = \|\psi\| - \lim_{t \to +\infty} \|\mathcal{M}_1(t)\psi\| < \|\psi\|.$$

Let us show our claim. Let $\varphi \in D(M_1)$ be an invariant density. Then

$$M_1\varphi + P\varphi = 0.$$

Let $\psi_\lambda := \lambda\varphi - M_1\varphi$ then

$$P\varphi = \lambda\varphi - M_1\varphi - \lambda\varphi = \psi_\lambda - \lambda\varphi$$

and

$$P\,(\lambda - M_1)^{-1}\,\psi_\lambda = \psi_\lambda - \lambda\varphi.$$

Since $\psi_\lambda \to -M_1\varphi\ (\lambda \to 0_+)$ and we know that $P\,(\lambda - M_1)^{-1} \to P\,(0_+ - M_1)^{-1}$ strongly then

$$P\,(0 - M_1)^{-1}\,\psi_0 = \psi_0$$

where $\psi_0 = -M_1\varphi \neq 0$ because M_1 is injective. $\square$

As a by-product of Corollary 3.3.1, we have that (3.3.3) implies that 1 is not an eigenvalue of $P\,(0 - M_1)^{-1}$. In the statement of Corollary 3.3.1, if $P\,(0 - M_1)^{-1}$ is irreducible then we can replace (3.3.3) by $\lim_{t\to+\infty}\|\mathcal{M}_1(t)\varphi\| \neq 0\ (\varphi > 0$ a.e.$)$. We are ready to show

Theorem 3.3.1 *Let $(\mathcal{M}_1(t))_{t\geq 0}$ be strongly stable. Suppose that the conservativity condition*

$$\int M_1\varphi + P\varphi = 0\ (\varphi \in D(M_1)).$$

is satisfied and that $(\mathcal{J}_1(t))_{t\geq 0}$ is honest (see Definition 2.11.2). Then
(i) $(\mathcal{J}_1(t))_{t\geq 0}$ and $P\,(0 - M_1)^{-1}$ are stochastic and $r(P\,(0 - M_1)^{-1}) = 1$.
(ii) If 1 is an eigenvalue of $P\,(0 - M_1)^{-1}$ (and $u \in L^1_+(\nu)$ is an eigenvector associated to 1) then we have the following alternative:
(a) If $v := (0 - M_1)^{-1}u \in L^1(\nu)$ then $v \in D(M_1)$ and $M_1 v + Pv = 0$, i.e., (after normalization) v is an invariant density of $(\mathcal{J}_1(t))_{t\geq 0}$. If additionally $(\mathcal{J}_1(t))_{t\geq 0}$ is irreducible and partially integral then v is asymptotically stable.
(b) Let $v := (0 - M_1)^{-1}u \notin L^1(\nu)$. If $(\mathcal{J}_1(t))_{t\geq 0}$ is genuinely honest and $(0 - M_1)^{-1}\left(\int_0^t \mathcal{M}_1(s)ds\right)u$ is finite a.e. then $(\mathcal{J}_1(t))_{t\geq 0}$ is sweeping from the sets of integrability of v.
(iii) If $r_{ess}(P(0 - M_1)^{-1}) < 1$ then $M_1 + P : D(M_1) \subset L^1(\nu) \to L^1(\nu)$ generates a stochastic C_0-semigroup $(\mathcal{J}_1(t))_{t\geq 0}$ (i.e., $(\mathcal{J}_1(t))_{t\geq 0}$ is genuinely honest) and 1 is an (isolated) eigenvalue of $P\,(0 - M_1)^{-1}$.

Proof (i) is a consequence of Lemma 3.3.1.
(ii) We know that $r(P\,(0 - M_1)^{-1}) \in \sigma\left(P\,(0 - M_1)^{-1}\right)$. Assume that 1 is an eigenvalue associated to a positive eigenvector u.
(a) If

$$v :=: (0 - M_1)^{-1}u = \lim_{\lambda \to 0}(\lambda - M_1)^{-1}u \text{ exists in } L^1(\nu)$$

i.e. $v := \lim_{\lambda \to 0}v_\lambda$ exists where $v_\lambda = (\lambda - M_1)^{-1}u$, (and as $M_1 v_\lambda \to -u$) then $v \in D(M_1)$ and $v_\lambda \to v$ for the graph norm so $M_1 v = -u$ and $Pv_\lambda \to Pv$. On the other hand

$$Pv_\lambda = P\,(\lambda - M_1)^{-1}u \to P\,(0 - M_1)^{-1}u = u$$

so $Pv = u$. Since $M_1 v = -u$ then $M_1 v + Pv = 0$. The asymptotic stability is a consequence of Theorem 2.10.3.
(b) By assumption

$$v := (0 - M_1)^{-1}u \notin L^1(\nu)$$

and

$$v_\lambda = (\lambda - M_1)^{-1} u \to v$$

pointwisely and monotonically and

$$P v_\lambda = P (\lambda - M_1)^{-1} u \to \text{``} P v \text{''} = u \quad \text{in } L^1 (\nu)$$

whence

$$\lambda v_\lambda - M_1 v_\lambda = u \geq P v_\lambda$$

and

$$M_1 v_\lambda + P v_\lambda \leq \lambda v_\lambda.$$

Since $J_1 := \overline{M_1 + P}$ then

$$J_1 v_\lambda \leq \lambda v_\lambda$$

therefore

$$\mathcal{J}_1(t) v_\lambda - v_\lambda = \int_0^t \mathcal{J}_1(s) J_1 v_\lambda ds \leq \lambda \int_0^t \mathcal{J}_1(s) v_\lambda ds$$

and

$$\mathcal{J}_1(t) v_\lambda \leq v_\lambda + \lambda \int_0^t \mathcal{J}_1(s) v_\lambda ds$$

$$= v_\lambda + \lambda \left(\int_0^t \mathcal{J}_1(s) ds \right) v_\lambda$$

$$\leq v + \lambda \left(\int_0^t \mathcal{J}_1(s) ds \right) v.$$

If

$$\left(\int_0^t \mathcal{J}_1(s) ds \right) v < +\infty \text{ a.e.} \tag{3.3.4}$$

then, by monotonic pointwise limit,

$$\mathcal{J}_1(t) v \leq v.$$

To check (3.3.4), note that $v_\lambda = (\lambda - M_1)^{-1} u$ so

$$\mathcal{J}_1(t) v_\lambda = \mathcal{M}_1(t) v_\lambda + \int_0^t \mathcal{J}_1(t - s) P \mathcal{M}_1(s) v_\lambda ds$$

$$= \mathcal{M}_1(t) (\lambda - M_1)^{-1} u + \int_0^t \mathcal{J}_1(t - s) P \mathcal{M}_1(s) (\lambda - M_1)^{-1} u ds$$

$$= (\lambda - M_1)^{-1} \, \mathcal{M}_1(t)u + \int_0^t \mathcal{J}_1(t-s)P\,(\lambda - M_1)^{-1}\,\mathcal{M}_1(s)u\,ds.$$

Since

$$\int_0^t \mathcal{J}_1(t-s)P\,(\lambda - T)^{-1}\,\mathcal{M}_1(s)ds \le \int_0^t \mathcal{J}_1(t-s)P\,(0-T)^{-1}\,\mathcal{M}_1(s)ds$$

is a bounded operator then (3.3.4) holds if and only if

$$(0 - M_1)^{-1} \left(\int_0^t \mathcal{M}_1(s)ds \right) u$$

is finite a.e.

Notice that u does not belong to the range of M_1 otherwise $u = M_1 z$ would imply that

$$\begin{aligned}
(0 - M_1)^{-1}\,u \;:\; &= \lim_{\lambda \to 0} (\lambda - M_1)^{-1}\,u \\
&= \lim_{\lambda \to 0} (\lambda - M_1)^{-1}\,M_1 z \\
&= \lim_{\lambda \to 0} M_1\,(\lambda - M_1)^{-1}\,z \\
&= -z + \lim_{\lambda \to 0} \lambda\,(\lambda - M_1)^{-1}\,z
\end{aligned}$$

exists in $L^1\,(\nu)$, i.e., $v \in L^1\,(\nu)$. Finally, by Proposition 2.10.2, $(\mathcal{J}_1(t))_{t \ge 0}$ has no invariant density and Theorem 2.10.7 ends the proof.

(iii) Note that $P\,(\lambda - M_1)^{-1}\,(\lambda \ge 0)$ is a contraction. Since $P\,(\lambda - M_1)^{-1} \le P\,(0_+ - M_1)^{-1}$ then $r_{ess}(P\,(\lambda - M_1)^{-1}) < 1$ (see Theorem 2.4.1). The generation is a consequence of Theorem 2.6.4 and $P\,(0_+ - M_1)^{-1}$ is stochastic. Since

$$r_{ess}(P\,(0_+ - M_1)^{-1}) < 1 = r(P\,(0_+ - M_1)^{-1})$$

then 1 is an isolated eigenvalue of $P\,(0 - M_1)^{-1}$ with finite algebraic multiplicitiy and is associated to a nonnegative eigenvector. $\square$

Remark 3.3.1 Note that if there exists $u \in L_+^1(\nu)$ such that $P\,(0 - M_1)^{-1}\,u = u$ then

$$P(\lambda - M_1)^{-1}u \le P\,(0 - M_1)^{-1}\,u = u \;\; (\lambda > 0)$$

so the trajectory $\{\mathcal{J}_1(t)u; \; t \ge 0\}$ is honest (Proposition 2.11.4 (i)). If additionally $u > 0$ a.e. then all trajectories are honest (Proposition 2.11.4 (iii)) and consequently (by Definition 2.11.2) $(\mathcal{J}_1(t))_{t \ge 0}$ is honest or equivalently $(\mathcal{J}_1(t))_{t \ge 0}$ is generated by $\overline{M_1 + P}$.

Remark 3.3.2 (*Open problem*) Let $(\mathcal{M}_1(t))_{t \ge 0}$ be strongly stable. In Theorem 3.3.1, we have explored the situation when $r(P\,(0 - M_1)^{-1})$ is an eigenvalue of

$P\left(0-M_1\right)^{-1}$. The analysis of the situation when this spectral radius is not an eigenvalue is an interesting open problem. We wonder whether the part (ii) of Theorem 3.3.1 remains true if 1 is a "generalized" eigenvalue of $P\left(0-M_1\right)^{-1}$ in the sense that there exists a measurable $u > 0$ a.e. such that

$$u \notin L^1\left(\nu\right), \ \ P\left(0-M_1\right)^{-1} u = u.$$

3.4 Scattering Theory

We have seen in Corollary 3.3.1 that the strong stability of $(\mathcal{M}_1(t))_{t\geq 0}$ is a necessary condition for $(\mathcal{J}_1(t))_{t\geq 0}$ to have an invariant density. We explore now the opposite direction where $(\mathcal{M}_1(t))_{t\geq 0}$ is not strongly stable, more precisely when

$$\|\mathcal{M}_1(t)\varphi\| \geq c \, \|\varphi\|, \ \ \varphi \in L^1_+(\Xi, \nu), \ t \geq 0. \tag{3.4.1}$$

Under an additional condition, we will fall into another time asymptotic regime related to scattering theory. Scattering theory for linear kinetic equations in L^1 spaces, including wave operators, traces back to [158, 337] and has since developed extensively, see [18, 75, 85, 118, 127, 128, 243, 309, 329, 363]. Further details appear in the monograph [119]. An abstract scattering theory in L^1 spaces with applications to Neutron Transport in the whole space is presented in [243]. In this section, under the dissipativity condition (3.2.1), we revisit this theory, removing some assumptions that become unnecessary within the Kato-Voigt perturbation framework considered here. The first result is

Theorem 3.4.1 *Suppose that* $(\mathcal{M}_1(t))_{t\geq 0}$ *extends as a positive* C_0*-group* $(\mathcal{M}_1(t))_{t\in\mathbb{R}}$ *on* $L^1(\Xi, \nu)$ *such that* $\sigma(M_1) \subset i\mathbb{R}$ *and the (left) strong limit*

$$P\left(0_- - M_1\right)^{-1} := s \lim_{\lambda \to 0_-} P\left(\lambda - M_1\right)^{-1} \ exists. \tag{3.4.2}$$

Then the wave operator

$$\Omega^+ := s \lim_{t\to+\infty} \mathcal{J}_1(t)\mathcal{M}_1(-t) \ exists. \tag{3.4.3}$$

Proof We need two steps.
   Step 1. Let us show

$$\|\mathcal{J}_1(t)\mathcal{M}_1(-t)\varphi\| \leq \|\varphi\| + \left\|P\left(0_- - M_1\right)^{-1}\varphi\right\|, \ \ \varphi \in L^1_+(\Xi, \nu). \tag{3.4.4}$$

Since $\mathcal{M}_1(t)\varphi \in D(M_1)$ if $\varphi \in D(M_1)$ then (3.3.1) gives

$$\mathcal{J}_1(t)\mathcal{M}_1(-t)\varphi = \varphi + \int_0^t \mathcal{J}_1(s)P\mathcal{M}_1(-s)\varphi ds, \ \ \varphi \in D(M_1).$$

Consider first the case $\varphi \in D_+(M_1)$. By the additivity of the norm on the positive cone

$$\left\| \int_0^t \mathcal{J}_1(s)P\mathcal{M}_1(-s)\varphi ds \right\| = \int_0^t \|\mathcal{J}_1(s)P\mathcal{M}_1(-s)\varphi\| \, ds$$

$$\leq \int_0^t \|P\mathcal{M}_1(-s)\varphi\| \, ds$$

$$\leq \int_0^{+\infty} \|P\mathcal{M}_1(-s)\varphi\| \, ds$$

$$= \left\| P \int_0^{+\infty} \mathcal{M}_1(-s)\varphi ds \right\|$$

$$= \left\| P\, (0_- - M_1)^{-1} \varphi \right\|$$

shows (3.4.4) for $\varphi \in D_+(M_1)$. Let $\varphi \in L^1_+(\Xi, \nu)$ and $\varphi_j := j \int_0^{j^{-1}} \mathcal{M}_1(s)\varphi ds$. Then $\varphi_j \in D_+(M_1)$ and $\varphi_j \to \varphi$ in $L^1(\Xi, \nu)$ as $j \to \infty$. Then

$$\left\| \mathcal{J}_1(t)\mathcal{M}_1(-t)\varphi_j \right\| \leq \|\varphi\| + \left\| P\, (0_- - M_1)^{-1} \varphi_j \right\|$$

implies

$$\left\| \mathcal{J}_1(t)\mathcal{M}_1(-t)\varphi \right\| \leq \|\varphi\| + \left\| P\, (0_- - M_1)^{-1} \varphi \right\| \quad (\varphi \in L^1_+(\Xi, \nu)).$$

Finally, since $\mathcal{J}_1(t)\mathcal{M}_1(-t) \geq 0$ then, for any $\varphi \in L^1(\Xi, \nu)$

$$\|\mathcal{J}_1(t)\mathcal{M}_1(-t)\varphi\|$$
$$\leq \|\mathcal{J}_1(t)\mathcal{M}_1(-t)\,|\varphi|\,\|$$
$$\leq \||\varphi|\| + \left\| P\, (0_- - M_1)^{-1}\,|\varphi|\,\right\|$$
$$\leq \left(1 + \left\| P\, (0_- - M_1)^{-1} \right\|\right) \|\varphi\|$$

and $\|\mathcal{J}_1(t)\mathcal{M}_1(-t)\| \leq 1 + \left\| P\,(0_- - M_1)^{-1} \right\| \quad (t \geq 0)$.

Step 2. Since $(\mathcal{J}_1(t)\mathcal{M}_1(-t))_{t\geq 0}$ is uniformly bounded, it suffices to show the convergence (3.4.3) for $\varphi \in D(T)$. Let $\varphi = \varphi_+ - \varphi_-$ be the canonical decomposition of φ into positive and negative parts. Note that $\varphi_\pm$ need not belong to $D(T)$ but

$$\varphi_\pm^{(j)} := j \int_0^{j^{-1}} \mathcal{M}_1(s)\varphi_\pm ds \to \varphi_\pm \text{ in } L^1(\Xi, \nu)$$

while

$$\varphi_+^{(j)} - \varphi_-^{(j)} = j \int_0^{j^{-1}} \mathcal{M}_1(s)\varphi ds \to \varphi \text{ in } D(T) \tag{3.4.5}$$

(endowed with the graph norm). On the other hand

$$\int_0^t \mathcal{J}_1(s) P \mathcal{M}_1(-s) \varphi_\pm^{(j)} ds \le \int_0^{+\infty} \mathcal{J}_1(s) P \mathcal{M}_1(-s) \varphi_\pm^{(j)} ds$$

and

$$\int_0^{+\infty} \left\| \mathcal{J}_1(s) P \mathcal{M}_1(-s) \varphi_\pm^{(j)} \right\| ds \le \int_0^{+\infty} \left\| P \mathcal{M}_1(-s) \varphi_\pm^{(j)} \right\| ds$$
$$= \left\| P \left(0_- - M_1 \right)^{-1} \varphi_\pm^{(j)} \right\|$$

so

$$\int_0^t \mathcal{J}_1(s) P \mathcal{M}_1(-s) \left(\varphi_+^{(j)} - \varphi_-^{(j)} \right) ds$$
$$\to \int_0^{+\infty} \mathcal{J}_1(s) P \mathcal{M}_1(-s) \left(\varphi_+^{(j)} - \varphi_-^{(j)} \right) ds \quad (t \to +\infty).$$

Finally (3.4.5) implies

$$\int_0^t \mathcal{J}_1(s) P \mathcal{M}_1(-s) \varphi \, ds$$
$$\to \int_0^{+\infty} \mathcal{J}_1(s) P \mathcal{M}_1(-s) \varphi \, ds \quad (t \to +\infty)$$

for all $\varphi \in D(M_1)$. This ends the proof. $\qquad\square$

Note that if $(\mathcal{M}_1(t))_{t \ge 0}$ extends as a positive C_0-group $(\mathcal{M}_1(t))_{t \in \mathbb{R}}$ on $L^1(\Xi, \nu)$, in general this is not the case for the perturbed C_0-semigroup $(\mathcal{J}_1(t))_{t \ge 0}$. If it is the case (e.g. if P is bounded), $\mathcal{J}_1(t)$ need not be positive for $t < 0$. Before dealing with a second wave operator, we need several preliminary results. We need first a useful definition.

Definition 3.4.1 We say that $P : D(P) \subset L^1(\nu) \to L^1(\nu)$ is "monotonically closed" if for any nondecreasing sequence $(\varphi_j) \subset D_+(M_1)$ such that $\sup_j \|\varphi_j\| + \|P \varphi_j\| < +\infty$ then $\varphi := \lim_j \varphi_j \in D(P)$ and $P\varphi = \lim_j P\varphi_j$.

By using Theorem 2.6.1, for any $0 \le r < 1$, let $\mathcal{J}_1^{(r)}(t)$ the C_0-semigroup with generator
$$J_1^{(r)} := M_1 + rP : D(M_1) \to L^1(\nu).$$

By construction, $\mathcal{J}_1^{(r)}(t) \to \mathcal{J}_1(t)$ $(r \to 1)$ strongly uniformly in $t \in [0, c]$ $(c > 0)$ and

$$\mathcal{J}_1^{(r)}(t)(t)\varphi = \mathcal{M}_1(t)\varphi + r \int_0^t \mathcal{M}_1(t - s) P \mathcal{J}_1^{(r)}(s)\varphi \, ds, \quad (\varphi \in D(M_1))$$

(see (2.6.2)).

Lemma 3.4.1 *Suppose that P is "monotonically closed" in the sense of Definition 3.4.1 Let there exist $c > 0$ such that $\|\mathcal{M}_1(t)\varphi\| \geq c\,\|\varphi\|$ for all $\varphi \in L^1_+(\Xi, \nu)$ and $t \geq 0$. Then P is J_1-bounded and the (right) strong limit*

$$P(0_+ - J_1)^{-1} := s \lim_{\lambda \to 0_+} P(\lambda - J_1)^{-1} \ exists.$$

Proof For $r < 1$, we have the Duhamel equation (2.6.2)

$$\mathcal{J}_1^{(r)}(t)\psi = \mathcal{M}_1(t)\psi + r \int_0^t \mathcal{M}_1(t - s) P \mathcal{J}_1^{(r)}(s)\psi ds \ \ \psi \in D(M_1).$$

The additivity of the norm on the positive cone and $\psi \in D_+(M_1)$ give

$$\|\mathcal{M}_1(t)\psi\| + r \int_0^t \left\| \mathcal{M}_1(t - s) P \mathcal{J}_1^{(r)}(s)\psi \right\| ds \leq \|\psi\|$$

so

$$\|\mathcal{M}_1(t)\psi\| + rc \int_0^t \left\| P \mathcal{J}_1^{(r)}(s)\psi \right\| ds \leq \|\psi\|$$

$$\|\mathcal{M}_1(t)\psi\| + rc \left\| P \int_0^t \mathcal{J}_1^{(r)}(s)\psi ds \right\| \leq \|\psi\|$$

and consequently

$$rc \left\| P \int_0^t \mathcal{J}_1^{(r)}(s)\psi ds \right\| \leq \|\psi\|, \ \ \psi \in D_+(M_1).$$

Let $\psi \in L^1_+(\nu)$ and $\psi_j := j \int_0^{j^{-1}} \mathcal{M}_1(s)\psi ds$. Then $\psi_j \in D_+(M_1)$ and

$$rc \left\| P \int_0^t \mathcal{J}_1^{(r)}(s)\psi_j ds \right\| \leq \|\psi_j\|.$$

Since $\psi_j \to \psi$ in $L^1(\nu)$ then

$$\int_0^t \mathcal{J}_1^{(r)}(s)\psi_j ds \to \int_0^t \mathcal{J}_1^{(r)}(s)\psi ds (j \to \infty)$$

in $D(M_1)$ endowed with the graph norm. Since P is M_1-bounded then

$$P \int_0^t \mathcal{J}_1^{(r)}(s)\psi_j ds \to P \int_0^t \mathcal{J}_1^{(r)}(s)\psi ds \ (j \to \infty)$$

and

$$rc \left\| P \int_0^t \mathcal{J}_1^{(r)}(s)\psi ds \right\| \leq \|\psi\| \quad (\psi \in L_+^1(\nu)).$$

Since P is "monotonically closed" then letting $r \to 1$ monotonically

$$\int_0^t \mathcal{J}_1(s)\psi ds \in D(P)$$

and

$$c \left\| P \int_0^t \mathcal{J}_1(s)\psi ds \right\| \leq \|\psi\|, \quad (\psi \in L_+^1(\nu)), \ (t \geq 0).$$

This shows that P is J_1-bounded and also

$$c \left\| P \int_0^{+\infty} \mathcal{J}_1(s)\psi ds \right\| \leq \|\psi\|, \quad (\psi \in L_+^1(\nu))$$

i.e., $P(0_+ - J_1)^{-1}$ exists. We can of course decompose any $\psi \in L^1(\nu)$ into positive and negative parts and get

$$c \left\| P \int_0^{+\infty} \mathcal{J}_1(s)\psi ds \right\| \leq \|\psi\|, \quad (\psi \in L^1(\nu)). \qquad \square$$

Thus, under suitable conditions, the second Duhamel relation (2.6.2) remains true in Kato-Voigt perturbation theory when appropriately interpreted.

Corollary 3.4.1 *Suppose that P is "monotonically closed" in the sense of Definition 3.4.1 and there exists $c > 0$ such that $\|\mathcal{M}_1(t)\varphi\| \geq c\|\varphi\|$ for all $\varphi \in L_+^1(\Xi, \nu)$ and $t \geq 0$. Then*

$$\mathcal{J}_1(t)\psi = \mathcal{M}_1(t)\psi + \int_0^t \mathcal{M}_1(t - s)P\mathcal{J}_1(s)\psi ds \quad (\psi \in D(J_1)) \tag{3.4.6}$$

We are ready to deal with a second wave operator.

Theorem 3.4.2 *Let P be "monotonically closed" in the sense of Definition 3.4.1. Suppose that $(\mathcal{M}_1(t))_{t\geq 0}$ extends as a positive C_0-group on $L^1(\Xi, \nu)$ such that $\sigma(M_1) \subset i\mathbb{R}$ and (3.4.2) holds. Let there exist $c > 0$ such that $\|\mathcal{M}_1(t)\varphi\| \geq c\|\varphi\|$ for all $\varphi \in L_+^1(\Xi, \nu)$ and $t \geq 0$. Then the wave operator*

$$\Omega^- := s \lim_{t \to +\infty} \mathcal{M}_1(-t)\mathcal{J}_1(t) \text{ exists.}$$

Proof By (3.4.6)

$$\mathcal{M}_1(-t)\mathcal{J}_1(t)\varphi = \varphi + \int_0^t \mathcal{M}_1(-s)P\mathcal{J}_1(s)\varphi ds \quad \varphi \in D(J_1).$$

By using Lemma 3.4.1 and arguing as in the proof of Theorem 3.4.1, one sees that $(\mathcal{M}_1(-t)\mathcal{J}_1(t))_{t\geq 0}$ is uniformly bounded and it suffices to show the convergence for $\varphi \in D(J_1)$. As previously, if $\varphi = \varphi_+ - \varphi_-$ is the canonical decomposition of φ into positive and negative parts and $\varphi_\pm^{(j)} := j \int_0^{j^{-1}} \mathcal{J}_1(s)\varphi_\pm ds$ then

$$\int_0^{+\infty} \left\| \mathcal{M}_1(-s)P\mathcal{J}_1(s)\varphi_\pm^{(j)} \right\| ds \leq \int_0^{+\infty} \left\| P\mathcal{J}_1(s)\varphi_\pm^{(j)} \right\| ds$$
$$= \left\| P\left(0_+ - J_1\right)^{-1}\varphi_\pm^{(j)} \right\|$$

so the dominated convergence theorem gives

$$\lim_{t\to+\infty} \int_0^t \mathcal{M}_1(-s)P\mathcal{J}_1(s)\varphi_\pm^{(j)} ds = \int_0^{+\infty} \mathcal{M}_1(-s)P\mathcal{J}_1(s)\varphi_\pm^{(j)} ds$$

and

$$\int_0^t \mathcal{M}_1(-s)P\mathcal{J}_1(s)\left(\varphi_+^{(j)} - \varphi_-^{(j)}\right) ds$$
$$\to \int_0^{+\infty} \mathcal{M}_1(-s)P\mathcal{J}_1(s)\left(\varphi_+^{(j)} - \varphi_-^{(j)}\right) ds \quad (t \to +\infty).$$

Finally,

$$\varphi_+^{(j)} - \varphi_-^{(j)} := j \int_0^{j^{-1}} \mathcal{J}_1(s)\left(\varphi_+ - \varphi_-\right) ds$$
$$= j \int_0^{j^{-1}} \mathcal{J}_1(s)\varphi ds \to \varphi \quad (j \to \infty)$$

in the graph norm of J_1 and this ends the proof. $\square$

One of the main features of scattering theory is the asymptotic $(t \to \pm\infty)$ comparison of *free* dynamics $(\mathcal{M}_1(t))_{t\in\mathbb{R}}$ and *interacting* dynamics $(\mathcal{J}_1(t))_{t\geq 0}$.

Theorem 3.4.3 *Let* $\Omega^- = s\lim_{t\to+\infty} \mathcal{M}_1(-t)\mathcal{J}_1(t)$ *exist. Then*

$$\left\| \mathcal{J}_1(t)\varphi - \mathcal{M}_1(t)\Omega^-\varphi \right\| \to 0 \ (t \to +\infty) \ (\varphi \in L^1(\Xi, \nu)), \tag{3.4.7}$$

i.e., any trajectory $(\mathcal{J}_1(t)\varphi)_{t\geq 0}$ *of the full dynamics (with initial data* φ*) behaves asymptotically (as* $t \to +\infty$*) as a free trajectory* $(\mathcal{M}_1(t)\psi)_{t\geq 0}$ *(with initial data* $\psi = \Omega^-\varphi$*).*

Proof By definition, for all $\varphi \in L^1(\Xi, \nu)$

$$\left\| \mathcal{M}_1(-t)\mathcal{J}_1(t)\varphi - \Omega^-\varphi \right\| \to 0 \ (t \to +\infty).$$

Since $\mathcal{M}_1(t)$ is a contraction for $t \geq 0$ then

$$\left\| \mathcal{M}_1(t)\mathcal{M}_1(-t)\mathcal{J}_1(t)\varphi - \mathcal{M}_1(t)\Omega^-\varphi \right\|$$
$$\leq \left\| \mathcal{M}_1(-t)\mathcal{J}_1(t)\varphi - \Omega^-\varphi \right\| \to 0 \ (t \to +\infty)$$

so $\left\| \mathcal{J}_1(t)\varphi - \mathcal{M}_1(t)\Omega^-\varphi \right\| \to 0 \ (t \to +\infty)$. $\qquad\square$

Corollary 3.4.2 *Let $\Omega^- = s \lim_{t \to +\infty} \mathcal{M}_1(-t)\mathcal{J}_1(t)$ exist. If $(\mathcal{M}_1(t))_{t \geq 0}$ is sweeping from a family of measurable subsets $\Lambda \subset \Xi$ then so is $(\mathcal{J}_1(t))_{t \geq 0}$.*

Proof Let $\Lambda \subset \Xi$ be a subset of sweeping for $(\mathcal{M}_1(t))_{t \geq 0}$, i.e.,

$$\int_\Lambda |\mathcal{M}_1(t)\psi| \, d\nu \to 0 \ (t \to +\infty), \ (\psi \in L^1(\Xi, \nu)).$$

In particular

$$\int_\Lambda \left| \mathcal{M}_1(t)\Omega^-\varphi \right| \, d\nu \to 0 \ (t \to +\infty), \ (\varphi \in L^1(\Xi, \nu))$$

and consequently, for all $\varphi \in L^1(\Xi, \nu)$,

$$\int_\Lambda |\mathcal{J}_1(t)\varphi| \, d\nu \leq \int_\Lambda \left| \mathcal{J}_1(t)\varphi - \mathcal{M}_1(t)\Omega^-\varphi \right| \, d\nu + \int_\Lambda \left| \mathcal{M}_1(t)\Omega^-\varphi \right| \, d\nu$$
$$\leq \left\| \mathcal{J}_1(t)\varphi - \mathcal{M}_1(t)\Omega^-\varphi \right\| + \int_\Lambda \left| \mathcal{M}_1(t)\Omega^-\varphi \right| \, d\nu$$

tends to zero as $t \to +\infty$. $\qquad\square$

3.5 Comments

The monograph [320] provides general tools on asymptotic stability and sweeping for stochastic C_0-semigroups, with notable applications in Mathematical Biology.

A scattering theory for Neutron Transport in L^p spaces ($1 < p < +\infty$), based on factorization of collision operators and inspired by Lin's work [205], is developed in [245, Chap. 13].

Two very different time asymptotic regimes are explored in this chapter: The asymptotic stability of $(\mathcal{J}_1(t))_{t \geq 0}$ when $(\mathcal{M}_1(t))_{t \geq 0}$ is strongly stable (Theorem 3.3.1 (ii) (a) and a scattering theory when $\|\mathcal{M}_1(t)\varphi\| \geq c \, \|\varphi\|$ ($\varphi \in L^1_+(\Xi, \nu), t \geq 0$) (Theorem 3.4.3). In between these two opposite cases (e.g. if $(\mathcal{M}_1(t))_{t \geq 0}$ is strongly stable on a proper closed subspace only) there is a room for much more complex time asymptotics of $(\mathcal{J}_1(t))_{t \geq 0}$ which are worth studying.

The "alternative" between asymptotic stability and scattering theory was observed in [18, 19, 127–129] in the context of particle swarms of weakly ionized gas. A systematic functional analytic treatment in this context is given in Sect. 10.13 as a consequence of the general theory in Chap. 8.

In connection with Theorem 3.3.1: It is known (see [269, Theorem 5]) that if there exists $v \neq 0$, $v \in D_+(M_1)$ such that $M_1 v + P v = 0$ then 1 is an eigenvalue of $P(0 - M_1)^{-1}$ associated to a nonnegative eigenvector. Is this conclusion still true if $(\mathcal{J}_1(t))_{t \geq 0}$ is honest but *not* genuinely honest (so $D(J_1) \neq D(M_1)$) and there exists $v \neq 0$, $v \in D_+(J_1)$, $v \notin D(M_1)$ such that $J_1 v = 0$?

In connection with Remark 3.3.2:

(i) Does Theorem 3.3.1 remain true if $r(P(0 - M_1)^{-1})$ is a "generalized" eigenvalue of $P(0 - M_1)^{-1}$ (in the sense given in Remark 3.3.2)?

(ii) Another challenging open problem concerns the situation where $r(P(0 - M_1)^{-1})$ is *not* a "generalized" eigenvalue of $P(0 - M_1)^{-1}$. A deeper investigation into the impact of the "spectral nature" of $r(P(0 - M_1)^{-1})$ could yield valuable insights, potentially enriching both the theoretical framework and its practical applications.

In connection with Sect. 3.4, the assumption that the perturbation P is "monotonically closed" (as per Definition 3.4.1) holds whenever P is an integral operator, or at least integral with respect to certain variables and local with respect to the others; this condition is satisfied by the kinetic equations treated in this monograph.

Chapter 4
Spectral Analysis of Desch Semigroups

4.1 Chapter Aims

In this chapter, we consider Desch's semigroups $(\mathcal{J}(t))_{t\geq 0}$ in L^1-spaces which govern abstract Cauchy problems of the form

$$\frac{df}{dt} = Mf + Pf, \quad f(0) = f_0$$

where M is a generator of a positive C_0-semigroup $(\mathcal{M}(t))_{t\geq 0}$ on $L^1(\nu)$ and $P : D(M) \subset L^1(\nu) \to L^1(\nu)$ is a positive operator such that $\lim_{\lambda \to +\infty} r(P(\lambda - M_1)^{-1}) < 1$ with $r(.)$ denoting a spectral radius. Notably, no dissipativity assumption is imposed. Using weak compactness arguments, we investigate the existence of a spectral gap for the semigroup $(\mathcal{J}(t))_{t\geq 0}$. We also show how this construction and the associated spectral properties can be extended to L^p-spaces.

4.2 Chapter Overview

In this chapter, we investigate general spectral gap properties in L^1-spaces for a class of perturbed C_0-semigroups that differs significantly from the one considered in Chap. 3. Specifically, we focus on the class of C_0-semigroups described in Theorem 2.6.1. Unlike the setting of Chap. 3, this class is not limited to contraction C_0-semigroups and offers a key technical advantage: the domain of the perturbed generator coincides with that of the unperturbed one. This class is analyzed using weak compactness arguments and finds applications in various settings discussed in the subsequent chapters. An additional advantage of the approach is that the construction simplifies considerably when the unperturbed semigroup is immediately norm-continuous, particularly when it is holomorphic. This situation arises in Chap. 6, where the unperturbed semigroup is a multiplication semigroup, as well

M. Mokhtar-Kharroubi, *Peripheral Spectra of Perturbed Positive Semigroups*, Lecture Notes in Mathematics 2388, https://doi.org/10.1007/978-3-032-11173-9_4

as in the context of Kolmogorov differential equations studied in Chap. 9, and the neutron diffusion equations addressed in Chap. 10. We also present two methods for extending the construction to L^p-spaces. To our knowledge, no analogous systematic construction exists in the literature.

Here is the general framework: Let $(\mathcal{M}_1(t))_{t\geq 0}$ be a positive, not necessarily contractive, C_0-semigroup on $L^1(\nu)$ with generator M_1 and let $P : D(M_1) \to L^1(\nu)$ be positive. According to Theorem 2.6.1,

$$J_1 := M_1 + P : D(M_1) \to L^1(\nu)$$

generates a C_0-semigroup $(\mathcal{J}_1(t))_{t\geq 0}$ on $L^1(\nu)$ if and only if

$$\lim_{\lambda \to +\infty} r(P(\lambda - M_1)^{-1}) < 1.$$

In this monograph, we refer to such a perturbation P as a Desch perturbation and to the C_0-semigroup $(\mathcal{J}_1(t))_{t\geq 0}$ as a Desch semigroup. In Sect. 4.3, we explore several situations relying on *weak compactness* assumptions where a Desch semigroup $(\mathcal{J}_1(t))_{t\geq 0}$ is well defined and has a spectral gap, (see Theorems 4.3.1, 4.3.2, 4.3.3 and 4.3.4). In Sect. 4.4, we show that if a C_0-semigroup $(\mathcal{J}(t))_{t\geq 0}$ on $L^1(\nu)$ with generator J has spectral gap and if $a \in L^\infty_+(\nu)$ then the (absorption) semigroup generated by $J - a$ has a spectral gap provided that the oscillation of a is less than the spectral gap of $(\mathcal{J}_1(t))_{t\geq 0}$, (see Theorem 4.4.1).

In Sect. 4.5, we explore the impact of introducing an absorption term on the spectral properties and long-time asymptotics of a stochastic Kato-Voigt, (see Theorem 4.5.1).

Finally, in Sect. 4.6 we investigate two approaches to extending a Desch semigroup to $L^p(\nu)$ preserving peripheral spectral properties independently of p. The first one is a predual approach, (see Theorem 4.6.1) relying on several preliminary results; additional spectral results are also given, (see Theorem 4.6.4). In this approach, P need not be M-bounded in $L^p(\nu)$. A more classical interpolation approach relies on the assumption that P is M-bounded in $L^p(\nu)$, (see Theorems 4.6.5 and 4.6.7). In the last section, we present a somewhat surprising result on the computation of the relative M-bound of P specific to $L^p(\nu)$ space with $p \geq 2$ (see Corollary 4.6.1). Some remarks, comments or open questions are relegated to the end of this chapter in the section "Comments".

4.3 Spectral Gap for Perturbed Semigroups in $L^1(\nu)$

This section is devoted to the spectral theory and long-time asymptotics of perturbed positive C_0-semigroups beyond the setting of contractive C_0-semigroups. We demonstrate how compactness methods can be employed effectively when dissipativity conditions are not satisfied. Such situations arise naturally in various applied contexts. We give first

Definition 4.3.1 A C_0-semigroup $(U(t))_{t\geq 0}$ on a Banach space X is said to be immediate norm-continuous if $t \in (0, +\infty) \to U(t) \in \mathcal{L}(X)$ is continuous in operator norm.

We start with

Theorem 4.3.1 *Let $(\mathcal{M}_1(t))_{t\geq 0}$ be a C_0-semigroup on $L^1(\nu)$ with generator M_1 and let $P : D(M_1) \to L^1(\nu)$ be positive. Suppose that $(\mathcal{M}_1(t))_{t\geq 0}$ is immediate norm-continuous. Let there exists some integer m such that $(P(\lambda - M_1)^{-1})^m$ is weakly compact. Then*

$$J_1 := M_1 + P : D(M_1) \to L^1(\nu)$$

generates a C_0-semigroup $(\mathcal{J}_1(t))_{t\geq 0}$ on $L^1(\nu)$ and

$$\omega_{ess}(\mathcal{J}_1) \leq \omega_{ess}(\mathcal{M}_1).$$

In particular $(\mathcal{J}_1(t))_{t\geq 0}$ has a spectral gap if $s(M_1 + P) > s(M_1)$. This occurs if $(\mathcal{J}_1(t))_{t\geq 0}$ is stochastic and $s(M_1) < 0$.

Proof The generation result is a consequence of Theorem 2.6.3. According to Theorem 2.6.2

$$t \in (0, +\infty) \to U_j(t) \in \mathcal{L}(L^1(\nu)) \quad (j \in \mathbb{N})$$

is continuous in operator norm. Since

$$\int_0^{+\infty} e^{-\lambda t} U_n(t)dt = (\lambda - M_1)^{-1} \left(P(\lambda - M_1)^{-1}\right)^n$$

is weakly compact then for any $s > 0$, $\varepsilon > 0$ and $n \geq m$

$$\int_s^{s+\varepsilon} e^{-\lambda t} U_n(t)dt \leq (\lambda - M_1)^{-1} \left(P(\lambda - M_1)^{-1}\right)^n$$

is weakly compact (by domination) and consequently

$$e^{-\lambda s} U_n(s) = \lim_{\varepsilon \to 0} \varepsilon^{-1} \int_s^{s+\varepsilon} e^{-\lambda t} U_n(t)dt$$

(convergence in $\mathcal{L}(L^1(\nu))$) is also weakly compact. Hence, for any $t > 0$ and $n \geq m$ $U_n(t)$ is weakly compact and then so is $\sum_{j=m}^{+\infty} U_j(t)$ for any $t > 0$ because the series converges in operator norm (see Theorem 2.6.1). According to Theorem 2.8.4, $\omega_{ess}(\mathcal{J}_1) \leq \omega_{ess}(\mathcal{M}_1)$. In particular

$$\omega_{ess}(\mathcal{J}_1) \leq s(M_1).$$

If $s(M_1 + P) > s(M_1)$ then

$$\omega(\mathcal{J}_1) = s(M_1 + P) > s(M_1) \geq \omega_{ess}(\mathcal{J}_1)$$

and $(\mathcal{J}_1(t))_{t \geq 0}$ has a spectral gap. Finally, $\omega(\mathcal{J}_1) = 0$ if $(\mathcal{J}_1(t))_{t \geq 0}$ is stochastic. $\square$

Remark 4.3.1 We have a more precise result when $m = 1$ in Theorem 4.3.1. Indeed,

$$(\lambda - M_1 - P)^{-1} - (\lambda - M_1)^{-1} = (\lambda - M_1)^{-1} \sum_{j=1}^{\infty} \left(P(\lambda - M_1)^{-1} \right)^j$$

is weakly compact. Since

$$(\lambda - M_1 - P)^{-1} - (\lambda - M_1)^{-1} = \int_0^{+\infty} e^{-\lambda t} \left(\mathcal{J}_1(t) - \mathcal{M}_1(t) \right) dt$$

and both $\mathcal{J}_1(t)$ and $\mathcal{M}_1(t)$ are continuous (in operator norm) in $t > 0$ then arguing as in the previous proof, we get that $\mathcal{J}_1(t) - \mathcal{M}_1(t)$ is weakly compact so $\mathcal{J}_1(t)$ and $\mathcal{M}_1(t)$ share the same essential spectrum and consequently the same essential radius or equivalently $\omega_{ess}(\mathcal{J}_1) = \omega_{ess}(\mathcal{M}_1)$.

In the statement above, the interest of immediate norm-continuity of $(\mathcal{M}_1(t))_{t \geq 0}$ is just to deduce the weak compactness of $U_n(t)$ from that of $\left(P(\lambda - M_1)^{-1} \right)^n$. More generally, we have the following statement which does not rely on this assumption.

Theorem 4.3.2 *Let $(\mathcal{M}_1(t))_{t \geq 0}$ be a C_0-semigroup on $L^1(\nu)$ with generator M_1 and let $P : D(M_1) \to L^1(\nu)$ be positive and $\lim_{\lambda \to +\infty} r(P(\lambda - M_1)^{-1}) < 1$. Let $(\mathcal{J}_1(t))_{t \geq 0}$ be the C_0-semigroup on $L^1(\nu)$ with generator*

$$J_1 := M_1 + P : D(M_1) \to L^1(\nu).$$

If there exists an integer m such that $U_n(t)$ is weakly compact for all $t > 0$ and $n \geq m$ then

$$\omega_{ess}(\mathcal{J}_1) \leq \omega_{ess}(\mathcal{M}_1).$$

In particular $(\mathcal{J}_1(t))_{t \geq 0}$ has a spectral gap if $s(M_1 + P) > s(M_1)$. This occurs if $(\mathcal{J}_1(t))_{t \geq 0}$ is stochastic and $s(M_1) < 0$.

Proof We have just to show $\omega_{ess}(\mathcal{J}_1) \leq \omega_{ess}(\mathcal{M}_1)$. Since $R_m(t) = \sum_{j=m}^{\infty} U_j(t)$ converges in operator norm (see Theorem 2.6.1) then $R_m(t)$ is weakly compact and Theorem 2.8.4 ends the proof. $\square$

If $(\mathcal{M}_1(t))_{t \geq 0}$ is immediate norm-continuous, we can weaken the assumptions.

Theorem 4.3.3 *Let $(\mathcal{M}_1(t))_{t \geq 0}$ be a C_0-semigroup on $L^1(\nu)$ with generator M_1 and let $P : D(M_1) \to L^1(\nu)$ be positive. Suppose that $(\mathcal{M}_1(t))_{t \geq 0}$ is* *immediate norm-continuous. Let $P = P^{(1)} + P^{(2)}$ $(P^{(1)},\ P^{(2)} \geq 0)$ such that $\lim_{\lambda \to +\infty} r\left(P^{(1)}(\lambda - M_1)^{-1}\right) < 1$ and $P^{(2)}(\lambda - M_1)^{-1}$ is weakly compact in $L^1(\nu)$. Then*

$$J_1 := M_1 + P : D(M_1) \to L^1(\nu)$$

generates a C_0-semigroup $(\mathcal{J}_1(t))_{t \geq 0}$ in $L^1(\nu)$ and $(\mathcal{J}_1(t))_{t \geq 0}$ has a spectral gap if $s(M_1 + P) > s(M_1 + P^{(1)})$. This occurs if $(\mathcal{J}_1(t))_{t \geq 0}$ is stochastic and $s(M_1 + P^{(1)}) < 0$.

Proof According to Theorem 2.6.2, $(\mathcal{J}_1(t))_{t \geq 0}$ is also immediate norm-continuous. On the other hand

$$(\lambda - M_1 - P^{(1)})^{-1} = (\lambda - M_1)^{-1} \sum_{j=0}^{\infty} \left(P^{(1)}(\lambda - M_1)^{-1}\right)^j \quad (\lambda > s(M_1))$$

shows that

$$P^{(2)}(\lambda - M_1 - P^{(1)})^{-1} \text{ is weakly compact in } L^1(\nu)$$

because the series converges in operator norm. The weak compactness of $P^{(2)}(\lambda - M_1)^{-1}$ implies

$$r\left(P^{(2)}(\lambda - M_1 - P^{(1)})^{-1}\right) \to 0 \ (\lambda \to +\infty)$$

(see [260]) so

$$(\lambda - M_1 - P)^{-1} = (\lambda - M_1 - P^{(1)})^{-1} \sum_{j=0}^{\infty} \left(P^{(2)}(\lambda - M_1 - P^{(1)})^{-1}\right)^j \quad (\lambda > s(M_1 + P^{(1)}))$$

and

$$(\lambda - M_1 - P)^{-1} - (\lambda - M_1 - P^{(1)})^{-1}$$

is weakly compact in $L^1(\nu)$.

Let $\left(\mathcal{M}_1^{(1)}(t)\right)_{t \geq 0}$ be the C_0-semigroup generated by $M_1 + P^{(1)}$. Then

$$\int_0^{+\infty} e^{-\lambda t}\left[\mathcal{J}_1(t) - \mathcal{M}_1^{(1)}(t)\right] dt = (\lambda - M_1 - P)^{-1} - (\lambda - M_1 - P^{(1)})^{-1}$$

is weakly compact in $L^1(\nu)$ and then so is

$$\int_s^{s+\varepsilon} e^{-\lambda t}\left[\mathcal{J}_1(t) - \mathcal{M}_1^{(1)}(t)\right] dt \leq \int_0^{+\infty} e^{-\lambda t}\left[\mathcal{J}_1(t) - \mathcal{M}_1^{(1)}(t)\right] dt$$

for any $s > 0$, $\varepsilon > 0$. Since

$$\varepsilon^{-1} \int_s^{s+\varepsilon} e^{-\lambda t} \left[\mathcal{J}_1(t) - \mathcal{M}_1^{(1)}(t) \right] dt \to e^{-\lambda s} \left[\mathcal{J}_1(s) - \mathcal{M}_1^{(1)}(s) \right] \quad (\varepsilon \to 0)$$

in operator norm because $(\mathcal{J}_1(t))_{t \geq 0}$ and $\left(\mathcal{M}_1^{(1)}(t) \right)_{t \geq 0}$ immediate norm-continuous. Hence $\mathcal{J}_1(s) - \mathcal{M}_1^{(1)}(s)$ is weakly compact for all $s > 0$ and $(\mathcal{J}_1(t))_{t \geq 0}$ and $\left(\mathcal{M}_1^{(1)}(t) \right)_{t \geq 0}$ have the same essential spectrum and the same essential type. Thus

$$\omega_{ess}(\mathcal{J}_1) = \omega_{ess}(\mathcal{M}_1^{(1)}) \leq \omega(\mathcal{M}_1^{(1)}) = s(M_1 + P^{(1)})$$

and $(\mathcal{J}_1(t))_{t \geq 0}$ has a spectral gap if $s(M_1 + P) > s(M_1 + P^{(1)})$. Finally, if $(\mathcal{J}_1(t))_{t \geq 0}$ is stochastic then $s(M_1 + P) = 0$. $\qquad\square$

We explore now a different situation occuring in Population Dynamics [39, 275, 277, 278, 280].

Theorem 4.3.4 *Let* $(\mathcal{M}_1(t))_{t \geq 0}$ *be a positive C_0-semigroup in $L^1(\nu)$ with generator M_1 and let P be a positive M_1-bounded operator such that*

$$\lim_{\lambda \to +\infty} r\left(P(\lambda - M_1)^{-1} \right) < 1.$$

Suppose that M_1 is resolvent compact and there exists a positive weakly compact operator Q such that $P\varphi \geq Q\varphi$, $\varphi \in D_+(M_1)$. If the C_0-semigroup $(\mathcal{J}_1(t))_{t \geq 0}$ generated by $M_1 + P$ is irreducible then $(\mathcal{J}_1(t))_{t \geq 0}$ has a spectral gap.

Proof Note that

$$(\lambda - M_1 - P)^{-1} = (\lambda - M_1)^{-1} \sum_{j=0}^{\infty} \left(P(\lambda - M_1)^{-1} \right)^j \quad (\lambda > s(M_1 + P))$$

shows that $M_1 + P$ is also resolvent compact. Let

$$\widetilde{P} := P - Q \geq 0.$$

Since $\widetilde{P} \leq P$ then $\widetilde{P}$ is M_1-bounded and $\lim_{\lambda \to +\infty} r\left(\widetilde{P}(\lambda - M_1)^{-1} \right) < 1$ so that $M_1 + \widetilde{P} : D(M_1) \subset L^1(\nu) \to L^1(\nu)$ generates a positive C_0-semigroup $\left(\widetilde{\mathcal{J}}_1(t) \right)_{t \geq 0}$ in $L^1(\nu)$. Since

$$M_1 + P = \left(M_1 + \widetilde{P} \right) + Q$$

then

$$\mathcal{J}_1(t) - \widetilde{\mathcal{J}}_1(t) = \int_0^t \widetilde{\mathcal{J}}_1(t - s) Q \mathcal{J}_1(s) ds$$

is weakly compact by Theorem 2.9.1. Hence

$$\omega_{ess}(\mathcal{J}_1) = \omega_{ess}(\widetilde{\mathcal{J}_1}) \le \omega(\widetilde{\mathcal{J}_1}).$$

On the other hand

$$(\lambda - M_1 - P)^{-1} = (\lambda - M_1 - \widetilde{P})^{-1} \sum_{j=0}^{\infty} \left(Q(\lambda - M_1 - \widetilde{P})^{-1} \right)^j \ \ (\lambda > s(M_1 + P))$$

shows that

$$(\lambda - M_1 - P)^{-1} \ge (\lambda - M_1 - \widetilde{P})^{-1}$$

and

$$(\lambda - M_1 - P)^{-1} \ne (\lambda - M_1 - \widetilde{P})^{-1}$$

According to Proposition 2.4.1, the irreducibility of $(\lambda - M_1 - P)^{-1}$ (since $(\mathcal{J}_1(t))_{t \ge 0}$ is) implies

$$r((\lambda - M_1 - P)^{-1}) > r((\lambda - M_1 - \widetilde{P})^{-1})$$

or equivalently

$$s(M_1 + P) > s(M_1 + \widetilde{P}) = \omega(\widetilde{\mathcal{J}_1})$$

whence

$$\omega(\mathcal{J}_1) = s(M_1 + P) > \omega(\widetilde{\mathcal{J}_1}) \ge \omega_{ess}(\mathcal{J}_1)$$

which ends the proof. $\square$

We note that all the aforementioned results on spectral gaps of perturbed C_0-semigroups also imply an asymptotic expansion for these semigroups, as described in Theorems 2.7.1 and 2.7.2.

4.4 On Absorptions with Small Oscillation

This section is devoted to studying the impact of introducing absorption terms with small oscillation on general positive C_0-semigroups $(\mathcal{J}(t))_{t \ge 0}$ on $L^1(\nu)$ that possess a spectral gap.

Theorem 4.4.1 *Let $(\mathcal{J}(t))_{t \ge 0}$ be a positive C_0-semigroup on $L^1(\nu)$ with generator J. Suppose that $(\mathcal{J}(t))_{t \ge 0}$ has a spectral gap, i.e.,*

$$\omega_{ess}(\mathcal{J}) < \omega(\mathcal{J}).$$

Let $\sigma_a \in L^{\infty}_+(\nu)$ and $(\mathcal{J}^a(t))_{t \ge 0}$ be the C_0-semigroup generated by $J - \sigma_a$. Then $(\mathcal{J}^a(t))_{t \ge 0}$ has a spectral gap if the oscillation of σ_a is less than $\omega(\mathcal{J}) - \omega_{ess}(\mathcal{J})$.

Proof Let $\underline{\sigma}_a = \inf \sigma_a$ and $\sigma_a = \underline{\sigma}_a + \zeta$ where

$$\zeta = \sigma_a - \underline{\sigma}_a.$$

Since the (essential) spectrum is just shifted by constant perturbations

$$\omega_{ess}(\mathcal{J}^a) = -\underline{\sigma}_a + \omega_{ess}(\mathcal{J}^\zeta)$$

where $\mathcal{J}^\zeta$ is the semigroup generated by $J - \zeta$. On the other hand, according to Theorem 2.4.1,

$$r_{ess}(\mathcal{J}^\zeta(t)) \leq r_{ess}(\mathcal{J}(t)),$$

i.e.,

$$\omega_{ess}(\mathcal{J}^\zeta) \leq \omega_{ess}(\mathcal{J})$$

so

$$\omega_{ess}(\mathcal{J}^a) \leq -\underline{\sigma}_a + \omega_{ess}(\mathcal{J}).$$

On the other hand

$$
\begin{aligned}
s(J - \sigma_a) &= s(J - \underline{\sigma}_a - \zeta) \\
&= -\underline{\sigma}_a + s(J - \zeta) \\
&\geq -\underline{\sigma}_a - \sup \zeta + s(J)
\end{aligned}
$$

whence

$$\omega(\mathcal{J}^a) \geq -\underline{\sigma}_a - \sup \zeta + \omega(\mathcal{J}).$$

Finally, $(\mathcal{J}^a(t))_{t \geq 0}$ has a spectral gap, i.e., $\omega(\mathcal{J}^a) > \omega_{ess}(\mathcal{J}^a)$, provided that

$$-\underline{\sigma}_a - \sup \zeta + \omega(\mathcal{J}) > -\underline{\sigma}_a + \omega_{ess}(\mathcal{J})$$

which ends the proof since $\sup \zeta$ is the oscillation of σ_a. $\qquad\square$

4.5   Kato-Voigt Semigroups with Absorption

In this section, we investigate the spectral effects of perturbing the generator of a stochastic Kato–Voigt C_0-semigroup by an absorption term. This analysis is carried out independently of the framework of Desch semigroups. Let $(\mathcal{M}_1(t))_{t \geq 0}$ be a substochastic C_0-semigroup. Suppose that $(\mathcal{M}_1(t))_{t \geq 0}$ is strongly stable

$$\mathcal{M}_1(t)\varphi \to 0 \; (t \to \infty) \; \varphi \in L^1(\nu).$$

Let P be M_1-bounded and positive such that $\lim_{\lambda \to +\infty} r\left(P(\lambda - M_1)^{-1}\right) < 1$ and

$$\int_\Omega (M_1\varphi + P\varphi)\,d\nu = 0 \quad (\varphi \in D(M_1)).$$

Let $(\mathcal{J}_1(t))_{t \geq 0}$ be the stochastic C_0-semigroup with generator $M_1 + P : D(M_1) \subset L^1(\nu) \to L^1(\nu)$. Let

$$\sigma_a : \Omega \to [0, +\infty]$$

be measurable, finite a.e. such that $D(\sigma_a) \cap D(M_1)$ is dense in $L^1(\nu)$. According to the theory of absorption semigroups (see Chap. 5 below for the details),

$$\mathcal{M}_1^a(t)\varphi := \lim_{j \to +\infty} e^{t\left(M_1 - \sigma_a^{(j)}\right)}\varphi$$

(where $\sigma_a^{(j)} := \sigma_a \wedge j$) defines a substochastic C_0-semigroup in $L^1(\nu)$ with generator

$$M_1^a \supset M_1 - \sigma_a.$$

Since $\mathcal{M}_1^a(t) \leq \mathcal{M}_1(t)$ then $\lim_{\lambda \to +\infty} r\left(P(\lambda - M_1^a)^{-1}\right) < 1$ and $M_1^a + P : D(M_1^a) \subset L^1(\nu) \to L^1(\nu)$ generates a substochastic C_0-semigroup $\left(\mathcal{J}_1^a(t)\right)_{t \geq 0}$ and

$$s(M_1^a + P) \leq 0.$$

Let us explore the situation where

$$s(M_1^a + P) = 0. \tag{4.5.1}$$

Since $s(M_1^a + P) \leq -\inf \sigma_a$, then (4.5.1) implies that $\inf \sigma_a = 0$. We start with

Lemma 4.5.1 *Suppose that $Pf \neq 0$ for any $f \in D_+(M_1)$ such that $f \neq 0$. If $\sigma_a \neq 0$ then*

(i) The contraction $P(0 - M_1^a)^{-1}$ is not stochastic.

(ii) The time evolution of the total mass $\int \mathcal{J}_1^a(t)f$ is given by

$$\lim_{t \to \infty} \int \mathcal{J}_1^a(t)f = \langle f, e^* \rangle$$

where $e^* = \lim_{t \to \infty} \left(\mathcal{J}_1^a(t)\right)^* 1 \in L_+^\infty(\nu)$ (non-increasing pointwise convergence) is such that $\left(P(0 - M_1^a)^{-1}\right)^* e^* = e^*$.

Proof We have

$$\left\| P(0 - M_1^a)^{-1}f \right\| = \int_0^{+\infty} \left\| P\mathcal{M}_1^a(s)f \right\| ds$$

Since $\sigma_a \neq 0$ then $(\mathcal{M}_1(t))_{t \geq 0} \neq \big(\mathcal{M}_1^a(t)\big)_{t \geq 0}$ so there exists $f \in L_+^1(\nu)$ such that

$$\mathcal{M}_1^a(s)f \leq \mathcal{M}_1(s)f \, , \, \mathcal{M}_1(s)f - \mathcal{M}_1^a(s)f \neq 0$$

at least on some interval so

$$\int_0^{+\infty} \big\| P\mathcal{M}_1^a(s)f \big\| \, ds < \int_0^{+\infty} \big\| P\mathcal{M}_1(s)f \big\| \, ds = \|f\| .$$

Note that for all $f \in L_+^1(\nu)$

$$t \rightarrow \int_\Omega \mathcal{J}_1^a(t)f \text{ is non increasing}$$

so that

$$l(f) := \lim_{t \to \infty} \int_\Omega \mathcal{J}_1^a(t)f \text{ exists.}$$

Note also that

$$\int_\Omega \mathcal{J}_1^a(t)f = \langle \mathcal{J}_1^a(t)f, 1 \rangle = \langle f, \big(\mathcal{J}_1^a(t)\big)^* 1 \rangle$$

and $t \rightarrow \big(\mathcal{J}_1^a(t)\big)^* 1$ is non increasing and admits a pointwise limit

$$e^* = \lim_{t \to \infty} \big(\mathcal{J}_1^a(t)\big)^* 1 \leq 1.$$

Hence

$$l(f) := \lim_{t \to \infty} \int_\Omega \mathcal{J}_1^a(t)f = \langle f, e^* \rangle.$$

Integrating

$$\mathcal{J}_1^a(t)f - \mathcal{M}_1^a(t)f = \int_0^t \mathcal{J}_1^a(t - s)P\mathcal{M}_1^a(s)f \, ds$$

for $f \in L_+^1(\nu)$ gives

$$\big\| \mathcal{J}_1^a(t)f \big\| - \big\| \mathcal{M}_1^a(t)f \big\| = \int_0^t \big\| \mathcal{J}_1^a(t - s)P\mathcal{M}_1^a(s)f \big\| \, ds$$

$$= \int_0^{+\infty} 1_{\{s \leq t\}} \big\| \mathcal{J}_1^a(t - s)P\mathcal{M}_1^a(s)f \big\| \, ds.$$

Note that $\big\| \mathcal{M}_1^a(t)f \big\| \rightarrow 0 \ (t \rightarrow \infty)$. Since

$$\lim_{t \to \infty} 1_{\{s \leq t\}} \big\| \mathcal{J}_1^a(t - s)P\mathcal{M}_1^a(s)f \big\| = l(P\mathcal{M}_1^a(s)f)$$

and

$$1_{\{s \leq t\}} \left\| \mathcal{J}_1^a(t-s) P \mathcal{M}_1^a(s) f \right\| \leq \left\| P \mathcal{M}_1^a(s) f \right\|$$

with

$$\int_0^{+\infty} \left\| P \mathcal{M}_1^a(s) f \right\| ds = \left\| \int_0^{+\infty} P \mathcal{M}_1^a(s) f \, ds \right\| = \left\| P(0 - M_1^a)^{-1} f \right\| < +\infty$$

then, by dominated convergence theorem,

$$\begin{aligned}
\langle f, e^* \rangle = \lim_{t \to \infty} \left\| \mathcal{J}_1^a(t) f \right\| &= \int_0^{+\infty} l(P \mathcal{M}_1^a(s) f) ds \\
&= l\left(\int_0^{+\infty} P \mathcal{M}_1^a(s) f ds \right) \\
&= l\left(P(0 - M_1^a)^{-1} f \right) \\
&= \langle P(0 - M_1^a)^{-1} f, e^* \rangle \\
&= \langle f, \left(P(0 - M_1^a)^{-1} \right)^* e^* \rangle
\end{aligned}$$

whence

$$\left(P(0 - M_1^a)^{-1} \right)^* e^* = e^*$$

because $f \in L_+^1(\nu)$ is arbitrary. $\qquad\qquad\qquad\qquad\qquad\qquad\qquad\qquad\qquad\square$

We are ready to state:

Theorem 4.5.1 *Suppose that $\sigma_a \neq 0$ and $Pf \neq 0$ for any $f \in D_+(M_1))$ such that $f \neq 0$. If $r(P(0 - M_1^a)^{-1}) < 1$ then $\left(\mathcal{J}_1^a(t) \right)_{t \geq 0}$ is strongly stable. Let $r(P(0 - M_1^a)^{-1}) = 1$ and let $P(0 - M_1^a)^{-1}$ be irreducible. then:*

(i) 1 belongs to the spectrum of $P(0 - M_1^a)^{-1}$.

(ii) 1 is not an eigenvalue of $P(0 - M_1^a)^{-1}$.

(iii) $\left(\mathcal{J}_1^a(t) \right)_{t \geq 0}$ is strongly stable if and only if the range of $I - P(0 - M_1^a)^{-1}$ is dense in $L^1(\nu)$, i.e., 1 is not in the residual spectrum of $P(0 - M_1^a)^{-1}$.

Proof If $r(P(0 - M_1^a)^{-1}) < 1$ then $r\left(\left(P(0 - M_1^a)^{-1} \right)^* \right) < 1$ and consequently $e^* = 0$. We assume now that $r(P(0 - M_1^a)^{-1}) = 1$.

(i) The fact that $r(P(0 - M_1^a)^{-1})$ belongs to the spectrum of $P(0 - M_1^a)^{-1}$ is a general property of positive operators.

(ii) We argue by contradiction. Let $f \in L_+^1(\nu)$ (with norm 1) such that

$$P(0 - M_1^a)^{-1} f = f.$$

Then $f > 0$ a.e. by the irreducibilty assumption. We have

$$1 = \int P(0 - M_1^a)^{-1} f = \langle f, \left(P(0 - M_1^a)^{-1} \right)^* 1 \rangle.$$

We know that $\left(P(0-M_1^a)^{-1}\right)^* 1 \leq 1$ and $\left(P(0-M_1^a)^{-1}\right)^* 1 \neq 1$ because $P(0-M_1^a)^{-1}$ is not stochastic. Hence

$$\langle f, \left(P(0-M_1^a)^{-1}\right)^* 1 \rangle < \langle f, 1 \rangle = 1$$

and we are done.

(iii) According to Lemma 4.5.1, $\left(\mathcal{J}_1^a(t)\right)_{t \geq 0}$ is strongly stable if and only if the kernel of $I - \left(P(0-M_1^a)^{-1}\right)^*$ is $\{0\}$ or equivalently if the range of $I - P(0-M_1^a)^{-1}$ is dense. $\qquad \square$

4.6 Extension to $L^p(\nu)$ Spaces

We give two approaches to the extension of $(\mathcal{J}_1(t))_{t \geq 0}$ to $L^p(\nu)$ $(1 \leq p \leq +\infty)$.

4.6.1 Predual Approach to $L^p(\nu)$ Extension

This first approach relies on three hypotheses that are readily verifiable in typical applications where the perturbation P is either a genuine integral operator (see, e.g., Sect. 6.7) or, as in the case of Neutron Transport, an operator that is integral with respect to some variables and local with respect to others.

Hyp 1. $(\mathcal{M}_1(t))_{t \geq 0}$ extends (in a consistent way) to all $L^p(\nu)$ as a semigroup $\left(\mathcal{M}_p(t)\right)_{t \geq 0}$ $(1 \leq p \leq \infty)$ with generator M_p, strongly continuous for $p < +\infty$ and symmetric in $L^2(\nu)$.

This symmetry assumption could be relaxed; see e.g. in the Chap. 14 how to develop a predual analysis for the case of Neutron Transport. Before stating the other two hypotheses, we introduce some material. Let $\mathcal{E}_+(\Omega)$ the set of nonnegative and measurable functions

$$f : \Omega \to [0, +\infty].$$

Note that the restriction of $(\lambda - M_1)^{-1}$ to $L_+^1(\nu)$ extends to $\mathcal{E}_+(\Omega)$ in the following way

$$\left((\lambda - M_1)^{-1}\right)_{ext} f := \lim_{j \to \infty} (\lambda - M_1)^{-1} f_j \ (f \in \mathcal{E}_+(\Omega))$$

where $\left(f_j\right)_j \subset L_+^1(\nu)$ is non decreasing and $f_j \to f$ a.e. (the limit is independent of the choice of the sequence $\left(f_j\right)_j$); we find this kind of *extension* of positive operators in [33, p. 169]. Similarly, the restriction of $(\lambda - M_\infty)^{-1}$ to $L_+^\infty(\nu)$ extends to $\mathcal{E}_+(\Omega)$ as

$$\left((\lambda - M_\infty)^{-1}\right)_{ext} f = \lim_{j \to \infty} (\lambda - M_\infty)^{-1} f_j \ (f \in \mathcal{E}_+(\Omega))$$

where $\left(f_j\right)_j \subset L_+^\infty(\nu)$ is non decreasing and $f_j \to f$ a.e.

Note that the duality

$$\left((\lambda - M_1)^{-1}\right)^* = (\lambda - M_\infty)^{-1},$$

in particular

$$\int_\Omega \left((\lambda - M_1)^{-1} f\right).g d\nu = \int_\Omega f.\left((\lambda - M_\infty)^{-1} g\right) d\nu, \ f \in L_+^1(\nu), \ g \in L_+^\infty(\nu),$$

implies, by monotone convergence, an extended duality

$$0 \le \int_\Omega \left((\lambda - M_1)^{-1}\right)_{ext} f.g d\nu = \int_\Omega f \left((\lambda - M_\infty)^{-1}\right)_{ext} g d\nu \le +\infty \quad (4.6.1)$$

where $f, g \in \mathcal{E}_+(\Omega)$. We recall (see e.g. [73]) that

$$P : D(M_1) \subset L^1(\nu) \to L^1(\nu)$$

has a dense domain and consequently admits a dual operator

$$P^* : D(P^*) \subset L^\infty(\nu) \to L^\infty(\nu)$$

where

$$D(P^*) = \left\{g \in L^\infty(\nu); \ \exists c > 0, \ \left|\langle Pf, g\rangle_{L^1,L^\infty}\right| \le c \, \|f\|_{L^1}, \ f \in D(M_1)\right\}$$

and P^*g is the unique element in $L^\infty(\nu)$ such that

$$\langle Pf, g\rangle_{L^1,L^\infty} = \langle f, P^*g\rangle_{L^1,L^\infty}, \quad f \in D(M_1).$$

Our second hypothesis is

Hyp 2. We suppose that

$$Pf = \mathcal{P}f \ \ f \in D(M_1) \cap L_+^1(\nu)$$

where

$$\mathcal{P} : \mathcal{E}_+(\Omega) \to \mathcal{E}_+(\Omega)$$

is additive and positively homogeneous, i.e.,

$$\begin{cases} \mathcal{P}(f + g) = \mathcal{P}f + \mathcal{P}g \ (f, g \in \mathcal{E}_+(\Omega)) \\ \mathcal{P}(\lambda f) = \lambda \mathcal{P}f \ (\lambda \ge 0, \ f \in \mathcal{E}_+(\Omega)), \end{cases}$$

Similarly, we suppose that the existence of

$$\mathcal{P}^* : \mathcal{E}_+(\Omega) \to \mathcal{E}_+(\Omega),$$

additive, positively homogeneous, equal to P^* on $D(P^*) \cap L_+^\infty(\nu)$ and such that

$$0 \le \int_\Omega (\mathcal{P}^* g). f \, d\nu = \int_\Omega g. (\mathcal{P} f) \, d\nu \le +\infty, \quad (f, g \in \mathcal{E}_+(\Omega)).$$

Let

$$\mathcal{E}(\Omega) = \{f : \Omega \to [-\infty, +\infty] \text{ measurable and finite a.e.}\}.$$

We define

$$\left((\lambda - M_\infty)^{-1}\right)^e : D\left[\left((\lambda - M_\infty)^{-1}\right)^e\right] \subset \mathcal{E}(\Omega) \to L^\infty(\nu)$$

by

$$\left((\lambda - M_\infty)^{-1}\right)^e f = \left((\lambda - M_\infty)^{-1}\right)_{ext} f_+ - \left((\lambda - M_\infty)^{-1}\right)_{ext} f_-$$

with

$$D\left[\left((\lambda - M_\infty)^{-1}\right)^e\right] = \left\{f \in \mathcal{E}(\Omega); \; \left((\lambda - M_\infty)^{-1}\right)_{ext} |f| \in L_+^\infty(\nu)\right\}.$$

Note that $O := \left((\lambda - M_\infty)^{-1}\right)^e$ is *linear*. Indeed, note first that if $f \in D(O)$ then $f_\pm \in D(O)$. Let $f, g \in D(O)$. We have

$$f + g = (f + g)_+ - (f + g)_-$$

and

$$f + g = f_+ - f_- + g_+ - g_-$$

hence

$$(f + g)_+ + (f_- + g_-) = f_+ + g_+ + (f + g)_-$$

and

$$O\,(f + g)_+ + Of_- + Og_- = Of_+ + Og_+ + O\,(f + g)_-$$

so

$$\begin{aligned}
O(f + g) &= O\,(f + g)_+ - O\,(f + g)_- \\
&= Of_+ - Of_- + Og_+ - Og_- \\
&= Of + Og.
\end{aligned}$$

Similarly, if $\lambda = -\beta < 0$ then

$$O(\lambda f) = O((-\beta f)_+) - O((-\beta f)_-)$$
$$= O(\beta f_-) - O(\beta f_+)$$
$$= \beta O(f_-) - \beta O(f_+)$$
$$= -\beta\,(Of_+ - Of_-) = \lambda O f.$$

The last condition is

Hyp 3. We suppose that

$$\begin{cases} \forall f \in L^\infty(\nu),\ \mathcal{P}|f| \in D\left[\left((\lambda - M_\infty)^{-1}\right)^e\right] \\ \left((\lambda - M_\infty)^{-1}\right)^e \mathcal{P} : L^\infty(\nu) \to L^\infty(\nu) \text{ is bounded} \end{cases}$$

where

$$\left((\lambda - M_\infty)^{-1}\right)^e \mathcal{P}f = \left((\lambda - M_\infty)^{-1}\right)^e \mathcal{P}f_+ - \left((\lambda - M_\infty)^{-1}\right)^e \mathcal{P}f_-.$$

As for O, we can check that $\left((\lambda - M_\infty)^{-1}\right)^e \mathcal{P}$ is linear. We introduce now an unbounded operator P_* on $L^1(\nu)$ as follows

$$D(P_*) = \left\{f \in L^1(\nu);\ \mathcal{P}^*|f| \in L^1(\nu)\right\}$$

$$P_* f = \mathcal{P}^* f_+ - \mathcal{P}^* f_- ,\ f \in D(P_*);$$

note that $f_\pm \in D(P_*)$ if $f \in D(P_*)$. As previously, we can check that P_* is linear. Two preliminary lemmas are necessary.

Lemma 4.6.1 *Let Hyp 1–3 be satisfied. Then $\mathcal{P}^*\,(\lambda - M_1)^{-1} \in \mathcal{L}(L^1\,(\nu))$ and its dual operator is given by $\left((\lambda - M_\infty)^{-1}\right)^e \mathcal{P}$.*

Proof For $f \in L^1_+\,(\nu)$ and $g \in L^\infty_+\,(\nu)$ we have

$$\int_\Omega \left(\mathcal{P}^*\,(\lambda - M_1)^{-1}\,f\right).g d\nu = \int_\Omega (\lambda - M_1)^{-1}\,f.(\mathcal{P}g)d\nu$$
$$= \int_\Omega f.\left((\lambda - M_\infty)^{-1}\right)_{ext}(\mathcal{P}g)d\nu$$
$$= \int_\Omega f.\left((\lambda - M_\infty)^{-1}\right)^e \mathcal{P}g d\nu$$

Now

$$\left|\int_\Omega \left(\mathcal{P}^*\,(\lambda - M_1)^{-1}\,f\right).g d\nu\right| \le \left\|\left((\lambda - M_\infty)^{-1}\right)^e \mathcal{P}\right\|_{\mathcal{L}(L^\infty(\nu))} \|f\|_{L^1(\nu)}\,\|g\|_{L^\infty(\nu)}$$

shows that $\mathcal{P}^*\,(\lambda - M_1)^{-1}\,f \in L^1\,(\nu)$ and

$$\left\| \mathcal{P}^* (\lambda - M_1)^{-1} f \right\|_{L^1(\nu)} \leq \left\| \left((\lambda - M_\infty)^{-1} \right)^e \mathcal{P} \right\|_{\mathcal{L}(L^\infty(\nu))} \| f \|_{L^1(\nu)}.$$

We extend $\mathcal{P}^* (\lambda - M_1)^{-1}$ to $L^1 (\nu)$ by

$$\mathcal{P}^* (\lambda - M_1)^{-1} f := \mathcal{P}^* (\lambda - M_1)^{-1} f_+ - \mathcal{P}^* (\lambda - M_1)^{-1} f_-$$

and check, as previously, that $\mathcal{P}^* (\lambda - M_1)^{-1}$ is linear. We have

$$\begin{aligned}
\left\| \mathcal{P}^* (\lambda - M_1)^{-1} f \right\|_{L^1(\nu)} &\leq \left\| \mathcal{P}^* (\lambda - M_1)^{-1} f_+ \right\|_{L^1(\nu)} + \left\| \mathcal{P}^* (\lambda - M_1)^{-1} f_- \right\|_{L^1(\nu)} \\
&\leq \left\| \left((\lambda - M_\infty)^{-1} \right)^e \mathcal{P} \right\|_{\mathcal{L}(L^\infty(\nu))} \left(\| f_+ \|_{L^1(\nu)} + \| f_- \|_{L^1(\nu)} \right) \\
&= \left\| \left((\lambda - M_\infty)^{-1} \right)^e \mathcal{P} \right\|_{\mathcal{L}(L^\infty(\nu))} \| f \|_{L^1(\nu)}.
\end{aligned}$$

This shows that

$$D(M_1) \subset D(P_*)$$

and finally, by decomposition into positive and negative parts, for all $f \in L^1 (\nu)$ and $g \in L^\infty (\nu)$

$$\int_\Omega \left(P_* (\lambda - M_1)^{-1} f \right) . g d\nu = \int_\Omega f . \left((\lambda - M_\infty)^{-1} \right)^e \mathcal{P} g d\nu,$$

i.e.,

$$\left(P_* (\lambda - M_1)^{-1} \right)^* = \left((\lambda - M_\infty)^{-1} \right)^e \mathcal{P}.$$

$\square$

We have also

Lemma 4.6.2 *Let Hyp 1–3 be satisfied and let*

$$\lim_{\lambda \to +\infty} r \left(P_* (\lambda - M_1)^{-1} \right) < 1. \tag{4.6.2}$$

Then, for all $\varphi \in L^1 (\nu)$ and $\psi \in L^1 (\nu) \cap L^\infty (\nu)$

$$\langle (\lambda - M_1 - P_*)^{-1} \varphi, \psi \rangle_{L^1, L^\infty} = \langle \varphi, (\lambda - M_1 - P)^{-1} \psi \rangle_{L^1, L^\infty}. \tag{4.6.3}$$

Proof The resolvent of the generator is given by

$$(\lambda - M_1 - P_*)^{-1} = \sum_{j=0}^{\infty} (\lambda - M_1)^{-1} \left(P_* (\lambda - M_1)^{-1} \right)^j$$

for λ large enough. For $\varphi \in L^1 (\nu)$, $\psi \in L^\infty (\nu)$

$$\left(P_*\,(\lambda - M_1)^{-1}\right)^j \varphi, \psi\rangle_{L^1,L^\infty} = \langle\varphi, \left(\left((\lambda - M_\infty)^{-1}\right)^e \mathcal{P}\right)^j \psi\rangle_{L^1,L^\infty}$$

we have

$$\langle(\lambda - M_1)^{-1}\left(P_*\,(\lambda - M_1)^{-1}\right)^j \varphi, \psi\rangle_{L^1,L^\infty}$$
$$= \langle\left(P_*\,(\lambda - M_1)^{-1}\right)^j \varphi, (\lambda - M_\infty)^{-1}\psi\rangle_{L^1,L^\infty}$$
$$= \langle\varphi, \left(\left((\lambda - M_\infty)^{-1}\right)^e \mathcal{P}\right)^j (\lambda - M_\infty)^{-1}\psi\rangle_{L^\infty,L^1}$$

so

$$\langle(\lambda - M_1 - P_*)^{-1}\varphi, \psi\rangle_{L^1,L^\infty}$$
$$= \sum_{j=0}^{\infty}\langle(\lambda - M_1)^{-1}\left(P_*\,(\lambda - M_1)^{-1}\right)^j \varphi, \psi\rangle_{L^1,L^\infty}$$
$$= \sum_{j=0}^{\infty}\langle\left(P_*\,(\lambda - M_1)^{-1}\right)^j \varphi, (\lambda - M_\infty)^{-1}\psi\rangle_{L^1,L^\infty}$$
$$= \sum_{j=0}^{\infty}\langle\varphi, \left(\left((\lambda - M_\infty)^{-1}\right)^e \mathcal{P}\right)^j (\lambda - M_\infty)^{-1}\psi\rangle_{L^\infty,L^1}.$$

If $\psi \in L^1(\nu) \cap L^\infty(\nu)$

$$\sum_{j=0}^{\infty}\left(\left((\lambda - M_\infty)^{-1}\right)^e \mathcal{P}\right)^j (\lambda - M_\infty)^{-1}\psi$$
$$= \sum_{j=0}^{\infty}\left(\left((\lambda - M_\infty)^{-1}\right)^e \mathcal{P}\right)^j (\lambda - M_1)^{-1}\psi$$
$$= \sum_{j=0}^{\infty}\left((\lambda - M_1)^{-1}\,P\right)^j (\lambda - M_1)^{-1}\psi$$
$$= \sum_{j=0}^{\infty}(\lambda - M_1)^{-1}\left(P\,(\lambda - M_1)^{-1}\right)^j \psi$$
$$= (\lambda - M_1 - P)^{-1}\psi.$$

Finally, for all $\psi, \varphi \in L^1(\nu) \cap L^\infty(\nu)$

$$\langle(\lambda - M_1 - P_*)^{-1}\varphi, \psi\rangle_{L^1,L^\infty} = \langle\varphi, (\lambda - M_1 - P)^{-1}\psi\rangle_{L^1,L^\infty}$$

which ends the proof. $\qquad\square$

Note that (4.6.2) implies that

$$J_{1*} : M_1 + P_* : D(M_1) \to L^1(\nu)$$

generates a C_0-semigroup $(\mathcal{J}_{1*}(t))_{t \geq 0}$ on $L^1(\nu)$. We are ready to state

Theorem 4.6.1 *Let Hyp 1–3 be satisfied. If the condition (4.6.2) is satisfied then the C_0-semigroup $(\mathcal{J}_1(t))_{t \geq 0}$ on $L^1(\nu)$ extends (in a consistent way) to $L^p(\nu)$ ($1 \leq p \leq +\infty$) as a semigroup $(\mathcal{J}_p(t))_{t \geq 0}$, strongly continuous if $p < +\infty$, with generator $J_p : D(J_p) \subset L^p(\nu) \to L^p(\nu)$. Moreover*

$$J_p h = M_1 h + P h \text{ for } h \in D(J_p) \cap D(M_1)$$

and $D(J_p) \cap D(M_1)$ is a core of $D(J_p)$.

Proof The identity (4.6.3) shows that $(\lambda - M_1 - P)^{-1}$ and $\left((\lambda - M_1 - P_*)^{-1}\right)^*$ coincide on $L^1(\nu) \cap L^\infty(\nu)$. Thus $(\lambda - M_1 - P)^{-1}$ extends as a bounded operator on $L^\infty(\nu)$. By Riesz-Thorin theorem, it extends to all $L^p(\nu)$ ($1 \leq p \leq +\infty$). Similarly

$$\langle \mathcal{J}_{1*}(t)\varphi, \psi \rangle_{L^1, L^\infty} = \langle \varphi, \mathcal{J}_1(t)\psi \rangle_{L^1, L^\infty} \ \ \psi, \varphi \in L^1(\nu) \cap L^\infty(\nu)$$

shows that $\mathcal{J}_1(t)$ coincides with $(\mathcal{J}_{1*}(t))^*$ on $L^1(\nu) \cap L^\infty(\nu)$ so $\mathcal{J}_1(t)$ extends as a bounded operator on $L^\infty(\nu)$ and, by Riesz-Thorin theorem, to all $L^p(\nu)$ ($1 \leq p \leq +\infty$). We denote by $\mathcal{J}_p(t)$ this extension. The semigroup property $\mathcal{J}_1(t + s) = \mathcal{J}_1(t)\mathcal{J}_1(s)$ extends to $L^p(\nu)$

$$\mathcal{J}_p(t + s) = \mathcal{J}_p(t)\mathcal{J}_p(s).$$

On the other hand, the convexity property

$$\left\| \mathcal{J}_p(t)f - f \right\|_{L^p(\nu)} \leq \| \mathcal{J}_1(t)f - f \|_{L^1(\nu)}^{\frac{1}{p}} \, \| \mathcal{J}_\infty(t))f - f \|_{L^\infty(\nu)}^{1 - \frac{1}{p}}$$

shows that the strong continuity in $L^p(\Omega, \nu)$ ($1 \leq p < +\infty$) is inherited from that in $L^1(\nu)$. Thus $(\mathcal{J}_p(t))_{t \geq 0}$ is a C_0-semigroup in $L^p(\nu)$ ($1 \leq p < +\infty$). We denote by

$$J_p : D(J_p) \subset L^p(\nu) \to L^p(\nu)$$

its generator. We note that the type of $(\mathcal{J}_\infty(t))_{t \geq 0}$ is given by that of $(\mathcal{J}_{1*}(t))_{t \geq 0}$. The convexity property

$$\left\| \mathcal{J}_p(t) \right\|_{\mathcal{L}(L^p(\nu))} \leq \| \mathcal{J}_1(t) \|_{\mathcal{L}(L^1(\nu))}^{\frac{1}{p}} \, \| \mathcal{J}_\infty(t)) \|_{\mathcal{L}(L^\infty(\nu))}^{1 - \frac{1}{p}} \qquad (4.6.4)$$

gives

$$\ln \left\| \mathcal{J}_p(t) \right\|_{\mathcal{L}(L^p(\nu))} \leq \frac{1}{p} \ln \| \mathcal{J}_1(t) \|_{\mathcal{L}(L^1(\nu))} + \left(1 - \frac{1}{p}\right) \ln \| \mathcal{J}_\infty(t)) \|_{\mathcal{L}(L^\infty(\nu))}$$

so

$$\omega(\mathcal{J}_p) \le \frac{1}{p}\omega(\mathcal{J}_1) + \left(1 - \frac{1}{p}\right)\omega(\mathcal{J}_\infty) \tag{4.6.5}$$

$$= \frac{1}{p}\omega(\mathcal{J}_1) + \left(1 - \frac{1}{p}\right)\omega(\mathcal{J}_{1*}).$$

For all $f \in L^1(\nu) \cap L^p(\nu)$ and $\lambda > \max\{\omega(\mathcal{J}_1),\ \omega(\mathcal{J}_{1*})\}$

$$\left(\lambda - J_p\right)^{-1} f = (\lambda - J_1)^{-1} f = (\lambda - M_1 - P)^{-1} f.$$

Let $\left(h,\, J_p h\right)$ be in the graph of J_p and let

$$f := \lambda h - J_p h.$$

Then $h = \left(\lambda - J_p\right)^{-1} f$. Let $\left(f_j\right)_j \subset L^1(\nu) \cap L^p(\nu)$ such that

$$f_j \to f \text{ in } L^p(\Omega, \nu)$$

and

$$h_j := \left(\lambda - J_p\right)^{-1} f_j = (\lambda - M_1 - P)^{-1} f_j \to h = \left(\lambda - J_p\right)^{-1} f \quad \text{in } L^p(\nu).$$

Hence

$$\lambda h_j - M_1 h_j - P h_j = f_j \to f \text{ in } L^p(\nu)$$

and

$$M_1 h_j + P h_j = J_p h_j \to \lambda h - f = J_p h$$

$$h_j \in D(J_p) \cap D(M_1) \to h \ \text{ in } L^p(\nu)$$

so

$$(h_j,\, M_1 h_j + P h_j) \to (h,\, J_p h),$$

i.e., $\left\{h \in D(J_p) \cap D(M_1)\right\}$ is a core of $D(J_p)$. This ends the proof. $\qquad\square$

Under the "symmetry" of $\mathcal{P}$

$$0 \le \int_\Omega (\mathcal{P}g).f\, d\nu = \int_\Omega g.(\mathcal{P}f)\, d\nu \le +\infty, \quad (f, g \in \mathcal{E}_+(\Omega)), \tag{4.6.6}$$

we obtain a more precise result in $L^2(\nu)$.

Theorem 4.6.2 *Suppose that $\left(\mathcal{M}_p(t)\right)_{t\ge 0}$ is contractive in all $L^p(\nu)$. Let Hyp 1–3 be satisfied and*

$$\delta := \lim_{\lambda \to +\infty} r\left(P\,(\lambda - M_1)^{-1}\right) < 1.$$

We assume the existence of a non decreasing sequence of symmetric operators $\left(P_j\right)_j \subset \mathcal{L}_+(L^2\,(\nu))$ *such that* $P_j\varphi \le P\varphi$ $(\varphi \in D_+(M_1))$ *and*

$$P_j\varphi \to \mathcal{P}\varphi \; (j \to \infty) \; a.e. \; (\varphi \in L^2_+\,(\nu)). \tag{4.6.7}$$

Then

   (i) $(\mathcal{J}_2(t))_{t\ge 0}$ *is self-adjoint.*
   (ii) For all $\varphi \in D(\sqrt{-M_2})$

$$[\mathcal{P}\varphi, \varphi] := \lim_{j \to \infty} (P_j\varphi, \varphi) \; exists \tag{4.6.8}$$

and

$$|[\mathcal{P}\varphi, \varphi]| \le \delta \left\| \sqrt{-M_2}\varphi \right\|^2 + \delta s(J_1)\,\|\varphi\|^2_{L^2(\nu)} \; \forall \varphi \in D(\sqrt{-M_2}).$$

   (i) If $c(\varphi, \psi)$ *is the hermitian form associated to the quadratic form* $-[\mathcal{P}\varphi, \varphi]$ *then the hermitian form*

$$a(\varphi, \psi) := \left(\sqrt{-M_2}\varphi, \sqrt{-M_2}\psi\right) + c(\varphi, \psi), \quad \varphi, \psi \in D(\sqrt{-M_2})$$

is closed and $J_2 = -A$ *where* A *is the self-adjoint operator associated to* $a(.,.)$.

Proof (i) We know that for all $\psi, \varphi \in L^1\,(\nu) \cap L^\infty\,(\nu)$

$$\langle (\lambda - M_1 - P_*)^{-1}\varphi, \psi \rangle_{L^1, L^\infty} = \langle \varphi, (\lambda - M_1 - P)^{-1}\psi \rangle_{L^1, L^\infty}.$$

Under the condition (4.6.6) we have $P_* = P$ and

$$\langle (\lambda - M_1 - P)^{-1}\varphi, \psi \rangle_{L^1, L^\infty} = \langle \varphi, (\lambda - M_1 - P)^{-1}\psi \rangle_{L^1, L^\infty}.$$

Since

$$\left(\lambda - J_p\right)^{-1}\varphi = (\lambda - M_1 - P)^{-1}\varphi$$

then

$$\langle (\lambda - J_2)^{-1}\varphi, \psi \rangle_{L^1, L^\infty} = \langle \varphi, (\lambda - J_2)^{-1}\psi \rangle_{L^1, L^\infty}.$$

By density of $L^1\,(\nu) \cap L^\infty\,(\nu)$ in $L^2\,(\nu)$ one sees that $(\lambda - J_2)^{-1}$ is self-adjoint and consequently the generator

$$J_2 : D(J_2) \subset L^2\,(\nu) \to L^2\,(\nu)$$

is self-adjoint.

(ii) According to (4.6.5)

$$\omega(\mathcal{J}_p) \leq \frac{1}{p}\omega(\mathcal{J}_1) + \left(1 - \frac{1}{p}\right)\omega(\mathcal{J}_{1*}).$$

Since $\mathcal{J}_1 = \mathcal{J}_{1*}$ then $\omega(\mathcal{J}_p) \leq \omega(\mathcal{J}_1)$ or equivalently

$$s(J_p) \leq s(J_1).$$

Since J_2 self-adjoint then

$$(J_2\varphi, \varphi) \leq s(J_2)\,\|\varphi\|^2_{L^2(\nu)} \quad (\varphi \in D(J_2)).$$

If we choose c arbitrarily such that $1 < c < \frac{1}{\delta}$ then

$$\lim_{\lambda \to +\infty} r_\sigma\left[cP(\lambda - M_1)^{-1}\right] = c\delta < 1.$$

The same construction with P_j in place of P shows that the self-adjoint operator in $L^2(\nu)$

$$J_2^j := M_2 + cP_j : D(M_2) \to L^2(\nu)$$

is such that

$$s(J_2^j) \leq s(J_1^j).$$

Since $P_j\varphi \leq P\varphi$ $(\varphi \in D_+(M_1))$ then $s(J_1^j) \leq s(J_1)$ and consequently the uniform estimate

$$s(J_2^j) \leq s(J_1)\ (j \geq 0).$$

Thus

$$(J_2^j\varphi, \varphi) \leq s(J_1)\,\|\varphi\|^2_{L^2(\nu)} \ \forall\varphi \in D(M_2), \ (j \geq 0),$$

i.e.,

$$(M_2\varphi, \varphi) + c(P_j\varphi, \varphi) \leq s(J_1)\,\|\varphi\|^2_{L^2(\nu)} \ \forall\varphi \in D(M_2), \ (j \geq 0).$$

Since $(\mathcal{M}_2(t))_{t\geq 0}$ is contractive then $M_2 \leq 0$ and then

$$(P_j\varphi, \varphi) \leq c^{-1}\left\|\sqrt{-M_2}\varphi\right\|^2 + c^{-1}s(J_1)\,\|\varphi\|^2_{L^2(\nu)} \ \forall\varphi \in D(M_2), \ (j \geq 0).$$

Since c such that $1 < c < \frac{1}{\delta}$ is arbitrary then

$$(P_j\varphi, \varphi) \leq \delta\left\|\sqrt{-M_2}\varphi\right\|^2 + \delta s(J_1)\,\|\varphi\|^2_{L^2(\nu)} \ \forall\varphi \in D(M_2), \ (j \geq 0).$$

The density of $D(M_2)$ in $D(\sqrt{-M_2})$ for the graph norm of $\sqrt{-M_2}$ (see e.g. [166, Theorem 4.1 p. 78]) gives

$$(P_j\varphi, \varphi) \leq \delta \left\| \sqrt{-M_2}\varphi \right\|^2 + \delta s(J_1) \|\varphi\|_{L^2(\nu)}^2 \quad \forall \varphi \in D(\sqrt{-M_2}), \quad (j \geq 0).$$

Since $\left(\mathcal{M}_p(t)\right)_{t \geq 0}$ is contractive in all $L^p(\nu)$ then $\left\| \sqrt{-M_2}\varphi \right\|^2$ is a Dirichlet form in $D(\sqrt{-M_2})$ and consequently

$$\varphi \in D(\sqrt{-M_2}) \implies |\varphi| \in D(\sqrt{-M_2}) \text{ and } \left\| \sqrt{-M_2}\,|\varphi| \right\| \leq \left\| \sqrt{-M_2}\varphi \right\|$$

(see [99, Theorem 1.3.2]). By the positivity of P_j

$$\left|(P_j\varphi, \varphi)\right| \leq (P_j|\varphi|, |\varphi|) \leq \delta \left\| \sqrt{-M_2}\varphi \right\|^2 + \delta s(J_1) \|\varphi\|_{L^2(\nu)}^2 \tag{4.6.9}$$

and

$$(P_j|\varphi|, |\varphi|) \leq \delta \left\| \sqrt{-M_2}\,|\varphi| \right\|^2 + \delta s(J_1) \|\varphi\|_{L^2(\nu)}^2$$
$$\leq \delta \left\| \sqrt{-M_2}\varphi \right\|^2 + \delta s(J_1) \|\varphi\|_{L^2(\nu)}^2 \quad \forall \varphi \in D(\sqrt{-M_2}), \quad (j \geq 0).$$

Since

$$(P_j|\varphi|, |\varphi|) \to \int \mathcal{P}|\varphi| \cdot |\varphi| \, d\nu \quad (j \to \infty)$$

then

$$\int \mathcal{P}|\varphi| \cdot |\varphi| \, d\nu \leq \delta \left\| \sqrt{-M_2}\varphi \right\|^2 + \delta s(J_1) \|\varphi\|_{L^2(\nu)}^2 \quad \forall \varphi \in D(\sqrt{-M_2}).$$

Note that

$$\int \mathcal{P}|\varphi| \cdot |\varphi| \, d\nu = \int \mathcal{P}(\varphi_+ + \varphi_-) \cdot (\varphi_+ + \varphi_-) \, d\nu$$
$$= \int \mathcal{P}\varphi_+ \cdot \varphi_+ d\nu + \int \mathcal{P}\varphi_- \cdot \varphi_- d\nu + 2 \int \mathcal{P}\varphi_+ \cdot \varphi_- d\nu$$

shows that $\mathcal{P}\varphi_\pm < +\infty$ a.e. and

$$\int \mathcal{P}\varphi_+ \cdot \varphi_- d\nu < +\infty.$$

For all $\varphi \in D(\sqrt{-M_2})$, we have φ_+ and $\varphi_- \in D(\sqrt{-M_2})$. We define

$$\mathcal{P}\varphi = \mathcal{P}\varphi_+ - \mathcal{P}\varphi_-, \quad \varphi \in D(\sqrt{-M_2})$$

and check that $\mathcal{P}$ is linear on $D(\sqrt{-M_2})$. Define

$$\begin{aligned}
[\mathcal{P}\varphi, \varphi] : &= \int \mathcal{P}\varphi_+.\varphi_+ d\nu + \int \mathcal{P}\varphi_-.\varphi_- d\nu - 2 \int \mathcal{P}\varphi_+.\varphi_- d\nu \\
&= \lim_{j\to\infty} \int P_j\varphi_+.\varphi_+ d\nu + \lim_{j\to\infty} \int P_j\varphi_-.\varphi_- d\nu - 2 \int P_j\varphi_+.\varphi_- d\nu \\
&= \lim_{j\to\infty} \int P_j\varphi.\varphi d\nu = \lim_{j\to\infty} (P_j\varphi, \varphi).
\end{aligned}$$

We have

$$[\mathcal{P}\varphi, \varphi] = \lim_{j\to\infty} (P_j\varphi, \varphi), \quad \varphi \in D(\sqrt{-M_2}) \tag{4.6.10}$$

and

$$|[\mathcal{P}\varphi, \varphi]| \leq \delta \left\| \sqrt{-M_2}\varphi \right\|^2 + \delta s(J_1) \|\varphi\|^2_{L^2(\nu)} \ \forall \varphi \in D(\sqrt{-M_2}). \tag{4.6.11}$$

We can define the hermitian form $c(.,.)$ associated to the quadratic form

$$\varphi \in D(\sqrt{-M_2}) \to -[\mathcal{P}\varphi, \varphi]$$

(we restrict ourselves to the real case for simplicity) via the polarisation identity

$$c(\varphi, \psi) = -\frac{1}{4} \left([P(\varphi + \psi), (\varphi + \psi)] - [P(\varphi - \psi), (\varphi - \psi)] \right), \ \varphi,\psi \in D(\sqrt{-M_2}).$$

Thus the hermitian form $a(.,.)$ on $D(\sqrt{-M_2})$

$$a(\varphi, \psi) = \left(\sqrt{-M_2}\varphi, \sqrt{-M_2}\psi \right) + c(\varphi, \psi)$$

is closed since

$$a(\varphi, \varphi) \geq (1 - \delta) \left\| \sqrt{-M_2}\varphi \right\|^2 - \delta s(J_1) \|\varphi\|^2_{L^2(\nu)}$$

and consequently

$$a(\varphi, \psi) + \delta s(J_1)(\varphi, \psi)$$

is coercive. There exists a unique self-adjoint operator A_2 on $L^2(\nu)$ defined by

$$(A_2\varphi, \psi) = a(\varphi, \psi), \quad \varphi \in D(A_2), \ \psi \in D(\sqrt{-M_2})$$

$$D(A_2) = \left\{ \varphi \in D(\sqrt{-M_2}); \ \exists c > 0, \ |a(\varphi, \psi)| \leq c \|\psi\|_{L^2(\nu)}, \ \forall \psi \in D(\sqrt{-M_2}) \right\}.$$

We can denote symbolically

$$(-M_2) \oplus (-\mathcal{P})$$

the operator A_2 because formally "$c(\varphi, \psi) = (-\mathcal{P}\varphi, \psi)$".

(iii) The last part of the proof is to identify $-J_2$ as a form-sum. Note that (4.6.7) and the fact that

$$P : D(M_1) \subset L^1(\nu) \to L^1(\nu)$$

imply by dominated convergence theorem

$$P_j\varphi \to P\varphi \ (j \to \infty) \text{ in } L^1(\nu) \ (\varphi \in D_+(M_1))$$

so

$$(\lambda - M_1 - P_j)^{-1}\varphi$$
$$= (\lambda - M_1)^{-1} \sum_{n=0}^{\infty} \left[P_j(\lambda - M_1)^{-1} \right]^n \varphi$$
$$\to (\lambda - M_1)^{-1} \sum_{n=0}^{\infty} \left[P(\lambda - M_1)^{-1} \right]^n \varphi \ (j \to \infty) \ (\varphi \in L^1_+(\nu)),$$

i.e., $(\lambda - M_1 - P_j)^{-1} \to (\lambda - M_1 - P_-)^{-1}$ $(j \to +\infty)$ strongly in $L^1(\nu)$. It follows by Riesz-Thorin's interpolation theorem that

$$(\lambda - J_2^j)^{-1} \to (\lambda - J_2)^{-1} \text{ strongly in } L^2(\nu) \tag{4.6.12}$$

as $j \to \infty$. On the other hand, since P_j is a bounded operator then $-J_2^j$ is also a form-sum

$$- J_2^j = (-M_2) \oplus \left(-P_j\right).$$

A key point is that the resolvent of the form-sum operator $(-M_2) \oplus \left(-P_j\right)$ is given by

$$(\lambda + (-M_2) \oplus \left(-P_j\right))^{-1} = (\lambda - M_2)^{-\frac{1}{2}}(I - C_j(\lambda))^{-1}(\lambda - M_2)^{-\frac{1}{2}}$$

(see [345, Theorem 6.25, p. 150]) where $C_j(\lambda)$ is the positive bounded self-adjoint operator on $L^2(\nu)$ defined by the positive bounded quadratic form

$$\varphi \in L^2(\nu) \to \left(P_j(\lambda - M_2)^{-\frac{1}{2}}\varphi, (\lambda - M_2)^{-\frac{1}{2}}\varphi \right)_{L^2(\nu)}$$

and (by (4.6.9)) $\left\| C_j(\lambda) \right\|_{\mathcal{L}(L^2(\nu))} \leq \delta$ for $\lambda > \frac{\delta s(J_1)}{\delta} = s(J_1)$ (see [345, Theorem 6.25, p. 150]). Similarly, the resolvent of the form-sum operator $(-M_2) \oplus (-\mathcal{P})$ is given by

$$(\lambda + (-M_2) \oplus (-\mathcal{P}))^{-1} = (\lambda - M_2)^{-\frac{1}{2}}(I - C(\lambda))^{-1}(\lambda - M_2)^{-\frac{1}{2}}$$

where $C(\lambda)$ is the positive bounded self-adjoint operator on $L^2(\nu)$ defined by the positive bounded quadratic form

$$\varphi \in L^2(\nu) \to \left[\mathcal{P}(\lambda - M_2)^{-\frac{1}{2}}\varphi, (\lambda - M_2)^{-\frac{1}{2}}\varphi\right]$$

and (by (4.6.11)) $\|C(\lambda)\|_{\mathcal{L}(L^2(\nu))} \le \delta$ for $\lambda > \frac{\delta s(J_1)}{\delta} = s(J_1)$ (see [345, Theorem 6.25, p. 150]).

The convergence of the quadratic forms

$$(C_j(\lambda)\varphi, \varphi) = \left(P_j(\lambda - M_2)^{-\frac{1}{2}}\varphi, (\lambda - M_2)^{-\frac{1}{2}}\varphi\right)_{L^2(\nu)}$$
$$\to (C(\lambda)\varphi, \varphi) = \left[\mathcal{P}(\lambda - M_2)^{-\frac{1}{2}}\varphi, (\lambda - M_2)^{-\frac{1}{2}}\varphi\right]$$

(see (4.6.10)) implies the strong convergence

$$(I - C_j(\lambda))^{-1} \to (I - C(\lambda))^{-1}(j \to +\infty)$$

(see e.g. [311, Theorem S. 14, p. 373]) and finally the strong convergence

$$(\lambda - M_2)^{-\frac{1}{2}}(I - C_j(\lambda))^{-1}(\lambda - M_2)^{-\frac{1}{2}} \to (\lambda - M_2)^{-\frac{1}{2}}(I - C(\lambda))^{-1}(\lambda - M_2)^{-\frac{1}{2}}$$

$(j \to +\infty)$ which, combined to (4.6.12), shows the equality

$$(\lambda - J_2)^{-1} = (\lambda + A_2)^{-1}$$

and finally $J_2 = -A_2$. $\qquad\square$

Even if it is convenient to denote A_2 as a form-sum $(-M_2) \oplus (-\mathcal{P})$ (formally "$[\mathcal{P}\varphi, \varphi] = (\mathcal{P}\varphi, \varphi)$"), this is purely symbolic because, a priori, $\mathcal{P}\varphi$ need not be in $L^2(\nu)$ if $\varphi \in D(\sqrt{-M_2})$.

4.6.2 *On p-Independence of "Asymptotic Spectra" I*

We show now that generators of the C_0-semigroups $\left(\mathcal{J}_p(t)\right)_{t \ge 0}$ introduced in Theorem 4.6.1 share the same "asymptotic spectra". Define the essential spectral bound of

$$\begin{cases} J_1 = M_1 + P \\ D(J_1) = D(M_1) \end{cases}$$

(if $s(J_1) > -\infty$) by

$$s_{ess}(J_1) := \inf\left\{\alpha \in (-\infty, s(J_1)) \,;\; \sigma(J_1) \cap \{\operatorname{Re}\lambda > \alpha\} \subset \sigma_{discr}(J_1)\right\}$$

where $\sigma_{discr}(J_1)$ is the set of isolated eigenvalues of J_1 with finite algebraic multiplicities, with the convention that $s_{ess}(J_1) = s(J_1)$ if the set above is empty. Similarly, If $s(J_{1*}) > -\infty$, define the essential spectral bound of

$$\begin{cases} J_{1*} = M_1 + P_* \\ D(J_{1*}) = D(M_1) \end{cases}$$

(if $s(J_{1*}) > -\infty$) by

$$s_{ess}(J_{1*}) := \inf\{\alpha \in (-\infty, s(J_{1*})); \ \sigma(J_{1*}) \cap \{\mathrm{Re}\,\lambda > \alpha\} \subset \sigma_{discr}(J_{1*})\}.$$

Theorem 4.6.3 *Let Hyp 1–3 be satisfied. Let the condition (4.6.2) be satisfied and*

$$\gamma := s_{ess}(J_{1*}) \vee s_{ess}(J_1) < s(J_{1*}) \wedge s(J_1). \tag{4.6.13}$$

Then

$$\sigma(J_p) \cap \{\mathrm{Re}\,\lambda > \gamma\}$$

is p-independent and consists of isolated eigenvalues with finite (p-independent) algebraic multiplicities. If additionally $(\mathcal{M}_1(t))_{t\geq 0}$ is immediately continuous then

$$\sigma\left(\mathcal{J}_p(t)\right) \cap \left\{\mu; \ |\mu| > e^{\gamma t}\right\}$$

is p-independent and consists of isolated eigenvalues with finite (p-independent) algebraic multiplicities.

Proof Let $(J_{1*})^*$ be the dual operator of J_{1*}. Note that $(J_{1*})^*$ and J_{1*} have the same isolated eigenvalues with finite algebraic multiplicities so

$$\sigma\left((J_{1*})^*\right) \cap \{\mathrm{Re}\,\lambda > \gamma\}$$

consists of isolated eigenvalues with finite algebraic multiplicities. We have seen in the proof of Theorem 4.6.1 that the restriction of $(\lambda - J_1)^{-1}$ to $L^1(\nu) \cap L^\infty(\nu)$ coincides with $(\lambda - (J_{1*})^*)^{-1}$. Then the first item follows from [160, Corollary 1.4] and the second one from [160, Remark 1.5 (a)]. $\square$

Remark 4.6.1 If $P_* = P$ then (4.6.13) reduces to $s_{ess}(J_1) < s(J_1)$.

We complement Theorem 4.6.3 by a more special statement.

Theorem 4.6.4 *Let Hyp 1–3 be satisfied. Suppose that $(\mathcal{M}_1(t))_{t\geq 0}$ is immediately continuous. If $P\,(\lambda - M_1)^{-1}$ is weakly compact in $L^1(\nu)$ then*

$$\mathcal{J}_p(t) - \mathcal{M}_p(t) \text{ is compact } (1 < p < \infty).$$

In particular $\left(\mathcal{J}_p(t)\right)_{t\geq 0}$ and $\left(\mathcal{M}_p(t)\right)_{t\geq 0}$ have the same essential spectrum and the same essential type.

Proof According to Remark 4.3.1 $\mathcal{J}_1(t) - \mathcal{M}_1(t)$ is weakly compact in $L^1(\nu)$ or equivalently (see [300]) $\mathcal{J}_1(t) - \mathcal{M}_1(t)$ is stricly singular in $L^1(\nu)$; it follows from [161, Theorem 4.2] that $\mathcal{J}_p(t) - \mathcal{M}_p(t)$ is compact $(1 < p < \infty)$. $\square$

4.6.3 Alternative Approach to $L^p(\nu)$ Extension

In the previous predual approach to $L^p(\nu)$ extension, P need *not* be M-bounded in $L^p(\nu)$. Another more classical strategy for extending $(\mathcal{J}_1(t))_{t\geq 0}$ to $L^p(\nu)$ is to extend the M-boundedness of P to $L^p(\nu)$. We have:

Theorem 4.6.5 *Suppose that* $(\mathcal{M}_1(t))_{t\geq 0}$ *acts in all* $L^p(\nu)$ *as a holomorphic semigroup* $(\mathcal{M}_p(t))_{t\geq 0}$ *with generator* M_p *and*

$$P(\lambda - M_\infty)^{-1}1 \in L^\infty(\nu).$$

Then $P(\lambda - M_p)^{-1}$ *is bounded* $L^p(\nu)$ $(1 \leq p \leq +\infty)$. *If* $r\left(P(\lambda - M_1)^{-1}\right) < 1$ *(for* λ *large enough) then* $M_p + P : D(M_p) \subset L^p(\nu) \to L^p(\nu)$ *generates a holomorphic semigroup* $(\mathcal{J}_p(t))_{t\geq 0}$ *in* $L^p(\nu)$ *for* p *close to 1 and for all* $p < \infty$ *if* $P(\lambda - M_\infty)^{-1}1 \leq 1$.

Proof By assumption $P(\lambda - M)^{-1}$ is bounded in $L^\infty(\nu)$ and consequently, by Riesz-Thorin theorem, in all $L^p(\nu)$ $(1 \leq p \leq +\infty)$ with the convexity estimate

$$\left\| P(\lambda - M_p)^{-1}\right\|_{\mathcal{L}(L^p(\nu))} \leq \left\| P(\lambda - M_1)^{-1}\right\|_{\mathcal{L}(L^1(\nu))}^{\alpha} \left\| P(\lambda - M)^{-1}\right\|_{\mathcal{L}(L^\infty(\nu))}^{1-\alpha}$$

where $\frac{1}{p} = \frac{\alpha}{1} + \frac{1-\alpha}{\infty}$ i.e.,

$$\left\| P(\lambda - M_p)^{-1}\right\|_{\mathcal{L}(L^p(\nu))} \leq \left\| P(\lambda - M_1)^{-1}\right\|_{\mathcal{L}(L^1(\nu))}^{\frac{1}{p}} \left\| P(\lambda - M)^{-1}\right\|_{\mathcal{L}(L^\infty(\nu))}^{1-\frac{1}{p}}.$$

Thus

$$\left\|(P(\lambda - M_p)^{-1})^n\right\|_{\mathcal{L}(L^p(\nu))}^{\frac{1}{n}} \leq \left(\left\|(P(\lambda - M_1)^{-1})^n\right\|_{\mathcal{L}(L^1(\nu))}^{\frac{1}{p}}\right)^{\frac{1}{n}} \left(\left\|(P(\lambda - M)^{-1})^n\right\|_{\mathcal{L}(L^\infty(\nu))}^{1-\frac{1}{p}}\right)^{\frac{1}{n}}$$

and taking the limit $n \to \infty$

$$r\left(P(\lambda - M_p)^{-1}\right) \leq \left(r\left(P(\lambda - M_1)^{-1}\right)\right)^{\frac{1}{p}} \left(r\left(P(\lambda - M_\infty)^{-1}\right)\right)^{1-\frac{1}{p}}.$$

Hence $r\left(P(\lambda - M_p)^{-1}\right) < 1$ for p close to 1 (and λ large enough). We conclude with Lemma 2.6.2. If $P(\lambda - M_\infty)^{-1}1 \le 1$ then $r\left(P(\lambda - M_\infty)^{-1}\right) \le 1$ and for all p

$$r\left(P(\lambda - M_p)^{-1}\right) \le \left(r\left(P(\lambda - M_1)^{-1}\right)\right)^{\frac{1}{p}} < 1$$

λ large enough which ends the proof with Lemma 2.6.2 again. $\square$

4.6.4 On p-Independence of "Asymptotic Spectra" II

We have much more precise (generation and spectral) results under a compactness assumption. We start with

Theorem 4.6.6 *Let $(\mathcal{M}_1(t))_{t \ge 0}$ acts in all $L^p(\nu)$ as a holomorphic C_0-semigroup $\left(\mathcal{M}_p(t)\right)_{t \ge 0}$ with generator M_p. Suppose that*

$$P(\lambda - M_\infty)^{-1}1 \in L^\infty(\nu).$$

and some power $\left(P(\lambda - M_1)^{-1}\right)^n$ is compact in $L^1(\nu)$. Then

$$M_p + P : D(M_p) \subset L^p(\nu) \to L^p(\nu)$$

generates a holomorphic C_0-semigroup $\left(\mathcal{J}_p(t)\right)_{t \ge 0}$ in $L^p(\nu)$ for all $p \in [1, +\infty)$.

Proof We know that P is M_p-bounded in all $L^p(\nu)$. By interpolation (see e.g. [99, Theorem 1.6.1]), $\left(P(\lambda - M_p)^{-1}\right)^n$ is compact in all $L^p(\nu)$ ($1 \le p < \infty$). Finally, the generation follows from Theorem 2.6.5. $\square$

We give now a very detailed spectral result.

Theorem 4.6.7 *Let $(\mathcal{M}_1(t))_{t \ge 0}$ acts in all $L^p(\nu)$ as a holomorphic C_0-semigroup $\left(\mathcal{M}_p(t)\right)_{t \ge 0}$ with generator M_p. Suppose that*

$$P(\lambda - M_\infty)^{-1}1 \in L^\infty(\nu).$$

and some power $\left(P(\lambda - M_1)^{-1}\right)^n$ is compact in $L^1(\nu)$. Then

(i) M_p and $M_p + P$ have the same essential bound, i.e., $s_{ess}(M_p) = s_{ess}(M_p + P)$.

(ii) The spectral bound $s(M_p + P)$ is p-independent.

(ii) If $s_{ess}(M_p)$ is p-independent then

$$\sigma(M_p + P) \cap \left\{\operatorname{Re}\lambda > s_{ess}(M_p)\right\}$$

is also p-independent and consists of isolated eigenvalues with finite (p-independent) algebraic multiplicities. Similarly,

$$\sigma\left(\mathcal{J}_p(t)\right) \cap \left\{\mu; \ |\mu| > e^{s_{ess}(M_p)t}\right\}$$

is p-independent and consists of isolated eigenvalues with finite (p-independent) algebraic multiplicities.

Proof (i) It follows by interpolation (see e.g. [99, Theorem 1.6.1]) that $\left(P(\lambda - M_p)^{-1}\right)^n$ is compact in all $L^p(\nu)$ and (see the proof of Theorem 2.6.5)

$$r\left(P\left(\lambda - M_p\right)^{-1}\right) \to 0 \ (\lambda \to +\infty).$$

According to Theorem 2.8.1

$$\sigma(M_p + P) \cap \left\{\mathrm{Re}\,\lambda > s_{ess}(M_p)\right\}$$

consists of isolated eigenvalues with finite algebraic multiplicities and consequently

$$s_{ess}(M_p + P) \leq s_{ess}(M_p).$$

Conversely, according to [364, Theorem 1.1] $\left(P(\lambda - M_p - P)^{-1}\right)^n$ is also compact and consequently (using again the proof of Theorem 2.6.5)

$$r\left(P\left(\lambda - M_p - P\right)^{-1}\right) \to 0 \ (\lambda \to +\infty).$$

Then, using again [364, Theorem 1.1] in the other direction

$$\sigma(M_p) \cap \left\{\mathrm{Re}\,\lambda > s_{ess}(M_p + P)\right\}$$

consists of isolated eigenvalues with finite algebraic multiplicities and consequently

$$s_{ess}(M_p) \leq s_{ess}(M_p + P).$$

(ii) Since $\left(P(\lambda - M_p)^{-1}\right)^n$ is compact in all $L^p(\nu)$ then the spectrum of $(P(\lambda - M_p)^{-1})^n$ is p-independent, (see e.g. [99, Corollary 1.6.2, p. 35]). In particular

$$r\left[\left(P(\lambda - M_p)^{-1}\right)^n\right]$$

is p-independent and

$$r\left(P(\lambda - M_p)^{-1}\right) = \left(r\left[\left(P(\lambda - M_p)^{-1}\right)^n\right]\right)^{\frac{1}{n}}$$

is also p-independent. Finally, the spectral bound of $M_p + P$, given by

$$s(M_p + P) = \inf\left\{\lambda > -\inf\sigma; \ r\left(P(\lambda - M_p)^{-1}\right) < 1\right\},$$

is also p-independent.

(iii) is a consequence of [160, Corollary 1.4 and Remark 1.5 (a)]. $\square$

4.6.5   On the M_p-Bound in $L^p(\nu)$ ($p \geq 2$)

It is well known (see e.g. [46, Theorem 13.3, p. 198]) that generation of positive contraction C_0-semigroups on Banach lattices in a perturbation context depends critically on estimate of relative operator bound and on a dispersivity condition. In this subsection, although it is not needed elsewhere in this monograph, we highlight an interesting result on the M_p-bound of P in $L^p(\nu)$ (see Corollary 4.6.1) which is peculiar to the case $p \geq 2$. To this end, we note that is well known that if X is a Banach space and

$$A : D(A) \subset X \to X$$

(with $\rho(A) \neq \varnothing$) then $B : D(A) \to X$ is A-bounded, i.e.,

$$\|Bx\| \leq a \|Ax\| + b \|x\| \quad (x \in D(A)) \tag{4.6.14}$$

(for some $a, b \geq 0$), if and only if $B(\lambda - A)^{-1} \in \mathcal{L}(X)$ for some $\lambda \in \rho(A)$; in this case, the A-bound of B, i.e.,

$$a_B := \inf \{a \geq 0; \ \exists b \geq 0; \ (4.6.14) \text{ is satisfied}\},$$

is less than or equal to $\inf_{\lambda \in \rho(A)} \left\| B(\lambda - A)^{-1} \right\|_{\mathcal{L}(X)}$ (see e.g. [33, Lemma 4.1 p. 116]). In particular, if A is a generator then

$$a_B \leq \lim_{\lambda \to +\infty} \inf \left\| B(\lambda - A)^{-1} \right\|_{\mathcal{L}(X)}. \tag{4.6.15}$$

We give here a result from [272] on the "computation" of the A-bound a_B of B in the context of positive C_0-semigroups and positive perturbations on Banach lattices. We first complement (4.6.15) by

Proposition 4.6.1 ([272, Lemma 2.1]) *Let X be Banach space and let $A : D(A) \subset X \to X$ be a generator of a C_0-semigroup. Let $B : D(A) \to X$ be A-bounded and let a_B be its A-bound. Then*

$$\lim_{\lambda \to +\infty} \sup \left\| B(\lambda - A)^{-1} \right\|_{\mathcal{L}(X)} \leq a_B \lim_{\lambda \to +\infty} \sup \left\| A(\lambda - A)^{-1} \right\|_{\mathcal{L}(X)}.$$

In particular, if

$$\lim_{\lambda \to +\infty} \sup \left\| A(\lambda - A)^{-1} \right\|_{\mathcal{L}(X)} \leq 1 \tag{4.6.16}$$

then $a_B = \lim_{\lambda \to +\infty} \left\| B(\lambda - A)^{-1} \right\|_{\mathcal{L}(X)}$.

Hence the "computation" of the A-bound of B relies on the key condition (4.6.16) which is the object of the following

Theorem 4.6.8 ([272, Theorem 4.13]) *Let X be a Banach lattice and let $(U(t))_{t \geq 0}$ be a positive contraction C_0-semigroup on X with generator A. Let there exist a duality map $\Phi : X \to X'$ (relative to some gauge $\zeta : \mathbb{R}_+ \to \mathbb{R}_+$) uniformly monotone on the positive cone X_+ in the sense*

$$\langle \Phi(x) - \Phi(y), x - y \rangle \geq \langle \Phi(x - y), x - y \rangle \ \forall x, y \in X_+. \tag{4.6.17}$$

Then $A(\lambda - A)^{-1}$ is a contraction.

It turns out that in Lebesgue spaces $L^p(\nu)$, the uniform monotony of the duality map (4.6.17) on $L^p_+(\nu)$ is true for $p \geq 2$ (see [272, Theorem 4.7]) and is false in general if $1 \leq p < 2$ (see [272, Remark 4.22]).

Corollary 4.6.1 *Suppose that $(\mathcal{M}(t))_{t \geq 0}$ acts in all $L^p(\nu)$ as a C_0-semigroup $(\mathcal{M}_p(t))_{t \geq 0}$ with generator M_p and*

$$P(\lambda - M_\infty)^{-1} 1 \in L^\infty(\nu).$$

Then P is M_p-bounded in $L^p(\nu)$ $(1 \leq p \leq +\infty)$ and the M_p-bound of P in $L^p(\nu)$ is equal to $\lim_{\lambda \to +\infty} \left\| P(\lambda - M_p)^{-1} \right\|_{\mathcal{L}(L^p(\nu))}$ if $p \geq 2$.

4.7 Comments

In this chapter, we have focused exclusively on the presence of spectral gaps. However, even in their absence, one can still obtain meaningful results using weak compactness arguments. Specifically, it is possible to show that the spectral bound $s(J_1)$ of the perturbed generator $J_1 := M_1 + P$ is an eigenvalue, though not necessarily a dominant or isolated one, and that the corresponding C_0-semigroup $(\mathcal{J}_1(t))_{t \geq 0}$ is partially integral. In this case, the spectral bound of the generator of the rescaled semigroup $\left(e^{-s(J_1)t} \mathcal{J}_1(t) \right)_{t \geq 0}$, which remains partially integral, is equal to zero. Whether this rescaled semigroup is *uniformly bounded*, and hence asymptotically stable by the general theory of [135], is a priori an open problem.

Theorem 4.6.8 is a key part of a general construction, given in [272], on various "contractivity" results (on the positive cone) for linear operators of the form $I - C$ where C are positive contractions on real ordered Banach spaces with applications to relative operator bounds, ergodic projections and conditional expectations. The proof that the key assumption in Theorem 4.6.8 is true in $L^p(\nu)$ spaces for $p \geq 2$ is generalized by E. Ricard to non commutative L^p spaces [316].

Chapter 5
Spectral Analysis of Absorption Semigroups

5.1 Chapter Aims

In this chapter, we study absorption semigroups $(U_V(t))_{t \geqslant 0}$, a class of C_0-semigroups on abstract L^1-spaces formally generated by $T_V = T - V$, where T is the generator of a substochastic C_0-semigroup $(U(t))_{t \geqslant 0}$ and V is an unbounded but measurable function bounded from below. By exploiting a specific contractive property of the operator $V(\lambda - T_V)^{-1}$, valid in L^1-spaces, along with the confining nature of the potential V and a "local weak compactness" condition on the resolvent $(\lambda - T)^{-1}$, we establish the existence of a *local* spectral gap at the boundary of the spectrum of T_V. We also present a more robust approach that applies in all L^p-spaces, does not rely on contractivity, and yields a stronger result: when the semigroup $(U(t))_{t \geqslant 0}$ is "locally compact", the absorption semigroup possesses $(U_V(t))_{t \geqslant 0}$ a spectral gap.

5.2 Chapter Overview

In the previous chapters, we dealt with spectral properties of positively perturbed positive C_0-semigroups. It is possible to capture positive C_0-semigroups by perturbing negatively positive C_0-semigroups on Lebesgue spaces provided that the perturbation is a (possibly unbounded) multiplication operator V. This is the subject of absorption semigroups we consider now. Let $(\Omega, \mathcal{E}, \nu)$ be a sigma-finite measure space. Let $(U(t))_{t \geqslant 0}$ be a positive C_0-semigroup on

$$L^p(\nu) := L^p(\Omega, \mathcal{E}, \nu)$$

with generator T and let

$$V : \Omega \to [-\infty, +\infty]$$

© The Author(s), under exclusive license to Springer Nature Switzerland AG 2026　　　113
M. Mokhtar-Kharroubi, *Peripheral Spectra of Perturbed Positive Semigroups*, Lecture Notes in Mathematics 2388, https://doi.org/10.1007/978-3-032-11173-9_5

be measurable. By absorption semigroup relative to V, we mean a C_0-semigroup $(U_V(t))_{t \geqslant 0}$ with (formal) generator

$$T_V = \text{``}T - V\text{''}$$

A general theory of absorption semigroups in L^p spaces is given in [370] for indefinite singular, i.e., unbounded, potentials V. We restrict ourselves here to potentials V bounded from below

$$V \geq -c$$

for some constant $c > 0$.

This chapter deals with peripheral spectral analysis of absorption semigroups $(U_V(t))_{t \geqslant 0}$ and their time asymptotics. A few historical remarks are in order before approaching the subject of this chapter. There exists a huge hilbertian literature on full discreteness of the spectrum of a Schrödinger operator in $L^2(\mathbb{R}^N)$

$$-\Delta \dotplus V \quad \text{(form-sum)} \tag{5.2.1}$$

which goes back to the 1930 and 1940s with K. Friedrichs, F. Rellich and pursued since then by A.M. Molchanov and many others; we refer to the introductory chapter in [271] for precise references. Much more developments in this direction, in terms of Wiener capacity of suitable sets, are given by Maz'ya and Shubin [227]. On the other hand, spectral gaps for (5.2.1), i.e., the location of its essential spectra, are dealt with e.g. in [229, 301]. More recently it was shown that $T - V$ is resolvent compact in $L^2(\mathbb{R}^N)$ when T is the relativistic α-stable operator

$$T = -(-\Delta + m^{\frac{2}{\alpha}})^{\frac{\alpha}{2}} + m$$

provided that $\lim_{|x| \to \infty} V(x) = +\infty$ [188]; various subsequent improvements are also given by others; see the introductory chapter in [271] for the references.

The objective of the monograph [271] was to address these questions in $L^1(\mathbb{R}^N)$ for a broader class of Schrödinger-type operators, and more generally, for generators of absorption semigroups in abstract Lebesgue spaces, by leveraging the confining effect of absorption terms. It is important to emphasize that, a priori, such L^1-based results cannot be derived from the existing L^2-literature.

The objective of the present chapter is to revisit and extend, in several directions, the construction developed in [271], with particular emphasis on spectral gap problems. To the best of the author's knowledge, no comparable systematic construction has been documented in the existing literature. It is important to note that this construction is meaningful only in cases where the operator T is not resolvent compact and when the semigroup $(U(t))_{t \geqslant 0}$ is not compact. Indeed, it is straightforward to show, using domination arguments, that if $(U(t))_{t \geqslant 0}$ is a compact semigroup then so is $(U_V(t))_{t \geqslant 0}$ and if T is resolvent compact then so is T_V. In this chapter, we address the following issues:

(1) Compactness of $(U_V(t))_{t\geqslant 0}$; (or at least resolvent compactness of T_V).
(2) T_V has a spectral gap, i.e.,

$$\sigma_{ess}(T_V) \subset \{\text{Re}\,\lambda \leq (s(T_V) - \varepsilon)\} \text{ for some } \varepsilon > 0.$$

(3) T_V has locally a spectral gap, i.e., there exists a function $c \to \varepsilon(c) > 0$ and

$$\sigma_{ess}(T_V) \cap \{|\text{Im}\,\lambda| \leq c\} \subset \{\text{Re}\,\lambda \leq (s(T_V) - \varepsilon(c))\}.$$

(4) $(U_V(t))_{t\geqslant 0}$ has a spectral gap, i.e., $\omega_{ess}(U_V) < \omega(U_V)$.
(5) Time asymptotics of $(U_V(t))_{t\geqslant 0}$.

In Sect. 5.3, we collect first some useful results from [370] on absorption C_0-semigroups, in particular a contraction property

$$\left\| V (\lambda - T_V)^{-1} \right\|_{\mathcal{L}(L^1(\nu))} \leq 1 \ (\lambda > 0) \tag{5.2.2}$$

which is specific to substochastic C_0-semigroup $(U(t))_{t\geqslant 0}$ in L^1 spaces.

Here are the results given in this chapter: We give a non conventional Dyson-Phillips expansion of the unperturbed semigroup $(U(t))_{t\geqslant 0}$ from the perturbed one $(U_V(t))_{t\geqslant 0}$, (see Theorem 5.4.1 and Corollary 5.4.1) which allows the study of irreducibility of absorption semigroups $(U_V(t))_{t\geqslant 0}$, (see Theorem 5.5.1). In Theorem 5.6.1, we recall a result from [271] according to which if the unperturbed resolvent $(\lambda - T)^{-1}$ is "locally" weakly compact and if the sublevel sets of V are "thin at infinity" (in a suitable sense) then T_V is resolvent compact in $L^1(\nu)$. We also address the case where the sublevel sets of V are not "thin at infinity"; indeed, in Theorem 5.7.1, we show that if the unperturbed resolvent $(\lambda - T)^{-1}$ is "locally" weakly compact and if

$$\lim_{``x\to\infty"} \inf V(x) > -s(T_V)$$

then T_V has locally a spectral gap, i.e., there exists a function $c \to \varepsilon(c) > 0$ such that

$$\sigma_{ess}(T_V) \cap \{|\text{Im}\,\lambda| \leq c\} \subset \{\text{Re}\,\lambda \leq (s(T_V) - \varepsilon(c))\}.$$

We note that $s(T_V) > -\infty$ if $(U(t))_{t\geqslant 0}$ is a Markov semigroup and is symmetric in $L^2(\nu)$, (see Theorem 5.7.2). We derive from the previous results that if $(U_V(t))_{t\geq 0}$ is partially integral then $\left(e^{-s(T_V)t} U_V(t)\right)_{t\geqslant 0}$ is stochastic and asymptotically stable in a suitable *weighted* L^1 space, (see Theorem 5.8.1). We show also how to extend the results to $L^p(\nu)$, (see Theorem 5.9.1).

In Theorem 5.10.1, we give a direct and completely different approach to spectral discreteness (or spectral gaps) for absorption semigroups provided that V is locally bounded; (see Remark 5.10.1 for a more general assumption). This approach is significantly more powerful than the previous one, as it does not rely on substochasticity of $(U(t))_{t\geqslant 0}$ and applies uniformly across all L^p-spaces. Moreover, it yields a global spectral gap for T_V, not merely a local one. Finally it provides spectral gap results directly at the semigroup level, not just for the generators. Indeed, we show that if $U(t)$ is "locally" weakly compact if $p = 1$ (resp. "locally" compact if $p > 1$) then $(U_V(t))_{t \geqslant 0}$ has a spectral gap in $L^p(\nu)$ provided that

$$\lim_{"x \to \infty"} \inf V(x) > s(T) - s(T_V),$$

(see Theorem 5.10.1 for the details). Similarly, if $U(t)$ is not necessarily "locally" weakly compact if $p = 1$ (resp. "locally" compact if $p > 1$) but $(\lambda - T)^{-1}$ is then T_V has a spectral gap, (see Theorem 5.10.2). Finally, we deal with Kato-Voigt perturbations of absorption semigroups and improve the results of Chap. 3 when the unperturbed semigroup itself is an absorption semigroup, (see Theorems 5.11.1, 5.11.2 and 5.11.3). Some remarks, comments or open questions are relegated to the end of this chapter in the section "Comments".

5.3 Reminders on Absorption Semigroups

Since the compactness (or spectral gap) properties are stable under a shift of the generator T then, by replacing V by $V + c$ if necessary, we may restrict ourselves, without loss of generality, to nonnegative potentials

$$V : \Omega \to [0, +\infty] . \tag{5.3.1}$$

In this section, we collect some key results on absorption semigroups [370]. Other developments are given in [9, 168, 221]. Let $V_n := V \wedge n$ and let $\left(e^{t(T-V_n)}\right)_{t \geqslant 0}$ be the C_0-semigroup generated by $T - V_n$. It is elementary to see that

$$e^{t(T-V_{n+1})} f \leq e^{t(T-V_n)} f \quad \forall f \in L^p_+(\Omega; \nu)$$

so that a monotone convergence in $L^p(\Omega; \nu)$

$$U_V(t)f := \lim_{n \to +\infty} e^{t(T-V_n)} f$$

defines a semigroup $(U_V(t))_{t \geqslant 0}$. This semigroup is a priori strongly continuous for $t > 0$ only (see e.g. [9]).

Definition 5.3.1 We say that V is *admissible* for $(U(t))_{t \geqslant 0}$ (or U-admissible for short) if $(U_V(t))_{t \geqslant 0}$ is a C_0-semigroup, i.e., is strongly continuous at $t = 0$.

Let V be U-admissible and let T_V be the generator of $(U_V(t))_{t \geqslant 0}$, then

$$T - V \subset T_V, \tag{5.3.2}$$

i.e., T_V is an *extension* of

$$T - V : D(T) \cap D(V) \to L^p(\Omega; \nu),$$

(see [370, Corollary 2.7]). Note that V is U-admissible if

$$D(T) \cap D(V) \text{ is dense in } L^p(\Omega; \nu),$$

(see [370, Proposition 2.9]). Note also that if V is U-admissible and $V \leq \widetilde{V}$ then $\widetilde{V}$ is also U-admissible admissible and $U_{\widetilde{V}}(t) \leq U_V(t)$, see ([370, Remark 2.3(b)]).

We resume now some key results from [370] we need here.

Lemma 5.3.1 (see [370, Definition 2.12 and Proposition 4.2, Corollary 4.3 (a)]) *Let V be U-admissible then for $\eta \in (0, 1]$, ηV is U-admissible and*

$$\begin{cases} T_{\eta V} = T_V + (1 - \eta)V \\ \quad D(T_{\eta V}) = D(T_V) \end{cases} \tag{5.3.3}$$

and

$$U_{(0,V)}(t)f := \lim_{\eta \to 0} U_{\eta V}(t)f \tag{5.3.4}$$

defines a C_0-semigroup $\left(U_{(0,V)}(t)\right)_{t \geqslant 0}$ with generator $T_{(0,V)}$ such that

$$T_{(0,V)} \supset T_V + V$$

and

$$U_V(t) \leq U_{(0,V)}(t) \leq U(t).$$

Moreover, V is $U_{(0,V)}$-admissible and

$$\left(U_{(0,V)}\right)_V (t) = U_V(t)$$

$$\begin{cases} T_V = \left(T_{(0,V)}\right)_V = T_{(0,V)} - V \\ D(T_V) = D(T_{(0,V)}) \cap D(V). \end{cases}$$

Definition 5.3.2 Let V be U-admissible. We say that V is regular for $(U(t))_{t \geqslant 0}$ (or U-regular for short) if $\left(U_{(0,V)}(t)\right)_{t \geqslant 0} = (U(t))_{t \geqslant 0}$.

Remark 5.3.1 Note that if V is U-regular then $D(T_V) = D(T) \cap D(V)$.

Here is a criterion of U-regularity.

Lemma 5.3.2 ([370, Propositions 2.13, 4.4]) *Let V be U-admissible. Then*
(i) If $D(T) \cap D(V)$ is a core for T then V is U-regular.
(ii) Let $p = 1$ and let $(U(t))_{t \geqslant 0}$ be stochastic. If V is U-regular then $D(T) \cap D(V)$ is a core for T.

Finally, absorption C_0-semigroups enjoy nice interpolation properties.

Proposition 5.3.1 ([370, Proposition 3.1]) *Let* $1 \leq p_0 < p_1 < +\infty$ *and (for* $p_0 \leq p \leq p_1$*) let* $\left(U^{(p)}(t)\right)_{t \geq 0}$ *be a positive* C_0*-semigroup on* $L^p(\nu)$ *with generator* $T^{(p)}$ *such that* $U^{(p)}(t)$ *coincides with* $U^{(p_0)}(t)$ *on* $L^{p_0}(\nu) \cap L^{p_1}(\nu)$*. Then*

(i) V *be* $U^{(p)}$*-admissible for all* $p \in [p_0, p_1]$ *if and only if* V *be* $U^{(p)}$*-admissible for some* $p \in [p_0, p_1]$*. In this case,* $U_V^{(p)}(t)$ *coincides with* $U_V^{(p_0)}(t)$ *on* $L^{p_0}(\nu) \cap L^{p_1}(\nu)$*.*

(ii) V *is* $U^{(p)}$*-regular for all* $p \in [p_0, p_1]$ *if and only if* V *is* $U^{(p)}$*-regular for some* $p \in [p_0, p_1]$*.*

Finally, we recall a key contraction estimate peculiar to L^1 spaces; (a result in the same spirit can be found in [292]).

Lemma 5.3.3 ([370]) *Let* $(U(t))_{t \geq 0}$ *be a substochastic* C_0*-semigroup in* $L^1(\nu)$ *with generator* T *and let*

$$V : (\Omega, \nu) \to [0, +\infty]$$

be U*-admissible. Let* $(U_V(t))_{t \geq 0}$ *be the absorption* C_0*-semigroup with generator* T_V*. Then* V *is* T_V*-bounded and*

$$\left\| V(\lambda - T_V)^{-1} \right\| \leq 1, \quad (\lambda > 0). \tag{5.3.5}$$

5.4 Absorption Semigroups Versus Kato-Voigt Perturbation Theory

We start with a non conventional *reverse* Dyson-Phillips expansion.

Theorem 5.4.1 *Let* $(U(t))_{t \geq 0}$ *be a substochastic* C_0*-semigroup on* $L^1(\nu)$ *with generator* T*. Let* V *be* U*-admissible. Then* $\left(U_{(0,V)}(t)\right)_{t \geq 0}$ *is the Kato-Voigt perturbation of* $(U_V(t))_{t \geq 0}$ *relative to the (generator) perturbation* V*. In particular,* $\left(U_{(0,V)}(t)\right)_{t \geq 0}$ *is given by the Dyson-Phillips expansion*

$$U_{(0,V)}(t)f = \sum_j U_j(t)f, \quad (f \in L^1(\nu)) \tag{5.4.1}$$

(a strongly convergent series) where

$$U_0(t) = U_V(t); \quad U_{j+1}(t)f = \int_0^t U_j(t-s)VU_V(s)f\,ds, \quad j \geq 0.$$

Proof We know that V is T_V-bounded and $\left\| V(\lambda - T_V)^{-1} \right\| \leq 1$, $(\lambda > 0)$. Since $T_{(0,V)}$ is the generator of a substochastic semigroup $\left(U_{(0,V)}(t)\right)_{t \geq 0}$ then

$$\int T_{(0,V)}\varphi \le 0 \quad (\varphi \in D(T_{(0,V)}) \cap L_+^1(\nu)).$$

Hence

$$\int T_V\varphi + V\varphi \le 0 \quad (\varphi \in D(T_V) \cap L_+^1(\nu)).$$

because $T_V + V \subset T_{(0,V)}$. Hence by Theorem 2.11.1 there exists a unique extension of $T_V + V$ which generates a minimal substochastic semigroup $(Z(t))_{t \geqslant 0}$. Le us identify $(Z(t))_{t \geqslant 0}$. The proof of Theorem 2.11.1 (via Desch's perturbation theorem) is as follows:

$$\left\| (1 - \eta)\,V\,(\lambda - T_V)^{-1} \right\| \le 1 - \eta < 1, \quad (\eta > 0)$$

shows that $T_V + (1 - \eta)\,V$ generates a positive semigroup $\big(Z_\eta(t)\big)_{t \geqslant 0}$ which must be contractive since

$$\int T_V\varphi + (1 - \eta)\,V\varphi \le 0 \quad (\varphi \in D(T_V) \cap L_+^1(\nu));$$

and $Z_\eta(t) \to Z(t) \quad (\eta \to 0)$ monotonically. Since $T_V + (1 - \eta)\,V = T_{\eta V}$ (see (5.3.3)) then $\big(Z_\eta(t)\big)_{t \geqslant 0}$ is nothing but $\big(U_{\eta V}(t)\big)_{t \geqslant 0}$ and by (5.3.4)

$$U_{(0,V)}(t) := s \lim_{\eta \to 0} U_{\eta V}(t).$$

Finally, the Dyson-Phillips expansion is given by Theorem 2.11.1.   □

In particular, we have

Corollary 5.4.1 *Let V be U-admissible. If V is U-regular (e.g. if $D(T) \cap D(V)$ is a core for T) then $(U(t))_{t \geqslant 0}$ is given by the Dyson-Phillips expansion*

$$U(t)f = \sum_j U_j(t)f, \quad (f \in L^1(\nu))$$

(a strongly convergent series) where

$$U_0(t) = U_V(t); \quad U_{j+1}(t)f = \int_0^t U_j(t - s)V U_V(s)f\,ds, \quad j \ge 0.$$

5.5 Irreducibility of Absorption Semigroups

We show now how the above Dyson-Phillips expansion can be used to address the irreducibility of absorption semigroups .

Theorem 5.5.1 *Let V be U-admissible. Then $\left(U_{(0,V)}(t)\right)_{t\geqslant 0}$ is irreducible if and only if $(U_V(t))_{t\geq 0}$ is. In particular, if V is U-regular (e.g. if $D(T) \cap D(V)$ is a core for T) and if $(U(t))_{t\geq 0}$ is irreducible then $(U_V(t))_{t\geq 0}$ is also irreducible.*

Proof (i) Since $U_V(t) \leq U_{(0,V)}$ then the irreducibility of $(U_V(t))_{t\geq 0}$ implies trivially that $\left(U_{(0,V)}(t)\right)_{t\geqslant 0}$ is also irreducible.

(ii) If $(U_V(t))_{t\geq 0}$ is not irreducible then there exists a measurable set $\Xi \subset \Omega$ such that $\nu(\Xi) > 0$ and $\nu(\Xi^c) > 0$ and $L^1(\Xi, \nu)$ is invariant under $(U_V(t))_{t\geq 0}$. Note that we identify $L^1(\Xi, \nu)$ to the closed subspace of $L^1(\Omega, \nu)$ of (class of) measurable functions vanishing almost everywhere on Ξ^c. Then $L^1(\Xi, \nu)$ is invariant under

$$U_1(t)f = \int_0^t U_V(t-s)VU_V(s)f\,ds$$

because the multiplication operator V does not augment the support. It follows by induction that all the operators $U_j(t)$ leave $L^1(\Xi, \nu)$ invariant and consequently, by (5.4.1) $\left(U_{(0,V)}(t)\right)_{t\geqslant 0}$ leaves $L^1(\Xi, \nu)$ invariant. This shows that $\left(U_{(0,V)}(t)\right)_{t\geqslant 0}$ is not irreducible. $\qquad\qquad\square$

5.6   Resolvent Compactness of Absorption Generator

In this section, we recall several known compactness results. It follows from (5.3.5) that singular, i.e., unbounded, absorptions V possess a confining effect which was fully exploited in [271] to address various spectral problems. We recall a key result from [271] relying on the sublevel sets of V

$$\Omega_c := \{V \leq c\} \quad (c > 0). \tag{5.6.1}$$

Theorem 5.6.1 *([271, Theorem 15]) Let $(U(t))_{t\geq 0}$ be a substochastic C_0-semigroup in $L^1(\Omega, \nu)$ with generator T and let $V : (\Omega, \nu) \to [0, +\infty]$ be an U-admissible. Let $(U_V(t))_{t\geq 0}$ be the absorption C_0-semigroup with generator T_V. Then T_V is resolvent compact in $L^1(\Omega, \nu)$ if and only if*

$$1_{\Omega_c}(\lambda - T_V)^{-1} : L^1(\Omega, \nu) \to L^1(\Omega_c, \nu) \quad \text{is weakly compact } (c > 0) \tag{5.6.2}$$

where Ω_c are the sublevel sets of V. In particular, (5.6.2) is satisfied if

$$1_{\Omega_c}(\lambda - T)^{-1} : L^1(\Omega, \nu) \to L^1(\Omega_c, \nu) \quad \text{is weakly compact } (c > 0). \tag{5.6.3}$$

Remark 5.6.1 Here is a useful application of Theorem 5.6.1. Let Ω be a locally compact metric space and let $V(x) \to +\infty$ $(x \to \infty)$. If $1_K(\lambda - T)^{-1}$ is weakly compact for any compact subset $K \subset \Omega$ then T_V is resolvent compact in $L^1(\Omega, \nu)$.

If $(U_V(t))_{t\geq 0}$ is norm continuous for $t > 0$, for instance if $(U(t))_{t\geq 0}$ is holomorphic (in which case, $(U_V(t))_{t\geq 0}$ is holomorphic too [169]) then the resolvent compactness of T_V implies that $(U_V(t))_{t\geq 0}$ is a compact semigroup, (see e.g. [271, Corollary 21]).

Remark 5.6.2 If $(U(t))_{t\geq 0}$ operates on all $L^p(\nu)$ as a positive C_0-semigroup $\left(U^{(p)}(t)\right)_{t\geq 0}$ (with generator $T^{(p)}$) then similarly $(U_V(t))_{t\geq 0}$ operates on all $L^p(\nu)$ as a positive C_0-semigroup $\left(U_V^{(p)}(t)\right)_{t\geq 0}$ (with generator $T_V^{(p)}$). In particular if T_V is resolvent compact in $L^1(\nu)$ then, for all $p \geq 1$, $T_V^{(p)}$ is resolvent compact in $L^p(\nu)$. Similarly, if $(U_V(t))_{t\geq 0}$ is a compact semigroup in $L^1(\nu)$ then, for all $p \geq 1$, $\left(U_V^{(p)}(t)\right)_{t\geq 0}$ is compact in $L^p(\nu)$.

The results above apply to convolution C_0-semigroups $(U(t))_{t\geq 0}$ on $L^p\left(\mathbb{R}^d\right)$ in terms of "thinness at infinity" of the sublevel sets of V [271, Chap. 4]; see e.g. Theorem 13.3.1.

5.7 Local Spectral Gaps for Absorption Generators

In the preceeding section, we gave sufficient conditions for the spectra $\sigma(T_V)$ or $\sigma(U_V(t))$ to consist (at most) of isolated eigenvalues with finite algebraic multiplicities. The object of the present section is to explore the possibility for $\sigma(T_V)$ or $\sigma(U_V(t))$ to exhibit (local) spectral gaps, i.e., a discrete "peripheral spectra".

Sufficient conditions for the existence of spectral gaps based on kernel estimates are provided in [271, Chap. 6]. However, these conditions are rather abstract and appear to have limited practical applicability. In this section, we present a more refined approach, based on verifiable assumptions, to establishing local spectral gaps for absorption generators. Here is a key result.

Theorem 5.7.1 *Let $(U(t))_{t\geq 0}$ be a substochastic C_0-semigroup in $L^1(\Omega, \nu)$ with generator T and let $V : (\Omega, \nu) \to [0, +\infty]$ be U-admissible. Let $(U_V(t))_{t\geq 0}$ be the absorption C_0-semigroup with generator T_V and let $s(T_V) > -\infty$. We assume the existence of an increasing sequence of measurable sets $\left(\Xi_j\right)_j$ such that $\cup_j \Xi_j = \Omega$ and*

$$1_{\Xi_j}(\lambda - T)^{-1} \text{ is weakly compact.} \tag{5.7.1}$$

Let

$$\eta := \lim_{j \to +\infty} \inf_{\Xi_j^c} V > -s(T_V). \tag{5.7.2}$$

Then:

(i) If $s(T_V) + i\beta \in \rho(T_V)$ for some $\beta \in \mathbb{R}$ then the complex numbers μ such that $\operatorname{Im} \mu = \beta$ and

$$s(T_V) - \left\|(s(T_V) + i\beta - T_V)^{-1}\right\|^{-1} < \operatorname{Re} \mu$$

belong to the resolvent set $\rho(T_V)$.

(ii) If $s(T_V) + i\beta \in \sigma(T_V)$ for some $\beta \in \mathbb{R}$ then any complex number μ such that $\mathrm{Im}\,\mu = \beta$ *and* $\mathrm{Re}\,\mu > -\eta$ *is either in* $\rho(T_V)$ *or is an isolated eigenvalue of* T_V *with finite algebraic multiplicity.*

Proof It consists in three steps.

Step 1. Let

$$\rho_b(T_V) := \rho(T_V) \cap \{s(T_V) + i\mathbb{R}\}$$

the (possibly empty) "boundary" resolvent set of T_V. It is elementary and well known to see that if $s(T_V) + i\beta \in \rho(T_V)$ for some $\beta \in \mathbb{R}$ then $\alpha + i\beta \in \rho(T_V)$ provided that

$$|\alpha - s(T_V)| < \left\| (s(T_V) + i\beta - T_V)^{-1} \right\|^{-1} ;$$

if $\alpha \leq s(T_V)$, this amounts to

$$\alpha > s(T_V) - \left\| (s(T_V) + i\beta - T_V)^{-1} \right\|^{-1} .$$

Hence the complex numbers $\alpha + i\beta$ such that

$$\begin{cases} s(T_V) + i\beta \in \rho(T_V) \\ s(T_V) - \left\| (s(T_V) + i\beta - T_V)^{-1} \right\|^{-1} < \alpha \leq s(T_V) \end{cases}$$

belong to the resolvent set $\rho(T_V)$.

Step 2. For $\mathrm{Re}\,\lambda > 0$ we decompose $(\lambda - T_V)^{-1}$ as

$$(\lambda - T_V)^{-1} = 1_{\Xi_j}\,(\lambda - T_V)^{-1} + 1_{\Xi_j^c}\,(\lambda - T_V)^{-1} .$$

Since $1_{\Xi_j}\,(\lambda - T_V)^{-1}$ is dominated by $1_{\Xi_j}\,(\mathrm{Re}\,\lambda - T)^{-1}$ then $1_{\Xi_j}\,(\lambda - T_V)^{-1}$ is weakly compact and

$$r_{ess}((\lambda - T_V)^{-1}) = r_{ess}(1_{\Xi_j^c}\,(\lambda - T_V)^{-1}),$$

in particular

$$r_{ess}((\lambda - T_V)^{-1}) \leq \left\| 1_{\Xi_j^c}\,(\lambda - T_V)^{-1} \right\|$$

$$\leq \left\| 1_{\Xi_j^c}\,(\mathrm{Re}\,\lambda - T_V)^{-1} \right\| .$$

By (5.7.2), V is bounded away from zero on Ξ_j^c for j large enough

$$\eta_j := \inf_{\Xi_j^c} V > 0.$$

Since

$$\int_\Omega V(x) \left| ((\mathrm{Re}\,\lambda - T_V)^{-1} f)(x) \right| \nu(dx) \leq \| f \|$$

then

$$\eta_j \int_{\Xi_j^c} \left| \left((\operatorname{Re}\lambda - T_V)^{-1} f \right)(x) \right| \nu(dx) \le \|f\|$$

so

$$\left\| 1_{\Xi_j^c} (\lambda - T_V)^{-1} \right\| \le \eta_j^{-1} \quad (\operatorname{Re}\lambda > 0)$$

and

$$r_{ess}((\lambda - T_V)^{-1}) \le \eta_j^{-1}.$$

Step 3. Let

$$\sigma_b(T_V) := \sigma(T_V) \cap \{s(T_V) + i\mathbb{R}\}$$

the "boundary" (peripheral) spectrum of T_V. Note that

$$s(T_V) \in \sigma_b(T_V).$$

Let $\beta \in \mathbb{R}$ be such that $s(T_V) + i\beta \in \sigma(T_V)$. Then any complex number

$$\lambda = \operatorname{Re}\lambda + i\beta \quad (\operatorname{Re}\lambda > 0)$$

belongs to $\rho(T_V)$ and

$$dis(\lambda, \sigma(T_V)) = \operatorname{Re}\lambda - s(T_V).$$

Since $r((\lambda - T_V)^{-1}) = \frac{1}{dis(\lambda, \sigma(T_V))}$ then

$$r((\lambda - T_V)^{-1}) = \frac{1}{\operatorname{Re}\lambda - s(T_V)}.$$

It follows that $(\lambda - T_V)^{-1}$ has spectral gap, i.e.,

$$r_{ess}((\lambda - T_V)^{-1}) < r((\lambda - T_V)^{-1}),$$

if

$$\eta_j^{-1} < \frac{1}{\operatorname{Re}\lambda - s(T_V)},$$

i.e.,

$$\operatorname{Re}\lambda < s(T_V) + \eta_j.$$

Since $\operatorname{Re}\lambda > 0$, this imposes the condition

$$\eta_j > -s(T_V).$$

We know that $\mu \in \rho(T_V)$ if and only if $\frac{1}{\lambda - \mu} \in \rho((\lambda - T_V)^{-1})$. Similarly, μ is an isolated eigenvalue of T_V with finite algebraic multiplicity if and only if $\frac{1}{\lambda - \mu}$ is an isolated eigenvalue of $(\lambda - T_V)^{-1}$ with finite algebraic multiplicity. Therefore if μ is such that

$$\eta_j^{-1} < \frac{1}{|\lambda - \mu|} \le \frac{1}{\mathrm{Re}\,\lambda - s(T_V)} \tag{5.7.3}$$

then either $\mu \in \rho(T_V)$ or μ is an isolated eigenvalue of T_V with finite algebraic multiplicity. If additionally μ is such that

$$\mathrm{Im}\,\mu = \beta$$

then (5.7.3) amounts to

$$\eta_j^{-1} < \frac{1}{\mathrm{Re}\,\lambda - \mathrm{Re}\,\mu} \le \frac{1}{\mathrm{Re}\,\lambda - s(T_V)},$$

i.e., $\mathrm{Re}\,\lambda - \mathrm{Re}\,\mu < \eta_j$ or

$$\mathrm{Re}\,\mu > \mathrm{Re}\,\lambda - \eta_j.$$

Since $\mathrm{Re}\,\lambda$ can be as close to 0 as we want then for all μ such that $\mathrm{Im}\,\mu = \beta$ and $\mathrm{Re}\,\mu > -\eta_j$ (5.7.3) is satisfied (for $\mathrm{Re}\,\lambda$ close to 0). Finally, letting $j \to +\infty$,

$$\{\mu;\ \mathrm{Im}\,\mu = \beta,\ \mathrm{Re}\,\mu > -\eta\} \cap \sigma(T_V)$$

(which contains $s(T_V) + i\beta$) consists of isolated eigenvalues with finite algebraic multiplicities. $\qquad\square$

Since $s(T_V)$ depends a priori on V, it is crucial that the condition (5.7.2) be readily verifiable. This is the object of the next result.

Theorem 5.7.2 *Let $(U(t))_{t\ge 0}$ be a symmetric Markov semigroup (i.e., acts in all $L^p(\nu)$ as a contraction C_0-semigroup and is symmetric in $L^2(\nu)$). We denote by $T^{(p)}$ its generator in $L^p(\nu)$. Let $V : (\Omega, \nu) \to [0, +\infty]$ be U-admissible and $\left(U_V^{(p)}(t)\right)_{t\ge 0}$ be the corresponding absorption C_0-semigroup in $L^p(\nu)$ with generator $T_V^{(p)}$. Then (5.7.2) is satisfied if η is larger than*

$$\inf_{\left\{\varphi \in D(\sqrt{-T^{(2)}}),\ \|\varphi\|_2 = 1\right\}} \left\{ \left\|\sqrt{-T^{(2)}}\varphi\right\|^2 + \int_\Omega V\,|\varphi(x)|^2\,\nu(dx) \right\}.$$

Proof It follows from [160, Corollary 1.4 and Remark 1.5 (a)] that the spectral bound $s(T_V^{(p)})$ is an isolated eigenvalue and is p-independent. Since $\left(U_V^{(2)}(t)\right)_{t\ge 0}$ is symmetric then $-s(T_V)$ is equal to the bottom of the spectrum of the self-adjoint generator $-T_V^{(2)}$

$$\inf_{\left\{\varphi \in D(\sqrt{-T^{(2)}}),\ \|\varphi\|=1\right\}} \left\{\left\|\sqrt{-T^{(2)}}\varphi\right\|^2 + \int_\Omega V(x)\,|\varphi(x)|^2\,\nu(dx)\right\}.$$

This ends the proof. $\qquad\square$

Remark 5.7.1 Note that (5.7.2) is satisfied if e.g. $\Omega = \mathbb{R}^d$ and if $\eta :=$ $\liminf_{|x|\to\infty} V(x)$ is larger than

$$\inf_{\left\{\varphi \in D(\sqrt{-T^{(2)}})\cap L^2(O),\ \|\varphi\|_2=1\right\}} \left\{\left\|\sqrt{-T^{(2)}}\varphi\right\|^2 + \int_O V(x)\,|\varphi(x)|^2\,\nu(dx)\right\}$$

for some bounded open set $O \subset \mathbb{R}^d$. This last quantity depends only on the values of V on the bounded set O.

5.8 Asymptotics of Absorption Semigroups

We show now how to derive time asymptotics from local spectral gaps.

Theorem 5.8.1 *Let $(U(t))_{t\geq 0}$ be a an irreducible substochastic C_0-semigroup in $L^1(\Omega, \nu)$ with generator T and let $V : \Omega \to [0, +\infty]$ be U-regular. Let $(U_V(t))_{t\geq 0}$ be the absorption C_0-semigroup with generator T_V. We assume the existence of an increasing sequence of measurable sets $\left(\Xi_j\right)_j$ such that $\cup_j \Xi_j = \Omega$ and (5.7.1) and (5.7.2) are satisfied. Then*

(i) The perturbed generator T_V has locally a spectral gap (in the sense of Theorem 5.7.1).

(ii) Let $\varphi \in D(T_V) \cap L^1_+(\nu)$ be such that $T_V\varphi = s(T_V)\varphi$. Let $(T_V)^*$ be the dual operator of T_V and $\varphi^* \in D((T_V)^*) \cap L^\infty_+(\nu)$ such that $(T_V)^*\varphi^* = s(T_V)\varphi^*$ with the normalization $\int_\Omega \varphi\varphi^* = 1$. Suppose that $(U_V(t))_{t\geq 0}$ is partially integral. Then $\left(e^{-s(T_V)t}U_V(t)\right)_{t\geq 0}$ is asymptotically stable in $L^1(\nu)$ if and only if $\left(e^{-s(T_V)t}U_V(t)\right)_{t\geq 0}$ is bounded; a sufficient condition is given by

$$\inf \varphi^* > 0. \tag{5.8.1}$$

(iii) Suppose that $(U_V(t))_{t\geq 0}$ is partially integral. Let $L^1_*(\nu) := L^1(\varphi^* d\nu)$ with norm

$$\|f\|_* := \int_\Omega |f(x)|\,\varphi^*(x)\nu(dx).$$

Regardless of (5.8.1), $L^1(\nu) \subset L^1_*(\nu)$ with continuous and dense imbedding and $\left(e^{-s(T_V)t}U_V(t)\right)_{t\geq 0}$ extends uniquely to $L^1_*(\nu)$ as an asymptotically stable stochastic C_0-semigroup with invariant density φ.

Proof (i) is given by Theorem 5.7.1.

(ii) Note that T_V and $(T_V)^*$ have the same spectrum and the same isolated eigenvalues with finite algebraic multiplicities. By Banach Steinhaus's theorem if $\left(e^{-s(T_V)t}U_V(t)\right)_{t\geq 0}$ is asymptotically stable in $L^1(\nu)$ then $\left(e^{-s(T_V)t}U_V(t)\right)_{t\geq 0}$ is bounded; the converse statement is given by Theorem 2.10.4. The sufficiency of (5.8.1) will be seen after.

(iii) Regardless of (5.8.1), we note that for $f \in L^1_+(\nu)$

$$\left\| e^{-s(T_V)t}U_V(t)f \right\|_* = e^{-s(T_V)t}\int_\Omega (U_V(t)f)\,\varphi^* d\nu \qquad (5.8.2)$$

$$= e^{-s(T_V)t}\langle U_V(t)f, \varphi^*\rangle$$

$$= e^{-s(T_V)t}\langle f, (U_V(t))^*\,\varphi^*\rangle$$

$$= e^{-s(T_V)t}\langle f, e^{s(T_V)t}\varphi^*\rangle$$

$$= \langle f, \varphi^*\rangle = \|f\|_*$$

(where $\langle .,.\rangle := \langle .,.\rangle_{L^1,L^\infty}$ is the usual bracket) so $\left(e^{-s(T_V)t}U_V(t)\right)_{t\geq 0}$ extends uniquely to $L^1_*(\nu)$ as a stochastic semigroup in $L^1(\nu^*)$ where $d\nu^* := \varphi^* d\nu$. Moreover, ψ is an invariant density of $\left(e^{-s(T_V)t}U_V(t)\right)_{t\geq 0}$ in $L^1(\nu^*)$. By assumption $(U_V(t))_{t\geq 0}$ is partially integral in $L^1(\nu)$. It is still partially integral in $L^1(\nu^*)$. Finally Theorem 2.10.3 shows that $\left(e^{-s(T_V)t}U_V(t)\right)_{t\geq 0}$ is strongly stable in $L^1(\nu^*)$. Note that under (5.8.1), the norm $\|\,\|_*$ in $L^1(\nu)$ is equivalent to the usual one and consequently $\left(e^{-s(T_V)t}U_V(t)\right)_{t\geq 0}$ is strongly stable in $L^1(\nu)$. Note that $L^1(\nu)$ is a proper subspace of $L^1_*(\nu)$ if (5.8.1) is not satisfied. $\square$

Note that if $1_{\Xi_j}U(t) : L^1(\Omega, \nu) \to L^1(\Omega_c, \nu)$ is weakly compact for all j then (5.7.1) is satisfied and $(U_V(t))_{t\geq 0}$ is partially integral. Indeed, $1_{\Xi_j}(\lambda - T)^{-1}$ is given by the strong integral

$$1_{\Xi_j}(\lambda - T)^{-1} = \int_0^{+\infty} e^{-\lambda t}1_{\Xi_j}U(t)dt$$

and (by Theorem 2.9.1)

$$\int_\varepsilon^{\varepsilon^{-1}} e^{-\lambda t}1_{\Xi_j}U(t)dt \text{ is weakly compact.}$$

The fact that $\int_\varepsilon^{\varepsilon^{-1}} e^{-\lambda t}1_{\Xi_j}U(t)dt \to \int_0^{+\infty} e^{-\lambda t}1_{\Xi_j}U(t)dt$ $(\varepsilon \to 0)$ in operator norm implies that $1_{\Xi_j}(\lambda - T)^{-1}$ is weakly compact. On the other hand, $1_{\Xi_j}U_V(t) \leq 1_{\Xi_j}U(t)$ shows that $1_{\Xi_j}U_V(t)$ is weakly compact and consequently is a kernel operator in $L^1(\nu)$ (see [115]) so $(U_V(t))_{t\geq 0}$ is partially integral.

5.9 Extension to L^p-Spaces

In this section, consider the case where $(U(t))_{t\geq 0}$ operates in a consistent way on all $L^p(\nu)$ as a positive C_0-semigroup $\left(U^{(p)}(t)\right)_{t\geq 0}$. If $V : \Omega \to [0, +\infty]$ is U-regular in $L^1(\nu)$ then it is so in all $L^p(\Omega, \nu)$. We show now how the L^1-theory extends to $L^p(\nu)$. We restrict ourselves to the symmetric case.

Theorem 5.9.1 *Let $(U(t))_{t\geq 0}$ be a substochastic C_0-semigroup in $L^1(\Omega, \nu)$ with generator T and let $V : (\Omega, \nu) \to [0, +\infty]$ be U-admissible. Suppose that $(U(t))_{t\geq 0}$ operates on all $L^p(\nu)$ as a positive C_0-semigroup $\left(U^{(p)}(t)\right)_{t\geq 0}$ with generator $T^{(p)}$ and that $\left(U^{(2)}(t)\right)_{t\geq 0}$ is symmetric. Let $(U_V(t))_{t\geq 0}$ be the absorption C_0-semigroup with generator T_V and let $s(T_V) > -\infty$. We assume the existence of an increasing sequence of measurable sets $\left(\Xi_j\right)_j$ such that $\cup_j \Xi_j = \Omega$ and $1_{\Xi_j}(\lambda - T)^{-1}$ is weakly compact. If $\eta := \lim_{j\to+\infty} \inf_{\Xi_j^c} V > -s(T_V)$ then $T_V^{(p)}$ shares the peripheral spectral structure given in Theorem 5.7.1.*

Proof According to Theorem 5.7.1, the statement is true for $p = 1$. Since $\left(U_V^{(2)}(t)\right)_{t\geq 0}$ is also symmetric then $\left(\lambda - T_V^{(1)}\right)^{-1}$ coincides with $\left(\lambda - \left(T_V^{(1)}\right)^*\right)^{-1}$ on $L^1(\nu) \cap L^\infty(\nu)$. Since $T_V^{(1)}$ and $\left(T_V^{(1)}\right)^*$ share the same resolvent set and the same isolated eigenvalues with finite algebraic multiplicities, hence the same essential resolvent set (see Definition 2.8.1), then [160, Corollary 1.4] shows that the peripheral spectral structure given in Theorem 5.7.1 is p-independent. $\qquad\square$

5.10   What About Locally Bounded Absorptions?

In this section, we give a new approach to spectral (gap) analysis of absorption semigroups. We explore them for locally bounded potentials V; (see Remark 5.10.1 below for more general assumptions). We build a general theory of full discreteness (or spectral gaps) for absorption generators T_V or absorption semigroups $(U_V(t))_{t\geq 0}$. This construction is significantly more general and more precise than the previous as it does not rely on Lemma 5.3.3. In particular, we can address (not necessarily contractive) C_0-semigroups $(U(t))_{t\geq 0}$ in any $L^p(\nu)$ space ($1 \leq p < +\infty$) and the spectral gap results are obtained at the semigroup level, not just for the generators. For the sake of simplicity, we suppose that Ω is a locally compact metric space and ν is a regular Borel measure. We give first

Definition 5.10.1 We say that $O \in \mathcal{L}(L^p(\nu))$ is "locally" weakly compact if $p = 1$ (resp. "locally" compact if $p > 1$) if for any compact set $\Xi \subset \Omega$, $1_\Xi O$ ($t > 0$) is a weakly compact if $p = 1$ (resp. compact if $p > 1$) where 1_Ξ is the indicator function of Ξ.

We start with

Lemma 5.10.1 *Let $(S(t))_{t\geqslant 0}$ be a positive C_0-semigroup on $L^p(\nu)$ $(1 \leq p < +\infty)$. Suppose that for all $t > 0$, $S(t)$ is "locally" weakly compact if $p = 1$ (resp. "locally" compact if $p > 1$). If $W \in L^\infty(\nu)$ has a compact support then $\omega_{ess}(S_W) = \omega_{ess}(S)$.*

Proof We have a Duhamel equation

$$S_W(t) = S(t) - \int_0^t S_W(t - s)W S(s)ds.$$

Since W is compactly supported then $W S(s)$ $(s > 0)$ is weakly compact in $L^1(\nu)$ (resp. compact in $L^p(\nu)$ if $1 < p < +\infty$)). By Theorem 2.9.1,

$$\int_\varepsilon^t S_W(t - s)W S(s)ds \text{ is weakly compact in } L^1(\nu)$$

or compact in compact in $L^p(\nu)$ if $1 < p < +\infty$) (see e.g. [372] for this last case). Since

$$\int_\varepsilon^t S_W(t - s)W S(s)ds \to \int_0^t S_W(t - s)W S(s)ds \ (\varepsilon \to 0)$$

in operator norm then $S_W(t) - S(t)$ is weakly compact in $L^1(\nu)$ (resp. compact in $L^p(\nu)$ if $1 < p < +\infty$)) so $(S(t))_{t\geqslant 0}$ and $(S_W(t))_{t\geqslant 0}$ have the same essential spectrum and consequenly the same essential type. $\square$

We are ready to prove

Theorem 5.10.1 *Let $(U(t))_{t\geqslant 0}$ be a positive C_0-semigroup on $L^p(\nu)$. Suppose that for all $t > 0$, $U(t)$ is "locally" weakly compact if $p = 1$ (resp. "locally" compact if $p > 1$). Let $V \in L^\infty_{loc}$ and*

$$\alpha := \lim_{x \to \infty} \inf V(x) > -\infty$$

$(\alpha = +\infty$ is allowed). Then

$$\omega_{ess}(U_V) \leq \omega(U) - \alpha \tag{5.10.1}$$

(with $\omega_{ess}(U_V) = -\infty$ if $\alpha = +\infty$). In particular, $(U_V(t))_{t\geqslant 0}$ has a spectral gap if

$$\alpha > \omega(U) - \omega(U_V) \tag{5.10.2}$$

or equivalently if $\alpha > s(T) - s(T_V)$; (this is meaningful provided that $s(T_V) > -\infty$).

Proof Note that there exists a sequence of compact sets $\left(\Xi_j\right)_j$ such that

$$\Xi_j \subset \Xi_{j+1}, \ \cup_j \Xi_j = \Omega.$$

The meaning of $\liminf_{x\to\infty} V(x) = \alpha$ is

$$\inf_{x \notin \Xi_j} V(x) \to \alpha \ (j \to +\infty).$$

We choose $\widetilde{\alpha}$ arbitrarily such that $\widetilde{\alpha} < \alpha$. There exists j such that

$$V(x) \geq \widetilde{\alpha} \ (x \notin \Xi_j).$$

Let

$$\widetilde{V}(x) := \begin{cases} V(x) \text{ if } x \notin \Xi_j \\ \widetilde{\alpha} \text{ if } x \in \Xi_j. \end{cases}$$

Then

$$V(x) = \widetilde{V}(x) + \widehat{V}(x)$$

where

$$\widehat{V}(x) := V(x) - \widetilde{V}(x)$$

is essentially bounded with compact support included in Ξ_j. Since

$$c := \inf \widetilde{V} > -\infty$$

then

$$U_{\widetilde{V}}(t) \leq U_c(t) = e^{-ct} U(t)$$

and $\left(U_{\widetilde{V}}(t)\right)_{t\geqslant 0}$ inherits the local weak compactness property of $(U(t))_{t\geqslant 0}$ if $p = 1$ (resp. the local compactness property of $(U(t))_{t\geqslant 0}$ if $p > 1$, see [4]). Since V is a "bounded and compactly supported" perturbation of $\widetilde{V}$ then Lemma 5.10.1 (with $(S(t))_{t\geqslant 0} := \left(U_{\widetilde{V}}(t)\right)_{t\geqslant 0}$) shows that

$$\omega_{ess}(U_V) = \omega_{ess}(U_{\widetilde{V}}).$$

Since

$$\widetilde{V}(x) = \left(\widetilde{V}(x) - \widetilde{\alpha}\right) + \widetilde{\alpha}$$

and $\widetilde{V} - \widetilde{\alpha} \geq 0$ then

$$U_{\widetilde{V}}(t) \leq U_{\widetilde{\alpha}}(t) = e^{-\widetilde{\alpha}t} U(t)$$

and

$$\omega(U_{\widetilde{V}}) \leq \omega(U_{\widetilde{\alpha}}) = \omega(U) - \widetilde{\alpha}.$$

Since

$$\omega_{ess}(U_V) = \omega_{ess}(U_{\widetilde{V}}) \leq \omega(U_{\widetilde{V}})$$

then

$$\omega_{ess}(U_V) \leq \omega(U) - \widetilde{\alpha}$$

for all $\widetilde{\alpha} < \alpha$, i.e.,

$$\omega_{ess}(U_V) \leq \omega(U) - \alpha.$$

Finally, $(U_V(t))_{t \geqslant 0}$ has a spectral gap if $\omega(U) - \alpha < \omega(U_V)$. $\qquad\square$

Note that when $\liminf_{x \to \infty} V(x) = +\infty$ it follows that $\omega_{ess}(U_V) = -\infty$; however, this does not, a priori, imply the compactness of the absorption semigroup $(U_V(t))_{t \geqslant 0}$. We give now a version of Theorem 5.10.1 that applies in the case where the semigroup $(U(t))_{t \geqslant 0}$ is *not* "locally" (weakly) compact but where the resolvent $(\lambda - T)^{-1}$ is. This result is useful in the treatment of growth equations with death, (see Theorem 11.6.4) and linearized non local Allen Cahn equations, (see Theorem 12.5.1).

Theorem 5.10.2 *Let* $(U(t))_{t \geqslant 0}$ *be a positive C_0-semigroup on $L^p(\nu)$ with generator T. Suppose that the resolvent $(\lambda - T)^{-1}$ is "locally" weakly compact if $p = 1$ (resp. "locally" compact if $p > 1$). Let $V \in L^{\infty}_{loc}$ and*

$$\alpha := \lim_{x \to \infty} \inf V(x) > -\infty,$$

($\alpha = +\infty$ is allowed). Then

$$\begin{cases} \sigma_{ess}(T_V) \subset \{\mathrm{Re}\,\lambda \leq s(T) - \alpha\} & \text{if } \alpha < +\infty \\ \sigma_{ess}(T_V) = \varnothing & \text{if } \alpha = +\infty. \end{cases}$$

In particular, T_V has a "spectral gap" if $\alpha > s(T) - s(T_V)$.

Proof Arguing as previously, let $\widetilde{\alpha} < \alpha$ and

$$\widetilde{V}(x) := \begin{cases} V(x) \text{ if } x \notin \Xi_j \\ \widetilde{\alpha} \text{ if } x \in \Xi_j. \end{cases}$$

Then $V(x) = \widetilde{V}(x) + \widehat{V}(x)$ where $\widehat{V}(x) := V(x) - \widetilde{V}(x)$ is bounded with compact support so that $\widehat{V}(\lambda - T)^{-1}$ is weakly compact in $L^1(\nu)$ if $p = 1$ (resp. compact in $L^p(\nu)$ if $p > 1$). Since $c := \inf V > -\infty$ then

$$(\lambda - T_V)^{-1} \leq (\lambda - T_c)^{-1} = (\lambda + c - T)^{-1}$$

and

$$\widehat{V}(\lambda - T_V)^{-1} = \widehat{V}_+(\lambda - T_V)^{-1} - \widehat{V}_-(\lambda - T_V)^{-1}$$

where

$$\widehat{V}_{\pm}(\lambda - T_V)^{-1} \leq \widehat{V}_{\pm}(\lambda + c - T)^{-1}$$

so $\widehat{V}_{\pm}\,(\lambda - T_V)^{-1}$ are weakly compact in $L^1\,(\nu)$ if $p = 1$ (resp. compact in $L^p\,(\nu)$ if $p > 1$, see [4]). Thus $\widehat{V}$ is T_V-weakly compact in $L^1\,(\nu)$ if $p = 1$ (resp. T_V -compact in $L^p\,(\nu)$ if $p > 1$) and consequently

$$\sigma_{ess}(T_V) = \sigma_{ess}(T_{\widetilde{V}}). \tag{5.10.3}$$

It follows from

$$\widetilde{V}(x) = \left(\widetilde{V}(x) - \widetilde{\alpha}\right) + \widetilde{\alpha}$$

(and $\widetilde{V} - \widetilde{\alpha} \geq 0$) that

$$\left(\lambda - T_{\widetilde{V}}\right)^{-1} \leq (\lambda - T_{\widetilde{\alpha}})^{-1}$$

so

$$s(T_{\widetilde{V}}) \leq s(T_{\widetilde{\alpha}}) = s(T) - \widetilde{\alpha}$$

because $\widetilde{\alpha}$ is a constant. Hence (5.10.3) gives

$$\sigma_{ess}(T_V) \subset \{\mathrm{Re}\,\lambda \leq s(T) - \widetilde{\alpha}\}$$

for *all* $\widetilde{\alpha} < \alpha$ and

$$\sigma_{ess}(T_V) \subset \{\mathrm{Re}\,\lambda \leq s(T) - \alpha\}.$$

Finally T_V has a "spectral gap" if $s(T) - \alpha < s(T_V)$. $\qquad\square$

Note that when $\liminf_{x\to\infty} V(x) = +\infty$, it follows that $\sigma_{ess}(T_V) = \varnothing$; however, this does not, a priori, imply that T_V is resolvent compact.

Remark 5.10.1 What we really need in the proof of Theorem 5.10.2 is just the following condition: for any compact set Ξ, $1_\Xi V\,(\lambda - T)^{-1}$ is weakly compact in $L^1\,(\nu)$ if $p = 1$ or compact in $L^p\,(\nu)$ if $p > 1$. For instance, if the imbedding of $D(T)$ in L^r_{loc} is compact for r such that $\frac{2}{r} = \frac{1}{p}$ (i.e., $r = 2p$) then Holder's inequality shows that it suffices to assume $V \in L^{2p}_{loc}\,(\nu)$. The assumption $V \in L^\infty_{loc}$ is made for the simplicity of the statement only.

5.11 Kato-Voigt Perturbations of Absorption Semigroups

In this chapter, we improve the spectral results on Kato-Voigt semigroups (given in Chap. 3) when the unperturbed C_0-semigroups are themselves absorption semigroups. Let $(U(t))_{t \geq 0}$ be a substochastic C_0-semigroup on $L^1(\nu)$ with generator T and let

$$V : \Omega \to [0, +\infty]$$

be measurable and finite a.e. Suppose that V is U-regular, e.g. $D(T) \cap D(V)$ is a core for T (see Lemma 5.3.2). Let $(U_V(t))_{t \geqslant 0}$ be the associated absorption semigroup with generator T_V. We know (see Remark 5.3.1) that $T_V = T - V$ with domain $D(T_V) = D(T) \cap D(V)$. Let

$$K : L^1(\nu) \to L^1(\nu)$$

be a subtochastic operator and let $P : D(V) \to L^1(\nu)$ be defined by

$$P\varphi = K(V\varphi), \quad \varphi \in D(V).$$

Then P is T_V-bounded and, for all $\varphi \in D_+(T_V)$,

$$\int T_V\varphi + P\varphi = \int T\varphi - \int V\varphi + \int P(V\varphi) \leq 0$$

since $\int P(V\varphi) \leq \int V\varphi$. It follows from Kato-Voigt perturbation theorem that there exists a unique extension

$$J \supset T_V + P$$

which generates a minimal substochastic C_0-semigroup $(\mathcal{J}(t))_{t \geq 0}$ on $L^1(\nu)$. For the sake of comparison with the notations in Chap. 4, the absorption semigroup $(U_V(t))_{t \geqslant 0}$ corresponds to $(\mathcal{M}_1(t))_{t \geqslant 0}$, its generator T_V corresponds to M_1 and $(\mathcal{J}(t))_{t \geq 0}$ corresponds to $(\mathcal{J}_1(t))_{t \geq 0}$.

5.11.1 Spectral Analysis and Time Asymptotics

We start with

Theorem 5.11.1 *Suppose that the substochastic operator K is quasi-compact. Then $(\mathcal{J}(t))_{t \geq 0}$ is genuinely honest (i.e., $J = T_V + P$).*

Proof In this statement, by quasi-compact we mean that $K = K_1 + K_2$ where K_1 is a strict contraction and K_2 is compact. We point out that K_i ($i = 1, 2$) need not be positive operators. We have

$$P = KV = K_1V + K_2V.$$

We note that any bounded operator O on $L^1(\Xi, \nu)$ admits a (bounded) *modulus* operator $|O| \geq 0$ satisfying the properties

$$\begin{cases} |Of| \leq |O|(|f|) \\ |O_1 + O_2| \leq |O_1| + |O_2| \\ \||O|\| = \|O\| \end{cases}$$

(see [86]) and $|O|$ is weakly compact if O is (see [3, Theorem 5.35, p. 299]). Hence

$$P \leq \widehat{P} := P_1 + P_2 = |K_1| V + |K_2| V$$

where $P_i = |K_i| V$, $\||K_1\|| < 1$ and $|K_2|$ is weakly compact. According to Lemma 5.3.3 $V (\lambda - T_V)^{-1}$ is contraction ($\lambda > 0$) so

$$\left\| P_1 (\lambda - T_V)^{-1} \right\| \leq \||K_1\|| = \|K_1\| < 1$$

while

$$P_2 (\lambda - T_V)^{-1} = |K_2| \left(V (\lambda - T_V)^{-1} \right)$$

is weakly compact. Hence, by Theorem 2.6.3, $T_V + \widehat{P} : D(T_V) \to L^1 (\Xi, \nu)$ generates a positive semigroup. This is equivalent to

$$\lim_{\lambda \to +\infty} r(\widehat{P} (\lambda - T_V)^{-1}) < 1$$

and consequently

$$\lim_{\lambda \to +\infty} r(P (\lambda - T_V)^{-1}) < 1$$

so $T_V + P : D(T_V) \to L^1 (\Xi, \nu)$ is also a generator, i.e., $(\mathcal{J}(t))_{t \geq 0}$ is genuinely honest. $\square$

We explore now some spectral properties when the substochastic operator

$$K : L^1(\nu) \to L^1(\nu) \text{ is weakly compact.} \tag{5.11.1}$$

The first spectral result is

Theorem 5.11.2 *Let (5.11.1) be satisfied. Then $(U_V(t))_{t \geqslant 0}$ and $(\mathcal{J}(t))_{t \geqslant 0}$ have the same essential spectrum. In particular:*

(i) If $(U_V(t))_{t \geqslant 0}$ has a spectral gap then $(\mathcal{J}(t))_{t \geq 0}$ has a (larger) spectral gap.

(ii) If $(U(t))_{t \geqslant 0}$ and K are stochastic and if $s(T_V) < 0$ then $(\mathcal{J}(t))_{t \geq 0}$ has a spectral gap.

Proof Let us show that $\mathcal{J}(t) - U_V(t)$ is weakly compact. We have the Duhamel equation on $D(T_V)$

$$\mathcal{J}(t)\varphi - U_V(t)\varphi = \int_0^t U_V(t-s)PU_V(s)\varphi ds.$$

Calculations similar to those in the proof of Desch's theorem show that

$$\left\| \int_0^t U_V(t-s)PU_V(s)\varphi ds \right\|$$

$$\le c_t \left\| P\,(\lambda - T_V)^{-1}\,\varphi \right\|$$

so

$$\left\| \int_0^t U_V(t-s)PU_V(s)ds \right\|$$
$$\le c_t \left\| P\,(\lambda - T_V)^{-1} \right\|$$
$$= c_t \left\| KV\,(\lambda - T_V)^{-1} \right\| \le c_t \|K\|.$$

Thus $\int_0^t U_V(t-s)PU_V(s)ds$ depends (linearly and) continuously on K for the norm operator topology. To show the weak compactness property, it suffices to proceed by approximation. Since K is weakly compact, we can approximate it (in operator norm) by operators dominated by rank one operators. Using a domination argument if necessary, we may assume that K has rank one

$$K\varphi = \left(\int_\Omega \varphi(x)g(x)\nu(dx) \right) f$$

where $f \in L^1_+(\Omega, \nu)$ and $g \in L^\infty_+(\Omega, \nu)$. By domination again, we may assume that K has the simpler form.

$$K\varphi = \left(\int_\Omega \varphi d\nu \right) f.$$

Thus, for $\varphi \in D_+(T_V)$

$$PU_V(s)\varphi = \left(\int_\Omega VU_V(s)\varphi d\nu \right) f$$

and

$$U_V(t-s)PU_V(s)\varphi$$
$$= U_V(t-s) \left[\left(\int_\Omega VU_V(s)\varphi d\nu \right) f \right]$$
$$= \left(\int_\Omega U_V(t-s)VU_V(s)\varphi d\nu \right) f.$$

Finally $\mathcal{J}(t)\varphi - U_V(t)\varphi$ is given by

$$\left(\int_0^t ds \int_\Omega U_V(t-s)VU_V(s)\varphi d\nu \right) f$$
$$= \left(\int_\Omega \left(\int_0^t U_V(t-s)VU_V(s)\varphi ds \right) d\nu \right) f.$$

Note that $\int_0^t U_V(t-s)VU_V(s)\varphi ds$ extends uniquely as a bounded operator on $L^1(\Omega,\nu)$ since

$$
\begin{aligned}
\left\| \int_0^t U_V(t-s)VU_V(s)\varphi ds \right\| \\
\leq \int_0^t \| U_V(t-s)VU_V(s)\varphi \|\, ds \\
\leq \int_0^t \| VU_V(s)\varphi \|\, ds \\
= \left\| V \int_0^t U_V(s)\varphi ds \right\|
\end{aligned}
$$

and $\psi \in L^1(\Omega,\nu) \to \int_0^t U_V(s)\psi ds \in D(T_V)$ is continuous. Thus $\mathcal{J}(t) - U_V(t)$ is a rank one operator. In particular, $(U_V(t))_{t\geq 0}$ and $(\mathcal{J}(t))_{t\geq 0}$ have the same essential spectrum and the same essential type

$$
\omega_{ess}(\mathcal{J}) = \omega_{ess}(U_V).
$$

(i) If $(U_V(t))_{t\geq 0}$ has a spectral gap, i.e., $\omega_{ess}(U_V) < \omega(U_V)$ then

$$
\omega_{ess}(\mathcal{J}) < \omega(U_V) \leq \omega(\mathcal{J}).
$$

(ii) In this case $(\mathcal{J}(t))_{t\geq 0}$ is stochastic and then $s(T_V + P) = 0$. If additionaly $\omega(U_V) = s(T_V) < 0$ then

$$
\omega_{ess}(\mathcal{J}) = \omega_{ess}(U_V) \leq \omega(U_V) < 0 = \omega(\mathcal{J})
$$

and we are done. $\qquad\square$

The second spectral result is:

Theorem 5.11.3 *Let $(U_V(t))_{t\geq 0}$ be strongly stable and $s(T_V) = 0$. Let $(\mathcal{J}(t))_{t\geq 0}$ be irreducible. Suppose that the conservativity condition*

$$
\int T_V\varphi + P\varphi = 0 \ (\varphi \in D(T_V))
$$

holds and that K is stochastic and weakly compact. Then $P\,(0 - T_V)^{-1}$ is stochastic and weakly compact. Let $u \in L^1_+(\nu)$ be an eigenvector of $P\,(0 - T_V)^{-1}$ associated to the eigenvalue 1. Then

(i) If $v := (0 - T_V)^{-1} u \in L^1(\nu)$ then $(\mathcal{J}(t))_{t\geq 0}$ has an invariant density and is asymptotically stable.

(ii) If $v := (0 - T_V)^{-1} u \notin L^1(\nu)$ then $(\mathcal{J}(t))_{t\geq 0}$ is sweeping from the sets

$$
\Xi_\varepsilon = \{V \geq \varepsilon\} \ (\varepsilon > 0).
$$

Proof The weak compactness assumption insures that $(\mathcal{J}(t))_{t \geq 0}$ is a Desch semigroup and then genuinely honest. We point out that $\inf V > 0$ would imply $s(T_V) \leq -\inf V < 0$ so

$$\inf V = 0.$$

According to Lemma 3.3.1 $P\,(0 - T_V)^{-1}$ is stochastic. In addition $P\,(0 - T_V)^{-1} = K\left(V\,(0 - T_V)^{-1}\right)$ is weakly compact since $V\,(0 - T_V)^{-1}$ is bounded. To end the proof, we appeal to Theorem 3.3.1. The sweeping from the sets Ξ_ε follows from the fact that v is integrable on the sets Ξ_ε since $P\,(0 - M_1)^{-1}\,u = u$ (with $P = KV$ and K stochastic) and

$$\|Vv\| = \left\|V\,(0 - M_1)^{-1}\,u\right\| = \left\|P\,(0 - M_1)^{-1}\,u\right\| = \|u\| < +\infty.$$

$\square$

5.12 Comments

In [271], peripheral spectral analysis of T_V relies on the key fact, specific to L^1 spaces, that V is always T_V-bounded in $L^1\,(\nu)$ spaces or equivalently the resolvent of the absorption generator T_V has a smoothing effect

$$(\lambda - T_V)^{-1} \in \mathcal{L}(L^1\,(\nu)\,, D(V)). \tag{5.12.1}$$

A priori, a smoothing effect for the absorption semigroup $(U_V(t))_{t \geq 0}$ of the form

$$U_V(t) \in \mathcal{L}(L^1\,(\nu)\,;\, D(V)) \quad (\forall t > 0) \tag{5.12.2}$$

does not hold in general. Then spectral results for $(U_V(t))_{t \geq 0}$ are derived in [271] by means of a spectral mapping argument under the assumption that $(U_V(t))_{t \geq 0}$ is immediately norm continuous; this assumption is satisfied when $(U(t))_{t \geq 0}$ is holomorphic as this implies that $(U_V(t))_{t \geq 0}$ is holomorphic too [169]. If the smoothing effect (5.12.2) were true, we would have had a "weak type" estimate for $t > 0$

$$\int_{\{V > M\}} (U_V(t)f)\nu(dx) \leq \frac{c_t\,\|f\|}{M}, \forall\, f \in L^1_+(\nu),\ \forall\, M > 0 \tag{5.12.3}$$

which is enough to derive directly spectral results for $(U_V(t))_{t \geq 0}$ without invoking e.g. the holomorphy of $(U(t))_{t \geq 0}$. Actually, it is shown in [271] that a sufficient condition for (5.12.2) to hold is

$$c_t := \lim_{\varepsilon \to 0_+} \inf \left\| \frac{U_V^*(t + \varepsilon)1 - U_V^*(t)1}{\varepsilon} \right\|_{L^\infty} < +\infty,\ (t > 0) \tag{5.12.4}$$

and a sufficient condition for (5.12.4) to hold is

$$\lim_{\varepsilon \to 0_+} \frac{U_V^*(t+\varepsilon)1 - U_V^*(t)1}{\varepsilon} \quad \text{exists in weak star topology,}$$

i.e., $U_V^*(t)1 \in D((T_V)^*)$, $\forall t > 0$ $((T_V)^*$, the dual of T_V, is the weak star generator of $\left(U_V^*(t)\right)_{t \geqslant 0})$ or equivalently

$$\forall f \in L^1(\nu), \ (0, +\infty) \ni t \to \int U_V(t)f \ \text{is differentiable.} \tag{5.12.5}$$

It is an interesting open problem to investigate the class of C_0-semigroups $(U(t))_{t \geqslant 0}$ (outside the realm of holomorphic C_0-semigroups) satisfying (5.12.2) or (5.12.5).

In Theorem 5.8.1, the boundedness of $\left(e^{-s(T_V)t}U_V(t)\right)_{t \geq 0}$ in $L^1(\nu)$ is an open question if $\inf \varphi^* = 0$.

In Theorem 5.9.1, if additionally we know that $\left(e^{-s(T_V)t}U_V(t)\right)_{t \geq 0}$ is asymptotically stable in $L^1(\nu)$ then $\left(e^{-s(T_V)t}U_V(t)\right)_{t \geq 0}$ is bounded in $L^1(\nu)$; by the symmetry of $\left(U_V^{(2)}(t)\right)_{t \geq 0}$ and an interpolation argument, $\left(e^{-s(T_V)t}U_V^{(p)}(t)\right)_{t \geq 0}$ is bounded in $L^p(\nu)$ and Theorem 2.10.4 shows that $\left(e^{-s(T_V)t}U_V^{(p)}(t)\right)_{t \geq 0}$ is asymptotically stable in $L^p(\nu)$ $(1 < p < +\infty)$.

Chapter 6
On Jump Semigroups in $L^1(\nu)$

6.1 Chapter Aims

This chapter focuses on well-posedness and spectral gap analysis in abstract L^1-spaces for semigroups $(\mathcal{J}(t))_{t\geq 0}$ that govern Cauchy problems of the form

$$\frac{df}{dt} = Mf + Pf, \quad f(0) = f_0,$$

where M is the generator of a multiplication C_0-semigroup $(\mathcal{M}(t))_{t\geq 0}$ on $L^1(\nu)$ and $P : D(M) \subset L^1(\nu) \to L^1(\nu)$ is a positive integral operator. The analysis relies on weak compactnes arguments. These equations are motivated by classical space-homogeneous linear kinetic equations and by Schrödinger equations in momentum space, as originally studied in the 1960s in the L^2-setting. We also address asymptotic stability or sweeping for $(\mathcal{J}(t))_{t\geq 0}$ when this semigroup possesses no spectral gap but satisfies a detailed balance condition.

6.2 Chapter Overview

In this chapter, we show how the abstract results developed in Chaps. 3 and 4 can be applied in the more concrete setting of jump equations of the form

$$\begin{cases} \frac{d\varphi}{dt} = -\sigma(x)\varphi(x,t) + \int_\Omega p(x,y)\varphi(y,t)\nu(dy) \\ \qquad\qquad \varphi(x,0) = \varphi_0(x). \end{cases} \tag{6.2.1}$$

where $\sigma = \sigma_a + \sigma_s$ and

$$p(.,.) : \Omega \times \Omega \to (0, +\infty) \quad \nu\text{-a.e.}$$

© The Author(s), under exclusive license to Springer Nature Switzerland AG 2026 139
M. Mokhtar-Kharroubi, *Peripheral Spectra of Perturbed Positive Semigroups*, Lecture Notes in Mathematics 2388, https://doi.org/10.1007/978-3-032-11173-9_6

$$\sigma_s(x) := \int_\Omega p(y, x)\nu(dy) < +\infty \quad \nu\text{-a.e.}$$

$$\sigma_a(.) : \Omega \to [0, +\infty) \quad \nu\text{-a.e.}$$

are measurable functions. Sometimes, it may be useful to formulate the jump equations (6.2.1) as

$$\begin{cases} \frac{d\varphi}{dt} = -\sigma(x)\varphi(x, t) + \int_\Omega p_s(x, y)\sigma_s(y)\varphi(y, t)\nu(dy) \\ \varphi(x, 0) = \varphi_0(x) \end{cases} \qquad (6.2.2)$$

where $p_s(x, y) := \frac{p(x,y)}{\sigma_s(y)}$ satisfies the stochastic property

$$\int_\Omega p_s(x, y)\nu(dx) = 1.$$

In the literature, we find also the formulation

$$\begin{cases} \frac{d\varphi}{dt} = -\sigma_a(x)\varphi(x, t) + \int_\Omega \left(p(x, y)\varphi(y, t) - p(y, x)\varphi(x, t)\right)\nu(dy) \\ \varphi(x, 0) = \varphi_0(x). \end{cases}$$

What we gain in this context is primarily the holomorphy of the unperturbed (multiplication) semigroup, which leads to several useful consequences. The equations considered here are mainly motivated by classical space-homogeneous linear kinetic equations, as originally studied in [92, 289, 333, 351, 352], and more recently revisited in [211, 264]. Similar types of equations also appear in the modeling of non-local dispersal in mathematical biology [93, 167] and in the study of Schrödinger equations in momentum space [351]. Historically, such kinetic equations have been treated in (weighted) L^2-spaces only. In this chapter, we revisit them in the L^1-setting, at an abstract level for now, focusing on well-posedness, peripheral spectral analysis, and long-time asymptotics. We emphasize that the existing L^2 spectral literature on (isotropic) space-homogeneous models in nuclear reactor theory (typically in one velocity dimension) is far more developed than the L^1-based peripheral spectral analysis we present here. We also show how the abstract L^p-extension theory developed in earlier chapters, particularly the predual approach, applies naturally and effectively in this setting. For brevity, we do not detail the numerous results of this chapter in this introduction; moreover, several of them follow directly from results established in previous chapters. In Sect. 10.6, we turn to more complex jump semigroups arising in the context of kinetic equations involving "downshift" and "Bragg" scatterings.

In the following chapter, we explore in depth the Hilbert space theory of jump C_0-semigroups, with a particular focus on their connection to detailed balance conditions. The case where Ω is countable, relevant for Kolmogorov differential equations and weighted graph models, will be treated separately in Chap. 9. The present chapter should be regarded as a preparatory step toward those developments. To the best of our knowledge, no comparable systematic construction for such jump semigroups exists in the current literature. The analysis will be further developed, at a more technical level, in the concrete setting of space-homogeneous linear kinetic equations, as discussed in Chap. 10. Some remarks, comments or open questions are relegated to the end of this chapter in the section "Comments".

6.3 First Generation Results and Irreducibility

We start with a standard observation for which we give a complete proof for reader's convenience.

Lemma 6.3.1 *Let* $1 \leq p < +\infty$. *The unbounded operator*

$$\begin{cases} M_p : \varphi \in D(M_p) \subset L^p(\nu) \to -\sigma\varphi \in L^p(\nu) \\ D(M_p) = \{\varphi \in L^p(\nu) \to \sigma\varphi \in L^p(\nu)\} \end{cases} \tag{6.3.1}$$

is densely defined and generates a positive (and holomorphic) contraction C_0-semigroup $\mathcal{M}_p = \left(\mathcal{M}_p(t)\right)_{t \geq 0}$ given by

$$\mathcal{M}_p(t)\varphi = e^{-t\sigma}\varphi \ \ (t \geq 0); \ \ \varphi \in L^p(\nu) \tag{6.3.2}$$

Proof Since $(\Omega, \mathcal{E}, \nu)$ is a sigma-finite measure space, there exists a sequence $\left(\Omega_j\right)_{j \in \mathbb{N}}$ of measurables subsets such that

$$\Omega_j \subset \Omega_{j+1}, \ \cup_{j \in \mathbb{N}}\Omega_j = \Omega \text{ and } \nu(\Omega_j) < \infty \ (j \in \mathbb{N}).$$

Let $h \in L^p(\Omega)$. We note first that

$$h_j := 1_{\Omega_j}h \to h \text{ in } L^p(\Omega) \ (j \to +\infty).$$

On the other hand, if

$$E_m^j := \left\{x \in \Omega_j; \ \sigma(x) \geq m\right\}$$

then, for any $j \in \mathbb{N}$, the sequence $\left(E_m^j\right)_{m \in \mathbb{N}}$ is nonincreasing and

$$\cap_{m \in \mathbb{N}} E_m^j = \left\{x \in \Omega_j; \ \sigma(x) = +\infty\right\}$$

has zero ν-measure since $\sigma(x) < +\infty$ a.e. Hence, for any fixed $j \in \mathbb{N}$,

$$\nu\left(E_m^j\right) \to 0 \ (m \to +\infty).$$

If we denote by $\left(E_m^j\right)^c$ the complement of E_m^j in Ω_j then, for any fixed $j \in \mathbb{N}$,

$$h_j^m := 1_{\left(E_m^j\right)^c} h \to h_j \text{ in } L^p(\Omega_j) \ (m \to +\infty).$$

Since $\sigma(x) \le m$ on $\left(E_m^j\right)^c$ then $\sigma h_j^m \in L^p(\Omega)$, i.e., $h_j^m \in D(M_p)$. Hence $D(M_p)$ is dense in $L^p(\Omega)$. It follows that M_p generates the positive contraction multiplication C_0-semigroup $\mathcal{M}_p = \left(\mathcal{M}_p(t)\right)_{t \ge 0}$ given by (6.3.2). The holomorphy of $\mathcal{M}_p$ follows from the estimate

$$\left\| (\lambda - M_p)^{-1} \right\|_{\mathcal{L}(L^p(\nu))} \le \frac{1}{|\lambda|} \ (\mathrm{Re}\,\lambda > 0)$$

(see e.g. [286, A-II-Theorem 1.14]). $\square$

Note that the spectrum of M_p is just the essential range of the measurable function $-\sigma$. In particular, the type $\omega(\mathcal{M}_p)$ of $\mathcal{M}_p$ and the spectral bound $s(M_p)$ of its generator are given by

$$\omega(\mathcal{M}_p) = s(M_p) = -\inf \sigma$$

where $\inf \sigma$ refers to the essential infimum of σ.

According to Theorem 2.11.1 there exists a unique extension

$$J_1 \supset M_1 + P : D(M_1) \subset L^1(\nu) \to -\sigma\varphi \in L^1(\nu) \tag{6.3.3}$$

which generates a sub-stochastic minimal C_0-semigroup $(\mathcal{J}_1(t))_{t \ge 0}$ in $L^1(\nu)$. We give now an irreducibility criterion.

Theorem 6.3.1 *The C_0-semigroup $(\mathcal{J}_1(t))_{t \ge 0}$ is irreducible if and only if $P(\lambda - M_1)^{-1}$ is irreducible for some $\lambda > 0$.*

Proof According to (2.11.3)

$$(\lambda - J_1)^{-1} f = (\lambda - M_1)^{-1} \sum_{j=0}^{\infty} \left(P(\lambda - M_1)^{-1}\right)^j f \ (\lambda > 0) \ (f \in L^1(\nu)).$$

The irreduciblity of $(\mathcal{J}_1(t))_{t \ge 0}$ is equivalent to the fact that $(\lambda - J_1)^{-1}$ is positivity improving in the sense that $(\lambda - J_1)^{-1}\varphi > 0$ a.e. for any non trivial nonnegative $\varphi \in L^1_+(\nu)$. This amounts to

$$\langle (\lambda - J_1)^{-1}\varphi, \psi \rangle > 0$$

for any non trivial nonnegative $\varphi \in L^1_+(\nu)$ and $\psi \in L^\infty_+(\nu)$, i.e.,

$$\sum_{j=0}^{\infty} \langle (\lambda - M_1)^{-1} \left(P(\lambda - M_1)^{-1}\right)^j \varphi, \psi \rangle$$

or equivalently for any non trivial nonnegative $\varphi \in L^1_+(\nu)$ and $\psi \in L^\infty_+(\nu)$ there exists $j \in \mathbb{N}$ such that

$$\left\langle \left(P(\lambda - M_1)^{-1}\right)^j \varphi, (\lambda - M_1^*)^{-1}\psi \right\rangle > 0 \qquad (6.3.4)$$

where

$$(\lambda - M_1^*)^{-1}\psi = \frac{\psi}{\lambda + \sigma}.$$

Since the support of $\frac{\psi}{\lambda+\sigma}$ coincides with the one of ψ then (6.3.4) amounts to

$$\left\langle \left(P(\lambda - M_1)^{-1}\right)^j \varphi, \psi \right\rangle > 0.$$

This amounts to irreducibility of $P(\lambda - M_1)^{-1}$. $\qquad\qquad\square$

Since $P(\lambda - M_1)^{-1}$ is an integral operator then a classical criterion (see e.g. [383]) gives

Proposition 6.3.1 $P(\lambda - M_1)^{-1}$ *is irreducible if and only if for any measurable subset* $\Xi \subset \Omega$ *such that* $\nu(\Xi) > 0$ *and* $\nu(\Xi^c) > 0$

$$\int_{\Xi^c \times \Xi} \frac{p(x, y)}{\lambda + \sigma(y)} \nu(dx)\nu(dy) > 0.$$

This property is λ-independent.

We point out that in general an irreducible semigroup need not be positivity improving; however, we have here.

Theorem 6.3.2 *If* $(\mathcal{J}_1(t))_{t\geq0}$ *is genuinely honest and irreducible then* $(\mathcal{J}_1(t))_{t\geq0}$ *is positivity improving, i.e., for any non trivial* $\varphi \in L^1_+(\nu)$

$$\mathcal{J}_1(t)\varphi > 0 \ a.e. \ \forall t > 0.$$

Proof We note first that C_0-semigroup $\left(\mathcal{M}_p(t)\right)_{t\geq0}$ is holomorphic. Since

$$r(P(\lambda - M_1)^{-1}) < 1 \ (\lambda > 0)$$

then, by Lemma 2.6.2, $(\mathcal{J}_1(t))_{t\geq0}$ is holomorphic too. Since $(\mathcal{J}_1(t))_{t\geq0}$ is irreducible then

$$\mathcal{J}_1(t)\varphi > 0 \text{ a.e. } \forall t > 0$$

for any non trivial $\varphi \in L^1_+(\nu)$ (see [286, Theorem 3.2 (b), p. 306]). $\qquad\square$

6.4   Honesty and Time Asymptotics Versus Detailed Balance

We consider the case in which a generalized detailed balance condition is satisfied

$$\omega(x)p(x, y) = \omega(y)p(y, x) \tag{6.4.1}$$

where $\omega : \Omega \to (0, +\infty)$ is a measurable function.

Theorem 6.4.1 *Let $\sigma_a = 0$ and let (6.4.1) be satisfied. Suppose that*

$$\sigma\omega^{-1} \in L^1(\nu). \tag{6.4.2}$$

Then
(i) $(\mathcal{J}_1(t))_{t\geq 0}$ is honest (or equivalently $J_1 = \overline{M_1 + P}$).
(ii) We have the following alternative:
(a) If $\omega^{-1} \in L^1(\nu)$ then (after normalization) ω^{-1} is an invariant density of $(\mathcal{J}_1(t))_{t\geq 0}$ and is asymptotically stable.
(b) Let $\omega^{-1} \notin L^1(\nu)$. If $(\mathcal{J}_1(t))_{t\geq 0}$ is genuinely honest then it is sweeping from the sets of integrability of ω^{-1}.

Proof Let $F = \omega^{-1}$ and $u := \sigma\omega^{-1}$. By writting (6.4.1) as

$$F(y)p(x, y) = F(x)p(y, x)$$

one sees that

$$\int_\Omega p(x, y)F(y)\nu(dy) = \sigma(x)F(x)$$

or

$$\int_\Omega \frac{p(x, y)}{\sigma(y)}u(y)\nu(dy) = u(x),$$

i.e.,

$$P(0 - M_1)^{-1}u = u.$$

Thus (i) follows from Remark 3.3.1. Now (ii) (a) follows from Theorem 3.3.1 (ii) (a); indeed, $(\mathcal{J}_1(t))_{t\geq 0}$ is partially integral since it is given by a Dyson-Phillips expansion (2.11.4) whose terms $U_j(t)$ $(j \geq 1)$ are integral operators because P is an integral operator. Finally, (ii) (b) follows from Theorem 3.3.1 (ii) (b) if we can check that

$$(0 - M_1)^{-1}\left(\int_0^t M_1(s)ds\right)u < +\infty \text{ a.e.}$$

To this end, note that

$$\left(\int_0^t M(s)ds\right)u = \left(\int_0^t e^{-s\sigma}ds\right)u = \frac{\left(1 - e^{-t\sigma}\right)u}{\sigma} = t\frac{\left(1 - e^{-t\sigma}\right)}{t\sigma}u$$

Since $\frac{1-e^{-t\sigma}}{t\sigma} \leq C$ then

$$\left(\int_0^t \mathcal{M}(s)ds\right) u \leq Ctu$$

and

$$0 - M_1)^{-1}\left(\int_0^t \mathcal{M}(s)ds\right) u \leq Ct(0 - M_1)^{-1}u.$$

Let $h := (0 - M_1)^{-1}u$. Since $Ph = u$ then

$$\int_\Omega \nu(dx) \int_\Omega p(x, y)h(y)\nu(dy) < +\infty,$$

i.e., $\int_\Omega \sigma(y)h(y)\nu(dy) < +\infty$ so $h(y) < +\infty$ a.e. This ends the proof. $\qquad\square$

Detailed balance will be investigated much more deeply in Chap. 7 in suitable weighted spaces.

6.5 Generation and Spectral Gap

À priori, all the results from Chap. 4 are applicable to this particular class of equations. However, we limit ourselves here to a few illustrative examples and instead complement them with several additional results specifically related to jump equations. In particular, Theorem 4.3.1 and Lemma 2.6.2 imply

Theorem 6.5.1 *If some power $\left(P(\lambda - M_1)^{-1}\right)^n$ is weakly compact in $L^1(\nu)$ then the C_0-semigroup $(\mathcal{J}_1(t))_{t\geq 0}$ is genuinely honest and holomorphic. Moreover, if $\sigma_a = 0$ and $\inf \sigma > 0$ then $(\mathcal{J}_1(t))_{t\geq 0}$ has a spectral gap.*

If we consider the formulation (6.2.2) of jump equations with $\sigma_a = 0$ then $P(0 - M_1)^{-1}$ is nothing but the stochastic operator P_s on $L^1(\nu)$ defined by

$$P_s\varphi = \int_\Omega p_s(x, y)\varphi(y)\nu(dy)$$

so $(\mathcal{J}_1(t))_{t\geq 0}$ has a spectral gap provided $\inf \sigma > 0$ and some power of P_s is compact. We can also complement Theorem 6.5.1 by noting that if we add an absorption $\sigma_a \in L^\infty_+(\nu)$ to the generator then the perturbed semigroup has a spectral gap provided that the oscillation of σ_a is less than the spectral gap of $(\mathcal{J}_1(t))_{t\geq 0}$, see Theorem 4.4.1.

If $P(\lambda - M_1)^{-1}$ is not weakly compact, we may attempt a decomposition approach to demonstrate that this operator is quasi-compact; this approach with be useful e.g. in Chap. 9 for Kolmogorov differential equations. We introduce a sequence $\Omega_j \subset \Omega$ of measurable sets such that

$$\Omega_j \subset \Omega_{j+1}, \quad \cup_j \Omega_j = \Omega \tag{6.5.1}$$

and identify $L^1(\Omega_j)$ with the closed subspace of $L^1(\Omega)$ of (class of) functions vanishing a.e. outside of Ω_j.

Theorem 6.5.2 *Let $\sigma_a = 0$ and $\inf \sigma > 0$. Let Ω_j be defined by (6.5.1). Suppose that for all $j \in \mathbb{N}$*

$$1_{\Omega_j} P(0 - M_1)^{-1} : L^1(\Omega, \nu) \to L^1(\Omega_j, \nu)$$

is weakly compact and

$$\lim_{j \to \infty} \sup_y \frac{1}{\sigma_s(y)} \int_{\Omega_j^c} p(x, y)\nu(dx) < 1.$$

Then $(\mathcal{J}_1(t))_{t \geq 0}$ is genuinely honest and has a spectral gap.

Proof Note first that

$$\frac{1}{\sigma_s(y)} \int_{\Omega} p(x, y)\nu(dx) = 1.$$

We decompose $p(x, y)$ as

$$p(x, y) = 1_{\Omega_j}(x)p(x, y) + 1_{\Omega_j^c}(x)p(x, y) =: p^{(1)}(x, y) + p^{(2)}(x, y).$$

By assumption

$$P^{(1)}(0 - M_1)^{-1} : \varphi \in L^1(\nu) \to 1_{\Omega_j}(x) \int_{\Omega} \frac{p(x, y)}{\sigma_s(y)} \varphi(y)\nu(dy) \in L^1(\nu)$$

is weakly compact while

$$\left\| P^{(2)}(0 - M_1)^{-1} \varphi \right\|_{L^1(\nu)} = \int_{\Omega} \frac{\int_{\Omega_j^c} p(x, y)\nu(dx)}{\sigma_s(y)} \varphi(y)\nu(dy)$$

shows that

$$\left\| P^{(2)}(0 - M_1)^{-1} \right\|_{\mathcal{L}(L^1(\nu))} \leq \sup_y \frac{1}{\sigma_s(y)} \int_{\Omega_j^c} p(x, y)\nu(dx) < 1$$

for j large enough so $s(M_1 + P^{(2)}) < 0$. Since $(\mathcal{J}_1(t))_{t \geq 0}$ is stochastic, we conclude with Theorem 4.3.3. $\qquad\square$

6.6 Asymptotics Without Spectral Gaps

In Theorem 6.4.1, the fact that 1 is an eigenvalue of $P(0 - M_1)^{-1}$ is given for free by the detailed balance condition (6.4.1). Here is another possibility combining also Theorem 3.3.1 (iii).

Theorem 6.6.1 *Let $\sigma_a = 0$ and $\inf \sigma_s = 0$. Suppose that $P(0 - M_1)^{-1}$ is quasi compact. Then*

(i) $M_1 + P : D(M_1) \subset L^1(\nu) \to L^1(\nu)$ *generates a stochastic semigroup* $(\mathcal{J}_1(t))_{t \geq 0}$.

(ii) 1 *is an eigenvalue of* $P(0 - M_1)^{-1}$ *associated to a nonnegative eigenvector u.*

(iii) If $v := (0 - M_1)^{-1}u \in L^1(\nu)$ then v is an invariant density of $(\mathcal{J}_1(t))_{t \geq 0}$, and is asymptotically stable if $v > 0$ a.e.

(iv) If $v := (0 - M_1)^{-1}u \notin L^1(\nu)$ and $v > 0$ a.e. then $(\mathcal{J}_1(t))_{t \geq 0}$ is sweeping from the integrability sets of v.

6.7 Predual Analysis in $L^1(\nu)$

The three conditions (*Hyp 1, Hyp 2* and *Hyp 3*) of the predual construction given in Sect. 4.6 are trivially satisfied by jump equations. This analysis consists in building a theory in $L^1(\nu)$ (nearly identical to the previous one) with

$$P_* : \varphi \to \int_\Omega p(y, x)\varphi(y)\nu(dy)$$

instead of P. This needs P_* to be M-bounded in $L^1(\nu)$, i.e.,

$$\sup_{x \in \Omega} \frac{\widehat{\sigma}_s(x)}{1 + \sigma(x)} < +\infty \tag{6.7.1}$$

where

$$\widehat{\sigma}_s(x) = \int_\Omega p(x, y)\nu(dy). \tag{6.7.2}$$

We observe that unlike $P(\lambda - M)^{-1}$, the operator $P_*(\lambda - M)^{-1}$ need not be a contraction. We start with

Proposition 6.7.1 *Let (6.7.1) be satisfied. Then*

(i)

$$P_* : \varphi \in D(M_1) \to \int_\Omega p(y, x)\varphi(y)\nu(dy) \in L^1(\nu)$$

is M_1-bounded on $L^1(\nu)$ and

$$\left\| P_* \left(\lambda - M_1\right)^{-1} \right\|_{\mathcal{L}(L^1(\nu))} = \sup_y \frac{\widehat{\sigma}_s(y)}{\lambda + \sigma(y)} \ (\lambda > 0).$$

(ii) If

$$\lim_{\lambda \to +\infty} r\left(P_*(\lambda - M_1)^{-1}\right) < 1 \tag{6.7.3}$$

then

$$M_1 + P_* : D(M_1) \to L^1(\nu) \tag{6.7.4}$$

generates an analytic C_0-semigroup $(\mathcal{J}_{1*}(t))_{t \geq 0}$ in $L^1(\nu)$. This occurs e.g. if

$$\lim_{c \to +\infty} \left(\sup_{\{\widehat{\sigma}_s(y) \geq c\}} \frac{\widehat{\sigma}_s(y)}{\sigma(y)} \right) < 1.$$

(iii) Let there exist $C > 0$ such that $\widehat{\sigma}_s(x) \leq C + \sigma(x)$ then there exists a C_0-semigroup $(\mathcal{J}_{1*}(t))_{t \geq 0}$ in $L^1(\nu)$ generated by an extension of

$$M_1 + P_* - C : D(M_1) \to L^1(\nu)$$

such that $\left(e^{-Ct} \mathcal{J}_{1*}(t)\right)_{t \geq 0}$ is substochastic.

Proof We have

$$P_* \left(\lambda - M_1\right)^{-1} \varphi = \int_\Omega \frac{p(y, x)}{\lambda + \sigma(y)} \varphi(y)\nu(dy)$$

so (for $\varphi \in L^1_+(\nu)$)

$$\left\| P_* \left(\lambda - M_1\right)^{-1} \varphi \right\|_{L^1(\nu)} = \int_\Omega \frac{\widehat{\sigma}_s(y)}{\lambda + \sigma(y)} \varphi(y)\nu(dy)$$

and

$$\left\| P_* \left(\lambda - M_1\right)^{-1} \right\|_{\mathcal{L}(L^1(\nu))} = \sup_y \frac{\widehat{\sigma}_s(y)}{\lambda + \sigma(y)}.$$

This shows (i).

Note that

$$\frac{\widehat{\sigma}_s(y)}{\lambda + \sigma(y)} = \frac{\widehat{\sigma}_s(y)}{\lambda + \sigma(y)} 1_{\{\widehat{\sigma}_s \geq c\}} + \frac{\widehat{\sigma}_s(y)}{\lambda + \sigma(y)} 1_{\{\widehat{\sigma}_s < c\}}$$

shows

$$\sup_y \frac{\widehat{\sigma}_s(y)}{\lambda + \sigma(y)} \leq \max \left\{ \sup_{\{\widehat{\sigma}_s(y) \geq c\}} \frac{\widehat{\sigma}_s(y)}{\sigma(y)}, \frac{c}{\lambda} \right\}$$

so

$$\lim_{\lambda \to +\infty} \sup_y \frac{\widehat{\sigma}_s(y)}{\lambda + \sigma(y)} \le \sup_{\{\widehat{\sigma}_s(y) \ge c\}} \frac{\widehat{\sigma}_s(y)}{\sigma(y)} \quad (\forall c > 0)$$

and

$$\lim_{\lambda \to +\infty} \sup_y \frac{\widehat{\sigma}_s(y)}{\lambda + \sigma(y)} \le \lim_{c \to +\infty} \left(\sup_{\{\widehat{\sigma}_s(y) \ge c\}} \frac{\widehat{\sigma}_s(y)}{\sigma(y)} \right).$$

Theorem 2.6.1 ends the proof of (ii).

Consider (iii).

By assumption

$$\left\| P_* (\lambda - M_1)^{-1} \right\|_{\mathcal{L}(L^1(\nu))} \le 1 \quad (\lambda \ge C).$$

Since $M_1^C := M_1 - C$ generates the C_0-semigroup $\left(e^{-Ct} \mathcal{M}_1(t) \right)_{t \ge 0}$ in $L^1(\nu)$ and

$$\left\| P_* \left(\lambda - M_1^C \right)^{-1} \right\|_{\mathcal{L}(L^1(\nu))} = \left\| P_* (\lambda + C - M_1)^{-1} \right\|_{\mathcal{L}(L^1(\nu))} \le 1 \quad (\lambda \ge 0)$$

then, by Theorem 2.11.1, there exists a unique extension of

$$M_1^C + P_* - C : D(M_1) \to L^1(\nu)$$

which generates a minimal substochastic semigroup. This ends the proof of (iii). $\square$

Remark 6.7.1 Unlike $(\mathcal{J}_1(t))_{t \ge 0}$, the C_0-semigroup $(\mathcal{J}_{1*}(t))_{t \ge 0}$ in $L^1(\nu)$ need not be contractive unless $\widehat{\sigma}_s(x) \le \sigma(x)$.

We have also

Proposition 6.7.2 *Let (6.7.1) be satisfied. If some power $\left(P_*(\lambda - M_1)^{-1} \right)^n$ is weakly compact in $L^1(\nu)$ then (6.7.4) generates a holomorphic C_0-semigroup $(\mathcal{J}_{1*}(t))_{t \ge 0}$ in $L^1(\nu)$ such that*

$$\omega_{ess}(\mathcal{J}_{1*}) \le \omega_{ess}(\mathcal{M}_1).$$

In particular $(\mathcal{J}_{1*}(t))_{t \ge 0}$ has a spectral gap if $s(M_1 + P_*) > s(M_1)$; this occurs if $\inf (\widehat{\sigma}_s - \sigma) > -\inf \sigma$.

Proof Apart from the last assertion, the statement is a direct consequence of Theorem 6.5.1. For the last assertion, it suffices to show that

$$s(M_1 + P_*) \ge \inf (\widehat{\sigma}_s - \sigma).$$

For any $\lambda > s(M_1 + P_*)$ and $g \in L_+^1(\nu)$ there exists a unique $f \in D(M_1) \cap L_+^1(\nu)$ such that

$$\lambda f - M_1 f - P_* f = g.$$

By integration, one sees that $\int_\Omega g(y)\nu(dy)$ is equal to

$$\lambda \int_\Omega f(y)\nu(dy) + \int_\Omega \sigma(y)\varphi(y)\nu(dy) - \int_\Omega \left(\int_\Omega p(y,x)\varphi(y)\nu(dy) \right) \nu(dx)$$

$$= \lambda \int_\Omega f(y)\nu(dy) - \int_\Omega (\widehat{\sigma}_s(y) - \sigma(y))\,\varphi(y)\nu(dy)$$

$$\leq (\lambda - \inf(\widehat{\sigma}_s - \sigma)) \int_\Omega f(y)\nu(dy)$$

so $\lambda - \inf(\widehat{\sigma}_s - \sigma) \geq 0$ for any $\lambda > s(M_1 + P_*)$ and finally $s(M_1 + P_*) \geq \inf (\widehat{\sigma}_s - \sigma)$. $\qquad\square$

Remark 6.7.2 In the Proposition above, if $P_*(\lambda - M_1)^{-1}$ is weakly compact in $L^1(\nu)$ then $\mathcal{J}_{1*}(t) - \mathcal{M}_1(t)$ is weakly compact, see Remark 4.3.1.

Similarly, as in Theorem 6.5.2, we have

Proposition 6.7.3 *Let $\sigma_a = 0$ and $\inf \sigma > 0$. Let Ω_j be defined by (6.5.1). Suppose that for all $j \in \mathbb{N}$*

$$1_{\Omega_j} P_*(0 - M_1)^{-1} : L^1(\Omega, \nu) \to L^1(\Omega_j, \nu)$$

is weakly compact and

$$\lim_{j \to \infty} \sup_y \frac{1}{\sigma_s(y)} \int_{\Omega_j^c} p(y,x)\nu(dx) < 1.$$

Then (6.7.4) generates a holomorphic C_0-semigroup $(\mathcal{J}_{1*}(t))_{t \geq 0}$ in $L^1(\nu)$.

6.8   Comments

The jump operators considered in this chapter represent a rather specific subclass of possible jump operators (see, e.g., [101, Example 13.6.4, p. 401]). In Sect. 10.6, we will explore the peripheral spectrum of certain complex jump semigroups arising in Kinetic Theory. It is worth noting that the Kato–Voigt perturbation theorem 2.11.1 can be formulated in general ordered spaces equipped with an additive norm on the positive cone. This general formulation allows one to address more general jump equations in spaces of measures, as developed in [347].

In Theorem 6.4.1, we wonder whether we can get rid of (6.4.2); see Remark 3.3.2.

The weak compactness assumption in Theorem 6.5.1 is a priori checkable. For example if $q(x) := \sup_y \frac{p(x,y)}{\sigma_s(y)} \in L^1(\nu)$ then a simple estimate shows a domination

$$P(\lambda - M_1)^{-1}\varphi \leq q(x) \int_\Omega \varphi(y)\nu(dy) \ (\varphi \geq 0)$$

so $P(\lambda - M_1)^{-1}$ is weakly compact. Actually, in situations of practical interest, things may be much more involved and, to recover the weak compactness of $P(\lambda - M_1)^{-1}$, we need to work in suitable weighted spaces $L^1(\omega\nu(dy))$ where the choice of the weight $\omega(.)$ is the fundamental issue as will be seen in the concrete contexts of Kolmogorov differential equations and Kinetic Theory we explore later.

As in Sect. 4.6, a priori the role of the predual construction in Sect. 6.7 is to allow the extension of $(\mathcal{J}_1(t))_{t\geq 0}$ from $L^1(\nu)$ to $L^p(\nu)$ if (6.7.3) is satisfied. Actually, for us the interest is above all to develop a theory in $L^1(\mu)$ for another measure μ, (see Chap. 7). Similarly, another L^p extension consists in assuming that $P(\lambda - M_1)^{-1}1 \in L^\infty(\nu)$, i.e.,

$$\sup_{x\in\Omega} \int_\Omega \frac{p(x, y)}{\lambda + \sigma(y)}\nu(dy) < +\infty \ (\lambda > 0); \tag{6.8.1}$$

and we could rewrite the spectral results of Sect. 4.6 for jump semigroups.

Chapter 7
Detailed Balance and Hilbert Space Theory

7.1 Chapter Aims

This chapter deals with the analysis in $L^2(\mu)$ of semigroups $(\mathcal{J}(t))_{t\geq 0}$ that govern Cauchy problems of the form

$$\frac{d\varphi}{dt} = -\sigma(x)\varphi(x,t) + \int_\Omega p(x,y)\varphi(y,t)\nu(dy), \quad \varphi(x,0) = \varphi_0(x),$$

under a detailed balance condition $\omega(x)p(x,y) = \omega(y)p(y,x)$, where $P : D(M) \subset L^1(\nu) \to L^1(\nu)$ is a positive integral operator, $\mu(dx) = \omega(x)\nu(dx)$ and $\omega(.)$ is an arbitrary nonnegative measurable function. A thorough spectral analysis, via form-perturbation theory, is provided.

7.2 Chapter Overview

This chapter continues the previous one in a hilbertian perspective. It is devoted to jump equations satisfying a generalized detailed balance condition

$$\omega(x)p(x,y) = \omega(y)p(y,x) \tag{7.2.1}$$

where $\omega : \Omega \to (0, +\infty)$ is a priori an arbitrary measurable function. We are going to present a general framework for the theory of jump equations in $L^2(\mu)$ where

$$\mu(dx) = \omega(x)\nu(dx)$$

and try to explore its links with the previous $L^1(\nu)$ theory in Chap. 6. Here $L^2(\mu)$ is endowed with scalar product

© The Author(s), under exclusive license to Springer Nature Switzerland AG 2026   153
M. Mokhtar-Kharroubi, *Peripheral Spectra of Perturbed Positive Semigroups*, Lecture Notes in Mathematics 2388, https://doi.org/10.1007/978-3-032-11173-9_7

$$(f, g) = \int_\Omega f(x)\overline{g(x)}\mu(dx) = \int_\Omega f(x)\overline{g(x)}\omega(x)\nu(dx)$$

and norm

$$\|f\|_{L^2(\mu)} = \sqrt{\int_\Omega |f(x)|^2\, \omega(x)\nu(dx)}.$$

In this chapter, unless explicitly stated otherwise, $\|f\|$ refers to $\|f\|_{L^2(\mu)}$. Note that $L^2(\mu) = L^2_\omega(\nu)$ where

$$L^2_\omega(\nu) := \left\{\varphi : \Omega \to \mathbb{R} \text{ measurable}; \sqrt{\omega}\,f \in L^2(\nu)\right\}$$

and

$$\|f\|_{L^2(\mu)} = \left\|\sqrt{\omega}\,f\right\|_{L^2(\nu)}.$$

To motivate the sequel note that formally the problem in $L^2(\mu)$

$$\begin{cases} \frac{d\varphi}{dt} = -\sigma(x)\varphi(x, t) + \int_\Omega p(x, y)\varphi(y, t)\nu(dy) \\ \qquad \varphi(., 0) = \varphi_0(.) \in L^2(\mu) \end{cases}$$

is unitarily equivalent to a problem in $L^2(\nu)$

$$\begin{cases} \frac{d\psi}{dt} = -\sigma(x)\psi(x, t) + \int_\Omega \frac{\sqrt{\omega(x)}p(x,y)}{\sqrt{\omega(y)}}\psi(y, t)\nu(dy) \\ \qquad \psi(., 0) = \psi_0(.) \in L^2(\nu) \end{cases}$$

via a unitary transformation

$$\psi(x, t) = \sqrt{\omega(x)}\varphi(x, t), \quad \psi_0 = \sqrt{\omega(x)}\varphi_0.$$

What we gain under (7.2.1) is a symmetry of the new kernel

$$\frac{\sqrt{\omega(x)}p(x, y)}{\sqrt{\omega(y)}} = \frac{\sqrt{\omega(y)}p(y, x)}{\sqrt{\omega(x)}}.$$

We point out that while the jump operator

$$\psi \to P\psi \tag{7.2.2}$$

is always σ-bounded in $L^1(\nu)$, this operator need not be σ-bounded in $L^2(\mu)$. The good news is that (7.2.2) is always σ-form-bounded in $L^2(\mu)$, i.e., in the sense of quadratic forms. This enables to build a rich *form-perturbation* theory in $L^2(\mu)$ in full generality. Thus "$-\sigma + P$" should be understood as a form-sum

$$-\sigma \dotplus P$$

in $L^2(\mu)$. In full generality, the corresponding nonnegative quadratic form need not be closed and the associated symmetric operator has a distinguished positive self-adjoint extension, the Friedrichs extension, which is the hilbertian counterpart of the (minus) Kato-Voigt generator in the $L^1(\nu)$-theory.

In nuclear reactor theory (see [65, 92, 289, 333, 351, 352]), the equilibrium function $\omega^{-1}(x) = M(x)$ is a Maxwellian, and the scattering kernel $p(x, y)$ endows the operator P with many favorable properties. In particular, P is a bounded operator on $L^2(\mu)$ for all types of moderators, liquid, solid, and gaseous, greatly simplifying the analysis. In this chapter, however, we deliberately set aside this fact and instead develop an abstract form-perturbation theory for its own sake. This general construction, going beyond the specific needs of nuclear reactor theory with inelastic scattering,[1] appears to be of independent interest and may have potential applications beyond the realm of Kinetic Theory.

The detailed balance condition (7.2.1) will be used in various forms, e.g.

$$\begin{cases} \omega^{-1}(y)p(x, y) = \omega^{-1}(x)p(y, x) \\ \dfrac{\sqrt{\omega(x)}p(x,y)}{\sqrt{\omega(y)}} = \dfrac{\sqrt{\omega(y)}p(y,x)}{\sqrt{\omega(x)}}. \end{cases}$$

We introduce the complex Hilbert space

$$V := \left\{ f \in L^2(\mu); \ \int_\Omega \sigma(x)\,|f(x)|^2\,\mu(dx) < +\infty \right\}$$

with norm

$$\|f\|_V = \sqrt{\int_\Omega (\sigma(x) + 1)\,|f(x)|^2\,\mu(dx)}.$$

Note that $V = L^2((\sigma + 1)\,d\mu)$ and V is dense in $L^2(\mu)$ since $1_{E_m} f \to f$ in $L^2(\mu)$ and $1_{E_m} f \in V$ where $E_m := \{\sigma \le m\}$. By identifying $L^2(\mu)$ to its anti-dual (i.e., the space of continuous anti-linear forms on $L^2(\mu)$) we have the continuous and dense imbeddings

$$V \subset L^2(\mu) \subset V'$$

where V' is the anti-dual of V (i.e., the space of continuous anti-linear forms on V) and we can identify $L^2(\mu)$ to a subspace of V'.

Without entering into the details of the numerous results, we outline them in rough terms. This long chapter is organized as follows.

In Sect. 7.3, we collect some useful preliminary results. In Sect. 7.4, we characterize the closedness of a suitable form c at the heart of form-perturbation theory. We explore the spectral properties of the self-adjoint operator associated to this closed form and the asymptotic stability of the associated contraction semigroup. We explore

---

[1] The framework presented here could likely be adapted to treat solid moderators (such as polycrystals) involving both elastic and inelastic collision operators [65]; we have not tried to explore this extension in this monograph.

also the link between the $L^2(\mu)$-theory and the previous $L^1(\nu)$-theory and show how our understanding of $L^1(\nu)$ -theory is improved by the existence of a detailled balance, (see Remark 7.4.1). In Sect. 7.5, we explore the critical situation of nonclosed form c. This form is always nonnegative and there exists a unique nonnegative *symmetric* (but not self-adjoint) operator in $L^2(\mu)$ associated to this nonclosed form. A construction similar to that of Kato-Voigt perturbation in $L^1(\nu)$ (see Theorem 2.11.1) shows that this symmetric operator has a unique nonnegative self-adjoint extension in $L^2(\mu)$ which turns out to coincide with its Friedrichs extension.

In Sect. 7.6, we explore jump C_0-semigroups in $L^1(\mu)$ in the spirit of the previous theory in $L^1(\nu)$. While

$$\sigma_s(y) := \int_\Omega p(x, y)\nu(dx) \leq \sigma(y)$$

is important in $L^1(\nu)$-theory, we need the additional condition

$$\frac{1}{\sigma(y)} \int_\Omega p(y, x)\nu(dx) \leq C$$

to develop a similar construction in $L^1(\mu)$. It turns out that $L^1(\mu)$-theory is nothing but a copy (via a suitable similarity transformation) of the predual theory given in $L^1(\nu)$ (see Chap. 6) to extend the theory to $L^p(\nu)$. Thus all the results of the predual theory given in $L^1(\nu)$ have their counterparts in $L^1(\mu)$. We show also how the results in $L^1(\mu)$ extend easily to $L^p(\mu)$ thanks to the "symmetry" of P with respect to the new measure μ. To the best of our knowledge, no comparable systematic construction exists in the literature. Some remarks, comments or open questions are relegated to the end of this chapter in the section "Comments".

7.3   Preliminary Results

We start with several preliminary results. Let

$$\sigma_\lambda(.) := \sigma(.) + \lambda \ (\lambda \geq 0).$$

Lemma 7.3.1 *For all $\lambda \geq 0$, the operator*

$$G_\lambda : \varphi \in L^2(\nu) \to \int_\Omega \frac{\sqrt{\omega(x)}p(x, y)}{\sqrt{\omega(y)}\sqrt{\sigma_\lambda(x)}\sqrt{\sigma_\lambda(y)}}\varphi(y)\nu(dy) \in L^2(\nu)$$

is a (self-adjoint) contraction in $L^2(\nu)$.

Proof By using Cauchy-Schwarz inequality

$$\left| \int_\Omega \frac{\sqrt{\omega(x)}\, p(x,y)}{\sqrt{\omega(y)}\sqrt{\sigma_\lambda(x)}\sqrt{\sigma_\lambda(y)}} \varphi(y)\nu(dy) \right|^2$$

$$= \left| \int_\Omega \frac{\sqrt{\omega(x)}\sqrt{p(x,y)}}{\sqrt{\omega(y)}\sqrt{\sigma_\lambda(x)}} \frac{\sqrt{p(x,y)}}{\sqrt{\sigma_\lambda(y)}} \varphi(y)\nu(dy) \right|^2$$

$$\leq \int_\Omega \frac{\omega(x)p(x,y)}{\omega(y)\sigma_\lambda(x)}\nu(dy) . \int_\Omega \frac{p(x,y)}{\sigma_\lambda(y)} |\varphi(y)|^2 \, \nu(dy)$$

and the detailed balance

$$\left| \int_\Omega \frac{\sqrt{\omega(x)}\, p(x,y)}{\sqrt{\omega(y)}\sqrt{\sigma_\lambda(x)}\sqrt{\sigma_\lambda(y)}} \varphi(y)\nu(dy) \right|^2$$

$$\leq \int_\Omega \frac{p(y,x)}{\sigma_\lambda(x)}\nu(dy) . \int_\Omega \frac{p(x,y)}{\sigma_\lambda(y)} |\varphi(y)|^2 \, \nu(dy)$$

$$\leq \frac{\sigma_s(x)}{\sigma_\lambda(x)} \int_\Omega \frac{p(x,y)}{\sigma_\lambda(y)} |\varphi(y)|^2 \, \nu(dy) \tag{7.3.1}$$

we get

$$\|G_\lambda\varphi\|_{L^2(\nu)}^2 \leq \int \frac{\sigma_s(x)}{\sigma_\lambda(x)}\nu(dx) \int_\Omega \frac{p(x,y)}{\sigma_\lambda(y)} |\varphi(y)|^2 \, \nu(dy)$$

$$\leq \left(\sup_y \frac{\sigma_s(y)}{\sigma_\lambda(y)} \right)^2 \|\varphi\|_{L^2(\nu)}^2 \leq \|\varphi\|_{L^2(\nu)}^2 . \qquad \square$$

Since $\varphi \in L^2(\nu)$ if and only if $\frac{\varphi}{\sqrt{\omega}} \in L^2(\mu)$ with the same norm

$$\int |\varphi(x)|^2 \, \nu(dx) = \int \left| \frac{\varphi(x)}{\sqrt{\omega(x)}} \right|^2 \mu(dx)$$

then $G_\lambda : \varphi \in L^2(\nu) \to G_\lambda \in L^2(\nu)$ is unitarily equivalent to $\mathcal{G}_\lambda : L^2(\mu) \to L^2(\mu)$ via

$$\psi = \frac{\varphi(y)}{\sqrt{\omega(y)}} \in L^2(\mu) \to \mathcal{G}_\lambda\psi = \frac{G_\lambda\varphi}{\sqrt{\omega(x)}} \in L^2(\mu),$$

i.e.,

$$\mathcal{G}_\lambda : \psi \in L^2(\mu) \to \int_\Omega \frac{p(x,y)}{\omega(y)\sqrt{\sigma_\lambda(x)}\sqrt{\sigma_\lambda(y)}} \psi(y)\mu(dy) \in L^2(\mu)$$

with $\|G_\lambda\|_{\mathcal{L}(L^2(\nu))} = \|\mathcal{G}_\lambda\|_{\mathcal{L}(L^2(\mu))} \leq 1$ ($\lambda \geq 0$). Note that $\mathcal{G}_\lambda$ is self-adjoint in $L^2(\mu)$ and G_λ is self-adjoint in $L^2(\nu)$. We have

Proposition 7.3.1 *If $f, g \in V$ then*

$$(Pf, g) = \left(\mathcal{G}_\lambda \sqrt{\sigma_\lambda} f, \sqrt{\sigma_\lambda} g\right) \quad (\lambda \geq 0).$$

Proof We note that if $f, g \in V$ then $\sqrt{\sigma_\lambda} f, \sqrt{\sigma_\lambda} g \in L^2(\mu)$ and (Pf, g) is given by

$$
\int_\Omega \int_\Omega p(x, y) f(y) \overline{g(x)} \omega(x) \nu(dy) \nu(dx)
$$
$$
= \int_\Omega \int_\Omega \frac{p(x, y)}{\omega(y) \sqrt{\sigma_\lambda(x)} \sqrt{\sigma_\lambda(y)}} \sqrt{\sigma_\lambda(y)} f(y) \sqrt{\sigma_\lambda(x)}\, \overline{g(x)} \omega(y) \nu(dy) \omega(x) \nu(dx)
$$
$$
= \int_\Omega \int_\Omega \frac{p(x, y)}{\omega(y) \sqrt{\sigma_\lambda(x)} \sqrt{\sigma_\lambda(y)}} \sqrt{\sigma_\lambda(y)} f(y) \sqrt{\sigma_\lambda(x)}\, \overline{g(x)} \mu(dy) \mu(dx)
$$
$$
= \left(\mathcal{G}_\lambda \sqrt{\sigma_\lambda} f, \sqrt{\sigma_\lambda} g\right). \qquad \qquad \square
$$

Definition 7.3.1 We say that a *continuous* sesquilinear and hermitian form $c(., .)$ on V is closed if, for $\lambda \geq 0$ large enough,

$$\sqrt{\lambda \|f\|^2 + c(f, f)} \text{ and } \|f\|_V \text{ are equivalent norms on } V.$$

Remark 7.3.1 We recall that if V is a complex Hilbert space *continuously* and *densely* imbedded in $L^2(\mu)$ and if $c(., .)$ is a *continuous* sesquilinear and hermitian form on V then there exists a unique unbounded *symmetric* operator

$$\mathcal{A} : D(\mathcal{A}) \subset L^2(\mu) \to L^2(\mu)$$

(i.e., $\mathcal{A} \subset \mathcal{A}^*$) such that

$$(\mathcal{A}u, v) = c(u, v), \quad u \in D(\mathcal{A}), \ v \in V$$

where
$$D(\mathcal{A}) = \{u \in V; \ \exists \alpha > 0, \ |c(u, v)| \leq \alpha \|v\| \ \ \forall v \in V\}.$$

If additionally, $c(., .)$ is λ-coercive on V for some $\lambda \geq 0$ in the sense that there exists $\alpha > 0$ such that
$$\lambda \|v\|^2 + c(v, v) \geq \alpha \|v\|_V^2 \quad (v \in V)$$

(i.e., if the form c is *closed* in the sense of Definition 7.3.1) then $\mathcal{A} = \mathcal{A}^*$, i.e., $\mathcal{A}$ is *self-adjoint.*

We introduce the continuous sesquilinear hermitian forms on V

$$c(f, g) = \int_\Omega \sigma(x) f(x) \overline{g(x)} \omega(x) \nu(dx) - (Pf, g) \tag{7.3.2}$$

and
$$c_\lambda(f, g) := \lambda(f, g) + c(f, g) \quad (\lambda \geq 0),$$

i.e.,
$$c_\lambda(f, g) = \left(\sqrt{\sigma_\lambda}f, \sqrt{\sigma_\lambda}g\right) - \left(\mathcal{G}_\lambda\sqrt{\sigma_\lambda}f, \sqrt{\sigma_\lambda}g\right) \quad (\lambda \geq 0).$$

We note that
$$\begin{aligned}
c_\lambda(f, f) &= \left(\sqrt{\sigma_\lambda}f, \sqrt{\sigma_\lambda}g\right) - \left(\mathcal{G}_\lambda\sqrt{\sigma_\lambda}f, \sqrt{\sigma_\lambda}g\right) \\
&= \left\|\sqrt{\sigma_\lambda}f\right\|^2 - \left(\mathcal{G}_\lambda\sqrt{\sigma_\lambda}f, \sqrt{\sigma_\lambda}f\right) \geq 0 \quad (\lambda \geq 0)
\end{aligned}$$

since $\|\mathcal{G}_\lambda\|_{\mathcal{L}(L^2(\mu))} \leq 1$. In particular $c(f, f) \geq 0$ $(f \in V)$. According to Remark 7.3.1, there exists a unique unbounded *symmetric* operator
$$\mathcal{A}_\lambda : D(\mathcal{A}_\lambda) \subset L^2(\mu) \to L^2(\mu)$$

such that
$$(\mathcal{A}_\lambda u, v) = c_\lambda(u, v), \quad u \in D(\mathcal{A}_\lambda), \ v \in V$$

where
$$D(\mathcal{A}_\lambda) = \{u \in V; \ \exists \alpha > 0, \ |c_\lambda(u, v)| \leq \alpha \|v\| \ \forall v \in V\}.$$

Note that $\mathcal{A}_\lambda$ is not a priori self-adjoint because c_λ is not a priori coercive (i.e., closed) on V. Note that the unbounded symmetric operator associated to the form $c(., .)$ is nothing but $\mathcal{A} := \mathcal{A}_\lambda - \lambda$ and is nonnegative. Let us characterize the domain of $\mathcal{A}$.

Theorem 7.3.1 *Let $f \in L^2(\mu)$. Then*
(i) If $f \in V$ then

$$\int_\Omega p(x, y) |f(y)| \nu(dy) < +\infty \ d\nu\text{-a.e.}$$

(ii) $f \in D(\mathcal{A})$ if and only if $f \in V$ and

$$\sigma(x) f(x) - \int_\Omega p(x, y) f(y) \nu(dy) \in L^2(\mu).$$

In this case

$$\mathcal{A}f = \sigma(x) f(x) - \int_\Omega p(x, y) f(y) \nu(dy).$$

Proof Note first that $V = L^2((\sigma + 1)\,d\mu)$ is a lattice. For any $f, g \in V_+$

$$\int_\Omega \int_\Omega p(x, y) f(y) g(x) \omega(x) \nu(dy) \nu(dx)$$
$$= (Pf, g) \leq \|G_\lambda\|_{\mathcal{L}(L^2(v))} \|f\|_\lambda \|g\|_\lambda < +\infty$$

where $\|f\|_\lambda := \sqrt{\int_\Omega (\sigma(x) + \lambda) |f(x)|^2 \mu(dx)}$ so

$$\int_\Omega \left(\int_\Omega p(x, y) f(y) \nu(dy) \right) g(x) \omega(x) \nu(dx) < +\infty, \quad (f, g \in V_+).$$

The choice $g > 0$ a.e. shows that

$$\int_\Omega p(x, y) f(y) \nu(dy) < +\infty \text{ a.e.}$$

Hence, for any $f \in V$

$$\left| \int_\Omega p(x, y) f(y) \nu(dy) \right| \leq \int_\Omega p(x, y) f(y) \nu(dy) < +\infty \text{ a.e}$$

(since $|f| \in V_+$). In particular $\sigma(x) f(x)$ and $\int_\Omega p(x, y) f(y) \nu(dy)$ are finite a.e so

$$\sigma(x) f(x) - \int_\Omega p(x, y) f(y) \nu(dy)$$

is well-defined a.e. Thus

$$c(f, g) = \int_\Omega \sigma(x) f(x) \overline{g(x)} \omega(x) \nu(dx) - \int_\Omega \int_\Omega p(x, y) f(y) \overline{g(x)} \omega(x) \nu(dy) \nu(dx)$$

can be written as

$$c(f, g) = \int_\Omega \left[\sigma(x) f(x) - \int_\Omega p(x, y) f(y) \nu(dy) \right] \overline{g(x)} \omega(x) \nu(dx)$$
$$= \int_\Omega \left[\sigma(x) f(x) - \int_\Omega p(x, y) f(y) \nu(dy) \right] \overline{g(x)} \mu(dx).$$

This shows that $g \in V \to c(f, g)$ extends uniquely as a continuous anti-linear functional on $L^2(\mu)$ if and only if

$$\sigma(x) f(x) - \int_\Omega p(x, y) f(y) \nu(dy) \in L^2(\mu).$$

In this case

$$\mathcal{A}f = \sigma(x)f(x) - \int_\Omega p(x, y) f(y) \nu(dy)$$

and

$$c(f, g) = \int_\Omega (\mathcal{A}f) \, \overline{g(x)} \mu(dx).$$ $\qquad\square$

7.4 The Case of Closed Form c

In general, we cannot assert that the sesquilinear form c is always closed on V. However, in nuclear reactor theory, for all kinds of moderators, P is a bounded operator on $L_\omega^2(\nu)$ [65] and consequently c is trivially closed on V. This section is devoted to the case of closed forms c; the critical non closed case is dealt with in Sect. 7.5. We start with

Theorem 7.4.1 *The sesquilinear form c given by (7.3.2) is closed on V if and only if*
$$\lim_{\lambda \to +\infty} \|G_\lambda\|_{\mathcal{L}(L^2(\nu))} < 1. \tag{7.4.1}$$

Proof It is easy to see that $\|f\|_V$ is equivalent to

$$\|f\|_\lambda := \sqrt{\int_\Omega (\sigma(x) + \lambda) \, |f(x)|^2 \, \mu(dx)} \quad (f \in V) \ (\forall \lambda > 0). \tag{7.4.2}$$

Note that
$$\left(\sqrt{\sigma_\lambda} f, \sqrt{\sigma_\lambda} g\right) \le \left\|\sqrt{\sigma_\lambda} f\right\| \left\|\sqrt{\sigma_\lambda} g\right\| = \|f\|_\lambda \|g\|_\lambda$$

and

$$\left(\mathcal{G}_\lambda \sqrt{\sigma_\lambda} f, \sqrt{\sigma_\lambda} g\right) \le \|\mathcal{G}_\lambda\|_{\mathcal{L}(L^2(\mu))} \left\|\sqrt{\sigma_\lambda} f\right\| \left\|\sqrt{\sigma_\lambda} g\right\| = \|\mathcal{G}_\lambda\|_{\mathcal{L}(L^2(\mu))} \|f\|_\lambda \|g\|_\lambda$$

so

$$\begin{aligned} c_\lambda(f, f) &= \left(\sqrt{\sigma_\lambda} f, \sqrt{\sigma_\lambda} f\right) - \left(\mathcal{G}_\lambda \sqrt{\sigma_\lambda} f, \sqrt{\sigma_\lambda} f\right) \\ &\ge \left(1 - \|\mathcal{G}_\lambda\|_{\mathcal{L}(L^2(\mu))}\right) \|f\|_\lambda^2 \\ &= \left(1 - \|G_\lambda\|_{\mathcal{L}(L^2(\nu))}\right) \|f\|_\lambda^2 \end{aligned}$$

since $\|\mathcal{G}_\lambda\|_{\mathcal{L}(L^2(\mu))} = \|G_\lambda\|_{\mathcal{L}(L^2(\nu))}$.

Now, if (7.4.1) is satisfied then c is closed on V. Conversely if $\lambda(f, g) + c(f, g)$ is coercive on V endowed with the norm $\|\ \|_V$ then it is coercive for the equivalent norm (7.4.2) so there exists $\alpha > 0$ such that

$$\left(\sqrt{\sigma_\lambda}f, \sqrt{\sigma_\lambda}f\right) - \left(\mathcal{G}_\lambda\sqrt{\sigma_\lambda}f, \sqrt{\sigma_\lambda}f\right) \geq \alpha \|f\|_\lambda^2 .$$

We may choose $\alpha < 1$ and consequently

$$\left(\mathcal{G}_\lambda\sqrt{\sigma_\lambda}f, \sqrt{\sigma_\lambda}f\right) \leq \|f\|_\lambda^2 - \alpha\|f\|_\lambda^2 = (1-a)\left\|\sqrt{\sigma_\lambda}f\right\|^2 .$$

Note that the unbounded self-adjoint operator on $L^2(\mu)$

$$f \in V \subset L^2(\mu) \to \sqrt{\sigma_\lambda}f \in L^2(\mu)$$

is injective and consequently has dense range. Hence

$$\|\mathcal{G}_\lambda\|_{\mathcal{L}(L^2(\mu))} \leq 1-a$$

since $\mathcal{G}_\lambda$ is a bounded self-adjoint operator on $L^2(\mu)$. This ends the proof since $\|\mathcal{G}_\lambda\|_{\mathcal{L}(L^2(\mu))} = \|G_\lambda\|_{\mathcal{L}(L^2(\nu))}$. $\qquad\square$

Theorem 7.4.2 *Let (7.4.1) be satisfied. Then* $-\mathcal{A}$ *is the generator of a self-adjoint (positive) contraction C_0-semigroup* $\left(\mathcal{U}_{(2)}(t)\right)_{t\geq 0}$ *on* $L^2(\mu)$.

Proof Since $-\mathcal{A}$ is a negative self-adjoint operator then it generates a self-adjoint contraction C_0-semigroup $\left(\mathcal{U}_{(2)}(t)\right)_{t\geq 0}$ on $L^2(\mu)$. To show that $\mathcal{U}_{(2)}(t)$ leaves invariant the cone $L_+^2(\mu)$ (i.e., $\left(\mathcal{U}_{(2)}(t)\right)_{t\geq 0}$ is a positive semigroup), it suffices to show that for λ large enough $(\lambda + \mathcal{A})^{-1} = \mathcal{A}_\lambda^{-1}$ leaves invariant the cone $L_+^2(\mu)$. To this end, consider the problem

$$\mathcal{A}_\lambda f = g \in L_+^2(\mu),$$

i.e., $f \in D(\mathcal{A})$ such that

$$\lambda f + \sigma(x)f(x) - \int_\Omega p(x,y)f(y)\nu(dy) = g,$$

i.e.,

$$\sqrt{\sigma_\lambda(x)}f(x) - \int_\Omega \frac{p(x,y)}{\sqrt{\sigma_\lambda(x)}\sqrt{\sigma_\lambda(y)}}\sqrt{\sigma_\lambda(y)}f(y)\nu(dy) = \frac{1}{\sqrt{\sigma_\lambda(x)}}g.$$

or

$$\sqrt{\sigma_\lambda(x)}f(x) - \int_\Omega \frac{p(x,y)}{\sqrt{\sigma_\lambda(x)}\sqrt{\sigma_\lambda(y)}\omega(y)}\sqrt{\sigma_\lambda(y)}f(y)\mu(dy) = \frac{1}{\sqrt{\sigma_\lambda(x)}}g.$$

This amounts to

$$\varphi := \sqrt{\sigma_\lambda}f \in L^2(\mu)$$

and

$$\varphi - \int_\Omega \frac{p(x, y)}{\sqrt{\sigma_\lambda(x)}\sqrt{\sigma_\lambda(y)}\omega(y)}\varphi(y)\mu(dy) = \frac{1}{\sqrt{\sigma_\lambda}}g,$$

i.e.,

$$\varphi - \mathcal{G}_\lambda\varphi = \frac{1}{\sqrt{\sigma_\lambda}}g.$$

We choose λ large enough so that $\|\mathcal{G}_\lambda\|_{\mathcal{L}(L^2(\mu))} < 1$ so

$$\varphi = \sum_{n=0}^\infty \mathcal{G}_\lambda^n\left(\frac{1}{\sqrt{\sigma_\lambda}}g\right) \geq 0$$

since $\mathcal{G}_\lambda \geq 0$ and $g \in L^2_+(\mu)$. Hence

$$f = (\lambda + \mathcal{A})^{-1}g = \frac{1}{\sqrt{\sigma_\lambda}}(I - \mathcal{G}_\lambda)^{-1}\left(\frac{1}{\sqrt{\sigma_\lambda}}g\right) \geq 0$$

ends the proof.   □

7.4.1   Sufficient Conditions of Closedness

Let us give some sufficient conditions of closedness of the form c. We start with:

Proposition 7.4.1 *The form c is closed in V if one of the following conditions is satisfied.*
  (i) $\lim_{m\to\infty} \inf_{\{\sigma_s(x)\geq m\}} \frac{\sigma_a(x)}{\sigma_s(x)} > 0$
  or
  (ii) $\lim_{m\to\infty} \sup_y \dfrac{\left[\int_{\{\sigma_s(x)\geq m\}} p(x,y)\nu(dx)\right]}{\sigma_s(y)} < 1.$

Proof According to (7.3.1) and φ real

$$|G_\lambda\varphi|^2 \leq \frac{\sigma_s(x)}{\sigma_\lambda(x)}\int_\Omega \frac{p(x, y)}{\sigma_\lambda(y)}\varphi^2(y)\nu(dy)$$

so for all $m > 0$

$$\|G_\lambda\varphi\|^2_{L^2(\nu)}$$
$$\leq \int_{\{\sigma_s(x)\geq m\}} \frac{\sigma_s(x)}{\sigma_\lambda(x)}\left(\int_\Omega \frac{p(x, y)}{\sigma_\lambda(y)}\varphi^2(y)\nu(dy)\right)\nu(dx)$$
$$+ \int_{\{\sigma_s(x)<m\}} \frac{\sigma_s(x)}{\sigma_\lambda(x)}\left(\int_\Omega \frac{p(x, y)}{\sigma_\lambda(y)}\varphi^2(y)\nu(dy)\right)\nu(dx).$$

We have

$$\int_{\{\sigma_s(x)<m\}} \frac{\sigma_s(x)}{\sigma_\lambda(x)} \left(\int_\Omega \frac{p(x,y)}{\sigma_\lambda(y)} \varphi^2(y)\nu(dy) \right) \nu(dx)$$

$$\leq \frac{m}{\lambda} \int_\Omega \frac{\sigma_s(y)}{\sigma_\lambda(y)} \varphi^2(y)\nu(dy) \leq \frac{m}{\lambda} \|\varphi\|^2_{L^2(v)}$$

and

$$\int_{\{\sigma_s(x)\geq m\}} \frac{\sigma_s(x)}{\sigma_\lambda(x)} \left(\int_\Omega \frac{p(x,y)}{\sigma_\lambda(y)} \varphi^2(y)\nu(dy) \right) \nu(dx)$$

$$\leq \sup_{\{\sigma_s(x)\geq m\}} \frac{\sigma_s(x)}{\sigma_a(x)+\sigma_s(x)}$$

$$\times \left(\int_\Omega \frac{\left[\int_{\{\sigma_s(x)\geq m\}} p(x,y)\nu(dx)\right]}{\sigma_\lambda(y)} \varphi^2(y)\nu(dy) \right)$$

$$\leq \frac{1}{\inf_{\{\sigma_s(x)\geq m\}} \frac{\sigma_a(x)}{\sigma_s(x)} + 1} \cdot \sup_y \frac{\left[\int_{\{\sigma_s(x)\geq m\}} p(x,y)\nu(dx)\right]}{\sigma_s(y)} \|\varphi\|^2_{L^2(v)}$$

whence $\lim_{\lambda\to+\infty} \|G_\lambda\|^2_{\mathcal{L}(L^2(v))}$ is less than or equal to

$$\frac{1}{\inf_{\{\sigma_s(x)\geq m\}} \frac{\sigma_a(x)}{\sigma_s(x)} + 1} \cdot \sup_y \frac{\left[\int_{\{\sigma_s(x)\geq m\}} p(x,y)\nu(dx)\right]}{\sigma_s(y)} \quad (\forall m > 0)$$

which ends the proof. $\qquad\square$

Here is a criterion of closedness in terms of compactness. In the context of nuclear reactor theory, this L^2 compactness property holds true for liquid and gaseous moderators but not for solid moderators [65, 104].

Proposition 7.4.2 *If G_λ is compact on $L^2(v)$ then the form c is closed.*

Proof Since G_λ is self-adjoint then

$$\|G_\lambda\|^2_{\mathcal{L}(L^2(v))} = \|G_\lambda^2\|_{\mathcal{L}(L^2(v))}.$$

Since G_λ is positive and non increasing λ then, for (a fixed) $\beta > 0$

$$\|G_\lambda^2\|_{\mathcal{L}(L^2(v))} \leq \|G_\lambda G_\beta\|_{\mathcal{L}(L^2(v))} \quad (\lambda \geq \beta).$$

Since $|G_\lambda\varphi|_{L^2(v)} \to 0 \ (\lambda \to \infty) \ \forall \varphi \in L^2(v)$ then the convergence is uniform on compact subsets of $L^2(v)$ and consequently

$$\sup_{|\varphi|_{L^2(\nu)} \leq 1} \left\| G_\lambda G_\beta \varphi \right\|_{\mathcal{L}(L^2(\nu))} \to 0 \ (\lambda \to \infty)$$

becaue G_β is compact. $\qquad\square$

We can relax the compactness assumption.

Proposition 7.4.3 *Suppose that* $P = P^{(1)} + P^{(2)}$ *(*$P^{(1)} \geq 0$, $i = 1, 2$*) such that there exists* $\beta \geq 0$ *with* $\left\| G_\beta^{(1)} \right\|_{\mathcal{L}(L^2(\nu))} < 1$ *and* $G_\beta^{(2)}$ *compact in* $L^2(\nu)$. *Then the form* c *is closed.*

Proof Since

$$\begin{aligned}
G_\lambda^2 &= \left(G_\lambda^{(1)} + G_\lambda^{(2)} \right)^2 = \left(G_\lambda^{(1)} \right)^2 + \left(G_\lambda^{(2)} \right)^2 + G_\lambda^{(1)} G_\lambda^{(2)} + G_\lambda^{(2)} G_\lambda^{(1)} \\
&\leq \left(G_\lambda^{(1)} \right)^2 + G_\lambda^{(2)} G_\lambda + G_\lambda G_\lambda^{(2)} + G_\lambda^{(2)} G_\lambda \\
&\leq \left(G_\beta^{(1)} \right)^2 + 2 G_\beta^{(2)} G_\lambda + G_\lambda G_\beta^{(2)} \ (\lambda \geq \beta)
\end{aligned}$$

then

$$\left\| G_\lambda^2 \right\|_{\mathcal{L}(L^2(\nu))} \leq \left\| G_\beta^{(1)} \right\|_{\mathcal{L}(L^2(\nu))}^2 + 2 \left\| G_\beta^{(2)} G_\lambda \right\|_{\mathcal{L}(L^2(\nu))} + \left\| G_\lambda G_\beta^{(2)} \right\|_{\mathcal{L}(L^2(\nu))}.$$

Since $G_\lambda \to 0$ strongly ($\lambda \to +\infty$) and its convergence is uniform on compact subsets of $L^2(\nu)$ then

$$\left\| G_\lambda G_\beta^{(2)} \right\|_{\mathcal{L}(L^2(\nu))} \to 0 \ (\lambda \to +\infty).$$

By duality $\left\| G_\beta^{(2)} G_\lambda \right\|_{\mathcal{L}(L^2(\nu))} = \left\| \left(G_\beta^{(2)} G_\lambda \right)^* \right\|_{\mathcal{L}(L^2(\nu))}$ is equal to

$$\left\| G_\lambda^* \left(G_\beta^{(2)} \right)^* \right\|_{\mathcal{L}(L^2(\nu))} = \left\| G_\lambda \left(G_\beta^{(2)} \right)^* \right\|_{\mathcal{L}(L^2(\nu))} \to 0 \ (\lambda \to +\infty)$$

because $\left(G_\beta^{(2)} \right)^*$ is compact and $G_\lambda \to 0$ strongly ($\lambda \to +\infty$). Thus

$$\lim_{\lambda \to +\infty} \left\| G_\lambda^2 \right\|_{\mathcal{L}(L^2(\nu))} < 1$$

since $\left\| G_\beta^{(1)} \right\|_{\mathcal{L}(L^2(\nu))}^2 < 1$. This ends the proof since $\| G_\lambda \|_{\mathcal{L}(L^2(\nu))}^2 = \left\| G_\lambda^2 \right\|_{\mathcal{L}(L^2(\nu))}$. $\square$

7.4.2 Spectral Analysis

We will show below (see Theorem 7.4.6) that if $\sigma_a = 0$ then

$$(\sigma + 1)\omega^{-1} \in L^1(\nu) \Longrightarrow s(-\mathcal{A}) = 0.$$

We complement this information by two kinds of spectral results. We start with:

Theorem 7.4.3 *Let there exist $\beta \geq 0$ such that $G_\beta : L^2(\nu) \to L^2(\nu)$ is compact then $(\lambda + \mathcal{A})^{-1} - (\lambda - M_2)^{-1}$ and $e^{-t\mathcal{A}} - e^{tM_2}$ are compact in $L^2(\mu)$. If additionally $(\sigma + 1)\omega^{-1} \in L^1(\upsilon)$, $\sigma_a = 0$ and $\inf \sigma_s > 0$ then $\big(\mathcal{U}_{(2)}(t)\big)_{t \geq 0}$ has a spectral gap in $L^2(\mu)$.*

Proof The compactness of G_β in $L^2(\nu)$ is equivalent to that of $\mathcal{G}_\beta$ in $L^2(\mu)$. Since $G_\lambda \leq G_\beta$ ($\lambda \geq \beta$) then, by domination, G_λ is compact for all $\lambda \geq \beta$ (see [4]). One can show (see (7.5.3) below with $r = 1$) that

$$(\lambda + \mathcal{A})^{-1} g = \frac{g}{\sigma_\lambda} + \frac{1}{\sqrt{\sigma_\lambda}} \sum_{n=1}^{\infty} \mathcal{G}_\lambda^n \left(\frac{1}{\sqrt{\sigma_\lambda}} g \right)$$

(where the series converges in operator norm) so the compactness of $(\lambda + \mathcal{A})^{-1} - (\lambda - M_2)^{-1}$. The compactness of $e^{-t\mathcal{A}} - e^{tM_2}$ follows from the holomorphy of self-adjoint C_0-semigroups. Indeed, the compactness of

$$\int_0^{+\infty} \left(e^{-t\mathcal{A}} - e^{tM_2} \right) e^{-\lambda t} dt = (\lambda + \mathcal{A})^{-1} - (\lambda - M_2)^{-1}$$

implies that of

$$\int_s^{s+\varepsilon} \left(e^{-t\mathcal{A}} - e^{tM_2} \right) e^{-\lambda t} dt$$

by a domination argument (see [4]) and consequently of that of

$$\lim_{\varepsilon \to 0} \varepsilon^{-1} \int_s^{s+\varepsilon} \left(e^{-t\mathcal{A}} - e^{tM_2} \right) e^{-\lambda t} dt = \left(e^{-s\mathcal{A}} - e^{sM_2} \right) e^{-\lambda s} \quad (s > 0)$$

because the convergence holds in operator norm. Since $e^{-t\mathcal{A}}$ and e^{tM_2} have the same essential spectrum and the essential radius of e^{tM_2} is < 1 then the essential radius of $e^{-t\mathcal{A}}$ is < 1 while its spectral radius is equal to 1 since $s(-\mathcal{A}) = 0$. $\quad\square$

If G_λ is not compact but quasi-compact only we have a more complex result.

Theorem 7.4.4 *Suppose that $P = P^{(1)} + P^{(2)}$ ($P^{(i)} \geq 0$, $i = 1, 2$) where $\left\| G_0^{(1)} \right\|_{\mathcal{L}(L^2(\nu))} < 1$ and $G_0^{(2)}$ is compact in $L^2(\nu)$. If $(\sigma + 1)\omega^{-1} \in L^1(\nu)$, $\sigma_a = 0$ and $\inf \sigma_s > 0$ then $\big(\mathcal{U}_{(2)}(t)\big)_{t \geq 0}$ has a spectral gap in $L^2(\mu)$.*

Proof Notice that even if G_0 is self-adjoint, a priori the operators $G_0^{(i)}$ ($i = 1, 2$) need not be so. We decompose the bilinear form

$$c_0(f, g) = (\sqrt{\sigma} f, \sqrt{\sigma} g) - (G_0 \sqrt{\sigma} f, \sqrt{\sigma} g)$$

as

$$c_0(f, g) = (\sqrt{\sigma} f, \sqrt{\sigma} g) - \left(G_0^{(1)} \sqrt{\sigma} f, \sqrt{\sigma} g\right) - \left(G_0^{(2)} \sqrt{\sigma} f, \sqrt{\sigma} g\right)$$

where the (a priori non hermitian) bilinear form

$$c_0^{(1)}(f, g) := (\sqrt{\sigma} f, \sqrt{\sigma} g) - \left(G_0^{(1)} \sqrt{\sigma} f, \sqrt{\sigma} g\right)$$

is coercive

$$c_0^{(1)}(f, f) = \left\|\sqrt{\sigma} f\right\|^2_{L^2(\mu)} - \left(G_0^{(1)} \sqrt{\sigma} f, \sqrt{\sigma} f\right) \geq (1 - \left\|G_0^{(1)}\right\|_{\mathcal{L}(L^2(\mu))}) \left\|\sqrt{\sigma} f\right\|^2_{L^2(\mu)}$$

because

$$\left\|G_0^{(1)}\right\|_{\mathcal{L}(L^2(\mu))} = \left\|G_0^{(1)}\right\|_{\mathcal{L}(L^2(\nu))} < 1.$$

In particular

$$c_0^{(1)}(f, f) \geq (1 - \left\|G_0^{(1)}\right\|_{\mathcal{L}(L^2(\mu))})(\inf \sigma) \|f\|^2_{L^2(\mu)} .$$

We introduce the (a priori non self-adjoint) operator

$$\mathcal{A}^{(1)} : D(\mathcal{A}^{(1)}) \subset L^2(\mu) \to L^2(\mu)$$

such that

$$\left(\mathcal{A}^{(1)} u, v\right) = c_0^{(1)}(u, v), \quad u \in D(\mathcal{A}^{(1)}), \ v \in V$$

where

$$D(\mathcal{A}^{(1)}) = \left\{u \in V; \ \exists \alpha > 0, \ \left|c_0^{(1)}(u, v)\right| \leq \alpha \|v\| \ \forall v \in V\right\}.$$

By the general theory, $-\mathcal{A}^{(1)}$ generates a positive semigroup of contractions. Its spectrum is a priori in the half plane $\{\text{Re } \lambda \leq s(-\mathcal{A}^{(1)})\}$. Since 0 is not in the spectrum of $-\mathcal{A}^{(1)}$ ($\mathcal{A}^{(1)}$ is invertible) then a neighborhood of 0 is in the resolvent set of $-\mathcal{A}^{(1)}$ and $s(-\mathcal{A}^{(1)}) < 0$. As in the proof of Theorem 7.4.2, for λ large enough

$$(\lambda + \mathcal{A})^{-1} g = \frac{1}{\sqrt{\sigma_\lambda}} (I - G_\lambda)^{-1} \left(\frac{1}{\sqrt{\sigma_\lambda}} g\right)$$

and

$$(\lambda + \mathcal{A}^{(1)})^{-1} g = \frac{1}{\sqrt{\sigma_\lambda}} \left(I - G_\lambda^{(1)}\right)^{-1} \left(\frac{1}{\sqrt{\sigma_\lambda}} g\right).$$

Note that $G_\lambda^{(2)} (\leq G_0^{(2)})$ is compact in $L^2(v)$ (see [4]) so $\mathcal{G}_\lambda^{(2)}$ is compact in $L^2(\mu)$. Since

$$\mathcal{G}_\lambda = \mathcal{G}_\lambda^{(1)} + \mathcal{G}_\lambda^{(2)}$$

and $\mathcal{G}_\lambda^{(2)}$ is compact in $L^2(\mu)$ then

$$(\alpha - \mathcal{G}_\lambda)^{-1} - \left(\alpha - \mathcal{G}_\lambda^{(1)}\right)^{-1} \text{ is compact } \left(\alpha > \max\left\{\|\mathcal{G}_\lambda\|, \left\|\mathcal{G}_\lambda^{(1)}\right\|\right\}\right),$$

in particular

$$(I - \mathcal{G}_\lambda)^{-1} - \left(I - \mathcal{G}_\lambda^{(1)}\right)^{-1} \text{ is compact.}$$

Hence

$$(\lambda + \mathcal{A})^{-1} - \left(\lambda + \mathcal{A}^{(1)}\right)^{-1} = \frac{1}{\sqrt{\sigma_\lambda}} \left[(I - \mathcal{G}_\lambda)^{-1} - \left(I - \mathcal{G}_\lambda^{(1)}\right)^{-1} \right] \frac{1}{\sqrt{\sigma_\lambda}}$$

is compact. Thus $-\mathcal{A}$ and $-\mathcal{A}^{(1)}$ share the same essential spectrum. In particular, the essential spectrum of the self-adjoint operator $-\mathcal{A}$ is $\leq s(-\mathcal{A}^{(1)}) < 0$. Since $s(-\mathcal{A}) = 0$ then $\left(\mathcal{U}_{(2)}(t)\right)_{t\geq 0}$ has a spectral gap in $L^2(\mu)$. $\square$

7.4.3 $L^2(\mu)$-Theory Versus $L^1(v)$-Theory

We explore now the connections of the $L^2(\mu)$ theory with the $L^1(v)$ theory given in Chap. 6. The connection is settled through the key condition

$$(\sigma + 1)\omega^{-1} \in L^1(v). \tag{7.4.3}$$

We need first two lemmas.

Lemma 7.4.1 *If $\omega^{-1} \in L^1(v)$ then $L^2(\mu)$ is continuously imbedded into $L^1(v)$; more precisely*

$$\|f\|_{L^1(v)} \leq \sqrt{\int_\Omega \frac{1}{\omega(x)} v(dx)} \, \|f\|_{L^2(\mu)} \quad \forall f \in L^2(\mu). \tag{7.4.4}$$

Moreover $L^2(\mu)$ is dense in $L^1(v)$.

Proof The first claim follows from

$$\int_\Omega |f(x)|\, \nu(dx) = \int_\Omega |f(x)|\, \sqrt{\omega(x)}\, \frac{1}{\sqrt{\omega(x)}} \nu(dx)$$

$$\leq \sqrt{\int_\Omega |f(x)|^2\, \omega(x)\nu(dx)}\sqrt{\int_\Omega \frac{1}{\omega(x)}\nu(dx)}$$

$$= \sqrt{\int_\Omega \frac{1}{\omega(x)}\nu(dx)}\, \|f\|_{L^2(\mu)}\, .$$

Now let $f \in L^1(\nu)$. Since $\nu(dx)$ is sigma finite, there exists an increasing sequence Ω_j of measurable subsets with finite ν-measure such that $\cup_j \Omega_j = \Omega$. Since ω is finite a.e. then the sets $\Xi_n = \{\omega \leq n\}$ are such that $\cup_n \Xi_n = \Omega$ and $\cup_j \cup_n \left(\Omega_j \cap \Xi_n\right) = \Omega$ so

$$1_{\Omega_j \cap \Xi_n} f \to f \to \text{ in } L^1(\nu) \ (j, n \to +\infty).$$

Since f is finite a.e. we can approximate each $1_{\Omega_j \cap \Xi_n} f$ by a sequence of essentially bounded functions

$$\left(1_{\Omega_j \cap \Xi_n} f\right) 1_{\{|f| \leq m\}} \to 1_{\Omega_j \cap \Xi_n} f \ \text{ in } L^1(v) \ (m \to +\infty).$$

Finally $\left(1_{\Omega_j \cap \Xi_n} f\right) 1_{\{|f| \leq m\}} \in L^2(\mu)$ ends the proof. $\qquad\square$

Lemma 7.4.2 *Let (7.4.3) be satisfied. Then V is included in $D(M_1)$, the domain of the multiplication operator par $-\sigma$ in $L^1(\nu)$.*

Proof Let $f \in V$ then $\sqrt{\sigma_\lambda} f \in L^2(\mu)$ and

$$\int_\Omega |f(x)|\, \sigma_\lambda(x)\nu(dx) = \int_\Omega |f(x)|\, \sqrt{\sigma_\lambda(x)}\sqrt{\omega(x)}\frac{\sqrt{\sigma_\lambda(x)}}{\sqrt{\omega(x)}}\nu(dx)$$

$$\leq \sqrt{\int_\Omega \left|\sqrt{\sigma_\lambda(x)}f(x)\right|^2 \omega(x)\nu(dx)}\sqrt{\int_\Omega \frac{\sigma_\lambda(x)}{\omega(x)}\nu(dx)}$$

$$= \sqrt{\int_\Omega \frac{\sigma_\lambda(x)}{\omega(x)}\nu(dx)}\, \left\|\sqrt{\sigma_\lambda} f\right\|_{L^2(\mu)}$$

and we are done. $\qquad\square$

We are now ready to state.

Theorem 7.4.5 *Let (7.4.3) be satisfied. Then the C_0-semigroup $\left(\mathcal{U}_{(2)}(t)\right)_{t \geq 0}$ on $L^2(\mu)$ extends uniquely as a positive contraction C_0-semigroup $\left(\widehat{U}_1(t)\right)_{t \geq 0}$ on $L^1(\nu)$. Moreover, if $\sigma_a(.) = 0$ then $\left(\widehat{U}_1(t)\right)_{t \geq 0}$ is stochastic, i.e., mass-preserving*

$$\left\|\widehat{U}_1(t)\varphi\right\|_{L^1(\nu)} = \|\varphi\|_{L^1(\nu)} \ (t \geq 0) \ \forall \varphi \in L^1_+(\mu).$$

The semigroup $\left(\widehat{U}_1(t)\right)_{t \geq 0}$ *coincides with the C_0-semigroup* $(\mathcal{J}_1(t))_{t \geq 0}$ *built in Chap. 6.*

Proof Let $\varphi \in L^2_+(\mu)$ and $F(t) = \mathcal{U}_{(2)}(t)\varphi$. The semigroup being self-adjoint has many smoothing properties, for instance

$$F \in C^0([0, +\infty), L^2(\mu)) \cap C^\infty((0, +\infty), L^2(\mu)) \cap C^0((0, +\infty), D(\mathcal{A}^j)) \ (j \geq 1)$$

(see [73]) and

$$\frac{dF(t)}{dt} = -\mathcal{A}F(t) \ (t > 0)$$

where the time derivative is in the sense of $L^2(\mu)$. Due to the continuous imbedding of $L^2(\mu)$ into $L^1(\nu)$, this time derivative holds also in the sense of $L^1(\nu)$. In particular

$$\int_\Omega \frac{dF(t)}{dt} \, d\nu = \frac{d}{dt}\left(\int_\Omega F(t)d\nu\right).$$

On the other hand, by Theorem 7.3.1

$$\mathcal{A}F(t) = \sigma(x)F(t, x) - \int_\Omega p(x, y)F(t, y)\nu(dy)$$

and Lemma 7.4.2 asserts that $\sigma F(t) \in L^1(\nu)$. For $t > 0$, $F(t) \in D(\mathcal{A}^\infty)$. In particular

$$\mathcal{A}F(t) \in D(\mathcal{A}) \subset V \subset D_1(M_1) \subset L^1(\nu)$$

so

$$\int_\Omega p(., y)F(t, y)\nu(dy) \in L^1(\nu).$$

It follows that

$$\frac{d}{dt}\left(\int_\Omega F(t)d\nu\right) = -\int_\Omega \sigma(x)F(t, x)\nu(dx) + \int_\Omega\left(\int_\Omega p(x, y)F(t, y)\nu(dy)\right)\nu(dx)$$

$$= -\int_\Omega (\sigma(x) - \sigma_s(x))\, F(t, x)\nu(dx).$$

Hence

$$\left\|\mathcal{U}_{(2)}(t)\varphi\right\|_{L^1(\nu)} \leq \|\varphi\|_{L^1(\nu)} \ \forall \varphi \in L^2_+(\mu)$$

and

$$\left\|\mathcal{U}_{(2)}(t)\varphi\right\|_{L^1(\nu)} \leq \|\varphi\|_{L^1(\nu)} \ \forall \varphi \in L^2(\mu)$$

because $\left(\mathcal{U}_{(2)}(t)\right)_{t\geq 0}$ is positive. We have also

$$\left\|\mathcal{U}_{(2)}(t)\varphi\right\|_{L^1(\nu)} = \|\varphi\|_{L^1(\nu)} \quad \forall \varphi \in L^2_+(\mu) \text{ if } \sigma_a = 0.$$

Since $L^2(\mu)$ is dense in $L^1(\nu)$ then $\mathcal{U}_{(2)}(t)$ extends uniquely as a contraction $\widehat{U}_1(t)$ in $L^1(\nu)$ which is stochastic when $\sigma_a = 0$. The semigroup property extends to $L^1(\nu)$ while the estimate (7.4.4) gives

$$\left\|\widehat{U}_1(t)f - f\right\|_{L^1(\nu)} \leq \sqrt{\int_\Omega \frac{1}{\omega(x)}\nu(dx)}\,\left\|\mathcal{U}_{(2)}(t)f - f\right\|_{L^2(\mu)} \quad (f \in L^2(\mu)).$$

The density of $L^2(\mu)$ in $L^1(\nu)$ shows that $\left(\widehat{U}_1(t)\right)_{t\geq 0}$ is strongly continuous on $L^1(\nu)$. The identification of $\left(\widehat{U}_1(t)\right)_{t\geq 0}$ with the semigroup $(\mathcal{J}_1(t))_{t\geq 0}$ in Chap. 6 is clear. $\square$

Remark 7.4.1   If (7.4.3) is satisfied then

(i) According to Theorem 6.4.1, the C_0-semigroup $(\mathcal{J}_1(t))_{t\geq 0}$ is honest and asymptotically stable in $L^1(\nu)$.

(ii) A priori $(\mathcal{J}_1(t))_{t\geq 0}$ need not be genuinely honest in the sense that the domain $D(\mathcal{B})$ of its generator $\mathcal{B}$ need not be equal to $D(M_1)$. Despite this fact, we observe that for $g \in L^2(\mu) \subset L^1(\nu)$

$$\lambda f - \mathcal{B}f = g$$

has its solution f in V so $\sigma f \in L^1(\nu)$, $Pf \in L^1(\nu)$ and consequently $\mathcal{B}f = -\sigma f + Pf$ is a sum of two functions in $L^1(\nu)$. Similarly, the solution to evolution equation

$$\frac{df}{dt} = \mathcal{B}f, \ f(0) = f_0 \in L^2(\mu) \subset L^1(\nu) \tag{7.4.5}$$

belongs to $D(\mathcal{A})$ (once $t > 0$) because $\left(\mathcal{U}_{(2)}(t)\right)_{t\geq 0}$ is analytic. In particular $f(t) \in V$ and $\mathcal{B}f(t) = -\sigma f(t) + Pf(t)$ is a sum of two functions in $L^1(\nu)$. Thus, if we consider the Cauchy problem (7.4.5) in $L^1(\nu)$ then, for any initial data $f_0 \in L^2(\mu)$, the solution $f(.)$ is such that $f(t) \in D(M_1)$ once $t > 0$. Since $D(M_1)$ is not invariant under $(\mathcal{J}_1(t))_{t\geq 0}$, (and there is no natural link between $L^2(\mu)$ and $D(\mathcal{B})$), this regularizing effect is not predictable a priori. One sees how a detailed balance condition improves our understanding of the $L^1(\nu)$ theory.

7.4.4   On Asymptotic Stability in $L^2(\mu)$

We have seen in Theorem 6.4.1 that if $\sigma_a = 0$, $\omega^{-1} \in L^1(\nu)$ and $\sigma\omega^{-1} \in L^1(\nu)$ then ω^{-1} (normalized in $L^1(\nu)$) is an invariant density of $(\mathcal{J}_1(t))_{t\geq 0}$ and is asymptotically stable in $L^1(\nu)$. Here is its counterpart in $L^2(\mu)$.

Theorem 7.4.6 *Let (7.4.3) be satisfied. If $\sigma_a = 0$ then $s(-\mathcal{A}) = 0$ and $s(-\mathcal{A})$ is an eigenvalue of $-\mathcal{A}$ associated to the eigenfunction ω^{-1} and the C_0-semigroup $\left(\mathcal{U}_{(2)}(t)\right)_{t \geq 0}$ is asymptotically stable in $L^2(\mu)$.*

Proof We already know that $s(-\mathcal{A}) \leq 0$. If $s(-\mathcal{A}) < 0$ then $\left\|\mathcal{U}_{(2)}(t)\right\|_{\mathcal{L}(L^2(\mu))} \to 0$ $(t \to +\infty)$. On the other hand, we know that

$$\left\|\mathcal{U}_{(2)}(t)\varphi\right\|_{L^1(\nu)} = \|\varphi\|_{L^1(\nu)} \quad (\varphi \in L^2_+(\mu))$$

and

$$\left\|\mathcal{U}_{(2)}(t)\varphi\right\|_{L^1(\nu)} \leq \sqrt{\int_\Omega \frac{1}{\omega(x)}\nu(dx)} \left\|\mathcal{U}_{(2)}(t)\varphi\right\|_{L^2(\mu)}.$$

This leads to the contradiction

$$\|\varphi\|_{L^1(\nu)} \leq \sqrt{\int_\Omega \frac{1}{\omega(x)}\nu(dx)} \left\|\mathcal{U}_{(2)}(t)\varphi\right\|_{L^2(\mu)} \to 0 \ (t \to +\infty).$$

Hence $s(-\mathcal{A}) = 0$.

Note that (7.4.3), i.e.,

$$\int_\Omega \frac{(1 + \sigma(x))}{\omega(x)}\nu(dx) < +\infty$$

means

$$\int_\Omega \left(\omega^{-1}(x)\right)^2 (1 + \sigma(x))\, \mu(dx) < +\infty$$

so $\omega^{-1} \in V$. On the other hand

$$\omega^{-1}(y)p(x, y) = \omega^{-1}(x)p(y, x)$$

implies

$$\sigma\omega^{-1} - \int_\Omega p(x, y)\omega^{-1}(y)\nu(dy) = 0.$$

Hence $\int_\Omega p(x, y)\omega^{-1}(y)\nu(dy)$ is finite a.e. According to the characterization of $D(\mathcal{A})$ (Theorem 7.3.1) $\omega^{-1} \in D(\mathcal{A})$ and $\mathcal{A}\omega^{-1} = 0$. The asymptotic stability in $L^2(\mu)$ follows from Theorem 2.10.4. $\qquad\square$

7.4.5 *Remark on Hyperboundedness*

We mention here how the existence of a spectral gap for $\left(\mathcal{U}_{(2)}(t)\right)_{t\geq 0}$ can be related to hyperboundedness of a suitable C_0-semigroup. Indeed, according to Theorem 7.4.6, if (7.4.3) is satisfied and $\sigma_a = 0$ then $s(-\mathcal{A}) = 0$ and $s(-\mathcal{A})$ is an eigenvalue of $-\mathcal{A}$ associated to the eigenfunction ω^{-1}. By normalizing ω^{-1} if necessary, we suppose that $\left\|\omega^{-1}\right\|_{L^2(\mu)} = 1$ and introduce the probability measure

$$\beta(dx) = \left(\omega^{-1}(x)\right)^2 \mu(dx).$$

We consider the unitary transformation

$$T_\omega : f \in L^2(\beta) \to T_\omega f = \omega^{-1} f \in L^2(\mu)$$

and $\mathcal{H} := T_\omega^{-1} \mathcal{A} T_\omega$. Then $\left(\mathcal{U}_{(2)}(t)\right)_{t\geq 0}$ is unitarily equivalent to $\left(e^{-t\mathcal{H}}\right)_{t\geq 0}$. Notice that $\mathcal{U}_{(2)}(t)\omega^{-1} = \omega^{-1}$ implies

$$e^{-t\mathcal{H}}1 = T_\omega^{-1} U_2(t) T_\omega 1 = 1$$

so $-\mathcal{H}$ generates a C_0-semigroup of contractions $\left(e^{-t\mathcal{H}}\right)_{t\geq 0}$ in all $L^p(\beta)$ ($1 \leq p < +\infty$), (see [311, Theorem X.55]). It turns out that a regularizing effect of $\left(e^{-t\mathcal{H}}\right)_{t\geq 0}$ (called hyperboundedness) in the sense that there exists $p > 2$ such that

$$e^{-t\mathcal{H}} \in \mathcal{L}\left(L^2(\beta), L^p(\beta)\right) \ (t > 0) \tag{7.4.6}$$

implies that $\left(e^{-t\mathcal{H}}\right)_{t\geq 0}$ has a spectral gap [232]; (see [138] and references therein for more recent information on hyperboundedness of operators). Thus the existence of the spectral gap for $\left(\mathcal{U}_{(2)}(t)\right)_{t\geq 0}$ has a connection with hyperboundedness of $\left(e^{-t\mathcal{H}}\right)_{t\geq 0}$.

7.4.6 *Is the Closed Form c a Dirichlet Form?*

We consider the case where the form c is closed on V and wonder whether c is a Dirichlet form in the sense of Beurling-Deny theory (see [99, Sect. 1.3]) or equivalently whether the C_0-semigroup $\left(\mathcal{U}_{(2)}(t)\right)_{t\geq 0}$ on $L^2(\mu)$ extends, in a consistent way, to all $L^p(\mu)$ spaces ($1 \leq p \leq \infty$) as a contraction C_0-semigroup $\left(\mathcal{U}_{(p)}(t)\right)_{t\geq 0}$ (with generator $\mathcal{A}_{(p)}$). To this end, we suppose that P is M_1-bounded in $L^1(\mu)$. We give a result under a *density* assumption which we do not explore further.

Proposition 7.4.4 *Suppose that the form c on V is closed and is a Dirichlet form. Let P be σ-bounded in $L^1(\mu)$. If*

$$D(\mathcal{A}) \cap D(\mathcal{A}_{(1)}) \cap D(M_1) \cap L^1_+(\mu)$$

is dense in $D(M_1) \cap L^1_+(\mu)$ *then*

$$\widehat{\sigma}_s(x) := \int_\Omega p(x, y) \upsilon(dy) \leq \sigma(x) \ \ (x \in \Omega).$$

Proof According to the general theory (see [99])

$$\mathcal{U}_{(2)}(t) : L^1(\mu) \cap L^\infty(\mu) \to L^1(\mu) \cap L^\infty(\mu)$$

and defines contractive C_0-semigroups $\left(\mathcal{U}_{(p)}(t)\right)_{t \geq 0}$ in $L^p(\mu)$ with generator $\mathcal{A}_{(p)}$. According to Theorem 7.3.1 if $f \in D(\mathcal{A})$

$$\mathcal{A} f = -\sigma f + Pf$$

is a sum of two measurable functions (finite a.e.) which belongs to $L^2(\mu)$. It follows that if $f \in D_1(\sigma) \cap D(\mathcal{A}) \cap D(\mathcal{A}_{(1)})$ then

$$\mathcal{A}_{(1)} f = -\sigma f + Pf$$

is a sum of two measurable functions (finite a.e.) which belongs to $L^2(\mu) \cap L^1(\mu)$. Then for $f \in D(\mathcal{A}_{(1)}) \cap L^1_+(\mu)$

$$\int \mathcal{U}_{(1)}(t) f \, d\mu \leq \int f \, d\mu$$

implies

$$\int \mathcal{A}_{(1)} f \, d\mu = \lim_{t \to 0} \int t^{-1} \left(\mathcal{U}_{(1)}(t)\right) f - f) \, d\mu \leq 0.$$

The choice

$$f \in D(\mathcal{A}) \cap D(\mathcal{A}_{(1)}), \ \ f \geq 0$$

gives

$$\int_\Omega (-\sigma f + Pf) \, d\mu \leq 0. \tag{7.4.7}$$

We know that P is σ-bounded in $L^1(\mu)$ if and only if there exists $C > 0$ such that $\widehat{\sigma}_s(y) \leq C(1 + \sigma(y))$ so, if additionally $f \in D(M_1) \subset L^1(\nu)$,

$$\int_{\Omega} \left(\int_{\Omega} p(x, y) f(y) \nu(dy) \right) \mu(dx)$$

$$= \int_{\Omega} \left(\int_{\Omega} p(x, y) f(y) \nu(dy) \right) \omega(x) \nu(dx)$$

$$= \int_{\Omega} \left(\int_{\Omega} p(x, y) \omega(x) \nu(dx) \right) f(y) \nu(dy)$$

$$= \int_{\Omega} \left(\int_{\Omega} p(y, x) \omega(y) \nu(dx) \right) f(y) \nu(dy)$$

$$= \int_{\Omega} \widehat{\sigma}_s(y) f(y) \omega(y) \nu(dy)$$

$$\leq C \int_{\Omega} (1 + \sigma(y)) f(y) \omega(y) \nu(dy) < +\infty.$$

We can integrate separately the two pieces of (7.4.7)

$$-\int_{\Omega} \sigma(x) f(x) \omega(x) \nu(dx) + \int_{\Omega} \left(\int_{\Omega} p(x, y) f(y) \nu(dy) \right) \omega(x) \nu(dx) \leq 0$$

so

$$-\int_{\Omega} \sigma(x) f(x) \omega(x) \nu(dx) + \int_{\Omega} \widehat{\sigma}_s(y) f(y) \omega(y) \nu(dy) \leq 0$$

and

$$\int_{\Omega} \left(\widehat{\sigma}_s(y) - \sigma(y) \right) f(y) \mu(dy) \leq 0$$

for all

$$f \in D(\mathcal{A}) \cap D(\mathcal{A}_{(1)}) \cap D_1(\sigma), \quad f \geq 0.$$

Since this set is (assumed to be) dense in $D(M_1) \cap L^1_+(\mu)$ then

$$\int_{\Omega} \left(\widehat{\sigma}_s(y) - \sigma(y) \right) f(y) \mu(dy) \leq 0 \ \forall f \in D(M_1) \cap L^1_+(\mu)$$

which implies $\widehat{\sigma}_s(.) \leq \sigma(.)$. $\qquad\square$

The lesson we draw from Proposition 7.4.4 is that we cannot hope c to be a Dirichlet form on V if the condition $\int_{\Omega} p(x, y) \nu(dy) \leq \sigma(x)$ is not satisfied. It is of course satisfied when $p(x, y) = p(y, x)$, i.e., $\omega(x) = 1$.

7.5　Critical Case: Non-closed Form c

According to Lemma 7.3.1, $\|G_\lambda\|_{\mathcal{L}(L^2(\nu))} \leq 1$ $(\lambda \geq 0)$. Moreover, according to Theorem 7.4.1, $\lim_{\lambda \to +\infty} \|G_\lambda\|_{\mathcal{L}(L^2(\nu))} < 1$ if and only if the form c defined by (7.3.2) is closed. We investigate now the critical case

$$\|G_\lambda\|_{\mathcal{L}(L^2(\nu))} = \|\mathcal{G}_\lambda\|_{\mathcal{L}(L^2(\mu))} = 1 \quad (\lambda \geq 0). \tag{7.5.1}$$

To this end, we introduce the approximating forms

$$c^{(r)}(f, g) = \int_\Omega \sigma(x) f(x) \overline{g(x)} \omega(x) \nu(dx) - r\,(Pf, g) \quad (0 < r \leq 1)$$

and

$$c_\lambda^{(r)}(f, g) = \lambda(f, g) + c^{(r)}(f, g).$$

Thus

$$c_\lambda^{(r)}(f, g) = \left(\sqrt{\sigma_\lambda} f, \sqrt{\sigma_\lambda} g\right) - r\left(\mathcal{G}_\lambda \sqrt{\sigma_\lambda} f, \sqrt{\sigma_\lambda} g\right), \quad (0 < r \leq 1). \tag{7.5.2}$$

We start with

Lemma 7.5.1 *The sesquilinear forms $c^{(r)}$ $(0 < r \leq 1)$ are bounded on V. If $0 < r < 1$ then $c^{(r)}$ is closed.*

Proof It is easy to see that $\|f\|_V$ is equivalent to

$$\|f\|_\lambda := \sqrt{\int_\Omega (\sigma(x) + \lambda)\,|f(x)|^2\,\mu(dx)} \quad (\forall \lambda > 0).$$

Since $\left(\sqrt{\sigma_\lambda} f, \sqrt{\sigma_\lambda} g\right) \leq \|f\|_\lambda \|g\|_\lambda$ and

$$r\left(\mathcal{G}_\lambda \sqrt{\sigma_\lambda} f, \sqrt{\sigma_\lambda} g\right) \leq r\,\|\mathcal{G}_\lambda\|_{\mathcal{L}(L^2(\mu))} \left\|\sqrt{\sigma_\lambda} f\right\| \left\|\sqrt{\sigma_\lambda} g\right\| = r\,\|\mathcal{G}_\lambda\|_{\mathcal{L}(L^2(\mu))} \|f\|_\lambda \|g\|_\lambda$$

then

$$c_\lambda^{(r)}(f, f) = \left(\sqrt{\sigma_\lambda} f, \sqrt{\sigma_\lambda} f\right) - r\left(\mathcal{G}_\lambda \sqrt{\sigma_\lambda} f, \sqrt{\sigma_\lambda} f\right)$$
$$\geq \left(1 - r\,\|\mathcal{G}_\lambda\|_{\mathcal{L}(L^2(\mu))}\right) \|f\|_\lambda \|g\|_\lambda \geq (1 - r)\,\|f\|_\lambda \|g\|_\lambda$$

since $\|\mathcal{G}_\lambda\|_{\mathcal{L}(L^2(\mu))} = \|G_\lambda\|_{\mathcal{L}(L^2(\nu))} \leq 1$. $\qquad\square$

According to Remark 7.3.1 then there exists a unique unbounded symmetric operator

$$\mathcal{A}^{(r)} : D(\mathcal{A}^{(r)}) \subset L^2(\mu) \to L^2(\mu), \quad (0 < r \leq 1)$$

(i.e., $\mathcal{A}^{(r)} \subset \left(\mathcal{A}^{(r)}\right)^*$) such that

$$\left(\mathcal{A}^{(r)} u, v\right) = c^{(r)}(u, v), \quad u \in D(\mathcal{A}^{(r)}), \ v \in V \ (0 < r \le 1)$$

where

$$D(\mathcal{A}^{(r)}) = \left\{u \in V; \ \exists \alpha > 0, \ \left|c^{(r)}(u, v)\right| \le \alpha \|v\| \ \forall v \in V\right\}, \ (0 < r \le 1).$$

In addition, $\mathcal{A}^{(r)}$ is self-adjoint for $r < 1$

$$\mathcal{A}^{(r)} = \left(\mathcal{A}^{(r)}\right)^* \ \text{if } 0 < r < 1$$

while $\mathcal{A} = \mathcal{A}^{(1)}$ is not self-adjoint but symmetric only

$$\mathcal{A} \subset \mathcal{A}^*.$$

As in Theorem 7.3.1 we can characterize the domain of $\mathcal{A}_\lambda^{(r)}$ for $0 < r \le 1$.

Proposition 7.5.1 *(i) Let $f \in V$. Then*

$$\int_\Omega p(x, y) |f(y)| \nu(dy) < +\infty \ \nu\text{-a.e.}$$

(ii) $f \in D(\mathcal{A}_\lambda^{(r)})$ if and only if $f \in V$ and

$$\sigma_\lambda(x) f(x) - r \int_\Omega p(x, y) f(y) \nu(dy) \in L^2(\mu).$$

In this case

$$\mathcal{A}_\lambda^{(r)} f = \sigma_\lambda(x) f(x) - r \int_\Omega p(x, y) f(y) \nu(dy).$$

7.5.1 Generation

Here is a kind of L^2 version of Theorem 2.11.1.

Theorem 7.5.1 *We consider the case (7.5.1). There exists a unique self-adjoint extension $\mathcal{W}$ of $-\mathcal{A}$ which generates a positive contraction semigroup in $L^2(\mu)$. Its resolvent is given by a strongly convergent series*

$$(\lambda - \mathcal{W})^{-1} g = \frac{1}{\sqrt{\sigma_\lambda}} \sum_{n=0}^{\infty} \mathcal{G}_\lambda^n \left(\frac{1}{\sqrt{\sigma_\lambda}} g\right) \ \forall g \in L^2(\mu).$$

Proof First, we look for the resolvent of $-\mathcal{A}^{(r)}$ $(0 < r < 1)$. Let $\lambda > 0$ and $g \in L^2(\mu)$; we look for $f \in D(\mathcal{A}^{(r)})$ such that

$$f = \left(\lambda + \mathcal{A}^{(r)}\right)^{-1} g.$$

According to Proposition 7.5.1 we have

$$\lambda f + \sigma(x) f(x) - r \int_\Omega p(x, y) f(y) \nu(dy) = g,$$

i.e.,

$$\sqrt{\sigma_\lambda(x)} f(x) - r \int_\Omega \frac{p(x, y)}{\sqrt{\sigma_\lambda(x)}\sqrt{\sigma_\lambda(y)}} \sqrt{\sigma_\lambda(y)} f(y) \nu(dy) = \frac{1}{\sqrt{\sigma_\lambda(x)}} g$$

or

$$\sqrt{\sigma_\lambda(x)} f(x) - r \int_\Omega \frac{p(x, y)}{\sqrt{\sigma_\lambda(x)}\sqrt{\sigma_\lambda(y)}\omega(y)} \sqrt{\sigma_\lambda(y)} f(y) \mu(dy) = \frac{1}{\sqrt{\sigma_\lambda(x)}} g.$$

This amounts to

$$\varphi := \sqrt{\sigma_\lambda} f \in L^2(\mu)$$

and

$$\varphi - r \int_\Omega \frac{p(x, y)}{\sqrt{\sigma_\lambda(x)}\sqrt{\sigma_\lambda(y)}\omega(y)} \varphi(y) \mu(dy) = \frac{1}{\sqrt{\sigma_\lambda}} g,$$

i.e.,

$$\varphi - r \mathcal{G}_\lambda \varphi = \frac{1}{\sqrt{\sigma_\lambda}} g$$

or

$$\varphi = \sum_{n=0}^{\infty} r^n \mathcal{G}_\lambda^n \left(\frac{1}{\sqrt{\sigma_\lambda}} g\right)$$

which converges in operator norm since $\|\mathcal{G}_\lambda\|_{\mathcal{L}(L^2(\mu))} \leq 1$ and $r < 1$. Thus

$$\left(\lambda + \mathcal{A}^{(r)}\right)^{-1} g = \frac{1}{\sqrt{\sigma_\lambda}} \sum_{n=0}^{\infty} r^n \mathcal{G}_\lambda^n \left(\frac{1}{\sqrt{\sigma_\lambda}} g\right). \tag{7.5.3}$$

The variational formulation shows that

$$\left\|\left(\lambda + \mathcal{A}^{(r)}\right)^{-1}\right\|_{\mathcal{L}(L^2(\mu))} \leq \frac{1}{\lambda} \quad (0 < r < 1).$$

Hence $-\mathcal{A}^{(r)}$ generates a positive contraction semigroup $\left(T^{(r)}(t)\right)_{t\geq 0}$ in $L^2(\mu)$ and

$$\frac{1}{\sqrt{\sigma_\lambda}}\sum_{n=0}^{\infty} r^n \mathcal{G}_\lambda^n\left(\frac{1}{\sqrt{\sigma_\lambda}}g\right) = \int_0^{+\infty} e^{-\lambda t}T^{(r)}(t)g\,dt. \qquad (7.5.4)$$

Since

$$r \to \left(\lambda + \mathcal{A}^{(r)}\right)^{-1} g \text{ is non decreasing } (g \geq 0)$$

then, by semigroup theory, so is $r \to T^{(r)}(t)g$. Since

$$\left\|T^{(r)}(t)g\right\| \leq \|g\| \quad (0 < r < 1)$$

then by monotone convergence theorem, the strong limit

$$T(t)g := \lim_{r\to 1} T^{(r)}(t)g$$

exists. One can check that $\mathcal{T} = (T(t))_{t\geq 0}$ is a positive C_0-semigroup of contractions in $L^2(\mu)$ and is self-adjoint. Let $\mathcal{W}$ be its generator. Then (7.5.4) shows, by passing to the limit $(r \to 1)$ that the series with positive terms

$$\frac{1}{\sqrt{\sigma_\lambda}}\sum_{n=0}^{\infty} \mathcal{G}_\lambda^n\left(\frac{1}{\sqrt{\sigma_\lambda}}g\right)$$

converges in $L^2(\mu)$ and

$$(\lambda - \mathcal{W})^{-1} g = \frac{1}{\sqrt{\sigma_\lambda}}\sum_{n=0}^{\infty} \mathcal{G}_\lambda^n\left(\frac{1}{\sqrt{\sigma_\lambda}}g\right) \quad \forall g \in L_+^2(\mu).$$

Let us show that $\mathcal{W}$ is an *extension* of $-\mathcal{A}$. For $g, h \in L^2(\mu)$, we have

$$\begin{aligned}
\left((\lambda - \mathcal{W})^{-1} g, h\right) &= \left(\frac{1}{\sqrt{\sigma_\lambda}}\sum_{n=0}^{\infty} \mathcal{G}_\lambda^n\left(\frac{1}{\sqrt{\sigma_\lambda}}g\right), h\right) \\
&= \left(\sum_{n=0}^{\infty} \mathcal{G}_\lambda^n\left(\frac{1}{\sqrt{\sigma_\lambda}}g\right), \frac{1}{\sqrt{\sigma_\lambda}}h\right) \\
&= \lim_{n\to\infty} \left(\sum_{j=0}^{n} \mathcal{G}_\lambda^j\left(\frac{1}{\sqrt{\sigma_\lambda}}g\right), \frac{1}{\sqrt{\sigma_\lambda}}h\right).
\end{aligned}$$

Since

$$\left(\sum_{j=0}^{n} \mathcal{G}_\lambda^j \left(\frac{1}{\sqrt{\sigma_\lambda}} g \right), \frac{1}{\sqrt{\sigma_\lambda}} h \right)$$

$$= \left(\frac{1}{\sqrt{\sigma_\lambda}} g, \frac{1}{\sqrt{\sigma_\lambda}} h \right) + \left(\sum_{j=1}^{n} \mathcal{G}_\lambda^j \left(\frac{1}{\sqrt{\sigma_\lambda}} g \right), \frac{1}{\sqrt{\sigma_\lambda}} h \right)$$

$$= \left(\frac{1}{\sqrt{\sigma_\lambda}} g, \frac{1}{\sqrt{\sigma_\lambda}} h \right) + \left(\sum_{j=1}^{n} \mathcal{G}_\lambda^{j-1} \mathcal{G}_\lambda \left(\frac{1}{\sqrt{\sigma_\lambda}} g \right), \frac{1}{\sqrt{\sigma_\lambda}} h \right)$$

$$= \left(\frac{1}{\sqrt{\sigma_\lambda}} g, \frac{1}{\sqrt{\sigma_\lambda}} h \right) + \left(\sum_{j=0}^{n-1} \mathcal{G}_\lambda^j \mathcal{G}_\lambda \left(\frac{1}{\sqrt{\sigma_\lambda}} g \right), \frac{1}{\sqrt{\sigma_\lambda}} h \right)$$

then

$$\left(\sum_{j=0}^{n} \mathcal{G}_\lambda^j \left(\frac{1}{\sqrt{\sigma_\lambda}} g \right), \frac{1}{\sqrt{\sigma_\lambda}} h \right)$$

$$= \left(\frac{1}{\sqrt{\sigma_\lambda}} g, \frac{1}{\sqrt{\sigma_\lambda}} h \right) + \left(\sum_{j=0}^{n-1} \mathcal{G}_\lambda^j \mathcal{G}_\lambda \left(\frac{1}{\sqrt{\sigma_\lambda}} g \right), \frac{1}{\sqrt{\sigma_\lambda}} h \right).$$

Letting $n \to \infty$

$$\left(\sum_{j=0}^{\infty} \mathcal{G}_\lambda^j \left(\frac{1}{\sqrt{\sigma_\lambda}} g \right), \frac{1}{\sqrt{\sigma_\lambda}} h \right)$$

$$= \left(\frac{1}{\sqrt{\sigma_\lambda}} g, \frac{1}{\sqrt{\sigma_\lambda}} h \right) + \left(\sum_{j=0}^{\infty} \mathcal{G}_\lambda^j \mathcal{G}_\lambda \left(\frac{1}{\sqrt{\sigma_\lambda}} g \right), \frac{1}{\sqrt{\sigma_\lambda}} h \right)$$

$$= \left(\frac{1}{\sqrt{\sigma_\lambda}} g, \frac{1}{\sqrt{\sigma_\lambda}} h \right) + \left(\sum_{j=0}^{\infty} \mathcal{G}_\lambda^j \frac{1}{\sqrt{\sigma_\lambda}} \sqrt{\sigma_\lambda} \mathcal{G}_\lambda \left(\frac{1}{\sqrt{\sigma_\lambda}} g \right), \frac{1}{\sqrt{\sigma_\lambda}} h \right),$$

i.e.,

$$\left((\lambda - \mathcal{W})^{-1} g, h \right)$$

$$= \left(\frac{1}{\sqrt{\sigma_\lambda}} g, \frac{1}{\sqrt{\sigma_\lambda}} h \right) + \left((\lambda - \mathcal{W})^{-1} \sqrt{\sigma_\lambda} \mathcal{G}_\lambda \left(\frac{1}{\sqrt{\sigma_\lambda}} g \right), h \right)$$

or

$$\left((\lambda - \mathcal{W})^{-1} \left[g - \sqrt{\sigma_\lambda} \mathcal{G}_\lambda \left(\frac{1}{\sqrt{\sigma_\lambda}} g \right) \right], h \right)$$
$$= \left(\frac{1}{\sqrt{\sigma_\lambda}} g, \frac{1}{\sqrt{\sigma_\lambda}} h \right) = \left(\frac{1}{\sigma_\lambda} g, h \right)$$

hence

$$(\lambda - \mathcal{W})^{-1} \left[g - \sqrt{\sigma_\lambda} \mathcal{G}_\lambda \left(\frac{1}{\sqrt{\sigma_\lambda}} g \right) \right] = \frac{1}{\sigma_\lambda} g.$$

Since $g \in L^2(\mu)$ then $\frac{1}{\sigma_\lambda} g = f \in V$ and therefore for all

$$f \in V \text{ such that } \sigma_\lambda f - \sqrt{\sigma_\lambda} \mathcal{G}_\lambda \left(\sqrt{\sigma_\lambda} f \right) \in L^2(\mu),$$

i.e., for all $f \in D(\mathcal{A})$

$$(\lambda - \mathcal{W})^{-1} \left[\sigma_\lambda f - \sqrt{\sigma_\lambda} \mathcal{G}_\lambda \left(\sqrt{\sigma_\lambda} f \right) \right] = f$$

so

$$\lambda f - \mathcal{W} f = \sigma_\lambda f - \sqrt{\sigma_\lambda} \mathcal{G}_\lambda \left(\sqrt{\sigma_\lambda} f \right) = \mathcal{A} f$$

and $\mathcal{W}$ is an extension of $-\mathcal{A}$. $\square$

7.5.2 Identification of the Generator

The key result of this subsection is

Theorem 7.5.2 *We consider the case (7.5.1). Then $-\mathcal{W}$ is nothing but the Friedrichs extension of the positive symmetric operator $\mathcal{A}$.*

Proof Consider the form on V

$$c(f, g) = \left(\sqrt{\sigma} f, \sqrt{\sigma} g \right) - \left(\mathcal{G}_\lambda \sqrt{\sigma} f, \sqrt{\sigma} g \right)$$

and

$$c_1(f, g) = c(f, g) + (f, g)$$
$$= \left(\sqrt{\sigma} f, \sqrt{\sigma} g \right) - \left(\mathcal{G}_\lambda \sqrt{\sigma} f, \sqrt{\sigma} g \right) + (f, g).$$

We know that

$$c(f, f) = \left\| \sqrt{\sigma} f \right\|^2 - \left(\mathcal{G}_\lambda \sqrt{\sigma} f, \sqrt{\sigma} f \right) \geq 0.$$

We endow V with the scalar product

$$(f, g)_1 = c_1(f, g).$$

The corresponding norm

$$\|f\|_1 = \sqrt{\|f\|^2 + c_1(f, f)}$$

is less strong than the initial norm on V

$$\|f\|_V = \sqrt{\int_\Omega (\sigma(x) + 1)\,|f(x)|^2\,\mu(dx)}.$$

By assumption V is not complete for the norm $\|.\|_1$, (i.e., the form $c(.,.)$ on V is not closed). Let $\widetilde{V}$ be the completion of $(V, \|.\|_1)$. Hence $V \subsetneq \widetilde{V}$ and is dense in $\widetilde{V}$. Since $\|f\|_1 \geq \|f\|$ then

$$\widetilde{V} \subset L^2(\mu).$$

We still denote by $(.,.)_1$ and $\|.\|_1$ the scalar product and norm in $\widetilde{V}$. There exists a unique self-adjoint operator $\mathcal{A}_c$ on $L^2(\mu)$ such that

$$(f, g)_1 = (\mathcal{A}_c f, g),\ f \in D(\mathcal{A}_c),\ g \in \widetilde{V}$$

$$D(\mathcal{A}_c) = \left\{ f \in \widetilde{V};\ \exists \alpha > 0,\ \left|(f, g)_1\right| \leq \alpha\,\|g\|\ \forall g \in \widetilde{V} \right\}.$$

Since $c_1(f, g) = c(f, g) + (f, g)$ and

$$c(f, g) = (\mathcal{A}f, g),\quad f \in D(\mathcal{A}),\ g \in V$$

with

$$D(\mathcal{A}) = \{ f \in V;\ \exists c > 0,\ |c\,(f, g)| \leq c\,\|g\|\ \forall g \in V \}$$

then $D(\mathcal{A}) \subset D(\mathcal{A}_c)$ and

$$\mathcal{A}_c f = \mathcal{A}f + f,\quad f \in D(\mathcal{A}).$$

Hence $\mathcal{A}_c - 1$ is a self-adjoint extension of $\mathcal{A}$. This is the Friedrichs extension of the positive symmetric operator $\mathcal{A}$. As $-\mathcal{A}_c + 1$, $\mathcal{W}$ is also an extension of $-\mathcal{A}$ which generates contraction C_0-semigroup whence $\mathcal{W} = -\mathcal{A}_c + 1$. $\square$

Remark 7.5.1 To sum up, if c is closed then $-\mathcal{A}$ generates a contraction C_0-semigroup in $L^2(\mu)$. If c is not closed then $\mathcal{A}$ is not self-adjoint and it is its *Friedrichs extension* $-\mathcal{W} \supset \mathcal{A}$ which is such that $\mathcal{W}$ generates a contraction C_0-semigroup in $L^2(\mu)$. Thus, in both cases, $\mathcal{W}$ is the $L^2(\mu)$ counterpart of the $L^1(\nu)$ Kato-Voigt generator of the $L^1(\nu)$-theory in Chap. 6.

7.5.3 Essential Self-adjointness in $L^2(\mu)$ Versus L^1 Honesty

In this subsection, we give a sufficient criterion of *essential* self-adjointness of $\mathcal{A}$ in the spirit of $L^1(\nu)$ honesty as characterized in Theorem 2.11.2. To this end, we need P to be σ-bounded in $L^2(\mu)$.

Theorem 7.5.3 *Let (7.5.1) be satisfied. Suppose that*

$$P \text{ is } M_2\text{-bounded in } L^2(\mu). \tag{7.5.5}$$

A sufficient condition for the symmetric operator $\mathcal{A}$ to be essentially self-adjoint (i.e., $\overline{\mathcal{A}}$ is self-adjoint) is given by

$$\left\| \left(P(\lambda + \sigma)^{-1} \right)^n g \right\| \to 0 \ (n \to \infty) \ (g \in L^2(\mu)).$$

Proof We already know that $-\mathcal{A} \subset \mathcal{W}$. We have just to show that any $(g, \mathcal{W}g)$ in the graph of $\mathcal{W}$ can be approximated by elements in the graph of $-\mathcal{A}$. Let $g \in L^2(\mu)$ and

$$g_\lambda = \lambda (\lambda - \mathcal{W})^{-1} g.$$

It is standard that $g_\lambda \to g$ in $L^2(\mu)$ $(\lambda \to +\infty)$ and

$$\mathcal{W}g_\lambda \to \mathcal{W}g \ (\lambda \to +\infty) \text{ if } g \in D(\mathcal{W}).$$

Thus, we can approximate any element $(g, \mathcal{W}g)$ in the graph of $\mathcal{W}$ by elements of this graph of the form $(g_\lambda, \mathcal{W}g_\lambda)$. Hence it suffices to approximate $(g_\lambda, \mathcal{W}g_\lambda)$ by elements in the graph of $-\mathcal{A}$. We have

$$g_\lambda = (\lambda - \mathcal{W})^{-1} (\lambda g) = \frac{1}{\sqrt{\sigma_\lambda}} \sum_{n=0}^{\infty} \mathcal{G}_\lambda^n \left(\frac{1}{\sqrt{\sigma_\lambda}} (\lambda g) \right).$$

Let

$$g_\lambda^n := \frac{1}{\sqrt{\sigma_\lambda}} \sum_{j=0}^{n} \mathcal{G}_\lambda^j \left(\frac{1}{\sqrt{\sigma_\lambda}} (\lambda g) \right).$$

We know that $g_\lambda^n \to g_\lambda \ (n \to \infty)$. Moreover

$$\sqrt{\sigma_\lambda} g_\lambda^n = \sum_{j=0}^{n} \mathcal{G}_\lambda^j \left(\frac{1}{\sqrt{\sigma_\lambda}} (\lambda g) \right)$$

shows that $g_\lambda^n \in V$. We have

$$\sqrt{\sigma_\lambda}g_\lambda^n = \sum_{j=0}^{n}\mathcal{G}_\lambda^j\left(\frac{1}{\sqrt{\sigma_\lambda}}(\lambda g)\right) = \frac{1}{\sqrt{\sigma_\lambda}}(\lambda g) + \sum_{j=1}^{n}\mathcal{G}_\lambda^j\left(\frac{1}{\sqrt{\sigma_\lambda}}(\lambda g)\right)$$

$$\sqrt{\sigma_\lambda}g_\lambda^n - \mathcal{G}_\lambda\sqrt{\sigma_\lambda}g_\lambda^n = -\mathcal{G}_\lambda\sqrt{\sigma_\lambda}g_\lambda^n + \sum_{j=0}^{n}\mathcal{G}_\lambda^j\left(\frac{1}{\sqrt{\sigma_\lambda}}(\lambda g)\right).$$

Since

$$\mathcal{G}_\lambda\sqrt{\sigma_\lambda}g_\lambda^n = \sum_{j=0}^{n}\mathcal{G}_\lambda^{j+1}\left(\frac{1}{\sqrt{\sigma_\lambda}}(\lambda g)\right) = \sum_{j=1}^{n+1}\mathcal{G}_\lambda^j\left(\frac{1}{\sqrt{\sigma_\lambda}}(\lambda g)\right)$$

then

$$-\mathcal{G}_\lambda\sqrt{\sigma_\lambda}g_\lambda^n + \sum_{j=0}^{n}\mathcal{G}_\lambda^j\left(\frac{1}{\sqrt{\sigma_\lambda}}(\lambda g)\right)$$
$$= \frac{1}{\sqrt{\sigma_\lambda}}(\lambda g) - \mathcal{G}_\lambda^{n+1}\left(\frac{1}{\sqrt{\sigma_\lambda}}(\lambda g)\right)$$

so

$$\sqrt{\sigma_\lambda}g_\lambda^n - \mathcal{G}_\lambda\sqrt{\sigma_\lambda}g_\lambda^n = \frac{1}{\sqrt{\sigma_\lambda}}(\lambda g) - \mathcal{G}_\lambda^{n+1}\left(\frac{1}{\sqrt{\sigma_\lambda}}(\lambda g)\right).$$

Since $\lambda g = \lambda g_\lambda - \mathcal{W}g_\lambda$ then

$$\sqrt{\sigma_\lambda}g_\lambda^n - \mathcal{G}_\lambda\sqrt{\sigma_\lambda}g_\lambda^n = \frac{1}{\sqrt{\sigma_\lambda}}(\lambda g_\lambda - \mathcal{W}g_\lambda) - \mathcal{G}_\lambda^{n+1}\left(\frac{1}{\sqrt{\sigma_\lambda}}(\lambda g)\right)$$

or

$$\sigma_\lambda g_\lambda^n - \sqrt{\sigma_\lambda}\mathcal{G}_\lambda\sqrt{\sigma_\lambda}g_\lambda^n - (\lambda g_\lambda - \mathcal{W}g_\lambda) = -\sqrt{\sigma_\lambda}\mathcal{G}_\lambda^{n+1}\left(\frac{1}{\sqrt{\sigma_\lambda}}(\lambda g)\right).$$

Since

$$\mathcal{G}_\lambda\psi = \int_\Omega \frac{p(x,y)}{\omega(y)\sqrt{\sigma_\lambda(x)}\sqrt{\sigma_\lambda(y)}}\psi(y)\mu(dy)$$

then

$$\sqrt{\sigma_\lambda}\mathcal{G}_\lambda\sqrt{\sigma_\lambda}g_\lambda^n = \int_\Omega \frac{p(x,y)}{\omega(y)}g_\lambda^n(y)\mu(dy)$$
$$= \int_\Omega p(x,y)g_\lambda^n(y)\nu(dy) = Pg_\lambda^n.$$

Hence

$$\sigma_\lambda g_\lambda^n - Pg_\lambda^n - (\lambda g_\lambda - \mathcal{W}g_\lambda) = -\sqrt{\sigma_\lambda}\mathcal{G}_\lambda^{n+1}\left(\frac{1}{\sqrt{\sigma_\lambda}}(\lambda g)\right)$$

or

$$\lambda g_\lambda^n + \left(\sigma g_\lambda^n - P g_\lambda^n\right) - \lambda g_\lambda + \mathcal{W} g_\lambda = -\sqrt{\sigma_\lambda} \mathcal{G}_\lambda^{n+1} \left(\frac{1}{\sqrt{\sigma_\lambda}}(\lambda g)\right).$$

According to Proposition 7.5.1

$$g_\lambda^n \in D(\mathcal{A}) \text{ and } \mathcal{A} g_\lambda^n = \sigma g_\lambda^n - P g_\lambda^n.$$

Thus

$$\lambda \left(g_\lambda^n - g_\lambda\right) + \left[\mathcal{A} g_\lambda^n + \mathcal{W} g_\lambda\right] = -\sqrt{\sigma_\lambda} \mathcal{G}_\lambda^{n+1} \left(\frac{1}{\sqrt{\sigma_\lambda}}(\lambda g)\right)$$

or

$$\mathcal{A} g_\lambda^n + \mathcal{W} g_\lambda = -\lambda \left(g_\lambda^n - g_\lambda\right) - \sqrt{\sigma_\lambda} \mathcal{G}_\lambda^{n+1} \left(\frac{1}{\sqrt{\sigma_\lambda}}(\lambda g)\right)$$

and

$$\left\|\mathcal{A} g_\lambda^n + \mathcal{W} g_\lambda\right\| \le \lambda \left\|g_\lambda^n - g_\lambda\right\| + \left\|\sqrt{\sigma_\lambda} \mathcal{G}_\lambda^{n+1} \left(\frac{1}{\sqrt{\sigma_\lambda}}(\lambda g)\right)\right\|.$$

It follows that if

$$\left\|\sqrt{\sigma_\lambda} \mathcal{G}_\lambda^{n+1} \left(\frac{1}{\sqrt{\sigma_\lambda}}(\lambda g)\right)\right\| \to 0 \ (n \to \infty)$$

then

$$(g_\lambda^n, -\mathcal{A} g_\lambda^n) \to (g_\lambda, \mathcal{W} g_\lambda) \ \ (n \to +\infty).$$

Since

$$\sqrt{\sigma_\lambda} \mathcal{G}_\lambda^{n+1} \left(\frac{g}{\sqrt{\sigma_\lambda}}\right)$$

$$= \sqrt{\sigma_\lambda} \mathcal{G}_\lambda \frac{1}{\sqrt{\sigma_\lambda}} \sqrt{\sigma_\lambda} \mathcal{G}_\lambda \frac{1}{\sqrt{\sigma_\lambda}} \sqrt{\sigma_\lambda} \mathcal{G}_\lambda ... \sqrt{\sigma_\lambda} \mathcal{G}_\lambda \left(\frac{g}{\sqrt{\sigma_\lambda}}\right)$$

$$= \left(\sqrt{\sigma_\lambda} \mathcal{G}_\lambda \frac{1}{\sqrt{\sigma_\lambda}}\right) \left(\sqrt{\sigma_\lambda} \mathcal{G}_\lambda \frac{1}{\sqrt{\sigma_\lambda}}\right) ... \sqrt{\sigma_\lambda} \mathcal{G}_\lambda \left(\frac{1}{\sqrt{\sigma_\lambda}}\right) g$$

$$= \left(\sqrt{\sigma_\lambda} \mathcal{G}_\lambda \frac{1}{\sqrt{\sigma_\lambda}}\right)^{n+1} g$$

and

$$\mathcal{G}_\lambda \psi = \int_\Omega \frac{p(x, y)}{\omega(y) \sqrt{\sigma_\lambda(x)} \sqrt{\sigma_\lambda(y)}} \psi(y) \mu(dy) \ \ (\psi \in L^2(\mu))$$

then

$$\sqrt{\sigma_\lambda}\mathcal{G}_\lambda \frac{1}{\sqrt{\sigma_\lambda}}h = \int_\Omega \frac{p(x,y)}{\omega(y)\sqrt{\sigma_\lambda(y)}}\frac{1}{\sqrt{\sigma_\lambda(y)}}h(y)\mu(dy)$$

$$= \int_\Omega \frac{p(x,y)}{\sigma_\lambda(y)}h(y)\nu(dy) = P(\lambda+\sigma)^{-1}h$$

and

$$\sqrt{\sigma_\lambda}\mathcal{G}_\lambda^{n+1}\left(\frac{g}{\sqrt{\sigma_\lambda}}\right) = \left(P(\lambda+\sigma)^{-1}\right)^{n+1}g$$

which ends the proof. $\qquad\square$

Remark 7.5.2 (Open question). We complement Remark 7.5.1 by: We wonder whether the symmetric operator A is not self-adjoint but is essentially self-adjoint if and only if the substochastic C_0-semigroup $(\mathcal{J}_1(t))_{t\geq 0}$ in $L^1(\nu)$ is honest but not genuinely honest.

7.6　On Jump Semigroups in $L^p(\mu)$

On the basis of a detailed balance condition, we have built a general form-perturbation theory in $L^2(\mu)$. We show now, under an additional assumption (see (7.6.1)), how to produce a general theory in $L^1(\mu)$ spaces. (It is possible to deal with other weighted L^1 spaces; this will be done in the specific context of kinetic equations, see Sect. 10.5.) We show also how the $L^1(\mu)$ theory extends to all $L^p(\mu)$ spaces.

7.6.1　$L^1(\mu)$ Theory Versus $L^1(\nu)$ Theory: Similarity Techniques

We start with

Lemma 7.6.1 *Let $\mu(dx) = \omega(x)\nu(dx)$. Then P is M-bounded in $L^1(\mu)$ if and only if*

$$\widehat{C} := \sup_y \frac{\widehat{\sigma}_s(y)}{1+\sigma(y)} < +\infty \qquad (7.6.1)$$

where $\widehat{\sigma}_s(y) = \int_\Omega p(y,x)\nu(dx)$.

Proof To check whether P is σ-bounded in $L^1(\mu)$ where $\mu(dx) = \omega(x)\nu(dx)$, note that

$$\int_\Omega \omega(x)\nu(dx) \int_\Omega \frac{p(x,y)}{\lambda+\sigma(y)}\varphi(y)\nu(dy)$$

$$= \int_\Omega \frac{\left[\int_\Omega \omega(x)p(x,y)\nu(dx)\right]}{\lambda+\sigma(y)}\varphi(y)\nu(dy)$$

$$= \int_\Omega \frac{\left[\int_\Omega \omega(x)p(x,y)\nu(dx)\right]}{(\lambda+\sigma(y))\,\omega(y)}\varphi(y)\omega(y)\nu(dy)$$

so P is σ-bounded in $L^1(\mu)$ if and only if

$$\sup_y \frac{\left[\int_\Omega \omega(x)p(x,y)\nu(dx)\right]}{(\lambda+\sigma(y))\,\omega(y)} < +\infty. \tag{7.6.2}$$

In this case

$$\left\| P(\lambda-\sigma)^{-1} \right\|_{\mathcal{L}(L^1(\mu))} = \sup_y \frac{\left[\int_\Omega \omega(x)p(x,y)\nu(dx)\right]}{(\lambda+\sigma(y))\,\omega(y)}.$$

By using the detailed balance

$$\int_\Omega \omega(x)p(x,y)\nu(dx) = \int_\Omega \omega(y)p(y,x)\nu(dx)$$

$$= \omega(y)\int_\Omega p(y,x)\nu(dx)$$

$$= \omega(y)\widehat{\sigma}_s(y)$$

where

$$\widehat{\sigma}_s(y) = \int_\Omega p(y,x)\nu(dx).$$

Hence (7.6.2) amounts to $\sup_y \frac{\widehat{\sigma}_s(y)}{\lambda+\sigma(y)} < +\infty$. $\qquad\square$

Notice that the condition (7.6.1) depends on σ only and not on the weight ω. The condition (7.6.1) is nothing but the condition introduced in Sect. 6.7 to extend the $L^1(\nu)$ theory to $L^p(\nu)$ via the predual approach. Actually, we have the fundamental observation.

Lemma 7.6.2 *P is M-bounded in $L^1(\mu)$ if and only if P_* is M-bounded in $L^1(\nu)$. The isometry*

$$\mathcal{I} : \varphi \in L^1(\mu) \to \omega\varphi \in L^1(\nu) \tag{7.6.3}$$

is such that $\mathcal{I}P = P_*\mathcal{I}$ or equivalently $P = \mathcal{I}^{-1}P_*\mathcal{I}$.

Proof Note that

$$\|\varphi\|_{L^1(\mu)} = \int_\Omega |\varphi(y)|\,\mu(dy) = \int_\Omega |\omega(y)\varphi(y)|\,\nu(dy) = \|\mathcal{I}\varphi\|_{L^1(\nu)}$$

and

$$\mathcal{I}^{-1} : \psi \in L^1(\nu) \to \omega^{-1}\psi \in L^1(\mu).$$

We have already seen in Sect. 6.7 that (7.6.1) is exactly the condition for P_* to be M-bounded in $L^1(\nu)$. Now

$$\omega(x) \int_\Omega p(x, y)\varphi(y)\nu(dy) = \int_\Omega p(y, x)\omega(y)\varphi(y)\nu(dy)$$

shows that $\mathcal{I}P = P_*\mathcal{I}$. This ends the proof. $\square$

For the simplicity of notations, we still denote by M_1 the multiplication operator by $-\sigma$ in the new space $L^1(\mu)$. We complement Lemma 7.6.2 by

Lemma 7.6.3 *Let (7.6.1) be satisfied. Then*

$$P(\lambda - M_1)^{-1} = \mathcal{I}^{-1} P_*(\lambda - M_1)^{-1}\mathcal{I} \qquad (7.6.4)$$

and

$$\left\| P(\lambda - M_1)^{-1} \right\|_{\mathcal{L}(L^1(\mu))} = \sup_y \frac{\widehat{\sigma}_s(y)}{\lambda + \sigma(y)} = \left\| P_*(\lambda - M_1)^{-1} \right\|_{\mathcal{L}(L^1(\nu))}.$$

Proof It suffices to check

$$\mathcal{I}P(\lambda - M_1)^{-1} = P_*(\lambda - M_1)^{-1}\mathcal{I}.$$

For all $\varphi \in L^1(\mu)$

$$P(\lambda - M_1)^{-1}\varphi = \int_\Omega \frac{p(x, y)}{\lambda + \sigma(y)}\varphi(y)\nu(dy)$$

so

$$\begin{aligned}
\mathcal{I}P(\lambda - M_1)^{-1}\varphi &= \int_\Omega \frac{\omega(x)p(x, y)}{\lambda + \sigma(y)}\varphi(y)\nu(dy) \\
&= \int_\Omega \frac{p(y, x)}{\lambda + \sigma(y)}\omega(y)\varphi(y)\nu(dy) \\
&= P_*(\lambda - M_1)^{-1}\mathcal{I}\varphi.
\end{aligned}$$

Hence

$$\left\| P_*(\lambda - M_1)^{-1}\mathcal{I}\varphi \right\|_{L^1(\nu)} = \left\| \mathcal{I}P(\lambda - M_1)^{-1}\varphi \right\|_{L^1(\nu)} = \left\| P(\lambda - M_1)^{-1}\varphi \right\|_{L^1(\mu)}$$

ends the proof since $\mathcal{I}$ is surjective. $\square$

Hence a theory in $L^1(\mu)$ for

$$U_1 := M_1 + P : D(M_1) \subset L^1(\mu) \to L^1(\mu)$$

turns out to be just a copy (via the similarity transformation (7.6.3)) of a previous construction in $L^1(\nu)$ (see Chap. 6) for

$$M_1 + P_* : D(M_1) \subset L^1(\nu) \to L^1(\nu).$$

Proposition 7.6.1 *Let (7.6.1) be satisfied. Then $r\left(P(\lambda - M_1)^{-1}\right)$ (spectral radius in $L^1(\mu)$) is equal to $r\left(P_*(\lambda - M_1)^{-1}\right)$ (spectral radius in $L^1(\nu)$). In particular*

$$U_1 := M_1 + P : D(M_1) \subset L^1(\mu) \to L^1(\mu) \tag{7.6.5}$$

is a generator of a positive C_0-semigroup $(\mathcal{U}_1(t))_{t\geq 0}$ in $L^1(\mu)$ if and only if

$$M_1 + P_* : D(M_1) \subset L^1(\nu) \to L^1(\nu)$$

is a generator of a positive C_0-semigroup $(\mathcal{J}_{1*}(t))_{t\geq 0}$ in $L^1(\nu)$.

Proof It follows from (7.6.4) that the spectrum of $P(\lambda - M_1)^{-1}$ in $L^1(\mu)$ is the same as that of $P_*(\lambda - M_1)^{-1}$ in $L^1(\nu)$. In particular they share the same spectral radius. Finally Theorem 2.6.1 ends the proof. $\qquad\qquad\qquad\qquad\qquad\qquad\qquad\qquad\qquad$ $\square$

All the criteria of generation of $(\mathcal{J}_{1*}(t))_{t\geq 0}$ in $L^1(\nu)$, given in Chap. 6 , are criteria of generation of $(\mathcal{U}_1(t))_{t\geq 0}$ in $L^1(\mu)$. Similarly, all spectral results concerning $(\mathcal{J}_{1*}(t))_{t\geq 0}$ in $L^1(\nu)$ have their counterparts for the C_0-semigroup $(\mathcal{U}_1(t))_{t\geq 0}$ in $L^1(\mu)$. In particular we can translate Proposition 6.7.2 as

Proposition 7.6.2 *Let (7.6.1) be satisfied. If some power $\left(P_*(\lambda - M_1)^{-1}\right)^n$ is weakly compact in $L^1(\nu)$ then (7.6.5) generates a holomorphic C_0-semigroup $(\mathcal{U}_1(t))_{t\geq 0}$ in $L^1(\mu)$ such that $\omega_{ess}(\mathcal{U}_1) \leq \omega_{ess}(\mathcal{M}_1)$. In particular $(\mathcal{U}_1(t))_{t\geq 0}$ has a spectral gap if $s(U_1) > s(M_1)$; this occurs if $\inf\left(\hat{\sigma}_s - \sigma\right) > -\inf \sigma$.*

On Asymptotic Stability in $L^1(\mu)$

Assume that $\sigma_a = 0$. We have seen in Theorem 6.4.1 that if ω^{-1}, $\sigma\omega^{-1} \in L^1(\nu)$ then (after normalization) ω^{-1} is an invariant density and $(\mathcal{J}_1(t))_{t\geq 0}$ is asymptotically stable in $L^1(\nu)$. (We have also seen in Theorem 7.4.6, under the same assumption, that the C_0-semigroup $\left(\mathcal{U}_{(2)}(t)\right)_{t\geq 0}$ is asymptotically stable in $L^2(\mu)$.) Let the condition (7.6.1) be satisfied. Since $-\sigma\omega^{-1} = P\omega^{-1}$ and P is bounded on $L^1(\mu)$ then $\sigma\omega^{-1} \in L^1(\mu)$ if $\omega^{-1} \in L^1(\mu)$ and consequently $\omega^{-1} \in D(M_1) \subset L^1(\mu)$ is an eigenvector of $M_1 + P$ in $L^1(\mu)$ associated to a zero eigenvalue. A priori, it is not clear that $(\mathcal{U}_1(t))_{t\geq 0}$ is asymptotically stable in $L^1(\mu)$ because we do not know whether $(\mathcal{U}_1(t))_{t\geq 0}$ is uniformly bounded in $L^1(\mu)$. Actually, $(\mathcal{U}_1(t))_{t\geq 0}$ is asymptotically stable in $L^1(\mu)$ if and only if $(\mathcal{J}_{1*}(t))_{t\geq 0}$ is asymptotically stable in $L^1(\nu)$ and we know that in general $(\mathcal{J}_{1*}(t))_{t\geq 0}$ is not contractive in $L^1(\nu)$ unless $\hat{\sigma}_s(.) \leq \sigma(.)$, see Remark 6.7.1. We can state

Theorem 7.6.1 *Let $\sigma_a = 0$ and let the condition (7.6.1) be satisfied. Suppose that U_1 generates a uniformly bounded semigroup $(\mathcal{U}_1(t))_{t\geq0}$ in $L^1(\mu)$ (or $\widehat{\sigma}_s(.) \leq \sigma(.)$ where $\widehat{\sigma}_s(y) = \int_\Omega p(y, x)\nu(dx)$). If $\omega^{-1} \in L^1(\mu)$ then $(\mathcal{U}_1(t))_{t\geq0}$ is asymptotically stable in $L^1(\mu)$.*

Proof We know that $(\mathcal{J}_{1*}(t))_{t\geq0}$ is a sub-stochastic in $L^1(\nu)$ if $\widehat{\sigma}_s(.) \leq \sigma(.)$, see Proposition 6.7.1 (iii) and the C_0-semigroup $(\mathcal{U}_1(t))_{t\geq0}$ in $L^1(\mu)$ is a copy of C_0-semigroup $(\mathcal{J}_{1*}(t))_{t\geq0}$ in $L^1(\nu)$ via the similarity transformation (7.6.3). Since ω^{-1} is an eigenvector of $M_1 + P$ in $L^1(\mu)$ associated to a zero eigenvalue then Theorem 2.10.4 ends the proof. $\qquad\square$

7.6.2   Extension to $L^p(\mu)$. Part I

Even if the $L^1(\mu)$-analysis is just a copy (via a similarity transformation) of the previous predual $L^1(\nu)$ -analysis, their extensions respectively to $L^p(\mu)$ and to $L^p(\nu)$ are not clearly a copy of each other. So it is more convenient to develop a direct construction in $L^p(\mu)$. Actually, because of the "symmetry" of P with respect to the new measure, the analogue of Theorem 4.6.1 has much simpler form and proof.

Proposition 7.6.3 *Let (7.6.1) be satisfied and*

$$\delta := \lim_{\lambda\to+\infty} r\left(P(\lambda - M_1)^{-1}\right) < 1$$

where $r\left(P(\lambda + \sigma)^{-1}\right)$ is the spectral radius in $L^1(\mu)$. Let $\mathcal{U}_1 = (\mathcal{U}_1(t))_{t\geq0}$ be the C_0-semigroup in $L^1(\mu)$ generated by

$$U_1 := M_1 + P; \quad D(U_1) = D(M_1)$$

given by Proposition 7.6.1 and let

$$\Xi_p := \left\{f \in L^p(\mu) \cap D(M_1)); \ M_1 f + Pf \in L^p(\mu)\right\}.$$

Then the unbounded operator in $L^p(\mu)$

$$f \in \Xi_p \to M_1 f + Pf \in L^p(\mu)$$

is closable and its closure generates a C_0-semigroup $\mathcal{U}_p = \left(\mathcal{U}_p(t)\right)_{t\geq0}$ on $L^p(\mu)$ which coincides with $\mathcal{U}_1 = (\mathcal{U}_1(t))_{t\geq0}$ on $L^1(\mu) \cap L^\infty(\mu)$.

Proof *Step 1.* If $f, g \in L^1(\mu) \cap L^\infty(\mu)$ then

$$\langle(\lambda - U_1)^{-1} f, g\rangle = (f, \lambda - U_1)^{-1}g\rangle.$$

Indeed, the resolvent of $U_1 = M_1 + P$ is given by

$$(\lambda - U_1)^{-1} f = \sum_{n=0}^{\infty} (\lambda - M_1)^{-1} \left[P(\lambda - M_1)^{-1} \right]^n f$$

$$= \sum_{n=0}^{\infty} \left[(\lambda - M_1)^{-1} P \right]^n (\lambda - M_1)^{-1} f.$$

Since

$$P(\lambda - M_1)^{-1} f = \int_{\Omega} p(x, y) \frac{f(y)}{\lambda + \sigma(y)} \nu(dy)$$

then

$$\langle P(\lambda - M_1)^{-1} f, g \rangle$$
$$= \int_{\Omega} g(x)\omega(x)\nu(dx) \int_{\Omega} p(x, y) \frac{f(y)}{\lambda + \sigma(y)} \nu(dy)$$
$$= \int_{\Omega} \frac{f(y)}{\lambda + \sigma(y)} \nu(dy) \int_{\Omega} \omega(y) p(y, x) g(x)\nu(dx)$$
$$= \int_{\Omega} f(y) \left(\frac{1}{\lambda + \sigma(y)} \int_{\Omega} p(y, x) g(x)\nu(dx) \right) \omega(y)\nu(dy)$$

shows that

$$\left(P(\lambda - M_1)^{-1} \right)^* g = \frac{1}{\lambda + \sigma(y)} \int_{\Omega} p(y, x) g(x)\nu(dx)$$
$$= (\lambda - M_1)^{-1} P g$$

so

$$\langle (\lambda - M_1)^{-1} \left[P(\lambda - M_1)^{-1} \right]^n f, g \rangle$$
$$= \langle \left[P(\lambda - M_1)^{-1} \right]^n f, (\lambda - M_1)^{-1} g \rangle$$
$$= \langle f, \left((\lambda - M_1)^{-1} P \right)^n (\lambda - M_1)^{-1} g \rangle$$

and

$$\langle (\lambda - U_1)^{-1} f, g \rangle = \langle \sum_{n=0}^{\infty} (\lambda - M_1)^{-1} \left[P(\lambda - M_1)^{-1} \right]^n f, g \rangle$$
$$= \langle f, \sum_{n=0}^{\infty} \left((\lambda - M_1)^{-1} P \right)^n (\lambda - M_1)^{-1} g \rangle.$$

Hence

$$(\lambda - (U_1)^*)^{-1} g = \sum_{n=0}^{\infty} \left((\lambda - M_1)^{-1} P \right)^n (\lambda - M_1)^{-1} g$$
$$= \sum_{n=0}^{\infty} (\lambda - M_1)^{-1} \left[P(\lambda - M_1)^{-1} \right]^n g$$
$$= (\lambda - U_1)^{-1} g.$$

It follows that

$$\langle \mathcal{U}_1(t) f, g \rangle = (f, \mathcal{U}_1(t) g), \quad f, g \in L^1(\mu) \cap L^{\infty}(\mu).$$

Step 2. It follows that $(\mathcal{U}_1(t))_{t \geq 0}$ interpolates to all $L^p(\mu)$ spaces ($1 \leq p \leq +\infty$). We denote it by $\mathcal{U}_p = (\mathcal{U}_p(t))_{t \geq 0}$ on $L^p(\mu)$. In particular, the semigroup property is preserved

$$\mathcal{U}_p(t + s) = \mathcal{U}_p(t) \mathcal{U}_p(s).$$

The convexity property shows that

$$\left\| \mathcal{U}_p(t) \right\|_{\mathcal{L}(L^p(\mu))} \leq \left\| \mathcal{U}_1(t) \right\|_{\mathcal{L}(L^1(\mu))}^{\frac{1}{p}} \left\| \mathcal{U}_1(t) \right\|_{\mathcal{L}(L^{\infty}(\mu))}^{1 - \frac{1}{p}} = \left\| \mathcal{U}_1(t) \right\|_{\mathcal{L}(L^1(\mu))} \qquad (7.6.6)$$

and

$$\left\| \mathcal{U}_p(t) f - f \right\|_{L^p(\mu)} \leq \left\| \mathcal{U}_1(t) f - f \right\|_{L^1(\mu)} \left\| \mathcal{U}_1(t) \right\| f - f \right\|_{L^{\infty}(\mu)}.$$

Hence, for $1 < p < +\infty$, the strong continuity of $(\mathcal{U}_p(t))_{t \geq 0}$ in $L^p(\mu)$ is inherited from the strong continuity of $(\mathcal{U}_1(t))_{t \geq 0}$ in $L^1(\mu)$. Note that $\mathcal{U}_2 = (\mathcal{U}_2(t))_{t \geq 0}$ is self-adjoint on $L^2(\mu)$. We denote by

$$U_p : D(U_p) \subset L^p(\mu) \to L^p(\mu)$$

the generator of $(\mathcal{U}_p(t))_{t \geq 0}$. The estimate (7.6.6) shows that the type of $(\mathcal{U}_p(t))_{t \geq 0}$ (or equivalently the spectral bound of its generator) is less than or equal to the type of $(\mathcal{U}_1(t)))_{t \geq 0}$ or equivalently the spectral bound $s(U_1)$ of its generator U_1. The fact that

$$\mathcal{U}_p(t) g = \mathcal{U}_1(t) g, \quad g \in L^1(\mu) \cap L^p(\mu)$$

shows that

$$(\lambda - M_1 - P)^{-1} g = (\lambda - U_p)^{-1} g, \quad g \in L^1(\mu) \cap L^p(\mu).$$

Let

$$f := (\lambda - U_p)^{-1} g, \quad g \in L^p(\mu);$$

then $f \in D(U_p)$ and $\lambda f - U_p f = g$. Let

$$(g_n)_n \subset L^1(\mu) \cap L^p(\mu), \quad g_n \to g \text{ in } L^p(\mu)$$

and

$$f_n := \left(\lambda - U_p\right)^{-1} g_n = (\lambda - M_1 - P)^{-1} g_n.$$

Then $f_n \to \left(\lambda - U_p\right)^{-1} g$ in the graph norm of U_p, i.e.,

$$f_n \to f, \quad U_p f_n \to U_p f.$$

Since

$$\lambda f_n - U_p f_n = \lambda f_n - M_1 f_n - P f_n = g_n \to g \text{ in } L^p(\mu)$$

then $M_1 f_n + P f_n \to U_p f$ in $L^p(\mu)$. Let us introducce

$$\Xi_p := \left\{ f \in L^p(\mu) \cap D(M_1); \ M_1 f + P f \in L^p(\mu) \right\}.$$

Hence the unbounded operator in $L^p(\mu)$

$$f \in \Xi_p \to M_1 f + P f \in L^p(\mu)$$

is closable and its closure is nothing but U_p. $\qquad\qquad\square$

Under a probably unnecessary technical condition (see (7.6.7)), we can show that the C_0-semigroup $\mathcal{U}_2 = (\mathcal{U}_2(t))_{t\geq 0}$ on $L^2(\mu)$ coincides with the C_0-semigroup $\mathcal{U}_{(2)} = (\mathcal{U}_{(2)}(t))_{t\geq 0}$ built in Theorem 7.4.2.

Proposition 7.6.4 *Let (7.6.1) be satisfied and let*

$$\delta := \lim_{\lambda \to +\infty} r \left(P(\lambda - M_1)^{-1} \right) < 1$$

where $r \left(P(\lambda + \sigma)^{-1} \right)$ is the spectral radius in $L^1(\mu)$. Let the condition

$$\sigma \omega^{-1} \in L^\infty(\nu) \tag{7.6.7}$$

be satisfied. Then for any $\varepsilon > 0$, there exists $C_\varepsilon > 0$ such that

$$\int_\Omega (Pf)\overline{f} \, d\mu \leq (\delta + \varepsilon) \int_\Omega \sigma \, |f|^2 \, d\mu + C_\varepsilon \, \|f\|^2_{L^2(\mu)} \ (f \in V).$$

In particular, the form

$$c(f, g) = \int_\Omega \sigma f \overline{g} \, d\mu - \int_\Omega (Pf)\overline{g} \, d\mu \tag{7.6.8}$$

is closed and the self-adjoint operator associated to (7.6.8) is nothing but (minus) the generator of the self-adjoint semigroup given by Proposition 7.6.3.

Proof Let $c > 0$ such that $c\delta < 1$ and let $P_c := cP$. Then

$$\lim_{\lambda \to +\infty} r\left(P_c(\lambda - M_1)^{-1}\right) = c\delta < 1$$

and

$$M_1 + P_c : D(M_1) \subset L^1(\mu) \to L^1(\mu)$$

is a generator. The same arguments as in the proof of Proposition 7.6.3 show that with

$$\Xi_2^c := \left\{ f \in L^2(\mu) \cap D(M_1); \ M_1 f + P_c f \in L^2(\mu) \right\}$$

then unbounded operator in $L^2(\mu)$

$$f \in \Xi_2^c \to M_1 f + P_c f \in L^2(\mu)$$

is closable and its closure generates a self-adjoint C_0-semigroup and the spectral bound of its (self-adjoint) generator is less than or equal to $s(M_1 + P_c)$ (the spectral bound of $M_1 + P_c$ in $L^1(\mu)$). Hence

$$\int_\Omega -\sigma |f|^2 \, d\mu + (P_c f, f) \le s(M_1 + P_c) \|f\|^2_{L^2(\mu)} \ (f \in \Xi_2^c),$$

i.e.,

$$\int_\Omega -\sigma |f|^2 \, d\mu + c \, (Pf, f) \le s(M_1 + P_c) \|f\|^2_{L^2(\mu)} \ (f \in \Xi_2^c)$$

so

$$(Pf, f) \le c^{-1} \int_\Omega \sigma |f|^2 \, d\mu + c^{-1} s(M_1 + P_c) \|f\|^2_{L^2(\mu)} \ (f \in \Xi_2^c).$$

Let us extend this estimate to all $f \in V$. Note first that $|(Pf, f)| \le (P\,|f|, |f|)$ so it suffices to restrict ourselves to nonnegative f. Let $(\Omega_n)_n$ a nondecreasing sequence of meurable sets such that $\cup \Omega_n = \Omega$ and $\nu(\Omega_n) < +\infty$. Let $f \in V$ and

$$f_n^m = 1_{\{\sigma \le n\}} 1_{\Omega_n} 1_{\{\omega \le n\}} f.$$

We note first that $f_n \in V$. Let us show that $f_n \in L^1(\mu)$. We have

$$\int_\Omega |f_n(x)|\,\mu(dx)$$

$$= \int_{\Omega_n \cap \{\omega \le n\}} |f(x)|\,\omega(x)\nu(dx)$$

$$\le \sqrt{\int_{\Omega_n \cap \{\omega \le n\}} \omega(x)\nu(dx)}\sqrt{\int_{\Omega_n \cap \{\omega \le n\}} |f_n(x)|^2\,\omega(x)\nu(dx)}$$

$$\le \sqrt{\int_{\Omega_n \cap \{\omega \le n\}} \omega(x)\nu(dx)}\,\,\|f\|^2_{L^2(\mu)}.$$

Finally $\sigma f_n \in L^1(\mu)$. Hence $f_n \in D(M_1)$ in $L^1(\mu) \cap D(M_1)$. Let us check that

$$Pf_n = \int_\Omega p(x,y)f_n(y)\nu(dy) = \int_{\Omega_n \cap \{\omega \le n\}} p(x,y)f_n(y)\nu(dy)$$

belongs to $L^2(\mu)$. We have

$$\int_\Omega \left| \int_{\Omega_n \cap \{\omega \le n\}} p(x,y)f_n(y)\nu(dy) \right|^2 \omega(x)\nu(dx)$$

$$\le \int_\Omega \left(\int_{\{\Omega_n \cap \{\omega \le n\}\}} p(x,y)\nu(dy) \right) \left(\int_{\Omega_n \cap \{\omega \le n\}} p(x,y)\,|f_n(y)|^2\,\nu(dy) \right) \omega(x)\nu(dx).$$

The detailed balance gives

$$\int_{\{\Omega_n \cap \{\omega \le n\}\}} p(x,y)\nu(dy) = \int_{\{\Omega_n \cap \{\omega \le n\}\}} \frac{\omega(y)}{\omega(x)} p(y,x)\nu(dy) \le n\frac{\sigma(x)}{\omega(x)} \in L^\infty(\nu)$$

by assumption (7.6.7) so, with $C = n\left\|\frac{\sigma}{\omega}\right\|_{L^\infty}$

$$\int_\Omega \left| \int_{\Omega_n \cap \{\omega \le n\}} p(x,y)f_n(y)\nu(dy) \right|^2 \omega(x)\nu(dx)$$

$$\le C \int_\Omega \left(\int_\Omega p(x,y)\,|f_n(y)|^2\,\nu(dy) \right) \omega(x)\nu(dx)$$

$$= C \int_\Omega \left(\int_\Omega \omega(x)p(x,y)\,|f_n(y)|^2\,\nu(dy) \right) \nu(dx)$$

$$= C \int_\Omega \left(\int_\Omega \omega(y)p(y,x)\,|f_n(y)|^2\,\nu(dy) \right) \nu(dx)$$

$$= C \left(\int_\Omega \omega(y) \left[\int_\Omega p(y,x)\nu(dx) \right] |f_n(y)|^2\,\nu(dy) \right)$$

$$= C \left(\int_\Omega \omega(y)\widehat{\sigma}_s(y)\,|f_n(y)|^2\,\nu(dy) \right)$$

$$= C \left(\int_\Omega \widehat{\sigma}_s(y)\, |f_n(y)|^2\, \mu(dy) \right).$$

According to (7.6.1) $\widehat{\sigma}_s(y) \leq \widehat{C}(1 + \sigma(y))$ hence

$$\int_\Omega \left| \int_{\Omega_n \cap \{\omega \leq n\}} p(x, y)\, f_n(y)\nu(dy) \right|^2 \omega(x)\nu(dx)$$

$$\leq C\widehat{C} \left(\int_\Omega (1 + \sigma(y))\, |f_n(y)|^2\, \mu(dy) \right) < +\infty$$

since $f_n \in V$. Finally

$$f_n \in L^2(\mu) \cap D(A_1) \text{ and } -\sigma f_n + P f_n \in L^2(\mu)$$

so $f_n \in \Xi_2^c$ and

$$(P f_n, f_n) \leq c^{-1} \int_\Omega \sigma\, |f_n|^2\, d\mu + c^{-1} s(-\sigma + P_c)\, \|f_n\|^2_{L^2(\mu)}.$$

Since $f_n \to f$ in $L^2(\mu)$ and monotonically then, using the monotone convergence theorem we get

$$(P f, f) \leq c^{-1} \int_\Omega \sigma\, |f|^2\, d\mu + c^{-1} s(-\sigma + P_c)\, \|f\|^2_{L^2(\mu)} \quad \forall f \in V.$$

Our constraint on c (i.e., $c\delta < 1$) allows c^{-1} to be as close to δ as we want, i.e., $c^{-1} = \delta + \varepsilon$ and

$$C_\varepsilon = (\delta + \varepsilon)\, s(-\sigma + (\delta + \varepsilon)^{-1}\, P).$$

This shows that the form-bound of $\int_\Omega (Pf)\, \overline{f}\, d\mu$ relative to $\int_\Omega \sigma f\overline{g}\, d\mu$ is $\leq \delta$. Finally, this explains why the form

$$c(f, g) = \int_\Omega \sigma f\overline{g}\, d\mu - \int_\Omega (Pf)\, \overline{f}\, d\mu$$

is closed on V since $\delta < 1$.

Let us check that the self-adjoint semigroup $\mathcal{U}_2 = (\mathcal{U}_2(t))_{t \geq 0}$ (given by Proposition 7.6.3) coincides with the semigroup obtained by form perturbation theory. We have for $g \in L^1(\mu) \cap L^\infty(\mu)$

$$f = (\lambda - U_2)^{-1} g = \sum_{n=0}^{\infty} (\lambda - M_2)^{-1} \left[P(\lambda - M_2)^{-1} \right]^n g$$

where the convergence holds in all $L^p(\mu)$. The resolvent obtained directely in $L^2(\mu)$ by form-perturbation theory is

$$f = (\lambda + \mathcal{A})^{-1} g = \frac{1}{\sqrt{\sigma_\lambda}} \sum_{n=0}^{\infty} \mathcal{G}_\lambda^n \left(\frac{1}{\sqrt{\sigma_\lambda}} g \right)$$

where the convergence holds in $L^2(\mu)$. Since

$$\mathcal{G}_\lambda : \psi \in L^2(\mu) \to \int_\Omega \frac{p(x,y)}{w(y)\sqrt{\sigma_\lambda(x)}\sqrt{\sigma_\lambda(y)}} \psi(y)\mu(dy) \in L^2(\mu)$$

then

$$\mathcal{G}_\lambda \frac{1}{\sqrt{\sigma_\lambda}} g = \int_\Omega \frac{p(x,y)}{w(y)\sqrt{\sigma_\lambda(x)}\sqrt{\sigma_\lambda(y)}} \psi(y)\mu(dy) = \int_\Omega \frac{p(x,y)}{\sqrt{\sigma_\lambda(x)}} \frac{g(y)}{\sigma_\lambda(y)} \nu(dy),$$

i.e.,

$$\sqrt{\sigma_\lambda}\mathcal{G}_\lambda \frac{1}{\sqrt{\sigma_\lambda}} g = P(\lambda + \sigma)^{-1} g$$

and

$$\left(\sqrt{\sigma_\lambda}\mathcal{G}_\lambda \frac{1}{\sqrt{\sigma_\lambda}} \right)^n g = \left(P(\lambda + \sigma)^{-1} \right)^n g.$$

We have seen in the proof of Theorem 7.5.3 that

$$\sqrt{\sigma_\lambda}\mathcal{G}_\lambda^n \left(\frac{g}{\sqrt{\sigma_\lambda}} \right) = \left(\sqrt{\sigma_\lambda}\mathcal{G}_\lambda \frac{1}{\sqrt{\sigma_\lambda}} \right)^n g$$

so

$$\sqrt{\sigma_\lambda}\mathcal{G}_\lambda^n \left(\frac{g}{\sqrt{\sigma_\lambda}} \right) = \left(P(\lambda + \sigma)^{-1} \right)^n g$$

and

$$\frac{1}{\sqrt{\sigma_\lambda}}\mathcal{G}_\lambda^n \left(\frac{g}{\sqrt{\sigma_\lambda}} \right)$$
$$= \frac{1}{\sigma_\lambda} \left(P(\lambda + \sigma)^{-1} \right)^n g = (\lambda + \sigma)^{-1} \left(P(\lambda + \sigma)^{-1} \right)^n g.$$

Hence for all $m \in \mathbb{N}$

$$\frac{1}{\sqrt{\sigma_\lambda}} \sum_{n=0}^{m} \mathcal{G}_\lambda^n \left(\frac{1}{\sqrt{\sigma_\lambda}} g\right)$$

$$= (\lambda + \sigma)^{-1} \sum_{n=0}^{m} \left(P(\lambda + \sigma)^{-1}\right)^n g, \quad \left(g \in L^1(\mu) \cap L^2(\mu)\right).$$

Since $(\lambda + \sigma)^{-1} = (\lambda - M_2)^{-1}$ then

$$f = (\lambda + \mathcal{A})^{-1} g = (\lambda - U_2)^{-1} g, \quad \left(g \in L^1(\mu) \cap L^2(\mu)\right)$$

and $(\lambda + \mathcal{A})^{-1} = (\lambda - U_2)^{-1}$ because $L^1(\mu) \cap L^2(\mu)$ is dense in $L^2(\mu)$. It follows that U_2 is nothing but $-\mathcal{A}$. $\square$

p-Independence of Asymptotic Spectra

Because of the "symmetry" of P with respect to the new measure, the analogue of Theorem 4.6.3 has now a much more transparent form.

Theorem 7.6.2 *Let (7.6.1) be satisfied. Let $s_{ess}(U_1) < s(U_1)$. Then $\mathcal{U}_p = \left(\mathcal{U}_p(t)\right)_{t \geq 0}$ has a spectral gap on $L^p(\mu)$ for all $p \in [1, +\infty)$ and*

$$\sigma\left(\mathcal{U}_p(t)\right) \cap \left\{\mu; \ |\mu| > e^{s_{ess}(U_1)t}\right\}$$

is p-independent and consists of isolated eigenvalues with finite (p -independent) algebraic multiplicities.

Proof The proof is similar to that of Theorem 4.6.3. Indeed, we have

$$s_{ess}((U_1)^*) = s_{ess}(U_1) \text{ and } s((U_1)^*) = s(U_1).$$

On the other hand, we have seen in the proof of Proposition 7.6.3 that $(\lambda - (U_1)^*)^{-1}$ coincides with $(\lambda - U_1)^{-1}$ on $L^1(\mu) \cap L^2(\mu)$. Finally, Hemple and Voigt [160, Corollary 1.4 and Remark 1.5 (a)] end the proof. $\square$

We keep the same notation (as in $L^p(\nu)$) for the multiplication operator on $L^p(\mu)$

$$\begin{cases} M_p : \varphi \in D(M_p) \subset L^p(\mu) \to -\sigma\varphi \in L^p(\mu) \\ \quad D(M_p) = \{\varphi \in L^p(\mu); \ \sigma\varphi \in L^p(\mu)\} \end{cases}$$

which generates the analytic C_0-(multiplication) semigroup $\left(\mathcal{M}_p(t)\right)_{t \geq 0}$ in $L^p(\mu)$ given by

$$\mathcal{M}_p(t)\varphi = e^{-t\sigma}\varphi \ (t \geq 0).$$

We note that the (essential) spectrum of M_p or $\mathcal{M}_p(t)$ is the same in the new space. We complement Theorem 7.6.2 by

Proposition 7.6.5 *Let (7.6.1) be satisfied. Then P is M_1-weakly compact in $L^1(\mu)$ if and only if P_* is M_1-weakly compact in $L^1(\nu)$. In this case*
 (i) $\mathcal{U}_1(t) - \mathcal{M}_1(t)$ is weakly compact in $L^1(\mu)$
 (ii) $\mathcal{U}_p(t)) - \mathcal{M}_p(t)$ is compact in $L^p(\mu)$ $(1 < p < +\infty)$. In particular $\left(\mathcal{U}_p(t)\right)_{t\geq 0}$ and $\left(\mathcal{M}_p(t)\right)_{t\geq 0}$ have the same essential spectrum.

Proof The similarity property (7.6.4)

$$P(\lambda - M_1)^{-1} = \mathcal{I}^{-1} P_*(\lambda - M_1)^{-1}\mathcal{I}$$

shows that the operator $P(\lambda - M_1)^{-1}$ in $L^1(\mu)$ and the operator $P_*(\lambda - M_1)^{-1}$ in $L^1(\nu)$ share the same properties, in particular the weak compactness property. In this case,

$$M_1 + P : D(M_1) \subset L^1(\mu) \to L^1(\mu)$$

generates $(\mathcal{U}_1(t))_{t\geq 0}$. The similarity property

$$\mathcal{U}_1(t) - \mathcal{M}_1(t) = \mathcal{I}^{-1}\left(\mathcal{J}_{1*}(t) - \mathcal{M}_1(t)\right)\mathcal{I}$$

and the weak compactness of $\mathcal{J}_{1*}(t) - \mathcal{M}_1(t)$ in $L^1(\nu)$ (see Remark 6.7.2) shows (i).

(ii) Since $\mathcal{U}_1(t) - \mathcal{M}_1(t)$ is weakly compact in $L^1(\mu)$ (or equivalently (see [300]) stricly singular in $L^1(\mu)$) then it follows from [161, Theorem 4.2] that $\mathcal{U}_p(t)) - \mathcal{M}_p(t)$ is compact in $L^p(\mu)$ $(1 < p < +\infty)$. $\square$

7.6.3 *Extension to $L^p(\mu)$. Part II*

In the above $L^p(\mu)$ construction from $L^1(\mu)$ theory, the operator P is M_1-bounded in $L^1(\mu)$ but need not be M_p-bounded in $L^p(\mu)$. We explore now this situation. We start with

Lemma 7.6.4 *Let (7.6.1) be satisfied. If*

$$\sup_{x\in\Omega} \int_\Omega \frac{p(x, y)}{\lambda + \sigma(y)}\nu(dy) < +\infty \; (\lambda > 0) \tag{7.6.9}$$

then P is M_p-bounded in all $L^p(\mu)$ $(1 \leq p \leq +\infty)$.

Proof Since μ and ν have the same negligeable sets then $L^\infty(\mu) = L^\infty(\nu)$ and

$$P(\lambda - M_1)^{-1}1 = \int_\Omega \frac{p(x, y)}{\lambda + \sigma(y)}\nu(dy) \in L^\infty(\mu)$$

and consequently Riesz-Thorin theorem ends the proof. $\square$

Note that (7.6.9) is nothing but the condition (6.8.1) which insures that P_* is M_p-bounded in all $L^p(\nu)$ $(1 \leq p \leq +\infty)$. We are ready to state

Theorem 7.6.3 *Suppose that $(\mathcal{U}_1(t))_{t \geq 0}$ is genuinely honest and*

$$\sup_{x \in \Omega} \int_\Omega \frac{p(x, y)}{\lambda + \sigma(y)} \nu(dy) < +\infty \ (\lambda > 0). \tag{7.6.10}$$

Let $r\left(P(\lambda - M_p)^{-1}\right)$ be a spectral radius in $L^p(\mu)$. Then

$$r\left(P(\lambda - M_p)^{-1}\right) \leq \left(r_1\left(P(\lambda - M_1)^{-1}\right)\right)^{\frac{1}{p}} \left(r\left(P(\lambda - M_\infty)^{-1}\right)\right)^{1 - \frac{1}{p}}.$$

In particular

$$U_p := M_p + P : D(M_p) \subset L^p(\mu) \to L^p(\mu) \tag{7.6.11}$$

generates positive holomorphic semigroup $(\mathcal{U}_p(t))_{t \geq 0}$ in $L^p(\mu)$ for p close to 1. If

$$\exists \lambda \geq 0; \ \sup_{x \in \Omega} \int_\Omega \frac{p(x, y)}{\lambda + \sigma(y)} \nu(dy) \leq 1 \tag{7.6.12}$$

then (7.6.11) generates a positive holomorphic semigroup $(\mathcal{U}_p(t))_{t \geq 0}$ in all $L^p(\mu)$ $(1 \leq p < +\infty)$.

Here is the analogue of Theorem 4.6.7 in $L^p(\mu)$ spaces

Theorem 7.6.4 *Let (7.6.1) be satisfied and let some power $\left(P(\lambda - M_1)^{-1}\right)^n$ be compact in $L^1(\mu)$. Then:*

(i) (7.6.11) generates a positive holomorphic C_0 -semigroup $(\mathcal{U}_p(t))_{t \geq 0}$ in all $L^p(\mu)$ $(1 \leq p < +\infty)$.

(ii) $(\mathcal{U}_p(t))_{t \geq 0}$ has a spectral gap in all $L^p(\mu)$ $(1 \leq p < +\infty)$ if and only if $(\mathcal{U}_1(t))_{t \geq 0}$ has a spectral gap in $L^1(\mu)$. In this case

$$\sigma\left(\mathcal{U}_p(t)\right) \cap \left\{\mu; \ |\mu| > e^{s_{ess}(M_p)t}\right\}$$

is p-independent and consists of isolated eigenvalues with finite (p -independent) algebraic multiplicities.

7.7 Comments

Several technical open questions remain in this chapter. We list them below.

The construction presented in this chapter could likely be adapted to incorporate, alongside an inelastic collision operator, an elastic operator that exhibits a lack of compactness (see Sect. 10.6). Such operators arise in solid moderators (poly-crystals) [65]. In particular, we expect that Theorem 7.4.4, which addresses quasi-compactness of G_λ, can be suitably extended to cover these cases.

In Subsect. 7.4.3, the condition (7.4.3) is assumed in order to investigate the connection between the $L^2(\mu)$-theory and $L^1(\nu)$-theory; see, for instance, Remark 7.4.1. We have seen that the hilbertian theory, based on the detailed balance condition, provides valuable insights into the $L^1(\nu)$ theory. However, a natural question arises: what becomes of this connection when the condition (7.4.3) is not satisfied?

In connection with Theorem 7.4.6, several natural questions arise when $\sigma_a = 0$ and (7.4.3) is not satisfied:

(i) Is $s(-A) = 0$ still true ?

(ii) If $s(-A) = 0$, can we assert that 0 is not an eigenvalue of A? If so then $\left(\mathcal{U}_{(2)}(t)\right)_{t \geq 0}$ is strongly stable in $L^2(\mu)$, (i.e., converges strongly to zero as $t \to +\infty$) by appealing e.g. to Theorem 2.10.4. In this case, what about $(\mathcal{J}_1(t))_{t \geq 0}$ in $L^1(\nu)$?

See also the open question given in Remark 7.5.2 which complements Remark 7.5.1.

In connection with Theorem 7.6.1, Is it possible to show that $\omega^{-1} \in L^1(\mu)$ implies that U_1 generates a uniformly bounded semigroup $(\mathcal{U}_1(t))_{t \geq 0}$ in $L^1(\mu)$?

We have seen in Theorem 7.4.1 that the sesquilinear form c given by (7.3.2), at the heart of the hilbertian theory, is closed on V if and only if

$$\lim_{\lambda \to +\infty} \|G_\lambda\|_{\mathcal{L}(L^2(\nu))} < 1$$

is satisfied. We have also seen that this condition is satisfied if G_λ is at least quasi-compact. This suggests that the borderline case where the sesquilinear form (7.3.2) is not closed is strongly connected to operators G_λ whose spectral radius is not an isolated eigenvalue. This observation is consistent with the $L^1(\nu)$ theory where, under the assumption that $P(0 - M_1)^{-1}$ is quasi compact,

$$M_1 + P : D(M_1) \subset L^1(\nu) \to L^1(\nu)$$

generates a stochastic genuinely honest C_0-semigroup $(\mathcal{J}_1(t))_{t \geq 0}$ (see e.g. Theorem 6.6.1) while, in general we have just an extension of $M_1 + P$ which generates a substochastic C_0-semigroup in $L^1(\nu)$ according to Kato-Voigt Theorem 2.11.1.

In this chapter, we have focused on a natural Hilbert space framework motivated by the detailed balance condition. However, in certain concrete problems, alternative choices of weighted spaces may be more appropriate. For instance, see Sect. 10.5 for a discussion on the significance of suitably chosen L^1-weighted spaces in the study of space-homogeneous linear Boltzmann equations.

Chapter 8
Perturbed Frobenius-Perron Semigroups

8.1 Chapter Aims

This chapter deals with the analysis of a C_0-semigroup $(V(t))_{t\geq 0}$ on $L^1(\Xi, \nu)$ generated by $T_\Sigma + K(\Sigma)$ where Σ is a nonnegative measurable function on a measure space (Ξ, ν), K is a substochastic operator on $L^1(\Xi, \nu)$, and T_Σ is the generator of the weighted Frobenius-Perron C_0-semigroup $(U_\Sigma(t))_{t\geq 0}$ defined by

$$(U_\Sigma(t)f)(x) = e^{-\int_0^t \Sigma(\pi_{-s}(x))ds} f(\pi_{-t}(x)),$$

and $\pi_t : \Xi \to \Xi$ $(t \in \mathbb{R})$ is a bijective, measure-preserving flow on Ξ. We address spectral gap analysis, asymptotic stability or sweeping of $(V(t))_{t\geq 0}$ in the case where K is stochastic and weakly compact. We also consider the scattering theory regime. This chapter is motivated by the theory of particle swarms in a weakly ionized gas and in semiconductor theory.

8.2 Chapter Overview

In Chap. 3, we investigated various time asymptotic regimes for an abstract class of Kato–Voigt semigroups. The aim here is to revisit that analysis in the specific setting where the unperturbed semigroup is a weighted Frobenius–Perron semigroup. This case is of particular significance in Transport Theory, especially in the context of particle swarm models, which we explore in Chap. 10.

Here is our general framework: Let $(\Xi, \mathcal{E}, \nu)$ be a sigma-finite measure space and let π be a measurable autonomous flow on Ξ, i.e.,

$$\pi : (-\infty, +\infty) \times \Xi \to \Xi$$

M. Mokhtar-Kharroubi, *Peripheral Spectra of Perturbed Positive Semigroups*, Lecture Notes in Mathematics 2388, https://doi.org/10.1007/978-3-032-11173-9_8

such that the transformations

$$\pi_t : \Xi \to \Xi \ \ (t \in \mathbb{R})$$

are bijective, measure-preserving and satisfy the group property

$$\pi_0(x) = x, \ \ \pi_{t+s}(x) = \pi_t(\pi_s(x)), \ \ t, s \in \mathbb{R}, \ x \in \Xi.$$

Let

$$U(t) : L^1(\Xi, \nu) \to L^1(\Xi, \nu)$$

be the *Frobenius-Perron* operatorassociated to the measure-preserving transformation π_t, i.e.,

$$(U(t)f)(x) = f(\pi_{-t}(x)).$$

It is well known (see e.g. [194] Theorem 3.2.1) that $(U(t))_{t \in \mathbb{R}}$ is a C_0-group on $L^1(\nu)$ and is stochastic.

Definition 8.2.1 We refer to the stochastic C_0-group $(U(t))_{t \in \mathbb{R}}$ on $L^1(\nu)$ as the *Frobenius-Perron C_0-group.*

We denote by T the generator of $(U(t))_{t \in \mathbb{R}}$. Let

$$\Sigma : (\Xi, \mathcal{E}, \ \nu) \to [0, +\infty]$$

be measurable and finite a.e. We consider the weighted Frobenius-Perron C_0-semigroup $(U_\Sigma(t))_{t \geq 0}$,defined by

$$(U_\Sigma(t)f)(x) = e^{-\int_0^t \Sigma(\pi_{-s}(x))ds} f(\pi_{-t}(x)),$$

(see e.g. [320, 367]) which is nothing but an absorption semigroup (see Chap. 5) associated to the absorption Σ. We denote by T_Σ the generator of $(U_\Sigma(t))_{t \geq 0}$.
Since $(U(t))_{t \geq 0}$ is a stochastic C_0-semigroup then, according to Lemma 5.3.2, Σ is U-regular (in the sense of Definition 5.3.2) if and only if

$$D(T) \cap D(\Sigma) \text{ is a core of } T. \tag{8.2.1}$$

We assume for the sequel that (8.2.1) is satisfied. It follows (see Remark 5.3.1) that

$$\begin{cases} T_\Sigma = T - \Sigma \\ D(T_\Sigma) = D(T) \cap D(\Sigma). \end{cases}$$

Let $K : L^1(\Xi, \nu) \to L^1(\Xi, \nu)$ be a substochastic operator and

$$P : D(\Sigma) \to L^1(\Xi, \nu)$$

be defined by

$$P\varphi = K(\Sigma\varphi), \quad \varphi \in D(\Sigma).\tag{8.2.2}$$

Since $(U(t))_{t\geq 0}$ is stochastic then $\int_{\Xi} T\varphi = 0$ for $\varphi \in D(T)$ and

$$\int_{\Xi} (T\varphi - \Sigma\varphi + P\varphi) \leq 0, \quad \varphi \in D_+(T_\Sigma) = D_+(T) \cap D(\Sigma).$$

According to Kato-Voigt perturbation Theorem 2.11.1, there exists a unique extension

$$G \supset T_\Sigma + P$$

which generates a minimal substochastic C_0-semigroup $(V(t))_{t\geq 0}$ on $L^1(\Xi, \nu)$. This chapter is devoted to the spectral analysis and scattering theory of such perturbed Frobenius–Perron semigroups $(V(t))_{t\geq 0}$. We provide complementary results to those presented in the more general settings of Chaps. 3 and 4. In Sect. 8.3, we "compute" the spectral bound of T_Σ, (see Theorem 8.3.1). The question of whether $s(T_\Sigma) < 0$ or $s(T_\Sigma) = 0$ constitutes a key issue; sufficient conditions are provided, (see Theorems 8.3.2 and 8.3.3). Genuine honesty of $(V(t))_{t\geq 0}$ holds when K is quasi-compact, (see Theorem 8.4.1). Let K be stochastic and weakly compact. Then the semigroup $(V(t))_{t\geq 0}$ has a spectral gap if $s(T_\Sigma) < 0$, (see Theorem 8.4.2). If $s(T_\Sigma) = 0$ then there is no spectral gap and $(V(t))_{t\geq 0}$ is either asymptotically stable or sweeping (see Theorem 8.4.3). We also explore a completely different time-asymptotic regime based on scattering theory (see Theorem 8.5.1). To the best of our knowledge, at this level of generality, such a construction, connecting spectral (gap) theory, time asymptotics, and scattering theory, for abstract C_0-semigroups generated by jump perturbations of (abstract) weighted shifts appears to be new. Nevertheless, in a similar spirit, various specific examples have already been treated in contexts such as Kinetic Theory, Population Dynamics, and related fields. Some remarks, comments or open questions are relegated to the end of this chapter in the section "Comments".

8.3 Type of Absorption Frobenius-Perron Semigroup

This section is devoted to different characterizations of the type of the absorption Frobenius-Perron semigroup $(U_\Sigma(t))_{t\geq 0}$ or equivalently the spectral bound of its generator T_Σ.

Theorem 8.3.1 *The spectral bound of T_Σ is given by*

$$s(T_\Sigma) = -\lim_{t\to +\infty} \inf_{y\in\Xi} t^{-1} \int_0^t \Sigma(\pi_s(y)))ds.$$

Proof We note that

$$(U_\Sigma(t)f)(x) = e^{-\int_0^t \Sigma(\pi_{-s}(x))ds} f(\pi_{-t}(x))$$

$$\|U_\Sigma(t)f\| = \int_\Xi e^{-\int_0^t \Sigma(\pi_{-s}(x))ds} |f(\pi_{-t}(x))| \nu(dx).$$

Since $\pi_t : \Xi \to \Xi$ $(t \in \mathbb{R})$ is measure-preserving then the change of variable $y = \pi_{-t}(x)$ gives $x = \pi_t(y)$ and $\|U_\Sigma(t)f\|$ is equal to

$$\int_\Xi e^{-\int_0^t \Sigma(\pi_{-s}(\pi_t(y)))ds} |f(y)| \nu(dy)$$

$$= \int_\Xi e^{-\int_0^t \Sigma(\pi_{t-s}(y))ds} |f(y)| \nu(dy)$$

$$= \int_\Xi e^{-\int_0^t \Sigma(\pi_s(y))ds} |f(y)| \nu(dy).$$

Thus

$$\|U_\Sigma(t)f\| = \int_\Xi e^{-\int_0^t \Sigma(\pi_s(y))ds} |f(y)| \nu(dy) \tag{8.3.1}$$

and

$$\|U_\Sigma(t)\| = \sup_{y \in \Xi} e^{-\int_0^t \Sigma(\pi_s(y))ds}$$

$$= e^{-\inf_{y \in \Xi} \int_0^t \Sigma(\pi_s(y))ds}$$

and finally the type $\omega(U_\Sigma)$ of $(U_\Sigma(t))_{t \geq 0}$ is given by

$$\omega(U_\Sigma) = \lim_{t \to +\infty} t^{-1} \ln \|U_\Sigma(t)\|$$

$$= -\lim_{t \to +\infty} \inf_{y \in \Xi} t^{-1} \int_0^t \Sigma(\pi_s(y))ds$$

which ends the proof since $s(T_\Sigma) = \omega(U_\Sigma)$. $\qquad\qquad\square$

The question of whether $s(T_\Sigma) < 0$ or $s(T_\Sigma) = 0$ is a key issue. We start with

Theorem 8.3.2 $s(T_\Sigma) < 0$ *if and only if there exist* $C_i > 0$ $(i = 1, 2)$ *such that*

$$\int_0^{C_1} \Sigma(\pi_s(x))ds \geq C_2 \ a.e. \tag{8.3.2}$$

In particular, let Ξ *be a metric space,* ν *be a Borel measure on* Ξ *and let* $\pi_t :$ *$\Xi \to \Xi$ $(t \in \mathbb{R})$ be continuous. If* Σ *is continuous on some characteristic curve* $\{\pi_s(\widehat{x}); \ s \geq 0\}$ *and vanishes identically on it then* $s(T_\Sigma) = 0$.

Proof $s(T_\Sigma) < 0$ if and only if

$$\delta := \lim_{t \to +\infty} \inf_{x \in \Xi} t^{-1} \int_0^t \Sigma(\pi_s(x))ds > 0.$$

In this case, for t large enough

$$\inf_{x \in \Xi} t^{-1} \int_0^t \Sigma(\pi_s(x))ds \geq \frac{\delta}{2}$$

and we can choose $C_1 = t$ and $C_2 = t\frac{\delta}{2}$.

Conversely, let (8.3.2) be satisfied. For $t > 0$, if $\left[\frac{t}{C_1}\right]$ is the integer part of $\frac{t}{C_1}$ then

$$\int_0^t \Sigma(\pi_s(x))ds = \int_0^{C_1 \frac{t}{C_1}} \Sigma(\pi_s(x))ds$$

$$\geq \int_0^{C_1\left[\frac{t}{C_1}\right]} \Sigma(\pi_s(x))ds$$

$$= \sum_{k=0}^{\left[\frac{t}{C_1}\right]-1} \int_{kC_1}^{(k+1)C_1} \Sigma(\pi_s(x))ds$$

$$= \sum_{k=0}^{\left[\frac{t}{C_1}\right]-1} \int_0^{C_1} \Sigma(\pi_{s+kC_1}(x))ds$$

$$= \sum_{k=0}^{\left[\frac{t}{C_1}\right]-1} \int_0^{C_1} \Sigma(\pi_s(\pi_{kC_1}(x)))ds$$

so

$$\int_0^t \Sigma(\pi_s(\zeta))ds \geq \sum_{k=0}^{\left[\frac{t}{C_1}\right]-1} \int_0^{C_1} \Sigma(\pi_s(\pi_{kC_1}\zeta))ds$$

$$\geq C_2 \left[\frac{t}{C_1}\right] \geq C_2 \left(\frac{t}{C_1} - 1\right)$$

whence

$$\lim_{t \to +\infty} t^{-1} \inf_x \int_0^t \Sigma(\pi_s(x))ds \geq C_2 C_1^{-1}.$$

Finally, if Ξ is a metric space, if $\pi_t : \Xi \to \Xi$ $(t \in \mathbb{R})$ are continous, if Σ is continuous on some characteristic curve $\{\pi_s(\widehat{x}), s \geq 0\}$ and vanishes on it then for each $t > 0$

$$\int_0^t \Sigma(\pi_s(x))ds \to \int_0^t \Sigma(\pi_s(\widehat{x}))ds = 0 \ (x \to \widehat{x})$$

so

$$\inf_{x \in \Xi} t^{-1} \int_0^t \Sigma(\pi_s(x)) ds = 0 \ (t > 0)$$

and consequently $s(T_\Sigma) = 0$. $\qquad\qquad\qquad\qquad\qquad\qquad\qquad\qquad\square$

According to Theorem 8.3.2, a necessary condition for $s(T_\Sigma) < 0$ is that there exists no characteristic curve $\{\pi_s(y); \ s \geq 0\}$ along which Σ vanishes identically. We complement Theorem 8.3.2 by additional results.

Theorem 8.3.3 *Let Ξ be a locally compact metric space and let $(t, x) \in [0, +\infty) \times \Xi \to \pi_t(x) \in \Xi$ be continuous. Suppose that Σ is continuous and for each $t \geq 0$*

$$\pi_t(x) \to \infty \ as \ x \to \infty \tag{8.3.3}$$

(i.e., for each compact set $\Lambda_1 \subset \Xi$ there exists a compact set $\Lambda_2 \subset \Xi$ such that $\pi_t(x) \notin \Lambda_1$ if $x \notin \Lambda_2$). Then:
(i) If $\lim_{x \to \infty} \Sigma(x) = 0$ then $s(T_\Sigma) = 0$
(ii) If there exists a compact set outside which $\Sigma(.)$ is bounded away from zero and if there is no characteristic curve $\{\pi_s(y); \ s \geq 0\}$ on which Σ vansishes identically then $s(T_\Sigma) < 0$.

Proof (i) In this case, Σ is bounded. According to Theorem 8.3.2 if $s(T_\Sigma) < 0$ then there exist $c_i > 0 \ (i = 1, 2)$ such that

$$\int_0^{c_1} \Sigma(\pi_s(x)) ds \geq c_2 \ \text{a.e.}$$

But for any $c_1 > 0$ and sequence $x_j \to \infty$ we have

$$\int_0^{c_1} \Sigma(\pi_s(x_j)) ds \to 0 \ (j \to \infty)$$

by Lebesgue dominated convergence theorem. Hence

$$\lim_{x \to \infty} \int_0^{c_1} \Sigma(\pi_s(x)) ds = 0$$

and $s(T_\Sigma) = 0$.

(ii) In this case, Σ need not be bounded. We argue by contradiction by assuming that for any $c > 0$

$$\inf_{x \in \Xi} \int_0^c \Sigma(\pi_s(x)) ds = 0.$$

There exists a sequence $(x_j)_j \subset \Xi$ such that

$$\int_0^c \Sigma(\pi_s(x_j))ds \to 0 \ (j \to \infty).$$

This implies that $(x_j)_j$ remains in a compact subset $\Delta \subset \Xi$. Indeed, if not then there exists a subsequence $(x_{j_n})_n$ such that

$$x_{j_n} \to \infty \ (n \to \infty)$$

and Fatou's lemma gives

$$\lim_{n \to \infty} \int_0^c \Sigma(\pi_s(x_{j_n}))ds \geq \int_0^c \left(\lim_{n \to \infty} \inf \Sigma(\pi_s(x_{j_n})) \right) ds.$$

By assumption, there exists a compact set $\Lambda \subset \Xi$ and $\alpha > 0$ such that $\Sigma(y) \geq \alpha \ (y \notin \Lambda)$. Then, for each s, $\pi_s(x_{j_n}) \notin \Lambda$ for n large enough so $\Sigma(\pi_s(x_{j_n})) \geq \alpha$ for n large enough and consequently

$$\lim_{n \to \infty} \inf \Sigma(\pi_s(x_{j_n})) \geq \alpha \ (s \geq 0)$$

which leads to a contradiction. Since $(x_j)_j \subset \Delta$ then, by extracting a subsequence if necessary, we may assume that $x_j \to y \ (j \to \infty)$ and

$$\int_0^c \Sigma(\pi_s(y))ds = 0.$$

A priori y depends of the choice of c so we denote it rather by y_c. Hence

$$\int_0^j \Sigma(\pi_s(y_j))ds = 0 \ (j \in \mathbb{N}).$$

In particular, for any fixed n

$$\int_0^n \Sigma(\pi_s(y_j))ds = 0 \ (j \geq n).$$

By using Fatou's lemma again, $(y_j)_j$ must remain in a compact set $\Lambda \subset \Xi$. By extracting a subsequence if necessary $y_j \to y$ such that

$$\int_0^n \Sigma(\pi_s(y))ds = 0 \ (n \in \mathbb{N})$$

so that Σ vanishes identically on a characteristic curve $\{\pi_s(y); \ s \geq 0\}$ and this is excluded by assumption. $\qquad\square$

Assumption (8.3.3) is always satisfied by flows induced by differential equations, see Sect. 10.13 on the theory of particle swarms.

8.4 Spectral Results and Time Asymptotics

The following statement is a particular case of Theorem 5.11.1.

Theorem 8.4.1 *Suppose that the substochastic operator K is quasi-compact. Let P be defined by (8.2.2). Then $(V(t))_{t\geq0}$ is stochastic and genuinely honest (i.e., $G = T_\Sigma + P$).*

Similarly, here is a particular case of Theorem 5.11.2.

Theorem 8.4.2 *Suppose that the substochastic operator K is weakly compact. Let P be defined by (8.2.2). Then $(V(t))_{t\geq0}$ and $(U_\Sigma(t))_{t\geq0}$ share the same essential spectrum; in particular $\omega_{ess}(V) = \omega_{ess}(U_\Sigma)$. If additionaly $s(T_\Sigma) < 0$ and K is stochastic then $(V(t))_{t\geq0}$ has a spectral gap, i.e., $\omega_{ess}(V) < \omega(V) = 0$.*

The presence of a spectral gap implies of course the existence of an invariant density. It is clear that $s(T_\Sigma) < 0$ if Σ is bounded away from zero. Actually, this sufficient condition is far from being necessary. It is possible to obtain much more precise results in locally compact metric spaces Ξ with suitable assumptions on the flow. We do not try to elaborate on this point in this too abstract setting. This will be investigated in details in the context of kinetic equations, see Chap. 10.

We deal now with the *critical* case $s(T_\Sigma) = 0$, i.e.,

$$\lim_{t\to+\infty}\ \inf_{y\in\Xi}\ t^{-1}\int_0^t \Sigma(\pi_s(y)))ds = 0 \tag{8.4.1}$$

where, a priori, no spectral gap exists. This case was also dealt with in a more general context in Theorem 5.11.3.

Theorem 8.4.3 *Let K be stochastic and weakly compact and let (8.4.1) be satisfied. Let P be defined by (8.2.2) and $P\,(0 - T_\Sigma)^{-1}$ be irreducible. Then 1 is an eigenvalue of $P\,(0 - T_\Sigma)^{-1}$ associated to a positive eigenfunction φ. Moreover:*
(i) If $\psi := (0 - T_\Sigma)^{-1}\varphi \in L^1(\Xi,\nu)$ then $\psi \in D(T_\Sigma)$ and (after normalization) is a unique invariant density of $(V(t))_{t\geq0}$ and is asymptotically stable.
(ii) If $\psi := (0 - T_\Sigma)^{-1}\varphi \notin L^1(\Xi,\nu)$ then $(V(t))_{t\geq0}$ is sweepingfrom the sets $\Xi_\varepsilon := \{\Sigma \geq \varepsilon\}$.

Proof The irreducibility of $P\,(0 - T_\Sigma)^{-1}$ implies that of $P\,(\lambda - T_\Sigma)^{-1}$ at least for $\lambda > 0$ small enough. It follows that $(\lambda - T_\Sigma - P)^{-1}$ is positivity improving (see Corollary 2.5.1) and consequently $(V(t))_{t\geq0}$ is irreducible. Finally Theorem 5.11.3 ends the proof. $\square$

Whether $(V(t))_{t\geq 0}$ has a spectral gap or not, there is an important hidden condition behind the existence of an invariant density.

Theorem 8.4.4 *Let $(V(t))_{t\geq 0}$ be stochastic and genuinely honest (i.e., $G = T_\Sigma + P$). Suppose that there exists no equilibrium point of the flow (i.e., x such that $\pi_{-t}(x) = x$; $\forall t \geq 0$) on which Σ vanishes. If $(V(t))_{t\geq 0}$ has an invariant density and is irreducible then*

$$\int_0^{+\infty} \Sigma(\pi_s(y)))ds = +\infty \ a.e. \tag{8.4.2}$$

Proof According to Corollary 3.3.1, $(U_\Sigma(t))_{t\geq 0}$ is strongly stable, i.e.,

$$\lim_{t\to+\infty} \|U_\Sigma(t)\varphi\| = 0 \ \ (\varphi \in L^1(\Xi, \nu)).$$

On the other hand, we know that

$$\|U_\Sigma(t)\varphi\| = \int_\Xi e^{-\int_0^t \Sigma(\pi_s(y)))ds} \, |\varphi(y)| \, \nu(dy)$$

so the strong stability is equivalent to (8.4.2). $\square$

Actually, we have a much more precise result.

Theorem 8.4.5 *Let P be defined by (8.2.2). If 1 is an eigenvalue of $P(0 - T_\Sigma)^{-1}$ associated to a positive eigenfunction φ then (8.4.2) holds.*

Proof We point out that if φ be an invariant density of $(V(t))_{t\geq 0}$ then

$$P(0 - T_\Sigma)^{-1}\varphi = \varphi \tag{8.4.3}$$

but the converse statement need not be true in general, see [269]. Let $\varphi(y) > 0$ a.e satisfying (8.4.3). Since K is stochastic then $\left\| P(0 - T_\Sigma)^{-1}\varphi \right\| = \left\| \Sigma(0 - T_\Sigma)^{-1}\varphi \right\|$ and

$$\left\| \Sigma(0 - T_\Sigma)^{-1}\varphi \right\| = \|\varphi\|.$$

On the other hand $\left\| \Sigma(0 - T_\Sigma)^{-1}\varphi \right\|$ is equal to

$$\int_\Xi \nu(dx)\Sigma(x) \int_0^{+\infty} e^{-\int_0^t \Sigma(\pi_{-s}(x))ds}\varphi(\pi_{-t}(x))dt$$

$$= \int_0^{+\infty} dt \int_\Xi \Sigma(x)e^{-\int_0^t \Sigma(\pi_{-s}(x))ds}\varphi(\pi_{-t}(x))\nu(dx).$$

Making the change of variable $y = \pi_{-t}(x)$ (i.e., $x = \pi_t(y)$) one sees that $\|\varphi\|$ is equal to

$$\int_0^{+\infty} dt \int_\Xi \Sigma(\pi_t(y))e^{-\int_0^t \Sigma(\pi_{-s}(\pi_t(y)))ds}\varphi(y)\nu(dy)$$

$$= \int_{\Xi} \left(\int_0^{+\infty} \Sigma(\pi_t(y)) e^{-\int_0^t \Sigma(\pi_s(y))ds} dt \right) \varphi(y)\nu(dy)$$

$$= \int_{\Xi} \left(\int_0^{+\infty} -\frac{d}{dt} e^{-\int_0^t \Sigma(\pi_s(y))ds} dt \right) \varphi(y)\nu(dy)$$

$$= \int_{\Xi} \left(1 - e^{-\int_0^{+\infty} \Sigma(\pi_s(y))ds} \right) \varphi(y)\nu(dy)$$

so $e^{-\int_0^{+\infty} \Sigma(\pi_s(y))ds} = 0$ a.e. and this ends the proof. $\square$

8.5 Scattering Theory

We explore now another regime for perturbed Frobenius-Perron semigroups in which an invariant density fails to exist. This construction is also motivated by different scattering problems arising in Kinetic Theory, (see Chap. 10). In this section, P is defined by (8.2.2) where K is an abstract substochastic operator that need not be an integral operator. We show here how the abstract scattering theory given in Sect. 3.4 applies to a broad class of perturbed Frobenius-Perron semigroups. To this end, we need that $(U_\Sigma(t))_{t \geq 0}$ extends as a positive C_0-group $(U_\Sigma(t))_{t \in \mathbb{R}}$ on $L^1(\Xi, \nu)$. Since

$$(U_\Sigma(-t)f)(x) = e^{-\int_0^{-t} \Sigma(\pi_{-s}(x))ds} f(\pi_t(x)) \quad (t > 0)$$

and

$$e^{-\int_0^{-t} \Sigma(\pi_{-s}(x))ds} = e^{\int_0^t \Sigma(\pi_s(x))ds}$$

then we introduce the condition

$$\sup_{x \in \Xi} \int_0^{+\infty} \Sigma(\pi_s(x))ds < +\infty$$

which ensures that $(U_\Sigma(t))_{t \in \mathbb{R}}$ is a bounded C_0-group. In particular $\sigma(T_\Sigma) \subset i\mathbb{R}$. Actually, we introduce the stronger condition of finite "optical length"

$$\sup_{x \in \Xi} \int_{-\infty}^{+\infty} \Sigma(\pi_s(x))ds < +\infty \tag{8.5.1}$$

which is needed for other purposes. We start with

Lemma 8.5.1 *Let (8.5.1) be satisfied. Then the (left) strong limit*

$$s \lim_{\lambda \to 0_-} P(\lambda - T_\Sigma)^{-1} \; exists. \tag{8.5.2}$$

Proof Since $-T_\Sigma$ generates a bounded positive C_0-semigroup $(U_\Sigma(-t))_{t \geq 0}$ then

$$(\mu + T_\Sigma)^{-1}\,\varphi = \int_0^{+\infty} e^{-\mu t}\, U_\Sigma(-t)\varphi\, dt \ (\mu > 0)$$

so

$$\Sigma\,(\mu + T_\Sigma)^{-1}\,\varphi = \Sigma \int_0^{+\infty} e^{-\mu t}\, U_\Sigma(-t)\varphi\, dt \ (\mu > 0)$$

and

$$\lim_{\mu \to 0_+} \Sigma\,(\mu + T_\Sigma)^{-1}\,\varphi = \Sigma \int_0^{+\infty} U_\Sigma(-t)\varphi\, dt$$

$$= \Sigma \int_0^{+\infty} e^{-\int_0^{-t} \Sigma(\pi_{-s}(x))ds}\,\varphi(\pi_t(x))dt.$$

Since π_t is measure-preserving then the change of variable $y = \pi_t(x)$ gives $x = \pi_{-t}(y)$ and for $\varphi \geq 0$

$$\int_\Xi \Sigma(x) \int_0^{+\infty} e^{-\int_0^{-t} \Sigma(\pi_{-s}(x))ds}\,\varphi(\pi_t(x))dt\nu(dx)$$

$$= \int_\Xi \Sigma(\pi_{-t}(y)) \int_0^{+\infty} e^{-\int_0^{-t} \Sigma(\pi_{-s}(\pi_{-t}(y)))ds}\,\varphi(y)dt\nu(dy)$$

$$= \int_\Xi \left(\int_0^{+\infty} \Sigma(\pi_{-t}(y))e^{-\int_0^{-t} \Sigma(\pi_{-s}(\pi_{-t}(y)))ds}dt \right) \varphi(y)\nu(dy).$$

Since

$$\int_0^{+\infty} \Sigma(\pi_{-t}(y))e^{-\int_0^{-t} \Sigma(\pi_{-s}(\pi_{-t}(y)))ds}dt$$

$$= \int_0^{+\infty} \Sigma(\pi_{-t}(y))e^{\int_0^{t} \Sigma(\pi_s(\pi_{-t}(y)))ds}dt$$

$$= \int_0^{+\infty} \Sigma(\pi_{-t}(y))e^{\int_0^{t} \Sigma(\pi_{-(t-s)}(y))ds}dt$$

$$= \int_0^{+\infty} \Sigma(\pi_{-t}(y))e^{\int_0^{t} \Sigma(\pi_{-s}(y))ds}dt$$

$$= \int_0^{+\infty} \frac{d}{dt}e^{\int_0^{t} \Sigma(\pi_{-s}(y))ds}dt$$

$$= e^{\int_0^{+\infty} \Sigma(\pi_{-s}(y))ds} - 1$$

then the norm of $\displaystyle\lim_{\mu \to 0_+} \Sigma\,(\mu_- + T_\Sigma)^{-1}\,\varphi$ is equal to

$$\int_\Xi \left(e^{\int_0^{+\infty} \Sigma(\pi_{-s}(y))ds} - 1 \right) \varphi(y)\nu(dy)$$

and therefore finite thanks to (8.5.1). Since K is stochastic then

$$\left\| P\,(\lambda - T_\Sigma)^{-1}\,\varphi \right\| = \left\| \Sigma\,(\lambda - T_\Sigma)^{-1}\,\varphi \right\|$$

and (8.5.2) is satisfied. $\qquad\qquad\square$

Note that (8.3.1) implies

$$\| U_\Sigma(t)\varphi \| \geq e^{-\sup_y \int_0^{+\infty} \Sigma(\pi_s(y)))ds}\,\|\varphi\|,$$

i.e., (3.4.1) is satisfied. Hence all the conditions in Theorems 3.4.1 and 3.4.2 are met. We are ready to state the main results.

Theorem 8.5.1 *Suppose that* $P : D(P) \subset L^1(\nu) \to L^1(\nu)$ *is "monotonically closed" (in the sense of Definition 3.4.1) If (8.5.1) is satisfied then the wave operators*

$$\begin{cases} \Omega^+ := s\,\lim_{t\to+\infty} V(t)U_\Sigma(-t) \\ \Omega^- := s\,\lim_{t\to+\infty} U_\Sigma(-t)V(t) \end{cases}$$

exist.

We point out that if K is stochastic then $P : D(P) \subset L^1(\nu) \to L^1(\nu)$ is automatically "monotonically closed". Indeed, for any nondecreasing sequence $(\varphi_j) \subset D_+(P) = D_+(\Sigma)$, we have $\left\| P\varphi_j \right\| = \left\| \Sigma\varphi_j \right\|$ so that $\sup_j \left\| \varphi_j \right\| + \left\| P\varphi_j \right\| < +\infty$ implies $\varphi := \lim_j \varphi_j \in D(\Sigma)$. If K is just substochastic but is an integral operator then, by using the monotone convergence theorem, we easily check that P is also "monotonically closed".

A useful consequence of Theorem 8.5.1 is

Corollary 8.5.1 *If (8.5.1) is satisfied then*

$$\left\| V(t)\varphi - U_\Sigma(t)\Omega^-\varphi \right\| \to 0\ (t \to +\infty)\ (\varphi \in L^1(\Xi, \nu)),$$

i.e., any trajectory $(V(t)\varphi)_{t\geq 0}$ *of the full dynamics (with initial data* φ*) behaves asymptotically (as* $t \to +\infty$*) as a free trajectory* $(U_\Sigma(t)\psi)_{t\geq 0}$ *(with initial data* $\psi = \Omega^-\varphi$*).*

The assumption (8.5.1), which is central to scattering theory, encodes important conditions on the flow π_t and on the collision frequency Σ; for example, the existence of periodic orbits is excluded, or at least Σ must vanish on them; (see the discussion, in Chap. 10, on (8.5.1) and (8.4.2) in the context of kinetic equations.)

Remark 8.5.1 Let (8.5.1) be satisfied. If $(U_\Sigma(t)(t))_{t\geq 0}$ is sweeping from a family of measurable subsets $\Lambda \subset \Xi$ then so is $(V(t))_{t\geq 0}$, (see Corollary 3.4.2).

8.6 Comments

In this chapter, we restricted ourselves to backward flows π_{-t} $(t \in \mathbb{R})$ even if some results can be obtained for backward semiflows π_{-t} $(t \geq 0)$ as well.

In the literature on ergodic theory (see e.g. [111, 194]), the spectral properties of the Frobenius-Perron operator $f \to f(\pi_{-t}(x))$ capture the *statistical* properties (ergodicity, mixing...) of the deterministic map π_t[194]. See [46] Chap. 16 for a smooth and pedagogical introduction to the subject.

Weighted shift C_0-semigroups of the form

$$f \to e^{-\int_0^t \Sigma(\pi_{-s}(x))ds} f(\pi_{-t}(x)) \ (t \geq 0)$$

arise naturally in Transport Theory. A notable feature of these semigroups is that the spectrum of their generators occupies an entire half-plane. This surprising phenomenon was first observed in 1955 in the seminal spectral analysis of the simplest physical model for Neutron Transport in slab geometry by Lehner and Wing [204]. More general spectral results on weighted shift C_0-semigroups associated to semiflows, motivated by Transport Theory, can be found in [210, 255, 373].

We conjecture that Theorem 8.4.2 and Theorem 8.4.3 remain true under the more general assumption that K is quasi-compact.

The computation of the type in Theorem 8.3.1 follows earlier results on weighted shift semigroups in [210, 367].

The characterization given in Theorem 8.3.2 is given in [266] for neutron transport fields on the torus; see also [53] for an earlier result in this direction.

Chapter 9
Jump Semigroups on Countable State Spaces

9.1   Chapter Aims

The structure of jump semigroups in Lebesgue spaces $L^1(E, \nu)$ significantly simplifies when the measure space is countable, such as $\mathbb{N}$ equipped with the counting measure. In this chapter, we address peripheral spectral analysis and time asymptotics for Kolmogorov differential equations on sequence spaces $l^1(\mathbb{N})$, equipped with (weighted) counting measures. We also address the hilbertian spectral gap analysis of Dirichlet forms associated with symmetric weighted graphs, with killing term, defined over abstract countable measure spaces.

9.2   Chapter Overview

In Chap. 6, we examined various spectral problems related to jump C_0-semigroups in abstract Lebesgue spaces $L^1(E, \nu)$. When the measure space is countable, such as $\mathbb{N}$ equipped with the counting measure, the structure of jump semigroups becomes significantly simpler, allowing for more straightforward formulations and, in some cases, stronger spectral results. In the present chapter, we build upon the framework developed in Chap. 6, focusing on the case where (E, ν) is a countable measure space. As in Chap. 6, our focus here is on peripheral spectral theory and time asymptotics. This chapter presents various (new) results in different contexts, organized into four sections. The first section, which is relatively long, is devoted to Kolmogorov differential equations on the standard sequence space $l^1(\mathbb{N})$ equipped with the counting measure. The second and third sections deal respectively with Kolmogorov differential equations on weighted (with respect to the counting measure) l^1-spaces and on moment spaces. Finally, the fourth section addresses symmetric weighted graphs defined over abstract countable measure spaces. More detailed information about the results in each case is provided in the introductions to the individual sections. Some

© The Author(s), under exclusive license to Springer Nature Switzerland AG 2026

M. Mokhtar-Kharroubi, *Peripheral Spectra of Perturbed Positive Semigroups*, Lecture Notes in Mathematics 2388, https://doi.org/10.1007/978-3-032-11173-9_9

remarks, comments or open questions are relegated to the end of this chapter in the section "Comments".

9.3 Kolmogorov Differential Equations on Standard sequence Space

This section is devoted to peripheral spectral analysis and time asymptotics of Kolmogorov differential equations. We recall that the transition probability matrix $(p_{jk}(t))$ of a homogeneous Markov process with denumerable states satisfies the properties

$$\sum_{k=0}^{\infty} p_{jk}(t) = 1 \tag{9.3.1}$$

and the Chapman-Kolmogorov identity

$$p_{ik}(t+s) = \sum_{j=0}^{\infty} p_{ij}(t)p_{jk}(t). \tag{9.3.2}$$

Under suitable conditions, $(p_{jk}(t))$ satisfies the Kolmogorov differential equations

$$\frac{dp_{ik}(t)}{dt} = \sum_{j=0}^{\infty} p_{ij}(t)q_{jk}, \quad \frac{dp_{ik}(t)}{dt} = \sum_{j=0}^{\infty} q_{ij}p_{jk}(t) \tag{9.3.3}$$

with the initial conditions

$$p_{ij}(0) = \begin{cases} 1, & j = k \\ 0, & j \neq k \end{cases}, \ \forall i, j \tag{9.3.4}$$

where the Q-matrix $Q = (q_{ij})$ is such that

$$\begin{cases} q_{jj} \leq 0, \ \forall j; \ q_{ij} \geq 0, \forall i \neq j \\ \sum_{j=0, \ j \neq i}^{\infty} q_{ij} = -q_{ii}, \ \forall i. \end{cases} \tag{9.3.5}$$

Actually, $(p_{jk}(t))$ is the matrix of the solution operator

$$P(t) : \varphi_0 \in l^1(\mathbb{N}) \to \varphi(t) \in l^1(\mathbb{N})$$

of the Cauchy problem

$$\frac{d\varphi(t)}{dt} = Q^t\varphi(t), \quad \varphi(0) = \varphi_0 \in l^1(\mathbb{N}),$$

(Q^t is the transpose of Q), relative to the canonical (topological) basis $\{e_i, \; i \in \mathbb{N}\}$ of $l^1(\mathbb{N})$, i.e.,

$$\{p_{jk}(t)\}_{k\in\mathbb{N}} = P(t)e_j \quad (\forall j \in \mathbb{N}).$$

This is a classical subject, tracing back at least to the 1950s (see [180, 203, 313]). As noted in the Introduction to this monograph, the seminal paper by Kato [180] plays a central role throughout and is used repeatedly. In particular, the well-posedness of the Cauchy problem discussed above is thoroughly analyzed in [180], as well as in more recent contributions (see, e.g., [20, 369]). The objective of this chapter is to study the L^1 peripheral spectral theory and long-time asymptotics of various classes of Kolmogorov differential equations. Despite the vast literature on continuous-time Markov chains (see, e.g., [64] and references therein), to the best of the author's knowledge, the results presented here are new.

We translate now some of the previous results into the case of

$$E = \mathbb{N} = \{0, 1, \ldots\}$$

endowed with the sigma algebra $\mathcal{P}(\mathbb{N})$ and the counting measure ν and use the abbreviation

$$l^1(\mathbb{N}, \nu) := l^1(\mathbb{N}, \mathcal{P}(\mathbb{N}), \nu).$$

We introduce a little change of notation. To avoid additional cumbersome technicalities, we suppose that

$$\sigma_i := -q_{ii} > 0 \; (\forall i \in \mathbb{N}). \tag{9.3.6}$$

Let $M : D(M) \subset l^1(\mathbb{N}, \nu) \to l^1(\mathbb{N}, \nu)$ be the multiplication operator

$$\begin{cases} \quad\quad (x_i) \in D(M) \to -(\sigma_i x_i) \in l^1(\mathbb{N}, \nu) \\ D(M) = l_\sigma^1(\mathbb{N}, \nu) := \{(x_i) \in l^1(\mathbb{N}, \nu); \; (\sigma_i x_i) \in l^1(\mathbb{N}, \nu)\} \end{cases} \tag{9.3.7}$$

and the (holomorphic) C_0-semigroup $(\mathcal{M}(t))_{t\geq0}$ generated by M given by

$$\mathcal{M}(t) : x = (x_i)_i \in l^1(\mathbb{N}, \nu) \to (x_i e^{-t\sigma_i})_i \in l^1(\mathbb{N}, \nu).$$

We introduce the matrix (b_{ij}) (the transpose of $Q - diag(q_{ii})$)

$$b_{ij} : \begin{cases} q_{ji} \text{ if } i \neq j \\ 0 \text{ if } i = j \end{cases}$$

and note that

$$\sum_{j\in\mathbb{N}} b_{ji} = \sigma_i \; (i \in \mathbb{N}).$$

Let $B : l_\sigma^1(\mathbb{N}, \nu) \to l^1(\mathbb{N}, \nu)$ be defined by

$$Bx = \left(\sum_{j=0}^{\infty} b_{ij} x_j \right)_i . \tag{9.3.8}$$

According to Theorem 2.11.1, there exists a unique extension

$$J \supset M + B : D(M) \to l^1(\mathbb{N}, \nu) \tag{9.3.9}$$

which generates a minimal positive contraction C_0-semigroup $(\mathcal{J}(t))_{t\geq 0}$. Actually, this result is very classical, see e.g. [20, 369] and references therein.

This section is concerned with (genuine) honesty and time asymptotics of these Kato-Voigt semigroups in $l^1(\mathbb{N}, \nu)$. As a consequence of the abstract Theorem 6.4.1, we show that under a (generalized) detailed balance condition, we capture an alternative between asymptotic stability and sweeping, (see Theorem 9.3.1). We show also that a (generalized) detailed balance condition is always fulfilled for general birth and death models, (see Theorem 9.3.2). We introduce also a class of Kolmogorov differential equations displaying quasi-compactness properties (see Theorem 9.3.3 when $\inf_{i\in\mathbb{N}} \sigma_i > 0$ and Theorem 9.3.4 when $\inf_{i\in\mathbb{N}} \sigma_i = 0$) and consequently falling within the setting of Desch semigroups and their spectral properties as discussed in Chap. 4. In addition, since any bounded operator on $l^1(\mathbb{N}, \nu)$ is "locally compact" then the spectral gap theory for absorption semigroups given in Chap. 5 applies to Kolmogorov C_0-semigroups perturbed by an (artificial) absorption at infinity, (see Theorem 9.3.5).

9.3.1 Honesty and Time Asymptotics Versus Detailed Balance

In this subsection, we consider the case where a generalized detailed balance condition holds

$$\omega_i b_{ij} = \omega_j b_{ji} \tag{9.3.10}$$

where $(\omega_i)_{i\in\mathbb{N}}$ is just a positive sequence. Then the translation of Theorem 6.4.1 within $l^1(\mathbb{N})$ gives

Theorem 9.3.1 *Let (9.3.10) be satisfied and*

$$(u_i)_{i\in\mathbb{N}} := \left(\sigma_i \omega_i^{-1} \right)_{\in\mathbb{N}} \in l^1(\mathbb{N}, \nu). \tag{9.3.11}$$

Then
  (i) $(\mathcal{J}(t))_{t\geq 0}$ is honest (or equivalently $J = \overline{M + B}$).
  (ii) We have the following alternative:
  (a) If $\left(\omega_i^{-1} \right)_{\in\mathbb{N}} \in l^1(\mathbb{N}, \nu)$ then (after normalization) $\left(\omega_i^{-1} \right)_{\in\mathbb{N}}$ is an invariant density of $(\mathcal{J}(t))_{t\geq 0}$ and is asymptotically stable.

(b) Let $\left(\omega_i^{-1}\right)_{i \in \mathbb{N}} \notin l^1(\mathbb{N}, \nu)$. If $(\mathcal{J}(t))_{t \geq 0}$ is genuinely honest then it is sweeping from the sets of summability of $\left(\omega_i^{-1}\right)_{i \in \mathbb{N}}$; (this amounts to $\lim_{t \to +\infty} (\mathcal{J}(t)\zeta)_i = 0$ for all $i \in \mathbb{N}$ and $\zeta = (\zeta_n)_{n \in \mathbb{N}} \in l^1(\mathbb{N}, \nu)$).

We refer to [320] for various results on the alternative between asymptotic stability and sweeping for quite general Kolmogorov differential equations.

On General Birth and Death Models

We show now how Theorem 9.3.1 applies to birth and death models, i.e., when

$$b_{i,j} = 0 \text{ for } i = j \text{ or } |i - j| \geq 2. \tag{9.3.12}$$

It is convenient to introduce λ_i and μ_i such that

$$\begin{cases} b_{i(i-1)} = \lambda_{i-1} & (i \geq 1) \\ b_{i(i+1)} = \mu_{i+1} & (i \geq 0) \end{cases} \tag{9.3.13}$$

so

$$b_{ij} = \begin{cases} \lambda_j \text{ if } i = j + 1, \ j \geq 0 \\ \mu_j \text{ if } i = j - 1, \ j \geq 1 \\ \quad = 0 \ \text{ otherwise} \end{cases}$$

and

$$\begin{cases} \sigma_i = b_{(i-1)i} + b_{(i+1)i} = \mu_i + \lambda_i \ (i \geq 1) \\ \qquad \sigma_0 = b_{10} = \lambda_0. \end{cases}$$

We recall that a birth and death semigroup is well understood since a long time, see e.g. its thorough spectral analysis [203]. In particular, it is honest if and only if

$$\sum_{n=0}^{\infty} \frac{1}{\lambda_n} \left(\sum_{i=0}^{\infty} \prod_{j=1}^{i} \frac{\mu_{n+j}}{\lambda_{n+j}} \right) = \infty, \tag{9.3.14}$$

(see e.g. [33] Theorem 7.11, p. 186). A sufficient condition of honesty is given by

$$\{\mu_n^{-1}\} \in l^1(\mathbb{N}) \text{ and } \lim_{n \to \infty} \frac{\lambda_n}{\mu_n} < 1 \tag{9.3.15}$$

(see [33] Proposition 7.20, p. 195). We provide here another (sufficient) approach of honesty relying on Theorem 9.3.1. In addition, we capture time asymptotics of birth and death C_0-semigroups. To this end, we look for a positive sequence $(\omega_i)_{i \in \mathbb{N}}$ such that

$$\omega_i b_{ij} = \omega_j b_{ji}.$$

Introducing $\pi_i := \omega_i^{-1}$ this amounts to looking for a positive sequence $(\pi_i)_{i \in \mathbb{N}}$ such that

$$b_{ji}\pi_i = b_{ij}\pi_j.$$

Thus

$$b_{(n-1)n}\pi_n = b_{n(n-1)}\pi_{n-1} \; (n \geq 1),$$

i.e.,

$$\mu_n \pi_n = \lambda_{n-1}\pi_{n-1} \; (n \geq 1).$$

Hence

$$\frac{\pi_n}{\pi_{n-1}} = \frac{\lambda_{n-1}}{\mu_n} \; \forall n \geq 1$$

and

$$\pi_n = \pi_0 \prod_{j=1}^{n} \frac{\lambda_{j-1}}{\mu_j} \; (n \geq 1)$$

where $\pi_0 > 0$ is arbitrary. Finally, such a sequence always exists and is unique up to a multiplicative factor π_0

$$\pi_n = \pi_0 \frac{\lambda_0 \cdots \lambda_{n-1}}{\mu_1 \mu_2 \cdots \mu_n} \; (n \geq 1). \tag{9.3.16}$$

One sees that the general assumption (9.3.11)

$$\left(\sigma_n \omega_n^{-1}\right)_{n \in \mathbb{N}} \in l^1(\mathbb{N}, \nu)$$

amounts to

$$\left((\mu_n + \lambda_n) \prod_{j=1}^{n} \frac{\lambda_{j-1}}{\mu_j}\right)_{n \geq 1} \in l^1(\mathbb{N}_*, \nu),$$

i.e.,

$$\sum_{n=3}^{+\infty} \frac{\lambda_0 \cdots \lambda_{n-1}}{\mu_1 \mu_2 \cdots \mu_{n-1}} + \sum_{n=1}^{+\infty} \frac{\lambda_0 \cdots \lambda_n}{\mu_1 \mu_2 \cdots \mu_n} < \infty$$

or

$$\sum_{n=1}^{+\infty} \frac{\lambda_1 \cdots \lambda_n}{\mu_1 \mu_2 \cdots \mu_n} < \infty.$$

The condition $\left(\omega_n^{-1}\right)_{n \in \mathbb{N}} \in l^1(\mathbb{N}, \nu)$ amounts to

$$\sum_{n=1}^{+\infty} \frac{\lambda_1 \cdots \lambda_{n-1}}{\mu_1 \mu_2 \cdots \mu_n} < \infty.$$

Hence Theorem 9.3.1 implies.

Theorem 9.3.2 *We consider the case (9.3.12) with the notations (9.3.13) and assume that*

$$\sum_{n=1}^{+\infty} \frac{\lambda_1 \cdots \lambda_n}{\mu_1 \mu_2 \cdots \mu_n} < \infty. \tag{9.3.17}$$

Then
(i) $(\mathcal{J}(t))_{t \geq 0}$ is honest (or equivalently $J = \overline{M + B}$).
(ii) We have the following alternative:
(a) If $\sum_{n=1}^{+\infty} \frac{\lambda_1 \cdots \lambda_{n-1}}{\mu_1 \mu_2 \cdots \mu_n} < \infty$ then (after normalization) $(\pi_n)_{n \in \mathbb{N}}$ defined by (9.3.16) is an invariant density and $(\mathcal{J}(t))_{t \geq 0}$ is asymptotically stable.
(b) If $\sum_{n=1}^{+\infty} \frac{\lambda_1 \cdots \lambda_{n-1}}{\mu_1 \mu_2 \cdots \mu_n} = \infty$ then $(\mathcal{J}(t))_{t \geq 0}$ is sweeping from the sets of summability of $(\pi_n)_{n \in \mathbb{N}}$.

In particular

Corollary 9.3.1 *We consider the case (9.3.12) with the notations (9.3.13). If*

$$\lim_{n \to \infty} \sup \frac{\lambda_n}{\mu_n} < 1 \tag{9.3.18}$$

then $(\mathcal{J}(t))_{t \geq 0}$ is honest.

Proof Note that (9.3.18) implies that $\left(\frac{\lambda_n}{\mu_n}\right)_n$ is bounded. Let $m_0 \in \mathbb{N}$ et $c < 1$ such that

$$\frac{\lambda_n}{\mu_n} \leq c \ (n \geq m_0).$$

Then

$$\frac{\lambda_1 \cdots \lambda_n}{\mu_1 \mu_2 \cdots \mu_n} = \frac{\lambda_1 \cdots \lambda_{m_0}}{\mu_1 \mu_2 \cdots \mu_{m_0}} \prod_{j=m_0+1}^{n} \frac{\lambda_j}{\mu_j} \leq \left(\sup \frac{\lambda_n}{\mu_n}\right)^{m_0} c^{n-m_0-1}$$

shows that (9.3.17) is satisfied. $\qquad\qquad\square$

This corollary shows that the additional condition $\{\mu_n^{-1}\} \in l^1(\mathbb{N})$ given in ([33] Proposition 7.20) is unnecessary. Note also that if we complement (9.3.18) by the assumption that $(\lambda_n)_{n \in \mathbb{N}}$ is bounded away from zero then (9.3.18) implies that $\sum_{n=1}^{+\infty} \frac{\lambda_1 \cdots \lambda_{n-1}}{\mu_1 \mu_2 \cdots \mu_n} < \infty$ and consequently the birth and death semigroup has an invariant density and is asymptotically stable.

9.3.2　Time Asymptotics Versus Quasi-Compactness

In this subsection, we explore a class of Kolmogorov differential equations displaying a quasi-compactness property. We recall first that

$$\frac{1}{\sigma_i} \sum_{j \in \mathbb{N}} b_{ji} = 1 \quad (i \in \mathbb{N}).$$

Our main assumption is

$$\lim_{m \to +\infty} \left(\sup_{i \geq m} \frac{1}{\sigma_i} \sum_{j=m}^{\infty} b_{ji} \right) < 1. \tag{9.3.19}$$

We start with a key observation.

Lemma 9.3.1 *If (9.3.19) is satisfied then $B(0 - M)^{-1}$ is quasi compact in $l^1(\mathbb{N}, \nu)$.*

Proof We fix an integer $m > 1 \in \mathbb{N}$ such that

$$\alpha := \sup_{i \geq m} \frac{1}{\sigma_i} \sum_{j=m}^{\infty} b_{ji} < 1.$$

Hence

$$\sum_{j=m}^{\infty} b_{ji} \leq \alpha \sigma_i \quad (i \geq m).$$

We decompose B as

$$B = C + F$$

where $C = \left(c_{ij} \right)$ and $F = \left(f_{ij} \right)$ are given by

$$c_{ij} = \begin{cases} b_{ij} & \text{if } i \geq m \text{ and } j \geq m \\ 0 & \text{otherwise} \end{cases}$$

and $f_{ij} = b_{ij} - c_{ij}$. We note that

$$\sum_j c_{ji} = \sum_{j=m}^{\infty} b_{ji} \leq \alpha \sigma_i \quad (i \geq m)$$

so (since $c_{ji} = 0$ if $i < m$)

$$\sum_{j=1}^{\infty} c_{ji} \leq \alpha \sigma_i \quad (i \in \mathbb{N})$$

and consequently

$$\left\| C(0-M)^{-1} \right\|_{\mathcal{L}(l^1)} \le \alpha < 1. \tag{9.3.20}$$

Since

$$f_{ij} \ne 0 \text{ if and only if } i < m \text{ or } j < m$$

then

$$F(0-M)^{-1} \text{ has finite rank.} \tag{9.3.21}$$

Finally

$$B(0-M)^{-1} = C(0-M)^{-1} + F(0-M)^{-1} \tag{9.3.22}$$

is quasi-compact. $\square$

The splitting (9.3.22) with the properties (9.3.20) (9.3.21) and the use of Theorem 4.3.3 give

Theorem 9.3.3 *Suppose that (9.3.19) is satisfied and* $\inf_{i\in\mathbb{N}} \sigma_i > 0$. *Then* $(\mathcal{J}(t))_{t\ge0}$ *is genuinely honest in* $l^1(\mathbb{N}, \nu)$ *and has a spectral gap.*

Similarly, the translation of Theorem 6.6.1 within $l^1(\mathbb{N}, \nu)$ gives

Theorem 9.3.4 *Let (9.3.19) be satisfied and* $\inf_{i\in\mathbb{N}} \sigma_i = 0$. *Then 1 is an (isolated) eigenvalue of* $B(0-M)^{-1}$ *associated to an eigenvector* $(u_i)_{i\in\mathbb{N}} \in l^1_+(\mathbb{N}, \nu)$ *and* $(\mathcal{J}(t))_{t\ge0}$ *is genuinely honest. Assuming that* $u_i > 0$ $(i \in \mathbb{N})$ *we have the alternative.*
(i) If $\left(u_i\sigma_i^{-1}\right)_{i\in\mathbb{N}} \in l^1(\mathbb{N}, \nu)$ *then (after normalization)* $\left(u_i\sigma_i^{-1}\right)_{\in\mathbb{N}}$ *is an invariant density of* $(\mathcal{J}(t))_{t\ge0}$ *and is asymptotically stable.*
(ii) If $\left(u_i\sigma_i^{-1}\right)_{i\in\mathbb{N}} \notin l^1(\mathbb{N}, \nu)$ *then* $(\mathcal{J}(t))_{t\ge0}$ *is sweeping from the sets of summability of* $\left(u_i\sigma_i^{-1}\right)_{i\in\mathbb{N}}$.

Remark 9.3.1 It is easy to check that (9.3.19) is *not* satisfied by birth and death models.

9.3.3 *Spectral Gaps Versus "artificial" Absorption*

In this section, we consider a general Kato-Voigt substochastic C_0-semigroup $(\mathcal{J}(t))_{t\ge0}$ with generator

$$J \supset M + B : D(M) \to l^1(\mathbb{N}, \nu).$$

We do *not* assume a priori that $(\mathcal{J}(t))_{t\ge0}$ is honest. Let $s(J)$ be the spectral bound of J. The main statement is

Theorem 9.3.5 *Let $\beta = (\beta_i)_{i\geq 0}$ be a sequence bounded from below and let $(\mathcal{Z}(t))_{t\geq 0}$ be the C_0-semigroup generated by $J_\beta = J - \beta$ (we denote also by β the multiplication operator by $(\beta_i)_{i\geq 0}$ on $l^1(\mathbb{N}, \nu)$). If*

$$\lim_{i\to+\infty} \inf \beta_i > s(J) - s(J_\beta) \tag{9.3.23}$$

then $(\mathcal{Z}(t))_{t\geq 0}$ has a spectral gap.

Proof A key point in $l^1(\mathbb{N}, \nu)$ is that any bounded operator is locally compact. In particular $(\mathcal{J}(t))_{t\geq 0}$ is locally compact. Finally, Theorem 5.10.1 or Theorem 5.10.2 end the proof. $\qquad\square$

Of course, we may wonder whether (9.3.23) is checkable. To this end, we complement Theorem 9.3.5 by

Theorem 9.3.6 *Let $(\mathcal{J}(t))_{t\geq 0}$ be a symmetric Markov (i.e., contractive in all $l^p(\mathbb{N}, \nu)$ and symmetric in $l^2(\mathbb{N}, \nu)$) C_0-semigroup. Then (9.3.23) is satisfied provided that $\lim\inf_{i\to+\infty} \beta_i$ is greater than*

$$s(J) + \inf_{\left\{\|x\|=1,\ x\in D(\sqrt{-J^2})\right\}} \left\{ \left\| \sqrt{-J^2}x \right\|_{l^2}^2 + \sum_i \beta_i x_i^2 \right\}. \tag{9.3.24}$$

Proof Note that J^2, the generator of $(\mathcal{J}(t))_{t\geq 0}$ in $l^2(\mathbb{N}, \nu)$, is a self-adjoint and nonpositive. Then $\left(\mathcal{J}_\beta(t)\right)_{t\geq 0}$ is also symmetric in $l^2(\mathbb{N}, \nu)$ and $s(J_\beta^2) \leq s(J_\beta)$ where $-s(J_\beta^2)$ is equal to

$$\inf_{\|x\|=1,\ x\in D(\sqrt{-J^2})} \left\{ \left\| \sqrt{-J^2}x \right\|_{l^2}^2 + \sum_i \beta_i x_i^2 \right\} < +\infty$$

and we are done. $\qquad\square$

We note that if

$$l_m^2(\mathbb{N}, \nu) = \left\{ x = (x_i)_{i\geq 0};\ x_i = 0\ (i > m) \right\}$$

then (9.3.23) is satisfied provided that $\lim\inf_{i\to+\infty} \beta_i$ is greater than the quantity

$$s(J) + \inf_{\left\{x\in l_m^2,\ \|x\|=1,\ x\in D(\sqrt{-J^2})\right\}} \left\{ \left\| \sqrt{-J^2}x \right\|_{l^2}^2 + \sum_{i\leq m} \beta_i x_i^2 \right\}$$

which is a priori independent of $\lim\inf_{i\to+\infty} \beta_i$.

In some sense, up to a suitable absorption "at infinity", any Kato-Voigt semigroup in $l^1(\mathbb{N}, \nu)$ without spectral gap can be "approximated" by a substochastic C_0-semigroup having a spectral gap. We will see in Sect. 9.6 below how this "artificial" absorption occurs (in the form of a killing term) in the theory of weighted graphs.

9.4 Kolmogorov Differential Equations on Weighted Spaces

It may happen that (quasi-)compactness properties are not accessible in $l^1(\mathbb{N}, \nu)$ where ν denotes the counting measure, but can be recovered in suitably weighted l^1-spaces. In this subsection, we restrict our attention to a specific construction inspired by Transport Theory, (see Sects. 10.3 and 10.5). We explore a class of jump equations with detailed balance (9.3.10) in weighted spaces

$$l^1_p(\mathbb{N}) := l^1(\mathbb{N}, p\nu)$$

where $p := (p_n)_n$ is a suitable weight; (we denote the weight by $p := (p_n)_n$ instead of $\omega := (\omega_n)_n$ to avoid the confusion with $(\omega_n)_n$ occurring in the detailed balance condition). Their spectral and time asymptotic properties are given in Theorem 9.4.1 below which complements Theorem 9.3.1 in the standard sequence space $l^1(\mathbb{N})$. For the sake of motivation, we recall that a Cauchy problem

$$\begin{cases} \frac{d\varphi_i}{dt} = -\sigma_i \varphi_i + \sum_{j=0}^{\infty} b_{ij} \varphi_j & (i \in \mathbb{N}) \\ \varphi_i(0) = \varphi_i^0 & (i \in \mathbb{N}) \end{cases}$$

posed in $l^1_p(\mathbb{N})$ amounts (after the substitution to $\psi_i = p_i \varphi_i$) to a Cauchy problem

$$\begin{cases} \frac{d\psi_i}{dt} = -\sigma_i \psi_i + \sum_{j=0}^{\infty} \frac{p_i}{p_j} b_{ij} \psi_j & (i \in \mathbb{N}) \\ \psi_i(0) = \psi_i^0 & (i \in \mathbb{N}) \end{cases}$$

in $l^1(\mathbb{N})$ where the correspondence $\varphi \in l^1_p(\mathbb{N}) \to \psi = p\varphi \in l^1(\mathbb{N})$ is an isometry

$$\|\psi\|_{l^1(\mathbb{N})} = \|\varphi\|_{l^1_p(\mathbb{N})}.$$

Let $M : D(M) \subset l^1(\mathbb{N}) \to l^1(\mathbb{N})$ be the multiplication operator

$$\begin{cases} (x_i) \in D(M) \to -(\sigma_i x_i) \in l^1(\mathbb{N}) \\ D(M) = \left\{ (x_i) \in l^1(\mathbb{N}); \ (\sigma_i x_i) \in l^1(\mathbb{N}) \right\} \end{cases}$$

and the (holomorphic) C_0-semigroup $(\mathcal{M}(t))_{t \geq 0}$ generated by M given by

$$\mathcal{M}(t) : x = (x_i)_i \in l^1(\mathbb{N}) \to (x_i e^{-t\sigma_i})_i \in l^1(\mathbb{N}).$$

Let K be defined by

$$Kx = \left(\sum_{j=0}^{\infty} k_{ij} x_j \right)_i$$

where $k_{ij} := \frac{p_i}{p_j} b_{ij}$. The analysis of space homogeneous kinetic equations in Sect. 10.3 suggests a particular choice for the weight $p = (p_i)_i$

$$p_i = \sqrt{\omega_i} \ (i \in \mathbb{N})$$

so that $k_{ij} = \frac{\sqrt{\omega_i}}{\sqrt{\omega_j}} b_{ij}$ and

$$\begin{cases} k_{ii} = 0 \ (i \in \mathbb{N}) \\ \quad k_{ij} = k_{ji}. \end{cases}$$

The main spectral results with time asymptotics in weighted spaces are given in Theorem 9.4.1 below. We give first a key preliminary result.

Lemma 9.4.1 *Suppose that the conditions*

$$\begin{cases} \sup_{j\in\mathbb{N}} \sum_{i\in\mathbb{N}} \frac{\sqrt{\omega_i}}{\sqrt{\omega_j}} b_{ij} < +\infty \\ \lim_{m\to+\infty} \sup_{j\in\mathbb{N}} \sum_{i\geq m} \frac{\sqrt{\omega_i}}{\sqrt{\omega_j}} b_{ij} = 0 \end{cases} \tag{9.4.1}$$

are satisfied. Then K is a compact operator on $l^1(\mathbb{N})$. This holds true e.g. if there exists a non increasing $\Psi \in l^1_+(\mathbb{N})$ such that

$$\frac{\sqrt{\omega_i}}{\sqrt{\omega_j}} b_{ij} \leq \Psi(|i^2 - j^2|) \ (i \neq j). \tag{9.4.2}$$

Proof The first part of (9.4.1) insures that K is bounded on $l^1(\mathbb{N})$ while the second part of (9.4.1) expresses that

$$\|\chi_{\{i\geq m\}} K\| \to 0 \ (m \to +\infty).$$

Since $\chi_{\{i<m\}} K$ has finite dimensional rank then

$$\|K - \chi_{\{i<m\}} K\| = \|\chi_{\{i\geq m\}} K\| \to 0 \ (m \to +\infty)$$

shows that K is a compact operator.

Let (9.4.2) be satisfied. Since $k_{ii} = 0 \ (i \in \mathbb{N})$ then

$$\begin{aligned} \sup_{j\in\mathbb{N}} \sum_{i\in\mathbb{N}} \frac{\sqrt{\omega_i}}{\sqrt{\omega_j}} b_{ij} &= \sup_{j\in\mathbb{N}} \sum_{i\neq j} \frac{\sqrt{\omega_i}}{\sqrt{\omega_j}} b_{ij} \\ &\leq \sup_{j\in\mathbb{N}} \sum_{i\neq j} \Psi(|i^2 - j^2|) \\ &= \sup_{j\in\mathbb{N}} \sum_{i\neq j} \Psi((i+j)|i-j|) \end{aligned}$$

$$\leq \sup_{j \in \mathbb{N}} \sum_{i \neq j} \Psi(i+j) \leq \sum_{i} \Psi(i)$$

shows the first part of (9.4.1). Similarly

$$\sum_{i \geq m} \frac{\sqrt{\omega_i}}{\sqrt{\omega_j}} b_{ij} = \sum_{i \geq m,\, i \neq j} \frac{\sqrt{\omega_i}}{\sqrt{\omega_j}} b_{ij}$$

$$\leq \sum_{i \geq m,\, i \neq j} \Psi\big(|i^2 - j^2|\big)$$

$$= \sum_{i \geq m,\, i \neq j} \Psi\big((i+j)\,|i-j|\big)$$

$$\leq \Psi(i+j)$$

$$\leq \sum_{i \geq m,\, i \neq j} \Psi(i) \leq \sum_{i \geq m} \Psi(i)$$

implies the second part of (9.4.1). $\qquad\square$

We are going to develop a peripheral spectral theory in the weighted space

$$l^1_{\sqrt{\omega}}(\mathbb{N}) = \left\{ (x_j)_j \,;\; \sum_j \sqrt{\omega_j}\,|x_j| < +\infty \right\}$$

with norm

$$\|x\|_{l^1_{\sqrt{\omega}}} = \sum_j \sqrt{\omega_j}\,|x_j|.$$

Let $M_{\sqrt{\omega}} : D(M_{\sqrt{\omega}}) \subset l^1_{\sqrt{\omega}}(\mathbb{N}) \to l^1_{\sqrt{\omega}}(\mathbb{N})$ be the multiplication operator

$$(x_i) \in D(M_{\sqrt{\omega}}) \to -(\sigma_i x_i) \in l^1_{\sqrt{\omega}}(\mathbb{N})$$

with domain

$$D(M_{\sqrt{\omega}}) = \left\{ (x_i) \in l^1_{\sqrt{\omega}}(\mathbb{N});\; (\sigma_i x_i) \in l^1_{\sqrt{\omega}}(\mathbb{N}) \right\}$$

and let $\mathcal{M}_{\sqrt{\omega}} = \big(\mathcal{M}(t)_{\sqrt{\omega}}\big)_{t \geq 0}$ be the C_0-semigroup with generator $M_{\sqrt{\omega}}$. We are now ready to give the main result of this Section.

Theorem 9.4.1 *Let the detailed balance condition (9.3.10) and (9.4.1) be satisfied. Then B is a compact operator on $l^1_{\sqrt{\omega}}(\mathbb{N})$ and the C_0-semigroup $\big(\mathcal{J}_{\sqrt{\omega}}(t)\big)_{t \geq 0}$ generated by $M_{\sqrt{\omega}} + B$ in $l^1_{\sqrt{\omega}}(\mathbb{N})$ is holomorphic and shares the essential spectrum of $\big(\mathcal{M}(t)_{\sqrt{\omega}}\big)_{t \geq 0}$. In particular $\omega_{ess}(\mathcal{J}_{\sqrt{\omega}}) \leq -\inf \sigma_i$. Let additionally*

$$\sum_j \frac{1}{\sqrt{\omega_j}} + \sum_j \frac{\sigma_i}{\sqrt{\omega_j}} < +\infty. \qquad (9.4.3)$$

Then $0 = s(M_{\sqrt{\omega}} + B)$ is an eigenvalue of both $M_{\sqrt{\omega}} + B$ in $l^1_{\sqrt{\omega}}(\mathbb{N})$ and its dual operator $M^*_{\sqrt{\omega}} + B^*$ in $l^\infty_{\sqrt{\omega}}(\mathbb{N})$[1] associated to the (same) eigenvector ω^{-1} and:

(i) $\left(\mathcal{J}_{\sqrt{\omega}}(t)\right)_{t \geq 0}$ has a spectral gap if $\inf \sigma_i > 0$.

(ii) If $\inf \sigma_i = 0$ then $\left(\mathcal{J}_{\sqrt{\omega}}(t)\right)_{t \geq 0}$ extends to $l^1_{\frac{1}{\sqrt{\omega}}}(\mathbb{N})$ as a stochastic asymptoticaly stable C_0-semigroup in $l^1_{\frac{1}{\sqrt{\omega}}}(\mathbb{N})$.

Proof We have seen that the unbounded operator $M_{\sqrt{\omega}} + B$ in $l^1_{\sqrt{\omega}}(\mathbb{N})$ is similar, through the (isometric) similarity transformation

$$\varphi \in l^1_{\sqrt{\omega}}(\mathbb{N}) \to \psi = \sqrt{\omega}\varphi \in l^1(\mathbb{N}),$$

to the unbounded operator $M + K$ in $l^1(\mathbb{N})$. Hence the two operators share all properties and it follows from Lemma 9.4.1 that B is a compact operator on $l^1_{\sqrt{\omega}}(\mathbb{N})$. Note that (9.4.3) shows that $\omega^{-1} \in D(M_{\sqrt{\omega}})$ and, by (9.3.10),

$$M_{\sqrt{\omega}}\left(\omega^{-1}\right) + B\left(\omega^{-1}\right) = 0,$$

i.e., 0 is an eigenvalue of $M_{\sqrt{\omega}} + B$. Actually, (9.3.10) shows that B coincides with B^* on $l^1_{\sqrt{\omega}}(\mathbb{N}) \cap l^\infty_{\sqrt{\omega}}(\mathbb{N})$; indeed, if $f, g \in l^1_{\sqrt{\omega}}(\mathbb{N}) \cap l^\infty_{\sqrt{\omega}}(\mathbb{N})$ then

$$\langle Bf, g \rangle_{l^1_{\sqrt{\omega}}, l^\infty_{\sqrt{\omega}}} = \sum_i \sqrt{\omega_i}\left(\sum_j b_{ij} f_j\right) g_i = \sum_j \left(\sum_i \sqrt{\omega_i} b_{ij} g_i\right) f_j$$

$$= \sum_j \left(\sum_i \frac{\omega_j}{\sqrt{\omega_i}} b_{ji} g_i\right) f_j = \sum_j \sqrt{\omega_j}\left(\sum_i \frac{\sqrt{\omega_j}}{\sqrt{\omega_i}} b_{ji} g_i\right) f_j$$

$$= \sum_j \sqrt{\omega_j}\left(\sum_i \frac{\sqrt{\omega_i}}{\sqrt{\omega_j}} b_{ij} g_i\right) f_j = \langle f, Bg \rangle_{l^1_{\sqrt{\omega}}, l^\infty_{\sqrt{\omega}}}.$$

Thus 0 is an eigenvalue of $M^*_{\sqrt{\omega}} + B^*$ associated to ω^{-1} and consequently

$$\mathcal{J}^*_{\sqrt{\omega}}(t)\omega^{-1} = \omega^{-1}.$$

---

[1] Note that $l^\infty_{\sqrt{\omega}}(\mathbb{N})$ is nothing but $l^\infty(\mathbb{N})$ and the duality pairing is given by $\langle f, g \rangle_{l^1_{\sqrt{\omega}}, l^\infty} = \sum_j \sqrt{\omega_j} f_j g_j$.

Note that 0 must be equal to $s(M_{\sqrt{\omega}} + B)$; otherwise $s(M_{\sqrt{\omega}} + B) > 0$ would be another eigenvalue of $M_{\sqrt{\omega}} + B$ associated to a positive eigenvector and a standard argument leads to a contradiction.

Since $\omega_{ess}(\mathcal{J}_{\sqrt{\omega}}) \leq -\inf \sigma_i$ then $\left(\mathcal{J}_{\sqrt{\omega}}(t)\right)_{t \geq 0}$ has a spectral gap if $\inf \sigma_i > 0$.

Let now $\inf \sigma_i = 0$. Note that $\varphi \in l^1_{\sqrt{\omega}}(\mathbb{N})$, $\varphi \geq 0$

$$
\begin{aligned}
\left\| \mathcal{J}_{\sqrt{\omega}}(t)\varphi \right\|_{l^1_{\frac{1}{\sqrt{\omega}}}} &= \sum_j \left(\mathcal{J}_{\sqrt{\omega}}(t)\varphi \right)_j \frac{1}{\sqrt{\omega_j}} \\
&= \sum_j \left(\mathcal{J}_{\sqrt{\omega}}(t)\varphi \right)_j \frac{1}{\omega_j} \sqrt{\omega_j} \\
&= \langle \mathcal{J}_{\sqrt{\omega}}(t)\varphi, \omega^{-1} \rangle_{l^1_{\sqrt{\omega}}, l^\infty_{\sqrt{\omega}}} \\
&= \langle \varphi, \mathcal{J}^*_{\sqrt{\omega}}(t)\omega^{-1} \rangle_{l^1_{\sqrt{\omega}}, l^\infty_{\sqrt{\omega}}} \\
&= \langle \varphi, \omega^{-1} \rangle_{l^1_{\sqrt{\omega}}, l^\infty_{\sqrt{\omega}}} = \sum_j \varphi_j \omega_j^{-1} \sqrt{\omega_j} \\
&= \sum_j \frac{\varphi_j}{\sqrt{\omega_j}} = \|\varphi\|_{l^1_{\frac{1}{\sqrt{\omega}}}} .
\end{aligned}
$$

In sequence spaces, all operators are integral operators and consequently, by Theorem 2.10.3, $\left(\mathcal{J}_{\sqrt{\omega}}(t)\right)_{t \geq 0}$ is a stochastic asymptoticaly stable C_0-semigroup in $l^1_{\frac{1}{\sqrt{\omega}}}(\mathbb{N})$. $\qquad\square$

9.5 Kolmogorov Differential Equations on Moment Spaces

This section continues the spectral properties developed in Sect. 9.4 by focusing on specific, though important, choices of weights

$$
p_j = 1 + j^s, \quad (s \geq 1). \tag{9.5.1}
$$

A general construction of a semigroup in the first moment space (i.e., for $s = 1$) is given in [347], (see also [224]). The first moment $\sum_j j x_j$ is particularly significant due to its probabilistic interpretation as the expected value.

To the best of our knowledge, l^1 peripheral spectral analysis and long-time asymptotics of Kolmogorov differential equations on moment spaces have not yet been addressed in the existing literature. The aim of this section is to present several results in this direction.(see Theorem 9.5.1 below). Since the weight (9.5.1) is prescribed a priori, our strategy here is to adopt a *"reverse"* construction to that of Sect. 9.4. In other words, we choose first (a non increasing) $\Psi \in l^1_+(\mathbb{N})$ and a positive sequence $(\omega_j)_j$ such that $\left(\omega_j^{-1}\right)_j$ is bounded and then define b_{ij} by

$$\begin{cases} \frac{\sqrt{\omega_i}}{\sqrt{\omega_j}} b_{ij} = \Psi(|i^2 - j^2|) \ (i \neq j) \\ \qquad\qquad b_{ii} = 0. \end{cases}$$

By construction, $\frac{\sqrt{\omega_i}}{\sqrt{\omega_j}} b_{ij}$ is symmetric and the detailed balance is satisfied. Finally, we define $(\sigma_j)_j$ by

$$\sigma_j := \sum_i b_{ij} = \sum_i \frac{\sqrt{\omega_j}}{\sqrt{\omega_i}} \Psi(|i^2 - j^2|).$$

A simple calculation shows that if we choose $(\omega_j)_j$ such that

$$\sum_j \frac{1}{\sqrt{\omega_j}} < +\infty \tag{9.5.2}$$

then (9.4.3) is satisfied. In this way, we derive a whole class of jump equations, indexed by $\Psi \in l^1_+(\mathbb{N})$ and $(\omega_j)_j$, which is amenable to the framework of Theorem 9.4.1 and specifically tailored to the particular weights of interest.

However, whether $\inf \sigma > 0$ or $\inf \sigma = 0$ is unclear a priori and requires further analysis; this distinction is crucial, as it determines the presence or absence of a spectral gap. If we are interested in the weight

$$\sqrt{\omega_j} = 1 + j^s$$

then the constraint (9.5.2) amounts to $s > 1$. In particular, we are unable to establish a peripheral spectral theory exactly in the first moment space (i.e., for $s = 1$) but can do so in higher moment spaces. The main result of this Section follows from Theorem 9.4.1.

Theorem 9.5.1 *Let $s > 1$ and $\sqrt{\omega_j} = 1 + j^s$. Let $\Psi \in l^1_+(\mathbb{N})$ be non increasing and*

$$\begin{cases} b_{ij} = \frac{1+j^s}{1+i^s} \Psi(|i^2 - j^2|) \ (i \neq j) \\ \qquad\qquad b_{ii} = 0 \end{cases}$$

and let

$$\sigma_j = \sum_i b_{ij} = \sum_i \frac{1 + j^s}{1 + i^s} \Psi(|i^2 - j^2|).$$

Then the detailed balance condition (9.3.10) and (9.4.1) are satisfied. Moreover B is a compact operator on $l^1_{\sqrt{\omega}}(\mathbb{N})$ and the C_0-semigroup $\left(\mathcal{J}_{\sqrt{\omega}}(t)\right)_{t \geq 0}$ generated by $M_{\sqrt{\omega}} + B$ in $l^1_{\sqrt{\omega}}(\mathbb{N})$ is holomorphic and shares the essential spectrum of $\left(\mathcal{M}(t)_{\sqrt{\omega}}\right)_{t \geq 0}$. In particular $\omega_{ess}(\mathcal{J}_{\sqrt{\omega}}) \leq -\inf \sigma_i$. Furthermore $0 = s(M_{\sqrt{\omega}} + B)$ is an eigenvalue of $M_{\sqrt{\omega}} + B$ associated to the eigenvector ω^{-1} and:

(i) $\left(\mathcal{J}_{\sqrt{\omega}}(t)\right)_{t \geq 0}$ *has a spectral gap if* $\inf \sigma_i > 0$.

(ii) If $\inf \sigma_i = 0$ *then* $\left(\mathcal{J}_{\sqrt{\omega}}(t)\right)_{t \geq 0}$ *extends to* $l^1_{\frac{1}{\sqrt{\omega}}}(\mathbb{N})$ *as a stochastic asymptoticaly stable* C_0*-semigroup in* $l^1_{\frac{1}{\sqrt{\omega}}}(\mathbb{N})$.

9.6 Weighted Graphs with Killing Term

This section is devoted to the spectral gap analysis of weighted graphs. There is a rich literature on this topic (see, e.g., [184, 185] and the numerous references therein). Notably, [380] establishes the equivalence between stochastic completeness and the concept of honesty in Kato-Voigt perturbation theory, as discussed in Sect. 2.11. Additionally, [184] provides lower bounds on the infimum of the spectrum of weighted graphs based on isoperimetric inequalities, with applications to the absence of essential spectrum. In this section, we present new spectral results that rely on different mathematical tools. Before presenting them, we introduce first some basic objects. Let E be a countable set and m be a measure on E of full support, i.e., $m(e) > 0$ for all $e \in E$. Let

$$l^p_m := l^p(E, m) \ \ (1 \leq p < +\infty)$$

endowed with the norm

$$\|u\|_p = \left(\sum_{x \in E} m(x) |u(x)|^p\right)^{\frac{1}{p}}.$$

In this section, the measure is denoted by m instead of ν to stick to the notations in the literature. We denote the duality between l^1_m and l^∞ by

$$\langle u, v \rangle_m = \sum_{x \in E} m(x) u(x) v(x) \ \ (u \in l^1_m, \ v \in l^\infty).$$

A symmetric weighted graph on E is a pair (b, c) consisting of two maps on E

$$\begin{cases} c : E \to [0, +\infty) \\ b : E \times E \to [0, +\infty) \end{cases}$$

such that

(i) $b(x, x) = 0$

(ii) $b(x, y) = b(y, x)$ for all $x, y \in E$

(iii) $\sum_{y \in E} b(x, y) < \infty$ for all $x \in E$.

The triple (E, b, c) represents a weighted graph with vertex set E and $b(x, y)$ the weight on the edge connecting the points $x, y \in E$; the weighted graphs need not be locally finite. If $c(x) > 0$, we think of x as connected to the point "infinity" by an edge with weight $c(x)$. The map c is called the killing term. We define the Dirichlet form on l_m^2

$$Q(u) = \frac{1}{2} \sum_{x \in E} \sum_{y \in E} b(x, y)\,(u(x) - u(y))^2 + \sum_{x \in E} c(x)u(x)^2$$

with domain

$$D(Q) = \left\{ u \in l_m^2;\ Q(u) < +\infty \right\}.$$

It is known (see [99, 184, 185] for the details) that Q is a closed form, i.e., $D(Q)$ endowed with the norm

$$\|u\|_Q^2 = \sqrt{\|u\|_{l_m^2}^2 + Q(u)}$$

is complete, and there exists a unique selfadjoint operator $L_2 \geq 0$ such that $D(\sqrt{L_2}) = D(Q)$ and

$$Q(u) = \left\| \sqrt{L_2}u \right\|_{l_m^2}^2.$$

The object of this section is to give a spectral gap result of L_2 relying on the killing term. The main result is an upper bound of the bottom of essential spectrum of L_2,

$$\sigma_{ess}(L_2) \subset \left[\lim \inf_{\mathcal{P}_c(E)} \frac{c}{m},\ +\infty \right)$$

where $\mathcal{P}_c(E)$ is (the basis of filter of) the subsets of E with finite complements and lim inf is taken along the directed sets $\mathcal{P}_c(E)$, (see Theorem 9.6.2). As far as we know, no similar result exists in the literature.

According to the general theory (see [99] for the details), $\left(e^{-tL_2}\right)_{t \geq 0}$ is a Markov semigroup, i.e., extends to all l_m^p as a contraction semigroup $\left(e^{-tL_p}\right)_{t \geq 0}$ (with generator $-L_p$) and is strongly continuous if $1 \leq p < +\infty$. Note that

$$s(-L_2) = -\inf_{\left\{ u \in D(\sqrt{L_2}),\ \|u\|_{l_m^2} = 1 \right\}} Q(u).$$

We define the multiplication operator M on l_m^1 by

$$Mu(x) = -\frac{1}{m(x)} \left(\sum_{y \in E} b(x, y) + c(x) \right) u(x)$$

with maximal domain

$$D(M) = \left\{ u \in l_m^1;\ Mu \in l_m^1 \right\}$$

and $B : D(M) \to l_m^1$ by

$$Bu(x) = \frac{1}{m(x)} \sum_{y \in E} b(x, y)u(y).$$

Then B is M-bounded and

$$\sum_{y \in E} \left(Mu(x) + Bu(x)\right) m(x) \leq 0 \quad (u \in D_+(M))$$

so, by Kato-Voigt pertubation theory, there exists a unique extension $J \supset M + B$ which generates a minimal substochastic C_0-semigroup $(\mathcal{J}(t))_{t \geq 0}$ on l_m^1. We start with a key information.

Theorem 9.6.1 ([380] Proposition 3.2) *The Kato-Voigt generator J is nothing but $-L_1$ or equivalently $(\mathcal{J}(t))_{t \geq 0} = \left(e^{-tL_1}\right)_{t \geq 0}$.*

As in Theorem 4.6.3 (see also Remark 4.6.1), if $\left(e^{-tL_1}\right)_{t \geq 0}$ has spectral gap then $\left(e^{-tL_p}\right)_{t \geq 0}$ has also a spectral gap. We explore here the role of a killing term in the production of a spectral gap. Let us investigate further $(\mathcal{J}(t))_{t \geq 0}$. Let $V(x) := \frac{c(x)}{m(x)}$ and

$$\tilde{M}u(x) = -\frac{1}{m(x)} \left(\sum_{y \in E} b(x, y)\right) u(x).$$

Since

$$\sum_{y \in E} \left(\tilde{M}u(x) + Bu(x)\right) m(x) = 0 \quad (u \in D_+(M)) \tag{9.6.1}$$

then, by Kato-Voigt perturbation theory, there exists a unique extension

$$\tilde{J} \supset \tilde{M} + B$$

which generates a minimal substochastic C_0-semigroup $\left(\tilde{\mathcal{J}}(t)\right)_{t \geq 0}$ on l_m^1. It turns out that

$$(\mathcal{J}(t))_{t \geq 0} = \left(\tilde{\mathcal{J}}_V(t)\right)_{t \geq 0},$$

i.e., $(\mathcal{J}(t))_{t \geq 0}$ is the absorption semigroup on l_m^1 obtained from $\left(\tilde{\mathcal{J}}(t)\right)_{t \geq 0}$ and the absorption V. Before stating the main result of this section, we need a

Definition 9.6.1 Let $\mathcal{P}_c(E)$ be the class of subsets of E whose complements are finite sets. Since $\mathcal{P}_c(E)$ is (downward) directed with respect to inclusion (and hence a basis of a filter on E), for any function

$$\zeta : E \to [-\infty, +\infty],$$

we define its lim inf along this filter, denoted by $\liminf_{\mathcal{P}_c(E)} \zeta$, as the supremum in $[-\infty, +\infty]$ of the set $\cap_{S \in \mathcal{P}_c(E)} \overline{\zeta(S)}$; (see e.g. [89] p. 123).

We are ready to state

Theorem 9.6.2 *We have*

(i) *If* $\lim_{\mathcal{P}_c(E)} \frac{c}{m} = +\infty$ *then* L_2 *is resolvent compact on* l_m^2.

(ii) *If* $\liminf_{\mathcal{P}_c(E)} \frac{c}{m} > -s(-L_2)$ *then* $\sigma_{ess}(L_2) \subset \left[\liminf_{\mathcal{P}_c(E)} \frac{c}{m}, +\infty\right)$.

Proof We point out that a priori we need no metric on E. However, the results given in Chap. 5 for locally compact metric spaces remain true (with the same proofs) where the *finite* sets play the role of compact sets and any bounded operator O on l_m^1 is "locally compact" in the sense that $1_F O$ is compact (actually has finite rank) where 1_F is the indicator function of a finite set $F \subset E$. Thus we can use freely the results of Chap. 5 in the present context.

(i) According to Remark 5.6.1 $\lim_{\mathcal{P}_c(E)} \frac{c}{m} = +\infty$ implies that $\tilde{J}_V$ is resolvent compact in l_m^1 so $J = \tilde{J}_V$ is resolvent compact in l_m^1. It follows from Remark 5.6.2 that $-L_2$ is resolvent compact in l_m^2.

(ii) Since e^{-tL_2} is symmetric then a convexity (interpolation) argument shows that

$$\left\| e^{-tL_2} \right\| \leq \left\| e^{-tL_1} \right\|$$

and

$$\lim_{t \to +\infty} t^{-1} \ln \left\| e^{-tL_2} \right\| \leq \lim_{t \to +\infty} t^{-1} \ln \left\| e^{-tL_1} \right\| ,$$

i.e., $s(-L_2) \leq s(J)$ since the spectral bound of the generator coincides with the type of the semigroup. Since

$$J = \tilde{J}_V$$

then $-s(\tilde{J}_V) \leq -s(-L_2)$ and $\lim \inf_{\mathcal{P}_c(E)} \frac{c}{m} > -s(\tilde{J}_V)$. Since $s(\tilde{J}) \leq 0$ ($s(\tilde{J}) = 0$ if $\left(\tilde{J}(t)\right)_{t \geq 0}$ is honest because of (9.6.1)) then

$$\lim \inf_{\mathcal{P}_c(E)} \frac{c}{m} > s(\tilde{J}) - s(\tilde{J}_V)$$

and consequently Theorem 5.10.1 or Theorem 5.10.2 show that

$$\sigma_{ess}(J) \subset \left(-\infty, -\lim \inf_{\mathcal{P}_c(E)} \frac{c}{m}\right].$$

Finally, by Theorem 5.9.1

$$\sigma_{ess}(-L_2) \subset \left(-\infty, -\lim \inf_{\mathcal{P}_c(E)} \frac{c}{m}\right]$$

and we are done. $\qquad\Box$

Note that the bottom of $\sigma(L_2)$ is $< \liminf_{\mathcal{P}_c(E)} \frac{c}{m}$. Note also that if a weighted graph without killing term (i.e., corresponding to $c(.) = 0$) has spectral gap (i.e., $\delta := \omega(\mathcal{J}) - \omega_{ess}(\mathcal{J}) > 0$) and if we add a killing term $c(x)$ such that $\frac{c(.)}{m(.)} \in l^{\infty}(E)$ then the new weighted graph has a spectral gap provided that the oscillation of $\frac{c(.)}{m(.)}$ is less than δ, see Theorems 4.4.1 and 5.9.1.

9.7 Comments

About Theorem 9.3.1:

(i) We can of course translate all the hilbertian results of Chap. 7 in the weighted space $l^2(\mathbb{N}, \mu)$ where $\mu = \omega\nu$.
(ii) In case $\left(\omega_i^{-1}\right)_{\in\mathbb{N}} \in l^1(\mathbb{N}, \nu)$ and $(\sigma_i)_{\in\mathbb{N}}$ is bounded away from zero, we wonder whether the convergence of $(\mathcal{J}(t))_{t\geq 0}$ to equilibrium is exponential?
(iii) We wonder whether this theorem remains true when $\left(\sigma_i\omega_i^{-1}\right)_{\in\mathbb{N}} \notin l^1(\mathbb{N}, \nu)$; see Remark 3.3.2.

We have seen in Theorems 9.3.3 and 9.3.4 the key role of Assumption (9.3.19). If this assumption fails, i.e., if

$$\lim_{m\to+\infty} \left(\sup_{i\geq m} \frac{1}{\sigma_i} \sum_{j=m}^{\infty} b_{ji} \right) = 1,$$

then the peripheral spectral analysis becomes much more involved and deserves a separate study.

In connection with Theorem 9.5.1: Let $\sqrt{\omega_j} = 1 + j^s$ $(s > 1)$ and

$$\sigma_j = \sum_i \frac{1 + j^s}{1 + i^s} \Psi(|i^2 - j^2|) \quad (j \in \mathbb{N}),$$

for which class of non increasing $\Psi \in l^1_+(\mathbb{N})$ do we have $\inf \sigma_i > 0$?

We ask whether the analysis in Sect. 9.6 can be extended to non-symmetric weighted graphs, that is, when the condition $b(x, y) = b(y, x)$ $(x, y \in E)$ is dropped.

Chapter 10
Transport Theory

10.1   Chapter Aims

This long chapter revisits and extends in differents directions and within the physical L^1-settings, many classical problems in spectral and scattering theory for Neutron Transport, while also exploring non-classical issues. In particular, we study neutron transport equations (with non-incoming boundary conditions) involving more complex scattering phenomena, including inelastic, downshift, and Bragg scattering, as well as space-inhomogeneous linear Boltzmann equations with power-like hard-sphere potentials. We also provide a spectral analysis for boundary conditions modeled by stochastic diffuse operators, along with a spectral gap analysis for neutron diffusion equations. Variational, Max-Inf and Inf-Sup, characterizations of the leading eigenvalue of neutron transport operators are also provided.

10.2   Chapter Overview

This chapter is dedicated to an in-depth exploration of Transport Theory, revisiting and extending numerous classical results, while also addressing non-classical problems, particularly in the context of Neutron Transport. Alongside the companion chapter on Population Dynamics, it contains many of the theoretical developments that have motivated the broader construction of this monograph. Each section within the chapter focuses on a specific topic and is preceded by its own introduction, offering context and motivation. The mathematical framework employed throughout is primarily the physical L^1-space. However, a predual analysis is also outlined in Chap. 14, allowing the derivation of results in more general L^p-spaces. In line with the overarching goals of this monograph, we consider peripheral spectral problems that admit only general results. A more detailed treatment of specific classes of neutron transport operators, those that permit a deeper analysis, is briefly

M. Mokhtar-Kharroubi, *Peripheral Spectra of Perturbed Positive Semigroups*, Lecture Notes in Mathematics 2388, https://doi.org/10.1007/978-3-032-11173-9_10

239

discussed in Sect. 14.6. Due to the breadth of material, we do not delve into technical details here. Instead, for the reader's convenience, we provide below a list of the key topics addressed in this chapter. Each section's introduction offers further detail and insight into its particular focus. In Sect. 10.3, we revisit several classical studies on space-homogeneous and isotropic equations arising in neutron thermalization for monoatomic gas models [92, 289, 333, 351, 352]. Due to isotropy, these equations reduce, via angular averaging, to one-dimensional problems. While the original analyses were conducted in weighted L^2-spaces, we demonstrate how to treat them within the more physically relevant framework of weighted L^1-spaces. In Sect. 10.4, we revisit the multidimensional physical free gas model studied by Suhadolc [343], focusing specifically on the space-homogeneous case and working within the weighted L^1-framework. We show that compactness results can in fact be obtained in the natural, unweighted space $L^1(dx)$ provided the absorption cross section is sufficiently strong. In Sect. 10.5, still within the framework of space-homogeneous models, we undertake a deeper analysis of Suhadolc's work [343], this time without invoking any additional absorption cross section. The study is conducted in the setting of space-homogeneous linear Boltzmann equations, considered in various weighted L^1-spaces. In Sect. 10.6, we continue the investigation of conservative space-homogeneous equations with additional more complex scattering phenomena ("downshift" and "Bragg" scattering) occuring in nuclear reactor theory [192]; we build a general spectral gap theory for these models. In Sect. 10.7, we demonstrate how generation and mean ergodicity results for space-homogeneous equations, with arbitrary velocity measures, extend to the space-nonhomogeneous neutron transport equations. Section 10.8 is devoted to a systematic peripheral spectral analysis of space-nonhomogeneous neutron transport semigroups (for a suitable class of abstract velocity measures) and to the study of its consequences for their long-time asymptotics. This program was initiated in [251, 266] for bounded collision operators in spatial domains of finite volume. Here, we extend the analysis to unbounded (Σ-bounded) collision operators, where Σ denotes the collision frequency. In Sect. 10.9, we show how the abstract results of Sect. 10.8 apply to space-nonhomogeneous linear Boltzmann semigroups for free gas models, and, in particular, how they resolve a problem left open in Suhadolc's work [343].

In Sect. 10.10, we show how the abstract Max-Inf and Inf-Sup variational characterizations of the leading eigenvalue, developed in Sect. 2.5, apply to neutron transport theory, and we also provide existence results for this leading eigenvalue. Section 10.11 is devoted to the criticality eigenvalue problem and to the time asymptotics of the corresponding semigroup. In Sect. 10.12, we consider stochastic neutron transport semigroups in the whole space $\mathbb{R}^d$. Despite the extension of several key compactness results to this setting, for cross sections vanishing at infinity, the existence of an invariant density remains an open question. Conversely, we develop a general scattering theory for cross sections that are not necessarily compactly supported in space. In Sect. 10.13, we present a general spectral and scattering theory of particle swarms in weakly ionized gases. In Sect. 10.14, we analyze the peripheral spectrum and time asymptotics of neutron transport semigroups with partly elastic collision operators on the d-dimensional torus. In contrast to the previous sections, which focus on vacuum or periodic boundary conditions, Sect. 10.15 is devoted to the spectral theory and time asymptotics of neutron transport semigroups with non-local (mass-preserving) boundary operators relating incoming and outgoing fluxes. The previous sections focus mainly on the so-called asymptotic (discrete) spectrum of neutron transport semigroups; in Sect. 10.16, we show how to characterize the L^1 essential spectrum for Neutron Transport. In Sect. 10.17, we explore the spectral theory of a class of diffusion models used in nuclear reactor theory, in which the spatial transport operator is replaced by a diffusion operator that depends on the velocity variable as a parameter. Many remarks, comments or open questions are relegated to the end of this chapter in the section "Comments".

10.3 Classical Works in Neutron Thermalization Theory

As noted in the introduction of this monograph, space-homogeneous equations for Neutron Transport were investigated early on for isotropic models; see e.g. [92, 289, 333, 351, 352]. Similar equations also appear in the study of dispersal evolution in Mathematical Biology; see [93, 167]. The isotropy assumption allows angular averaging, reducing the equations to a one-dimensional form in the speed variable. This one-dimensional structure significantly simplifies their analysis. In Sect. 10.5, we consider the multidimensional case. These equations have been studied extensively in suitable weighted L^2-spaces. Our aim here is to revisit them in the framework of weighted L^1-spaces, focusing exclusively on peripheral spectral analysis. We also show that the results extend to weighted L^p-spaces. The main results are given in Theorem 10.3.1 below.

By using the notations in [352], the equation reads as

$$\frac{\partial \psi(t, x)}{\partial t} + \Sigma_s(x) x \psi(t, x) = \int_0^{+\infty} \Sigma_s(x, x') x' \psi(t, x') dx' \qquad (10.3.1)$$

$(x = |\zeta| \ (\zeta \in \mathbb{R}^3)$ denotes a speed) where

$$\Sigma_s(x) = \int_0^{+\infty} \Sigma_s(x', x) dx'.$$

Remark 10.3.1 It follows from

$$\int_{\mathbb{R}^3} f(\zeta) d\zeta = \int_0^{+\infty} x^2 dx \int_{S^2} f(x\omega) d\omega \quad (f \in L^1(\mathbb{R}^3))$$

that $h(x) := \int_{S^2} f(x\omega) d\omega$ belongs to $L^1(\mathbb{R}_+, x^2 dx)$. Thus, a priori, Eq. (10.3.1) should be investigated in $L^1(\mathbb{R}_+, x^2 dx)$.

The scattering kernel $\Sigma_s(x, x')$ satisfies the detailed balance

$$\Sigma_s(x, x')x'M(x') = \Sigma_s(x', x)xM(x) \tag{10.3.2}$$

where

$$M(x) = \frac{4}{\sqrt{\pi}}x^2 \exp(-x^2).$$

We focus here more specifically on the works of Shizuta [333] and Ukaï [352]. Following them, the substitution

$$\psi(x) = \varphi(x)x \exp(-\frac{x^2}{2}) \tag{10.3.3}$$

transforms formally Eq. (10.3.1) into

$$\frac{\partial \varphi(t, x)}{\partial t} + \Sigma_s(x)x\varphi(t, x) = \int_0^{+\infty} k(x, x')\varphi(t, x')dx' \tag{10.3.4}$$

where

$$k(x, x') = x^{-1}x'^2 \Sigma_s(x, x') \exp(-\frac{1}{2}\left(x'^2 - x^2\right))$$

is symmetric because of (10.3.2). It is Eq. (10.3.4) which is considered in the papers above in the Hilbert space

$$H := L^2(\mathbb{R}_+, dx).$$

In the case of mono-atomic gas model, explicit expressions of $\Sigma_s(x)$ and $k(x, x')$ are given e.g. in [333, 352]. We do not need here the details of these expressions; it is enough for us to know (see [333, 352]) that there exist positive constants α, β, C such that

$$\begin{cases} \int_0^{+\infty} k(x, x')dx' \le C \ (x \in \mathbb{R}_+) \\ k(x, x') \le \alpha e^{-\beta|x^2-x'^2|} \ (x, x' \in \mathbb{R}_+) \\ \inf_{x \in \mathbb{R}_+} \Sigma_s(x)x > 0. \end{cases} \tag{10.3.5}$$

Let V_p be the multiplication operator on $L^p(\mathbb{R}_+, dx)$ with maximal domain

$$V_p : \varphi \in D(V_p) \subset L^p(\mathbb{R}_+, dx) \to -\Sigma_s(x)x\varphi \in L^p(\mathbb{R}_+, dx).$$

Many results are given in [333, 352]. Among them, we just mention here that

$$K : \varphi \in L^2(\mathbb{R}_+, dx) \to \int_0^{+\infty} k(x, x')\varphi(x')dx' \tag{10.3.6}$$

is bounded on $L^2(\mathbb{R}_+, dx)$ and that $K(\lambda - V_2)^{-1}$ is a Hilbert-Schmidt operator and is therefore compact on $L^2(\mathbb{R}_+, dx)$. The authors explore finely the spectrum of the self-adjoint operator $M + K$. Our object here is to give a general peripheral spectral theory of these equations in

$$L^p(\mathbb{R}_+, dx), \quad (1 \le p < +\infty).$$

The key mathematical ingredient is an L^1 compactness result.

Lemma 10.3.1 *Let (10.3.5) be satisfied. Then the operator K extends to all $L^p(\mathbb{R}_+, dx)$ $(1 \le p \le +\infty)$ and is compact in all them.*

Proof (i) It follows from $\int_0^{+\infty} k(x, x')dx' \le C$ $(x \in \mathbb{R}_+)$ and the symmetry of $k(x, x')$, by using Hölder inequality, that K is bounded in all $L^p(\mathbb{R}_+, dx)$ $(1 \le p \le +\infty)$.

(ii) Consider the case $p = 1$. The estimate

$$\int_c^{+\infty} dx \int_0^{+\infty} k(x, x') \left|\varphi(x')\right| dx'$$

$$\le \alpha \int_c^{+\infty} dx \int_0^{+\infty} e^{-\beta(x+x')|x-x'|} \left|\varphi(x')\right| dx'$$

$$\le \alpha \int_c^{+\infty} dx \int_0^{+\infty} e^{-\beta c|x-x'|} \left|\varphi(x')\right| dx'$$

$$\le \alpha \int_0^{+\infty} dx \int_0^{+\infty} e^{-\beta c|x-x'|} \left|\varphi(x')\right| dx'$$

$$\le \alpha \int_0^{+\infty} e^{-\beta c|z|} dz \, \|\varphi\|_{L^1}$$

shows that the integral operator $1_{\{x \ge c\}} K$ tends to zero in operator norm as $c \to +\infty$. By the closedness of the set of compact operators, it suffices to show that $1_{\{x < c\}} K$ is compact. To this end, let us show that for a *given* $c > 0$, the norm of $1_{\{x < c\}} K 1_{\{x' > c'\}}$ tends to zero in operator norm as $c' \to +\infty$. Indeed, the estimate for $c' > c$

$$\int_0^c dx \int_{c'}^{+\infty} k(x, x') \left|\varphi(x')\right| dx'$$

$$\le \alpha \int_0^c dx \int_{c'}^{+\infty} e^{-\beta(x+x')|x-x'|} \left|\varphi(x')\right| dx'$$

$$\le \alpha \int_0^c dx \int_{c'}^{+\infty} e^{-\beta c'|x-x'|} \left|\varphi(x')\right| dx'$$

$$\le \alpha \int_0^c dx \int_{c'}^{+\infty} e^{-\beta c'|(c'-c)|} \left|\varphi(x')\right| dx'$$

$$\le \alpha c e^{-\beta c'|(c'-c)|} \|\varphi\|_{L^1}$$

shows our claim. Finally, it suffices to show that $1_{\{x < c\}} K 1_{\{x' < c'\}}$ is compact. This follows easily from the fact that its kernel

$$1_{\{x<c\}}k(x, x')1_{\{x'<c'\}} \leq \alpha 1_{\{x<c\}}e^{-\beta|x^2-x'^2|}1_{\{x'<c'\}} \quad (x, x' \in \mathbb{R}_+)$$

is smooth and has a bounded support.

(iii) Since K is compact in $L^1(\mathbb{R}_+, dx)$ then, by duality (and symmetry of K), it is also compact in $L^\infty(\mathbb{R}_+, dx)$. Finally, the compactness in $L^p(\mathbb{R}_+, dx)$ $(1 < p < \infty)$ follows by interpolation (see e.g. [99, Theorem 1.6.1]). $\qquad\qquad\square$

Let $\left(T_p(t)\right)_{t\geq0}$ be the holomorphic C_0-semigroup with generator V_p. It follows from (10.3.2) that

$$\int_0^{+\infty} \Sigma_s(x, x')x'M(x')dx' = \Sigma_s(x)xM(x).$$

Since $M \in L^p(\mathbb{R}_+, dx)$ and $\Sigma_s(x)xM \in L^p(\mathbb{R}_+, dx)$ then $M \in D(V_p)$ and

$$V_pM + KM = 0, \tag{10.3.7}$$

i.e., 0 is an eigenvalue of $V_p + K$. We are ready to state the main result of this section.

Theorem 10.3.1 *The unbounded operator*

$$A_p := V_p + K : D(V_p) \subset L^p(\mathbb{R}_+, dx) \to L^p(\mathbb{R}_+, dx)$$

generates a holomorphic positive C_0-semigroup $\left(Z_p(t)\right)_{t\geq0}$ and $\sigma_{ess}(T_p) = \sigma_{ess}(Z_p)$. In particular $\left(Z_p(t)\right)_{t\geq0}$ has a spectral gap. Moreover

$$\sigma(A_p) \cap \left\{\mathrm{Re}\,\lambda > -\inf_x \Sigma_s(x)x\right\}$$

is p-independent.

Proof The holomorphy of $\left(Z_p(t)\right)_{t\geq0}$ is a consequence e.g. of Theorem 2.6.5. According to Proposition 2.5.1

$$(\lambda - A_p)^{-1} = (\lambda - V_p)^{-1}\sum_{j=0}^{\infty}\left(K(\lambda - V_p)^{-1}\right)^j \quad (\lambda > s(A_p))$$

so

$$(\lambda - A_p)^{-1} - (\lambda - V_p)^{-1} = (\lambda - V_p)^{-1}\sum_{j=1}^{\infty}\left(K(\lambda - V_p)^{-1}\right)^j$$

is compact in $L^p(\mathbb{R}_+, dx)$ $(1 \leq p < +\infty)$. On the other hand

$$(\lambda - A_p)^{-1} - (\lambda - V_p)^{-1} = \int_0^{+\infty} e^{-\lambda t}(Z_p(t) - T_p(t))dt$$

$$\geq \int_{\bar{t}}^{\bar{t}+\varepsilon} e^{-\lambda t}(Z_p(t) - T_p(t))dt \ (\bar{t} > 0, \ \varepsilon > 0)$$

shows, by a domination argument, that

$$\int_{\bar{t}}^{\bar{t}+\varepsilon} e^{-\lambda t}(Z_p(t) - T_p(t))dt$$

is weakly compact in $L^1(\mathbb{R}_+, dx)$ and (see [4]) compact in $L^p(\mathbb{R}_+, dx)$ $(1 < p < +\infty)$. It follows that

$$e^{-\lambda \bar{t}}(Z_p(\bar{t}) - T_p(\bar{t})) = \lim_{\varepsilon \to 0} \varepsilon^{-1} \int_{\bar{t}}^{\bar{t}+\varepsilon} e^{-\lambda t}(Z_p(t) - T_p(t))dt \qquad (10.3.8)$$

is weakly compact in $L^1(\mathbb{R}_+, dx)$ and compact in $L^p(\mathbb{R}_+, dx)$ $(1 < p < +\infty)$ because the convergence (10.3.8) holds in operator norm since

$$t \in (0, +\infty) \to Z_p(t) - T_p(t)$$

is continuous in operator norm. Hence $\sigma_{ess}(T_p) = \sigma_{ess}(Z_p)$. In particular

$$\omega_{ess}(Z_p) = \omega_{ess}(T_p).$$

Since $s(V_p) < 0$ then
$$\omega_{ess}(Z_p) \leq s(V_p) < 0.$$

Actually $s(A_p) = 0$. Indeed, if not, i.e., if $s(A_p) > 0$, then $s(A_p)$ would be an isolated eigenvalue of A_p associated to a positive eigenvector and similarly $s(A_p^*) = s(A_p)$ would be an isolated eigenvalue of A_p^* (the dual operator in $L^{p^*}(\mathbb{R}_+, dx)$) associated to a positive eigenvector h and finally, by using (10.3.7),

$$0 = \langle A_p M, h \rangle_{L^p, L^{p^*}} = \langle M, A_p^* h \rangle_{L^p, L^{p^*}} = s(A_p^*)\langle M, h \rangle_{L^p, L^{p^*}}$$

we get a contradiction since $\langle M, h \rangle_{L^p, L^{p^*}} > 0$ and $s(A_p^*) = s(A_p) > 0$. Hence

$$\omega_{ess}(Z_p) < 0 = s(A_p) = \omega(Z_p),$$

i.e., $\left(Z_p(t)\right)_{t \geq 0}$ has a spectral gap. Finally, the last claim is a consequence of Theorem 4.6.7. $\qquad \square$

Remark 10.3.2 Consider the case $p = 1$. If we add an absorption $\Sigma_a(x)x$ to Eq. (10.3.4), i.e., if we consider

$$\frac{\partial\varphi(t,x)}{\partial t} + \Sigma_a(x)x + \Sigma_s(x)x\varphi(t,x) = \int_0^{+\infty} k(x,x')\varphi(t,x')dx'$$

where $\Sigma_a(x)x$ is bounded with oscillation less than $\inf_x \Sigma_s(x)x$ then the perturbed semigroup has a spectral gap, see Theorem 4.4.1. In particular, there exists a negative leading (isolated) eigenvalue of the generator associated to a positive eigenfunction which is of course different from the maxwellian function M. We point out that in all the classical literature [92, 289, 333, 351, 352], $\Sigma_a(x)x$ is a constant; in the physical terminology, this corresponds to the so-called $\frac{1}{x}$ law for Σ_a.

If we consider the case $p = 1$ then the inspection of (10.3.3) shows that instead of studying Eq. (10.3.1) in the natural space $L^1(\mathbb{R}_+, x^2dx)$ (see Remark 10.3.1) defined by the condition

$$\int_0^{+\infty} |\psi(x)|\, x^2dx = \int_0^{+\infty} \left|\frac{\psi(x)}{x}\right| x^3dx < +\infty,$$

we have developed a theory in the smaller space defined by

$$\int_0^{+\infty} \left|\frac{\psi(x)}{x}\right| \exp(\frac{x^2}{2})dx < +\infty.$$

It is possible to develop a general well posedness theory in $L^1(\mathbb{R}_+, x^2dx)$; see Sect. 10.4 below.

10.4   On a Paper by A. Suhadolc

The earliest works on the spectral theory of Neutron Transport date back to the 1950s and 1960s and were formulated in (weighted) L^2-spaces. The first attempt to study the spectrum of neutron transport operators in L^1-spaces with physically relevant cross sections is due to Suhadolc [343] in 1971 (see also [294] for related work in the context of linearized Boltzmann operators). Suhadolc's analysis is based on estimates of cross sections for free gas models given in [104]. Disregarding the notations introduced in the previous section, the equation studied in [343] takes the form

$$\frac{\partial\varphi}{\partial t} = -\zeta.\nabla_x\varphi - \Sigma_s(\zeta)\varphi(t,x,\zeta) + \int_{\mathbb{R}^3} c(\zeta,\zeta')\varphi(t,x,\zeta')d\zeta' \qquad (10.4.1)$$

with nonincoming boundary condition

$$\varphi_{|\Gamma_-} = 0$$

when $D \subset \mathbb{R}^3$ is a smooth bounded open subset. The scattering kernel is of the form

$$c(\zeta, \zeta') = \alpha \frac{e^{-|\zeta|^2 + |\zeta'|^2}}{|\zeta - \zeta'|} \exp\left\{ -\frac{1}{2}\left[\frac{|\zeta - \zeta'|^2}{\beta} + \beta\frac{\left(|\zeta|^2 - |\zeta'|^2\right)^2}{|\zeta - \zeta'|^2} \right]\right\}$$

where α, β are positive constants and

$$\Sigma_s(\zeta) := \int_{\mathbb{R}^3} c(\zeta', \zeta)d\zeta' \geq a_1 + a_2\,|\zeta| \qquad (10.4.2)$$

for some positive constants a_1 and a_2. It is noted in [343], as a consequence of (10.4.2), that the collision operator

$$C : \psi \to \int_{\mathbb{R}^3} c(\zeta, \zeta')\psi(x, \zeta')d\zeta'$$

is not bounded on the "natural" Lebesgue space

$$L^1(D \times \mathbb{R}^3) := L^1(D \times \mathbb{R}^3;\ dx \otimes d\zeta) \qquad (10.4.3)$$

but is M_s-bounded where M_s is the (unbounded) multiplication operator by the function $-\Sigma_s$ or equivalently, for $\lambda > 0$, $C(\lambda - M_s)^{-1}$ is a bounded operator on $L^1(\mathbb{R}^3;\ d\zeta)$. If we denote by

$$T\psi = -\zeta.\nabla_x\psi$$

(with suitable domain in L^1) the streaming generator and by T_{Σ_s} the corresponding absorption generator, then C is T_{Σ_s}-bounded, more precisely $C(\lambda - T_{\Sigma_s})^{-1}$ is a contraction in $L^1(D \times \mathbb{R}^3)$. We know today, according to Kato-Voigt perturbation theorem (see Theorem 2.11.1), that there exists a unique extension of

$$T_{\Sigma_s} + C : D(T_{\Sigma_s}) \subset L^1(D \times \mathbb{R}^3) \to L^1(D \times \mathbb{R}^3)$$

which generates a substochastic C_0-semigroup $(V(t))_{t \geq 0}$ on the natural space (10.4.3); this was observed first in [369] (see also [17] for a different approach and more general vector fields). However, at that time (in 1971), Kato's foundational work [180] was not yet known to researchers in Transport Theory. Suhadolc [343] disregarded this Lebesgue space and opted for weighted L^1 spaces on which C is bounded

$$L^1(D \times \mathbb{R}^3;\ dx\nu(d\zeta)),$$

where

$$\nu(d\zeta) = e^{\alpha|\zeta|^2}d\zeta \ (0 < \alpha < 2).$$

It is known that $C(\lambda - T_{\Sigma_s})^{-1}$ is power compact in the weighted spaces considered in [343]. However, the spectral properties of the corresponding semigroup remained unresolved until Weis clarified them in 1988 [375]. In Sect. 10.9, we revisit this topic in larger weighted L^1-spaces and for more general linear Boltzmann equations with power-like hard-sphere potentials. This analysis is developed as a consequence of a general construction, valid for abstract velocity measures, presented in Sect. 10.8.

In the present section, our object is to show that it is possible to obtain some quite general new spectral results in the natural (unweighted) Lebesgue space (10.4.3). The main result is given in Theorem 10.4.1 below.

We remark first that the scattering kernel $c(\zeta, \zeta')$ obeys a detailed balance condition

$$c(\zeta, \zeta')m(\zeta') = c(\zeta', \zeta)m(\zeta)$$

where $m(\zeta)$ is a maxwellian function (see e.g. [65, 104]) so $\Sigma_s m \in L^1(\mathbb{R}^3; \, d\zeta)$ and

$$- \Sigma_s m + Cm = 0$$

and consequently (see Proposition 2.11.4) $(V(t))_{t\geq 0}$ is a honest stochastic C_0-semigroup generated by the closure of $-\Sigma_s + C$; moreover (after normalization) m is an invariant density of $(V(t))_{t\geq 0}$ and finally $(V(t))_{t\geq 0}$ is asymptotically stable, see Theorem 6.4.1. Similarly, we can deal also with the full neutron transport Eq. (10.4.1) but with periodic boundary conditions (i.e., on the three dimensional torus T^3) and get a honest asymptotically stable stochastic C_0-semigroup $(V(t))_{t\geq 0}$ on

$$L^1(T^3 \times \mathbb{R}^3; \, dx \otimes d\zeta),$$

see [264]. Finally, the full neutron transport Eq. (10.4.1) with nonincoming boundary condition $\varphi_{|\Gamma_-} = 0$ is also governed by a honest sub-stochastic C_0-semigroup $(V(t))_{t\geq 0}$ on $L^1(D \times \mathbb{R}^3)$, see [264].

In this section, we show that we can recover a key compactness result if we add a sufficiently strong absorption cross section $\Sigma_a(v)$ (at infinity), i.e., we replace Σ_s by

$$\Sigma := \Sigma_s + \Sigma_a$$

in (10.4.1). We consider here just space-homogeneous equations. Let M be the multiplication operator by the function $-\Sigma$ and let $\mathcal{M}$ be the holomorphic C_0-semigroup generated by M. Note that

$$s(M) = - \inf \Sigma = \omega(\mathcal{M}) = \omega_{ess}(\mathcal{M}).$$

The main result of this section is

Theorem 10.4.1 *Let Σ_a be strong at infinity in the sense that*

$$\delta := \lim_{|\zeta| \to \infty} \inf \frac{\Sigma_a(\zeta)}{\Sigma_s(\zeta)} > 0.$$

Then $C(\lambda - M)^{-1}$ is quasi-compact in $L^1(\mathbb{R}^3; \, d\zeta)$, more precisely

$$r_{ess}(C(\lambda - M)^{-1}) \leq \frac{1}{1 + \delta} \quad (\lambda \geq 0).$$

In particular,

$$M + C : D(M) \subset L^1(\mathbb{R}^3; \, d\zeta) \to L^1(\mathbb{R}^3; \, d\zeta) \tag{10.4.4}$$

generates a positive holomorphic C_0-semigroup $V = (V(t))_{t \geq 0}$. If $\delta = +\infty$ then $C(\lambda - M)^{-1}$ is weakly compact and $\omega_{ess}(V) = \omega_{ess}(M) = -\inf \Sigma$.

Proof We decompose $C = C^{(1)} + C^{(2)}$ by decomposing its kernel $c(\zeta, \zeta')$ as

$$c(\zeta, \zeta') = c^{(1)}(\zeta, \zeta') + c^{(2)}(\zeta, \zeta')$$

where

$$c^{(1)}(\zeta, \zeta') = c(\zeta, \zeta')1_{\{|\zeta'| \leq j\}}, \quad c^{(2)}(\zeta, \zeta') = c(\zeta, \zeta')1_{\{|\zeta'| > j\}}$$

where $j > 0$ is a priori arbitrary. Then

$$\left\| C^{(2)}(0 - M)^{-1}\varphi \right\|_{L^1(\mathbb{R}^3; \, d\zeta)}$$

$$= \int_{\mathbb{R}^3} d\zeta \int_{\{|\zeta'| > j\}} \frac{c(\zeta, \zeta')}{\Sigma(\zeta')} \varphi(\zeta') d\zeta'$$

$$= \int_{\{|\zeta'| > j\}} \left[\int_{\mathbb{R}^3} \frac{c(\zeta, \zeta')}{\Sigma(\zeta')} d\zeta \right] \varphi(\zeta') d\zeta'$$

$$= \int_{\{|\zeta'| > j\}} \frac{\Sigma_s(\zeta')}{\Sigma(\zeta')} \varphi(\zeta') d\zeta'$$

$$\leq \sup_{\{|\zeta'| > j\}} \frac{\Sigma_s(\zeta')}{\Sigma(\zeta')} \|\varphi\|_{L^1}$$

so

$$\left\| C^{(2)}(0 - M)^{-1} \right\|_{\mathcal{L}(L^1(\mathbb{R}^3))} \leq \sup_{\{|\zeta'| > j\}} \frac{\Sigma_s(\zeta')}{\Sigma(\zeta')}$$

$$= \frac{1}{1 + \inf_{\{|\zeta'| > j\}} \frac{\Sigma_a(\zeta')}{\Sigma_s(\zeta')}}.$$

One sees that for any $\delta' < \delta$

$$\left\| C^{(2)}(0 - M)^{-1} \right\|_{\mathcal{L}(L^1(\mathbb{R}^3))} \leq \frac{1}{1 + \delta'}$$

for j is large enough. On the other hand, for a fixed j we have

$$C^{(1)}(0 - M)^{-1}\varphi = \int_{\{|\zeta'|\leq j\}} \frac{c(\zeta, \zeta')}{\Sigma(\zeta')}\varphi(\zeta')d\zeta'$$

where

$$c(\zeta, \zeta') = \alpha \frac{e^{-|\zeta|^2+|\zeta'|^2}}{|\zeta - \zeta'|} \exp\left\{-\frac{1}{2}\left[\frac{|\zeta - \zeta'|^2}{\beta} + \beta\frac{\left(|\zeta|^2 - |\zeta'|^2\right)^2}{|\zeta - \zeta'|^2}\right]\right\}$$

is such that

$$\lim_{|\zeta|\to+\infty} \sup_{|\zeta'|\leq j} c(\zeta, \zeta') = 0.$$

We decompose $c^{(1)}(\zeta, \zeta')$ as

$$c^{(1)}(\zeta, \zeta') = c_1^{(1)}(\zeta, \zeta') + c_2^{(1)}(\zeta, \zeta')$$

where

$$c_1^{(1)}(\zeta, \zeta') = 1_{\{|\zeta|\leq R\}}c(\zeta, \zeta')1_{\{|\zeta'|\leq j\}}$$

and

$$c_2^{(1)}(\zeta, \zeta') = 1_{\{|\zeta|> R\}}c(\zeta, \zeta')1_{\{|\zeta'|\leq j\}}.$$

One sees that $\left\|C_2^{(1)}(0 - M)^{-1}\right\|_{\mathcal{L}(L^1(\mathbb{R}^3))}$ is as small as we want provided that we choose R large enough. Since $C_2^{(1)}(0 - M)^{-1}$ has a bounded support and a bounded kernel then $C_2^{(1)}(0 - M)^{-1}$ is weakly compact. Thus $C(\lambda - M)^{-1}$ is a weakly compact perturbation of $C_2^{(1)}(0 - M)^{-1} + C^{(2)}(0 - M)^{-1}$ and

$$r_{ess}(C(\lambda - M)^{-1}) \leq \frac{1}{1 + \delta'} + \left\|C_2^{(1)}(0 - M)^{-1}\right\|_{\mathcal{L}(L^1(\mathbb{R}^3))} \quad (\forall\delta' < \delta);$$

this shows the first claim and $C(\lambda - M)^{-1}$ is quasi-compact. It follows from Theorem 2.6.4 that (10.4.4) is a generator. If $\delta = +\infty$ then $C(\lambda - M)^{-1}$ is as close to a weakly compact operator as we want and consequently is weakly compact. The stability of the essential type follows from Remark 4.3.1. $\square$

Note that if $\delta = +\infty$ then, due to the presence of Σ_a, $\mathcal{V} = (V(t))_{t\geq 0}$ is not stochastic and a priori the only information we have on $\omega(\mathcal{V})$ is $s(M + C) = \omega(\mathcal{V}) \leq 0$. Thus $\mathcal{V}$ has a spectral gap (i.e., $\omega(\mathcal{V}) > \omega_{ess}(\mathcal{V})$) if and only if $s(M + C) > -\inf \Sigma$ or equivalently if and only if $\lim_{\lambda\to-\inf \Sigma} r\left(C(\lambda - M)^{-1}\right) > 1$.

10.5   Space Homogeneous Linear Boltzmann Equations

In this section, we revisit Suhadolc's work [343] in the context of more general linear Boltzmann equations with power-like hard-sphere potentials, without relying on an additional absorption cross section. Our analysis remains within the framework of space-homogeneous equations, considered in various weighted spaces and arbitrary spatial dimensions. The space-nonhomogeneous case is addressed separately in Sect. 10.9. We deal here with linear Boltzmann equations

$$\frac{df}{dt} = \mathcal{Q}(f, M)$$

where $\mathcal{Q}(f, g)$ denotes the bilinear Boltzmann operator (M is a suitable maxwellian function) given by

$$\int_{\mathbb{R}^d \times S^{d-1}} B(\zeta - \zeta_*, \sigma) \left(f(\zeta')g(\zeta'_*) - f(\zeta)g(\zeta_*) \right) d\zeta_* d\sigma$$

where ζ' and ζ'_* are the pre-collisional velocities that result respectively in ζ and ζ_* after elastic collision

$$\zeta' = \frac{\zeta + \zeta_*}{2} + \frac{|\zeta - \zeta_*|}{2}\sigma, \quad \zeta'_* = \frac{\zeta + \zeta_*}{2} - \frac{|\zeta - \zeta_*|}{2}\sigma.$$

Here f and g are functions of the velocity $\zeta \in \mathbb{R}^d$.

Remark 10.5.1 It is well known (see e.g. [211]) that $\mathcal{Q}(M, M) = 0$.

The interaction collision kernels $B(\zeta - \zeta_*, \sigma)$ are nonnegative and have the form

$$B(\zeta - \zeta_*, \sigma) = \Phi(|\zeta - \zeta_*|)b(\cos\theta), \quad \cos\theta = \langle \frac{\zeta - \zeta_*}{|\zeta - \zeta_*|}, \sigma \rangle \tag{10.5.1}$$

with $\Phi : \mathbb{R}^+ \to \mathbb{R}^+$ given by

$$\Phi(r) = r^\gamma \ (\gamma > 0) \ \text{(hard sphere potentials)} \tag{10.5.2}$$

and $b : [-1, 1] \to \mathbb{R}^+$ such that

$$\int_{-1}^{1} (1 - s^2)^{\frac{d-3}{2}} b(s)ds < +\infty \text{ (angular cut-off).} \tag{10.5.3}$$

Thanks to the cut-off assumption, $\mathcal{Q}(f, M)$ splits into two parts

$$\mathcal{Q}(f, M) = -\Sigma f + Cf$$

where there exist two positive constants c_i $(i = 1, 2)$ such that

$$c_1 \left(1 + |\zeta|\right)^\gamma \leq \Sigma(\zeta) \leq c_2 \left(1 + |\zeta|\right)^\gamma \tag{10.5.4}$$

for hard sphere potentials and (in the so-called Carleman representation)

$$Cf = \int_{\mathbb{R}^d} c(\zeta, \zeta') f(\zeta') d\zeta'$$

and $\Sigma(\zeta) = \int_{\mathbb{R}^d} c(\zeta', \zeta) d\zeta'$ where the kernel $c(\zeta, \zeta')$ admits suitable estimates we give thereafter. We note that according to Remark 10.5.1

$$- \Sigma M + C M = 0. \tag{10.5.5}$$

We consider two weighted spaces

$$X = L^1(\mathbb{R}^d; \; m^{-1}(\zeta) d\zeta)$$

where

$$m(\zeta) = \exp\left(-a \, |\zeta|^s\right), \;\; s \in (0, 2) \quad \text{(exponential weight)}$$

(for $a > 0$) or

$$m(\zeta) = \left(1 + |\zeta|^\beta\right)^{-1}, \;\; \beta > 0 \;\; \text{(algebraic weight)}.$$

Note that $X \subset L^1(\mathbb{R}^d; \; d\zeta)$. Let

$$\begin{cases} M : D(M) \subset X \to X \\ \quad\quad f \to -\Sigma f \end{cases}$$

where (thanks to (10.5.4))

$$D(M) = D(\Sigma) = L^1(\mathbb{R}^d; \; (1 + |\zeta|)^\gamma \, m^{-1}(\zeta) d\zeta).$$

The purpose of the present section is to revisit the compactness results from [211] as a preparation for the peripheral spectral analysis of space-nonhomogeneous Boltzmann equations with hard-sphere potentials, which will be addressed in Sect. 10.9. The main result in this section is Theorem 10.5.1 below. Let

$$Y = Y_\gamma := L^1(\mathbb{R}^d; \; (1 + |\zeta|)^\gamma \, m^{-1}(\zeta) d\zeta).$$

We start with a useful observation.

Lemma 10.5.1 ([211, Lemma 3.1]) *The collision operator $C \in \mathcal{L}(Y, X)$ (i.e., C is Σ-bounded in X) and depends (linearly and) continuously on $b(.)$ for the norm (10.5.3).*

We recall now some key estimates.

Lemma 10.5.2 ([211, Proposition A.1]) *If $b(.) \in L^{\infty}([-1, 1])$ then C is dominated by $\|b\|_{L^{\infty}} C_{\gamma}$ where $C_{\gamma} \in \mathcal{L}(Y, X)$, given by*

$$C_{\gamma} f = \int_{\mathbb{R}^d} c_{\gamma}(\zeta, \zeta') f(\zeta') d\zeta',$$

is such that its kernel satisfies the following estimates:
(i) If $\gamma = d - 2$ then

$$c_{d-2}(\zeta, \zeta') = 2^{d-1} (2\pi)^{-\frac{1}{2}} |\zeta - \zeta'|^{-1} \exp\left(-\frac{1}{8}\left[|\zeta - \zeta'| + \frac{|\zeta|^2 - |\zeta'|^2}{|\zeta - \zeta'|}\right]^2\right).$$

(ii) If $\gamma < d - 2$ then

$$c_{\gamma}(\zeta, \zeta') \leq |\zeta - \zeta'|^{\gamma - (d-2)} c_{d-2}(\zeta, \zeta')$$

$$= 2^{d-1} (2\pi)^{-\frac{1}{2}} |\zeta - \zeta'|^{\gamma - (d-1)} \exp\left(-\frac{1}{8}\left[|\zeta - \zeta'| + \frac{|\zeta|^2 - |\zeta'|^2}{|\zeta - \zeta'|}\right]^2\right).$$

(iii) If $m(\zeta) = \exp\left(-a |\zeta|^s\right), \quad s \in (0, 2)$ then

$$\int_{\mathbb{R}^d} c_{\gamma}(\zeta, \zeta') m^{-1}(\zeta) d\zeta \leq c \left(1 + |\zeta'|^{\gamma - s}\right) m^{-1}(\zeta'). \qquad (10.5.6)$$

(iv) If $m(\zeta) = \left(1 + |\zeta|^{\beta}\right)^{-1} \; (\beta > 0)$ then

$$\int_{\mathbb{R}^d} c_{\gamma}(\zeta, \zeta') m^{-1}(\zeta) d\zeta \leq c \left(1 + |\zeta'|^{\gamma - 2}\right) m^{-1}(\zeta'). \qquad (10.5.7)$$

Remark 10.5.2 Actually, in ([211, Proposition A.1]), Estimate (10.5.6) is given for $s \in (0, 1]$ only. For reader's convenience, we indicate here briefly what should be changed (in the main step of the proof) to capture the case $s \in (1, 2)$. The author thanks B. Lods for providing him the arguments. We refer the interested reader to the proof of ([211, Proposition A.1]) just after (A.4) where

$$H_{\gamma}(\omega) \leq C_d \left|S^{d-2}\right| \int_0^{+\infty} d\rho \int_0^{\pi} G(\rho, \theta) d\theta$$

and

$$G(\rho, \theta) = \rho^{\gamma} \exp\left\{-\frac{1}{2}\left(\rho + |\omega|\cos\theta\right)^2 + a\left(\rho^2 + |\omega|^2 + 2\rho|\omega|\cos\theta\right)^{\frac{s}{2}}\right\}(\sin\theta)^{d-2}.$$

Since $s < 2$ then $(\alpha + \beta)^{\frac{s}{2}} \leq \alpha^{\frac{s}{2}} + \beta^{\frac{s}{2}}$ so that

$$\exp\left\{a\left(\rho^2 + |\omega|^2 + 2\rho|\omega|\cos\theta\right)^{\frac{s}{2}}\right\} \leq \exp\left\{a|\omega|^s\right\}\exp\left\{a\left(\rho^2 + 2\rho|\omega|\cos\theta\right)^{\frac{s}{2}}\right\}$$

and $\rho^2 + 2\rho|\omega|\cos\theta \leq (\rho + |\omega|\cos\theta)^2$ so that

$$a\left(\rho^2 + 2\rho|\omega|\cos\theta\right)^{\frac{s}{2}} \leq a\left(\rho + |\omega|\cos\theta\right)^s.$$

By using Young's inequality (with $p = \frac{2}{s}$ and $p' = \frac{s}{2-s}$), for all $\varepsilon > 0$ there is $C_\varepsilon > 0$ such

$$a\left(\rho + |\omega|\cos\theta\right)^s \leq C_\varepsilon a^{\frac{s}{s-2}} + \varepsilon\left(\rho + |\omega|\cos\theta\right)^2$$

and, choosing $\varepsilon = \frac{1}{4}$, we have

$$G(\rho, \theta) \leq C_\alpha \rho^{\gamma} \exp\left\{-\frac{1}{4}\left(\rho + |\omega|\cos\theta\right)^2\right\}\exp\left\{a|\omega|^s\right\}(\sin\theta)^{d-2}$$

where $C_\alpha = \exp(C_\varepsilon a^{\frac{s}{s-2}})$ with $\varepsilon = \frac{1}{4}$. From here, the proof ends as for $s \in (0, 1]$.

Remark 10.5.3 The inspection of (10.5.6) (resp. (10.5.7)) shows that C need not be a bounded operator on $L^1(\mathbb{R}^d; \ m^{-1}(\zeta)d\zeta)$ if $s < \gamma$ (resp. if $2 < \gamma$).

We are ready to state

Theorem 10.5.1 ([211, Corollary 3.6]) *Suppose that the interaction kernel is of the form (10.5.1) with angular cut-off (10.5.3) and hard sphere potential (10.5.2) such that $\gamma \in (0, d - 2]$. Then C is M-weakly compact in X.*

Proof We give here a proof slightly different from that given in [211]. We start as in [211] by using an approximation argument, in operator norm (Lemma 10.5.1), which allows us to restrict ourselves to the case $b(.) \in L^\infty([-1, 1])$. In this case, thanks to the domination $C \leq \|b\|_{L^\infty} C_\gamma$, it suffices to show that C_γ is M-weakly compact in X. To this end, we decompose $C_\gamma = C_\gamma^{(1)} + C_\gamma^{(2)}$ by decomposing its kernel $c_\gamma(\zeta, \zeta')$ as

$$c_\gamma(\zeta, \zeta') = c_\gamma^{(1)}(\zeta, \zeta') + c_\gamma^{(2)}(\zeta, \zeta')$$

where

$$c_\gamma^{(1)}(\zeta, \zeta') = c_\gamma(\zeta, \zeta')1_{\{|\zeta'|\leq j\}}, \quad c_\gamma^{(2)}(\zeta, \zeta') = c_\gamma(\zeta, \zeta')1_{\{|\zeta'|>j\}}$$

where $j > 0$ is a priori arbitrary. Without loss of generality, we use nonnegative functions φ. Then

$$\left\|C_\gamma^{(2)}(0 - M)^{-1}\varphi\right\|_{L^1(\mathbb{R}^d; \ m^{-1}(\zeta)d\zeta)}$$

$$= \int_{\mathbb{R}^d} m^{-1}(\zeta)d\zeta \int_{\{|\zeta'|>j\}} \frac{c_\gamma(\zeta,\zeta')}{\Sigma(\zeta')} \varphi(\zeta')d\zeta'$$

$$= \int_{\mathbb{R}^d} m^{-1}(\zeta)d\zeta \int_{\{|\zeta'|>j\}} \frac{c_\gamma(\zeta,\zeta')}{\Sigma(\zeta')} m(\zeta')\varphi(\zeta')m^{-1}(\zeta')d\zeta'$$

$$= \int_{\{|\zeta'|>j\}} \left[\int_{\mathbb{R}^d} \frac{c_\gamma(\zeta,\zeta')}{\Sigma(\zeta')} m^{-1}(\zeta)d\zeta\right] m(\zeta')\varphi(\zeta')m^{-1}(\zeta')d\zeta'.$$

Case 1. Consider first the weight

$$m(\zeta) = \exp\left(-a\,|\zeta|^s\right),\quad s \in (0,2).$$

It follows from (10.5.6) and (10.5.4) that

$$\left\|C_\gamma^{(2)}(0-M)^{-1}\varphi\right\|_{L^1(\mathbb{R}^d;\ m^{-1}(\zeta)d\zeta)}$$

$$\le c\int_{\{|\zeta'|>j\}} \frac{\left(1+|\zeta'|^{\gamma-s}\right)}{\Sigma(\zeta')}\varphi(\zeta')m^{-1}(\zeta')d\zeta'$$

$$\le \frac{c}{c_1}\int_{\{|\zeta'|>j\}} \frac{\left(1+|\zeta'|^{\gamma-s}\right)}{(1+|\zeta'|)^\gamma}\varphi(\zeta')m^{-1}(\zeta')d\zeta'$$

$$= \frac{c}{c_1}\sup_{|\zeta'|>j} \frac{\left(1+|\zeta'|^{\gamma-s}\right)}{(1+|\zeta'|)^\gamma}\int_{\{|\zeta'|>j\}} \varphi(\zeta')m^{-1}(\zeta')d\zeta'$$

$$\le \frac{c}{c_1}\sup_{|\zeta'|>j} \frac{\left(1+|\zeta'|^{\gamma-s}\right)}{(1+|\zeta'|)^\gamma} \|\varphi\|_{L^1(\mathbb{R}^d;\ m^{-1}(\zeta)d\zeta)}$$

so, due to $s>0$,

$$\left\|C_\gamma^{(2)}(0-M)^{-1}\right\|_{\mathcal{L}(L^1(\mathbb{R}^d;\ m^{-1}(\zeta)d\zeta))} \le \frac{c}{c_1}\sup_{|\zeta'|>j} \frac{\left(1+|\zeta'|^{\gamma-s}\right)}{(1+|\zeta'|)^\gamma} \to 0\ (j\to\infty).$$

Thus $C_\gamma^{(2)}(0-M)^{-1}$ is as small as we want provided that j is large enough. Let j be fixed. Note that we can treat the cases (i) and (ii) simultaneously by observing that for both cases there exists a constant $c>0$ such that

$$c_\gamma(\zeta,\zeta') \le c\,|\zeta-\zeta'|^{\gamma-(d-1)} \exp\left(-\frac{1}{8}\left[|\zeta-\zeta'| + \frac{|\zeta|^2-|\zeta'|^2}{|\zeta-\zeta'|}\right]^2\right).$$

Up to multiplicative constants (we can ignore), we have

$$C_\gamma^{(1)}(0-M)^{-1}\varphi$$

$$= \int_{\{|\zeta'|\leq j\}} \frac{c_\gamma(\zeta,\zeta')}{\Sigma(\zeta')}\varphi(\zeta')d\zeta'$$

$$\leq \int_{\{|\zeta'|\leq j\}} \frac{|\zeta-\zeta'|^{\gamma-(d-1)}\exp\left(-\frac{1}{8}\left[|\zeta-\zeta'|+\frac{|\zeta|^2-|\zeta'|^2}{|\zeta-\zeta'|}\right]^2\right)}{(1+|\zeta'|)^\gamma}\varphi(\zeta')d\zeta'.$$

For $|\zeta| > j$ we have

$$\int_{\{|\zeta'|\leq j\}} \frac{|\zeta-\zeta'|^{\gamma-(d-1)}\exp\left(-\frac{1}{8}\left[|\zeta-\zeta'|+\frac{|\zeta|^2-|\zeta'|^2}{|\zeta-\zeta'|}\right]^2\right)}{(1+|\zeta'|)^\gamma}\varphi(\zeta')d\zeta$$

$$\leq \int_{\{|\zeta'|\leq j\}} \frac{|\zeta-\zeta'|^{\gamma-(d-1)}\exp\left(-\frac{1}{8}\left[|\zeta-\zeta'|^2+\frac{\left(|\zeta|^2-|\zeta'|^2\right)^2}{|\zeta-\zeta'|^2}\right]\right)}{(1+|\zeta'|)^\gamma}\varphi(\zeta')d\zeta'$$

and

$$\exp\left(-\frac{1}{8}\left[\frac{\left(|\zeta|^2-|\zeta'|^2\right)^2}{|\zeta-\zeta'|^2}\right]\right)$$

$$=\exp\left(-\frac{1}{8}\left[\frac{\left(|\zeta|-|\zeta'|\right)^2\left(|\zeta|+|\zeta'|\right)^2}{|\zeta-\zeta'|^2}\right]\right)$$

$$\leq\exp\left(-\frac{1}{8}\left[\frac{\left(|\zeta|-|\zeta'|\right)^2\left(|\zeta|+|\zeta'|\right)^2}{\left(|\zeta|+|\zeta'|\right)^2}\right]\right)$$

$$=\exp\left(-\frac{1}{8}\left(|\zeta|-|\zeta'|\right)^2\right)\leq\exp\left(-\frac{1}{8}\left(|\zeta|-j\right)^2\right)$$

so, for $|\zeta| > j$, $\left(C_\gamma^{(1)}(0-M)^{-1}\varphi\right)(\zeta)$ is less than

$$\exp\left(-\frac{1}{8}\left(|\zeta|-j\right)^2\right)\int_{\{|\zeta'|\leq j\}}|\zeta-\zeta'|^{\gamma-(d-1)}\exp\left(-\frac{1}{8}\left[|\zeta-\zeta'|^2\right]\right)\varphi(\zeta')d\zeta'$$

$$\leq\exp\left(-\frac{1}{8}\left(|\zeta|-j\right)^2\right)\int_{\{|\zeta'|\leq j\}}|\zeta-\zeta'|^{\gamma-(d-1)}\exp\left(-\frac{1}{8}\left[|\zeta-\zeta'|^2\right]\right)m^{-1}(\zeta')\varphi(\zeta')d\zeta'$$

since $m^{-1}(\zeta') \geq 1$. For $R > j$

$$\int_{\{|\zeta\geq R|\}}\left(C_\gamma^{(1)}(0-M)^{-1}\varphi\right)(\zeta)m^{-1}(\zeta)d\zeta$$

$$\leq \sup_{|\zeta| \geq R} \left(\exp\left(-\frac{1}{8} \left(|\zeta| - j \right)^2 \right) \exp(a\,|\zeta|^s) \right) \times$$

$$\int_{\mathbb{R}^d} d\zeta \int_{\mathbb{R}^d} |\zeta - \zeta'|^{\gamma - (d-1)} \exp\left(-\frac{1}{8} \left[|\zeta - \zeta'|^2 \right] \right) m^{-1}(\zeta') \varphi(\zeta') d\zeta'.$$

Since

$$\int_{\mathbb{R}^d} d\zeta \int_{\mathbb{R}^d} |\zeta - \zeta'|^{\gamma - (d-1)} \exp\left(-\frac{1}{8} \left[|\zeta - \zeta'|^2 \right] \right) m^{-1}(\zeta') \varphi(\zeta') d\zeta'$$

$$= \left(\int_{\mathbb{R}^d} |z|^{\gamma - (d-1)} \exp\left(-\frac{1}{8} |z|^2 \right) dz \right) \int_{\mathbb{R}^d} m^{-1}(\zeta') \varphi(\zeta') d\zeta'$$

$$= \left(\int_{\mathbb{R}^d} |z|^{\gamma - (d-1)} \exp\left(-\frac{1}{8} |z|^2 \right) dz \right) \|\varphi\|_{L^1(\mathbb{R}^d;\, m^{-1}(\zeta)d\zeta)}$$

and

$$\sup_{|\zeta| \geq R|} \left(\exp\left(-\frac{1}{8} \left(|\zeta| - j \right)^2 \right) \exp(a\,|\zeta|^s) \right) \to 0 \ \ (R \to \infty) \ \ (s < 2)$$

then

$$\lim_{R \to \infty} \sup_{\|\varphi\|_{L^1(\mathbb{R}^d;\, m^{-1}(\zeta)d\zeta)} \leq 1} \int_{\{|\zeta| \geq R|\}} \left(C_\gamma^{(1)}(0 - M)^{-1}\varphi \right)(\zeta) m^{-1}(\zeta) d\zeta = 0.$$

Since the kernel of $C_\gamma^{(1)}(0 - M)^{-1}$ for $|\zeta| \leq R$ and $|\zeta'| \leq j$ is bounded then $1_{\{|\zeta| \leq R\}} C_\gamma^{(1)}(0 - M)^{-1}$ is weakly compact and finally $C_\gamma(0 - M)^{-1}$ is as close to a weakly compact operator as we want.

Case 2. Consider now the weight

$$m(\zeta) = \left(1 + |\zeta|^\beta \right)^{-1} \ \ (\beta > 0).$$

We know that

$$\left\| C_\gamma^{(2)}(0 - M)^{-1}\varphi \right\|_{L^1(\mathbb{R}^d;\, m^{-1}(\zeta)d\zeta)}$$

$$= \int_{\{|\zeta'| > j\}} \left[\int_{\mathbb{R}^d} \frac{c_\gamma(\zeta, \zeta')}{\Sigma(\zeta')} m^{-1}(\zeta) d\zeta \right] m(\zeta') \varphi(\zeta') m^{-1}(\zeta') d\zeta'.$$

Similarly, it follows from (10.5.7) and (10.5.4) that

$$\left\| C_\gamma^{(2)}(0 - M)^{-1}\varphi \right\|_{L^1(\mathbb{R}^d;\, m^{-1}(\zeta)d\zeta)}$$

$$\leq \int_{\{|\zeta'| > j\}} \frac{c \left(1 + |\zeta'|^{\gamma - 2} \right)}{\Sigma(\zeta')} \varphi(\zeta') m^{-1}(\zeta') d\zeta'$$

$$\leq \frac{c}{c_1} \int_{\{|\zeta'|>j\}} \frac{\left(1+|\zeta'|^{\gamma-2}\right)}{(1+|\zeta'|)^\gamma} \varphi(\zeta')m^{-1}(\zeta')d\zeta'$$

$$\leq \frac{c}{c_1} \sup_{|\zeta'|>j} \frac{\left(1+|\zeta'|^{\gamma-2}\right)}{(1+|\zeta'|)^\gamma} \|\varphi\|_{L^1(\mathbb{R}^d;\ m^{-1}(\zeta)d\zeta)}$$

and

$$\left\| C_\gamma^{(2)}(0-M)^{-1} \right\|_{\mathcal{L}(L^1(\mathbb{R}^d;\ m^{-1}(\zeta)d\zeta))} \leq \frac{c}{c_1} \sup_{|\zeta'|>j} \frac{\left(1+|\zeta'|^{\gamma-2}\right)}{(1+|\zeta'|)^\gamma} \to 0\ (j \to \infty).$$

As previously, for $R > j$

$$\int_{\{|\zeta \geq R|\}} \left(C_\gamma^{(1)}(0-M)^{-1}\varphi \right)(\zeta)m^{-1}(\zeta)dv\zeta$$

$$\leq \sup_{|\zeta|\geq R} \left(\exp\left(-\frac{1}{8}(|\zeta|-j)^2\right)\left(1+|\zeta|^\beta\right) \right) \times$$

$$\int_{\mathbb{R}^d} d\zeta \int_{\mathbb{R}^d} |\zeta-\zeta'|^{\gamma-(d-1)} \exp\left(-\frac{1}{8}\left[|\zeta-\zeta'|^2\right]\right) m^{-1}(\zeta')\varphi(\zeta')d\zeta'$$

$$\leq \sup_{|\zeta|\geq R} \left(\exp\left(-\frac{1}{8}(|\zeta|-j)^2\right)\left(1+|\zeta|^\beta\right) \right)$$

$$\times \left(\int_{\mathbb{R}^d} |z|^{\gamma-(d-1)} \exp\left(-\frac{1}{8}|z|^2\right) dz \right) \|\varphi\|_{L^1(\mathbb{R}^d;\ m^{-1}(\zeta)d\zeta)}$$

and

$$\lim_{R\to\infty} \sup_{\|\varphi\|_{L^1(\mathbb{R}^d;\ m^{-1}(\zeta)d\zeta)}\leq 1} \int_{\{|\zeta\geq R|\}} \left(C_\gamma^{(1)}(0-M)^{-1}\varphi \right)(\zeta)m^{-1}(\zeta)d\zeta = 0.$$

We can end the proof as in the case 1. $\qquad\square$

As in Remark 10.3.2, it follows from Theorem 10.5.1 that $-\Sigma + C$ generates a holomorphic C_0-semigroup with a spectral gap (equal to $\inf \Sigma(\zeta)$) in the above spaces $X = L^1(\mathbb{R}^d;\ m^{-1}(\zeta)d\zeta)$. Moreover, if we add a bounded absorption $\Sigma_a(.)$ with oscillation less than $\inf \Sigma(\zeta)$ then the perturbed semigroup has a spectral gap.

10.6 Downshift and Bragg Scatterings for Neutron Transport

We continue the investigation of space-homogeneous equations, now focusing on more complex scattering phenomena arising in nuclear reactor theory (see [192]), including "*downshift*" and "*Bragg*" scattering operators (see below). The purpose of this section is to provide a systematic peripheral spectral analysis, along with corresponding time asymptotics, for this class of equations. To our knowledge, this represents the first general attempt in this direction. The main results are stated in Lemma 10.6.1, Theorems 10.6.1 and 10.6.3.

In this section, the velocity variable is denoted by x instead of ζ. We work in $L^1(\mathbb{R}^3)$ endowed with Lebesgue measure $\nu(dx) = dx$. Following [192], let

$$E_m = \frac{1}{2}N\omega_n^2(x) - \frac{1}{2}N|x|^2 \quad (m = 1, \ldots, \overline{m})$$

(N is the mass of the neutron) so $m|x|^2 + 2E_m = N\omega_n^2(|x|)$ and

$$\begin{cases} \omega_n(|x|) = \sqrt{|x|^2 + \frac{2E_m}{m}} \\ |x| = \sqrt{N\omega_n^2(|x|) - 2E_m}. \end{cases}$$

Besides the usual inelastic scattering operator K_i (see (10.6.1)), we deal here with additional "downshift" scattering operator K_d

$$K_d\varphi(x) = \sum_{m=1}^{\overline{m}} K_d^{(m)}\varphi(x) = \sum_{m=1}^{\overline{m}} \int_{\{|y|=\omega_n(|x|)\}} k_d^{(m)}(x, y)\varphi(y)dS_{\omega_n(|x|)}(y)$$

and "Bragg" (elastic) scattering operator K_b

$$K_b\varphi(x) = \int_{\{|y|=|x|\}} k_b(x, y)\varphi(y)dS_{|x|}(y)$$

where $dS_{|x|}(y)$ and $dS_{\omega_n(|x|)}(y)$ are suitable Lebesgue surface measures on spheres (see below). We denote their sum by

$$K := K_d + K_b.$$

The key point is that K_d and K_b induce a loss of compactness; in particular, they give rise to spectral curves within the essential spectrum [192]. The purpose of this section is to show how a well-behaved peripheral spectral theory can still be established, thanks to the presence of the inelastic scattering operator K_i. The main results are given in Theorem 10.6.1 below. Note that

$$\omega_n : \rho \in (0, +\infty) \to \sqrt{\rho^2 + \frac{2E_m}{m}} \in \left(\sqrt{\frac{2E_m}{m}}, +\infty\right)$$

with $\inf \omega_n = \sqrt{\frac{2E_m}{m}}$ while

$$\omega_n^{-1} : \rho \in \left(\sqrt{\frac{2E_m}{m}}, +\infty\right) \to \sqrt{N\rho^2 - 2E_m}.$$

Note that if $\gamma : (0, +\infty) \to (0, +\infty)$ is an increasing C^1 function then the change of variable $\rho = \gamma(s)$ ($\rho > \inf \gamma$) is such that $d\rho = \gamma'(s)ds$ so that for any measurable positive function

$$G : (u, v) \in (0, +\infty)^2 \to (0, +\infty)$$

we have

$$\int_0^{+\infty} G(s, \gamma(s))ds = \int_{\inf \gamma}^{+\infty} G(\gamma^{-1}(\rho), \rho)\frac{d\rho}{\gamma'(\gamma^{-1}(\rho))} = \int_{\inf \gamma}^{+\infty} c_\alpha(\rho)G(\gamma^{-1}(\rho), \rho)d\rho$$

where $c_\gamma(\rho) = \frac{1}{\gamma'(\gamma^{-1}(\rho))}$. Let

$$S_\rho = \{x \in \mathbb{R}^3; \ |x| = \rho\}$$

be the sphere centered at zero with radius $\rho > 0$ endowed with Lebesgue surface measure dS_ρ. Since for any locally integrable function $h \in L^1_{loc}(\mathbb{R}^3)$ we have

$$\int_{\mathbb{R}^3} h(x)\varphi(x)dx = \int_0^{+\infty} d\rho \int_{S_\rho} (h\varphi)(x)dS_\rho(x) \ \ (\varphi \in C_c^\infty(\mathbb{R}^3))$$

then we can identify h to a field of functions defined on the spheres S_ρ

$$\begin{cases} \rho \in (0, +\infty) \to h_\rho \\ \quad h_\rho : S_\rho \to \mathbb{R}. \end{cases}$$

In particular, if $h \in L^1(\mathbb{R}^3)$ then

$$\int_{\mathbb{R}^3} |h(x)|\, dx = \int_0^{+\infty} d\rho \int_{S_\rho} |h_\rho(x)|\, dS_\rho(x)$$

and

$$\|h\|_{L^1(\mathbb{R}^3)} = \int_0^{+\infty} \|h_\rho\|_{L^1(S_\rho, dS_\rho)} d\rho.$$

For $\varphi \geq 0$, the norm of $K_b\varphi$ is equal to

$$\int_0^{+\infty} d\rho \int_{\{|x|=\rho\}} dS_\rho(x) \left(\int_{\{|y|=|x|\}} k_b(x, y)\varphi(y)dS_{|x|}(y)\right)$$
$$= \int_0^{+\infty} d\rho \int_{\{|x|=\rho\}} dS_\rho(x) \left(\int_{\{|y|=\rho\}} k_b(x, y)\varphi(y)dS_\rho(y)\right)$$

$$= \int_0^{+\infty} d\rho \int_{\{|y|=\rho\}} \left(\int_{\{|x|=\rho\}} k_b(x,y) dS_\rho(x) \right) \varphi(y) dS_\rho(y)$$

so

$$\|K_b\|_{\mathcal{L}(L^1(\mathbb{R}^3))} = \sup_{\rho>0} \ \sup_{\{|y|=\rho\}} \int_{\{|x|=\rho\}} k_b(x,y) dS_\rho(x).$$

For $\varphi \geq 0$, the norm of

$$K_d^{(m)}\varphi = \int_{\{|y|=\omega_n(|x|)\}} k_d^{(m)}(x,y)\varphi(y) dS_{\omega_n(|x|)}(y)$$

is equal to

$$\int_0^{+\infty} d\rho \int_{S_\rho} dS_\rho(x) \left(\int_{\{|y|=\omega_n(|x|)\}} k_d^{(m)}(x,y)\varphi(y) dS_{\omega_n(|x|)}(y) \right)$$
$$= \int_0^{+\infty} d\rho \int_{S_\rho} dS_\rho(x) \left(\int_{\{|y|=\omega_n(\rho)\}} k_d^{(m)}(x,y)\varphi(y) dS_{\omega_n(\rho)}(y) \right).$$

By using the formula

$$\int_0^{+\infty} G(\rho, \gamma(\rho)) d\rho = \int_{\inf \gamma}^{+\infty} c_\gamma(\rho) G(\gamma^{-1}(\rho), \rho) d\rho$$

with $\gamma(\rho) = \omega_n(\rho) > \sqrt{\frac{2E_m}{m}}$

$$\int_0^{+\infty} d\rho \int_{S_\rho} dS_\rho(x) \left(\int_{\{|y|=\omega_n(\rho)\}} k_d^{(m)}(x,y)\varphi(y) dS_{\omega_n(\rho)}(y) \right)$$
$$= \int_{\sqrt{\frac{2E_m}{m}}}^{+\infty} c_{\omega_n}(\rho) d\rho \int_{S_{\omega_n^{-1}(\rho)}} dS_{\omega_n^{-1}(\rho)}(x) \left(\int_{\{|y|=\rho\}} k_d^{(m)}(x,y)\varphi(y) dS_\rho(y) \right)$$
$$= \int_{\sqrt{\frac{2E_m}{m}}}^{+\infty} d\rho \int_{\{|y|=\rho\}} \left(\int_{S_{\omega_n^{-1}(\rho)}} c_{\omega_n}(\rho) k_d^{(m)}(x,y) dS_{\omega_n^{-1}(\rho)}(x) \right) \varphi(y) dS_\rho(y)$$
$$= \int_0^{+\infty} d\rho \int_{\{|y|=\rho\}} \left(1_{\{\rho \geq \sqrt{\frac{2E_m}{m}}\}} \int_{S_{\omega_n^{-1}(\rho)}} c_{\omega_n}(\rho) k_d^{(m)}(x,y) dS_{\omega_n^{-1}(\rho)}(x) \right) \varphi(y) dS_\rho(y)$$

so $\|K_d\varphi\|$ is given by

$$\int_0^{+\infty} d\rho \int_{\{|y|=\rho\}} \left(\sum_{m=1}^{\overline{m}} 1_{\{\rho \geq \sqrt{\frac{2E_m}{m}}\}} \int_{S_{\omega_n^{-1}(\rho)}} c_{\omega_n}(\rho) k_d^{(m)}(x,y) dS_{\omega_n^{-1}(\rho)}(x) \right) \varphi(y) dS_\rho(y).$$

Finally $\|K\varphi\| = \|K_d\varphi\| + \|K_b\varphi\|$ is given by

$$\int_0^{+\infty} d\rho \int_{\{|y|=\rho\}} \left(\sum_{m=1}^{\overline{m}} 1_{\left\{\rho \geq \sqrt{\frac{2E_m}{m}}\right\}} \int_{S_{\omega_n^{-1}(\rho)}} c_{\omega_n}(\rho) k_d^{(m)}(x, y) dS_{\omega_n^{-1}(\rho)}(x) \right) \varphi(y) dS_\rho(y)$$

$$+ \int_0^{+\infty} d\rho \int_{\{|y|=\rho\}} \left(\int_{\{|x|=\rho\}} k_b(x, y) dS_\rho(x) \right) \varphi(y) dS_\rho(y)$$

$$= \int_0^{+\infty} d\rho \int_{\{|y|=\rho\}} \sigma_b^d(\rho, y) \varphi(y) dS_\rho(y)$$

where $\sigma_b^d(\rho, y)$ is given by

$$\sum_{m=1}^{\overline{m}} 1_{\left\{\rho \geq \sqrt{\frac{2E_m}{m}}\right\}} \int_{S_{\omega_n^{-1}(\rho)}} c_{\omega_n}(\rho) k_d^{(m)}(x, y) dS_{\omega_n^{-1}(\rho)}(x) + \int_{\{|x|=\rho\}} k_b(x, y) dS_\rho(x).$$

Let

$$K_i : \varphi \to \int_{\mathbb{R}^3} k_i(x, y) \varphi(y) dy \tag{10.6.1}$$

be the usual inelastic collision operator. We consider equations of the form

$$\begin{cases} \frac{d\varphi(t)}{dt} = -\sigma\varphi(t) + Cf(t) \\ \varphi(0) = \varphi_0 \in L^1(\mathbb{R}^3) \end{cases}$$

with C as the full collision operator

$$C = K + K_i = K_b + K_d + K_i$$

where

$$\sigma = \sigma_a + \sigma_i + \sigma_b^d \tag{10.6.2}$$

and $\sigma_i(y) = \int_{\mathbb{R}^3} k_i(x, y) dx$. Note that $K\varphi$ is given by

$$\int_{\{|y|=|x|\}} k_b(x, y) \varphi(y) dS_{|x|}(y) + \sum_{m=1}^{\overline{m}} \int_{\{|y|=\omega_n(|x|)\}} k_d^{(m)}(x, y) \varphi(y) dS_{\omega_n(|x|)}(y)$$

and, for $\varphi \geq 0$,

$$\|C\varphi\|_{L^1(\mathbb{R}^3)} = \int_0^{+\infty} d\rho \int_{S_\rho} \left(\sigma_i(\rho, y) + \sigma_b^d(\rho, y) \right) \varphi(\rho, y) dS_\rho(y)$$

$$\leq \int_0^{+\infty} d\rho \int_{S_\rho} \left(\sigma_a(\rho, y) + \sigma_i(\rho, y) + \sigma_b^d(\rho, y) \right) \varphi(\rho, y) dS_\rho(y).$$

If we introduce the multiplication operator

$$M : D(M) \subset \varphi \to -\sigma\varphi,$$

one sees that C is M-bounded in $L^1(\mathbb{R}^3)$. We are ready to provide a general peripheral spectral theory when K is M-small in the sense (10.6.3). We start with

Lemma 10.6.1 *Let* $\inf \sigma > 0$. *Suppose that* K *is* M-*small in the sense that*

$$r\left(K(0 - M)^{-1}\right) < 1, \tag{10.6.3}$$

e.g. $\inf \frac{\sigma_i + \sigma_a}{\sigma_b^d} > 0$. *Then*

$$\left\| K(\lambda - M)^{-1} \right\|_{\mathcal{L}(L^1(\mathbb{R}^3))} \le \sup \frac{\sigma_b^d}{\sigma_a + \sigma_i + \sigma_b^d} < 1 \ (\lambda \ge 0)$$

and $M + K$ *generates a holomorphic* C_0-*semigroup* $(U(t))_{t\ge 0}$ *in* $L^1(\mathbb{R}^3)$ *such that*

$$U(t) \le \left(I - K(0 - M)^{-1}\right)^{-1} \ (t \ge 0).$$

Moreover, $s(M + K) < 0$.

Proof Note that

$$\left\| K(\lambda - M)^{-1}\varphi \right\|$$
$$= \int_0^{+\infty} d\rho \int_{\{|y|=\rho\}} \frac{\sigma_b^d(\rho, y)}{\lambda + \sigma_a(\rho, y) + \sigma_i(\rho, y) + \sigma_b^d(\rho, y)} \varphi(y) dS_\rho(y)$$

shows

$$\left\| K(\lambda - M)^{-1} \right\|_{\mathcal{L}(L^1(\mathbb{R}^3))} \le \sup \frac{\sigma_b^d}{\sigma_a + \sigma_i + \sigma_b^d}$$
$$= \frac{1}{1 + \inf \frac{\sigma_a + \sigma_i}{\sigma_b^d}} \ (\lambda \ge 0)$$

so $\left\| K(0 - M)^{-1} \right\|_{\mathcal{L}(L^1(\mathbb{R}^3))} < 1$ if $\inf \frac{\sigma_a + \sigma_i}{\sigma_b^d} > 0$. The generation follows from Theorem 2.6.1 and the semigroup is given by a series

$$\sum_{j=0}^{\infty} S_j(t)\varphi$$

with $S_0(t)\varphi = e^{-t\sigma}\varphi$ and

$$S_{j+1}(t)\varphi = \int_0^t S_j(t-s)K S_0(s)\varphi ds$$

Let us show the domination

$$S_j(t)\varphi \leq \left(K(0-M)^{-1}\right)^j \varphi \ \ (\varphi \geq 0).$$

We have

$$S_1(t)\varphi = \int_0^t S_0(t-s)K S_0(s)\varphi ds$$
$$\leq \int_0^t K S_0(s)\varphi ds$$
$$\leq \int_0^{+\infty} K S_0(s)\varphi ds = K(0-M)^{-1}\varphi.$$

By induction

$$S_{j+1}(t)\varphi = \int_0^t S_j(t-s)K S_0(s)\varphi ds$$
$$\leq \int_0^t \left(K(0-M)^{-1}\right)^j K S_0(s)\varphi ds$$
$$= \left(K(0-M)^{-1}\right)^j \int_0^t K S_0(s)\varphi ds$$
$$\leq \left(K(0-M)^{-1}\right)^j \int_0^t K S_0(s)\varphi ds$$
$$= \left(K(0-M)^{-1}\right)^{j+1}.$$

Hence

$$U(t)\varphi \leq \sum_{j=0}^{\infty} \left(K(0-M)^{-1}\right)^j \varphi = \left(I - K(0-M)^{-1}\right)^{-1} \varphi.$$

By Proposition 2.5.1

$$s(M+K) = \inf\left\{\lambda > -\inf \sigma, \ r(K(0-M)^{-1}) < 1\right\}.$$

Since $r\left(K(0-M))^{-1}\right) < 1$ then the upper semi-continuity of the spectral radius (see e.g. [183]), $r(K(\lambda - M)^{-1}) < 1$ pour $\lambda < 0$ small enough and $s(M+K) < 0$. $\qquad\square$

Here are the main results when $\inf \sigma > 0$.

Theorem 10.6.1 *Suppose that* $\inf \sigma > 0$*. Let* K *be* M*-small in the sense that* $r\left(K(0-M))^{-1}\right) < 1$*, (e.g.* $\inf \frac{\sigma_i + \sigma_a}{\sigma_b^d} > 0$*), and let* K_i *be* M*-weakly compact. Then:*
(i) K_i *is* $(M+K)$*-weakly compact.*
(ii) $M+C$ *generates a holomorphic* C_0*-semigroup* $(V(t))_{t\geq 0}$ *in* $L^1(\mathbb{R}^3)$
(iii) $V(t) - U(t)$ *weakly compact.*
(iv) If $\sigma_a = 0$ *then* $(V(t))_{t\geq 0}$ *has a spectral gap.*

Proof Since

$$(\lambda - M - K)^{-1} = (\lambda - M)^{-1} \sum_{j=0}^{\infty} \left(K(\lambda - M)^{-1}\right)^j \quad (\lambda > s(M+K))$$

then

$$K_i\,(\lambda - M - K)^{-1} = K_i(\lambda - M)^{-1} \sum_{j=0}^{\infty} \left(K(\lambda - M)^{-1}\right)^j$$

is weakly compact since $K_i(\lambda - M)^{-1}$ is. The generation follows from Theorem 2.6.3. We have also

$$(\lambda - M - C)^{-1} = (\lambda - M - K)^{-1} \sum_{j=0}^{\infty} \left(K_i\,(\lambda - M - K)^{-1}\right)^j \quad (\lambda > s(M+P))$$

so $K_i\,(\lambda - M - C)^{-1}$ is weakly compact since $K_i\,(\lambda - M - K)^{-1}$ is. By using Lemma 10.6.1

$$
\begin{aligned}
e^{-\lambda t} V(t)\varphi - e^{-\lambda t} U(t)\varphi &= \int_0^t e^{-\lambda(t-s)} U(t-s) K_i \left(e^{-\lambda s} V(s)\right) \varphi ds \\
&\leq \int_0^t \left(I - K(0-M)^{-1}\right)^{-1} K_i \left(e^{-\lambda s} V(s)\right) \varphi ds \\
&= \left(I - K(0-M)^{-1}\right)^{-1} \int_0^t K_i \left(e^{-\lambda s} V(s)\right) \varphi ds \\
&\leq \left(I - K(0-M)^{-1}\right)^{-1} \int_0^{+\infty} K_i \left(e^{-\lambda s} V(s)\right) \varphi ds \\
&= \left(I - K(0-M)^{-1}\right)^{-1} K_i \, (\lambda - M - P)^{-1}
\end{aligned}
$$

shows that $e^{-\lambda t} V(t) - e^{-\lambda t} U(t)$ is weakly compact since it is dominated by a weakly compact operator.

Finally if $\sigma_a = 0$ then $(V(t))_{t\geq 0}$ is stochastic and its type is equal to zero

$$\omega(V) = 0.$$

Since $(V(t))_{t\geq 0}$ and $(U(t))_{t\geq 0}$ have the same essential spectrum and therefore the same essential type then

$$\omega_{ess}(V) = \omega_{ess}(U) \leq \omega(U) < 0 = \omega(V)$$

ends the proof. $\square$

It follows from Theorem 4.4.1 that we can preserve the existence of a spectral gap for $\sigma_a \neq 0$ provided that the oscillation of σ_a is small enough:

Theorem 10.6.2 *Let* $\inf \sigma > 0$. *Suppose that K is M-small in the sense that* $r\left(K(0-M))^{-1}\right) < 1$, *(e.g.* $\inf \frac{\sigma_c + \sigma_a}{\sigma_b^d} > 0$*). Let K_i be M-weakly compact.Then* $s(-\sigma_c - \sigma_b^d - K) < 0$ *and* $(V(t))_{t\geq 0}$ *has a spectral gap provided that*

$$\sup \sigma_a - \inf \sigma_a < -s(-\sigma_c - \sigma_b^d - K).$$

We note that all the results of this section can be stated in $\mathbb{R}^d$ rather than just $\mathbb{R}^3$. For the analysis of the full space-nonhomogeneous neutron transport equation with non-incoming boundary conditions, we refer to [324]. In Sect. 10.14, we consider this full equation on the torus, but without downshift scattering.

We give now the main results when $\inf \sigma = 0$.

Theorem 10.6.3 *Let* $\sigma_a = 0$ *and* $\inf \sigma = 0$. *We suppose that K be M-small in the sense that* $r\left(K\sigma^{-1}\right) < 1$ *(e.g.* $\inf \frac{\sigma_i}{\sigma_b^d} > 0$*) and that $K_i\sigma^{-1}$ is weakly compact. Then* $M + C : D(M) \subset L^1(\nu) \to L^1(\nu)$ *generates a stochastic holomorphic C_0-semigroup* $(V(t))_{t\geq 0}$. *Moreover, $C\sigma^{-1}$ is stochastic and* $r_{ess}\left(C\sigma^{-1}\right) < 1$. *In particular, 1 is an eigenvalue of $C\sigma^{-1}$ associated to a positive eigenvector $u \in L^1(\nu)$ and we have the following alternative:*

(i) If $v := \sigma^{-1}u \in L^1(\nu)$ *then* $v \in D(M)$ *and (after normalization) v is an invariant density of* $(V(t))_{t\geq 0}$. *Finally,* $(V(t))_{t\geq 0}$ *is asymptotically stable.*

(ii) If $v := \sigma^{-1}u \notin L^1(\nu)$ *then* $(V(t))_{t\geq 0}$ *has is sweeping from the sets* $\{\sigma \geq \varepsilon; \ \varepsilon > 0\}$.

Proof The generation follows from Theorem 2.6.3. The stochasticity follows from $\sigma_a = 0$. Note that

$$r_{ess}\left(C\sigma^{-1}\right) = r_{ess}\left(K\sigma^{-1} + K_i\sigma^{-1}\right) = r_{ess}\left(K\sigma^{-1}\right)$$

so $r_{ess}\left(C\sigma^{-1}\right) < 1$. Since K_i is integral then the Dyson Phillips expansion shows that $(V(t))_{t\geq 0}$ is partially integral and finally Theorem 3.3.1 ends the proof. $\square$

10.7 Neutron Transport Versus Space Homogeneous Equations

Neutron transport equations combine a free transport operator $\zeta.\nabla_x$ (acting on the spatial variable x) and a jump operator (acting on the velocity variable ζ). The purpose of this section is to show how various properties, such as generation, mean ergodicity, and sweeping, of space-homogeneous equations are inherited by the space-nonhomogeneous neutron transport equations. We mainly report results previously established in [264]. The main findings are presented in Theorems 10.7.1, 10.7.2, 10.7.3 and 10.7.4.

We define first the basic objects for the subsequent sections. Let $D \subset \mathbb{R}^d$ be a smooth (say with $\mathcal{C}^1$ boundary ∂D) open subset and let dx be the Lebesgue measure on D. The smoothness assumption on the spatial domain is made for the sake of simplicity in order to avoid additional cumbersome discussions. Let $\nu(d\zeta)$ be a general Borel measure on $\mathbb{R}^d$ such that

$$\nu\{0\} = 0$$

and let

$$L^p(D \times \mathbb{R}^d) := L^p(D \times \mathbb{R}^d; \ dx \otimes \nu(d\zeta)) \ \ (1 \le p < +\infty)$$

with norm

$$\|\varphi\|_{L^p(D \times \mathbb{R}^d)} = \left(\int_{\Omega \times \mathbb{R}^d} |\varphi(x,\zeta)|^p \, dx\nu(d\zeta) \right)^{\frac{1}{p}}.$$

Let

$$\tau(x,\zeta) := \inf\{s > 0; \ x - t\zeta \notin D\}$$

be the *exit time* function from D. Let

$$\Sigma : (x,\zeta) \in D \times \mathbb{R}^d \to \Sigma(x,\zeta) \in [0,+\infty]$$

be measurable and finite a.e. Suppose that

$$\int_0^t \Sigma(x - s\zeta, \zeta)ds < +\infty \ \ (x,\zeta)\text{-a.e. } (t < \tau(x,\zeta)),$$

then

$$U_\Sigma(t) : \varphi \in L^1(D \times \mathbb{R}^d) \to 1_{\{t<\tau(x,\zeta)\}} e^{-\int_0^t \Sigma(x - s\zeta,\zeta)ds} \varphi(x - t\zeta, \zeta)$$

defines a so-called streaming C_0-semigroup $(U_\Sigma(t))_{t \ge 0}$ on $L^p(D \times \mathbb{R}^d)$, see e.g. [367, 368]. We denote its generator by T_Σ. Let T_0 be the generator of the free streaming semigroup $(U_0(t))_{t \ge 0}$ (corresponding to $\Sigma = 0$) and let $D(T_0)$ be the domain of T_0.

We denote by $C_c(D \times \mathbb{R}^d)$ the continuous functions φ on $D \times \mathbb{R}^d$ with compact support such that $\varphi(x, \zeta) = 0$ if $|\zeta|$ is small enough and

$$C_c^{0,1}(D \times \mathbb{R}^d) = \left\{ \varphi \in C_c(D \times \mathbb{R}^d), \; \frac{\partial \varphi}{\partial x} \in C(D \times \mathbb{R}^d) \right\}.$$

Then $C_c^{0,1}(D \times \mathbb{R}^d) \subset D(T_0)$,

$$T_0\varphi = -\zeta.\nabla_x\varphi, \quad (\varphi \in D(T_0))$$

and

$$D_0 := lin \left\{ \cup_{t \geq 0} U_0(t) \left(C_c^{0,1}(D \times \mathbb{R}^d) \right) \right\}$$

($lin\ \{\}$ means linear span of $\{\}$) is dense in $L^p(D \times \mathbb{R}^d)$ and is invariant by $(U_0(t))_{t \geq 0}$ so that D_0 is a core of T_0, i.e., T_0 is the closure of its restriction to D_0, see [368].

Even if the (rigorous) proof is quite involved, the domain $D(T_0)$ has a "natural" characterization. Indeed, $\varphi \in D(T_0)$ if and only if $\varphi \in L^p(D \times \mathbb{R}^d)$ and, for almost all $(x, \zeta) \in D \times \mathbb{R}^d$,

$$t \in (-\tau(x, \zeta), \; \tau(x, -\zeta)) \rightarrow \varphi(x + t\zeta, \zeta)$$

is absolutely continuous, its derivative (at $t = 0$) $\zeta.\frac{\partial \varphi}{\partial x} \in L^p(D \times \mathbb{R}^d)$ and $\varphi_{|\Gamma_-} = 0$ in the sense

$$\varphi(x - \tau(x, \zeta)\zeta, \zeta) = 0; \tag{10.7.1}$$

in this case $T_0 = -\zeta.\frac{\partial \varphi}{\partial x}$, see [365] Chap. 2, p. 26 for the details.

We point out that $(U_\Sigma(t))_{t \geq 0}$ is nothing but the absorption semigroup (see Chap. 5) from $(U_0(t))_{t \geq 0}$ and the absorption Σ. Note that the elements of D_0 are finite linear combinations of functions of the form

$$\psi(x, \zeta) = 1_{\{t < \tau(x,\zeta)\}} e^{-\int_0^t \Sigma(x - s\zeta, \zeta) ds} \varphi(x - t\zeta, \zeta) \quad (\varphi \subset C_c^{0,1}(D \times \mathbb{R}^d))$$

which are bounded functions with bounded supports. It follows that if

$$\Sigma \in L_{loc}^p(D \times \mathbb{R}^d) \tag{10.7.2}$$

then $D_0 \subset D(\Sigma)$ (the domain of Σ). Since $D_0 \cap D(\Sigma) = D_0$ is a core of T_0 then so is $D(T_0) \cap D(\Sigma)$. Then Lemma 5.3.2 implies that Σ is U_0-regular and (see Remark 5.3.1)

$$\begin{cases} T_\Sigma = T_0 - \Sigma \\ D(T_\Sigma) = D(T_0) \cap D(\Sigma). \end{cases} \tag{10.7.3}$$

In all the sequel of this chapter, we restrict ourselves to the case

$$p = 1$$

and assume tacitly (10.7.2), i.e., Σ is locally integrable.

Let $C : D(C) \to L^1(D \times \mathbb{R}^d)$ be a collision operator defined by

$$C\varphi = \int_{\mathbb{R}^d} c(x, \zeta, \zeta')\varphi(x, \zeta')\nu(d\zeta').$$

We do not assume a priori a subcriticality condition

$$\int_{\mathbb{R}^d} c(x, \zeta', \zeta)\nu(d\zeta') \leq \Sigma(x, \zeta)$$

which would imply, by Kato-Voigt perturbation theory, that an extension of $T_\Sigma + C$ is a generator. Let

$$\begin{cases} \widehat{\Sigma}(\zeta) := \inf_{x \in D} \Sigma(x, \zeta) \\ \widehat{c}(\zeta, \zeta') = \sup_{x \in D} c(x, \zeta, \zeta') \end{cases}$$

and let

$$\widehat{C} : D(\widehat{C}) \subset L^1(\mathbb{R}^d_\zeta) \to L^1(\mathbb{R}^d_\zeta)$$

be a collision operator defined by

$$\widehat{C}\varphi = \int_{\mathbb{R}^d} \widehat{c}(\zeta, \zeta')\varphi(\zeta')\nu(d\zeta')$$

where $L^1(\mathbb{R}^d_\zeta) := L^1(\mathbb{R}^d, \nu(d\zeta))$.

10.7.1   Hereditary Generation

The main result of this subsection is

Theorem 10.7.1 *Let $\widehat{C}$ be $\widehat{\Sigma}$-bounded in $L^1(\mathbb{R}^d_\zeta)$ and assume that*

$$- \widehat{\Sigma} + \widehat{C} : D(\widehat{\Sigma}) \subset L^1(\mathbb{R}^d_\zeta) \to L^1(\mathbb{R}^d_\zeta)$$

generates a positive C_0-semigroup on $L^1(\mathbb{R}^d_\zeta)$. Then C is Σ-bounded in $L^1(D \times \mathbb{R}^d)$ and

$$T_\Sigma + C : D(T_\Sigma) \subset L^1(D \times \mathbb{R}^d) \to L^1(D \times \mathbb{R}^d)$$

generates a positive C_0-semigroup on $L^1(D \times \mathbb{R}^d)$.

This result is already given in [264]. We recall here the ingredients of the proof. Note that

$$(\lambda - T_\Sigma)^{-1} \varphi = \int_0^{\tau(x,\zeta)} e^{-\lambda t} e^{-\int_0^t \Sigma(x-s\zeta,\zeta)ds} \varphi(x - t\zeta, \zeta)dt \quad (\lambda > 0)$$

is dominated by

$$\left(\lambda - T_{\widehat{\Sigma}}\right)^{-1} \varphi = \int_0^{\tau(x,\zeta)} e^{-\lambda t} e^{-t\widehat{\Sigma}(\zeta)} \varphi(x - t\zeta, \zeta)dt \quad (\lambda > 0).$$

We still denote by $\widehat{C}$ the $\widehat{\Sigma}$-bounded operator on $L^1(D \times \mathbb{R}^d)$ given by

$$\widehat{C}\varphi = \int_{\mathbb{R}^d} \widehat{c}(\zeta, \zeta')\varphi(x, \zeta')\nu(d\zeta')$$

and note the domination

$$C\left(\lambda - T_\Sigma\right)^{-1} \leq \widehat{C}\left(\lambda - T_{\widehat{\Sigma}}\right)^{-1}. \tag{10.7.4}$$

The proof of Theorem 10.7.1 is based on the following key estimates.

Lemma 10.7.1 ([264]) *For any $\varphi \in L_+^1(D \times \mathbb{R}^d)$ and any $n \in \mathbb{N}$*

$$\left\|\left(\widehat{C}\left(\lambda - T_{\widehat{\Sigma}}\right)^{-1}\right)^n \varphi\right\|_{L^1(D \times \mathbb{R}^d)} \leq \left\|\left(\widehat{C}\left(\lambda + \widehat{\Sigma}\right)^{-1}\right)^n \psi\right\|_{L^1(\mathbb{R}_\zeta^d)} \tag{10.7.5}$$

where $\psi(\zeta) = \int_D \varphi(x, \zeta)dx$.

Proof (*of Theorem* 10.7.1) It follows from the estimate (10.7.5) that

$$\left\|\left(\widehat{C}\left(\lambda - T_{\widehat{\Sigma}}\right)^{-1}\right)^n\right\|_{\mathcal{L}(L^1(D \times \mathbb{R}^d))} \leq \left\|\left(\widehat{C}\left(\lambda + \widehat{\Sigma}\right)^{-1}\right)^n\right\|_{\mathcal{L}(L^1(\mathbb{R}_\zeta^d))}$$

and consequently

$$r\left(\widehat{C}\left(\lambda - T_{\widehat{\Sigma}}\right)^{-1}\right) \leq r\left(\widehat{C}\left(\lambda + \widehat{\Sigma}\right)^{-1}\right).$$

By assumption $-\widehat{\Sigma} + \widehat{C}$ is a generator on $L^1(\mathbb{R}_\zeta^d)$ so, by Theorem 2.6.1,

$$\lim_{\lambda \to +\infty} r\left(\widehat{C}\left(\lambda + \widehat{\Sigma}\right)^{-1}\right) < 1$$

and consequently $\lim_{\lambda \to +\infty} r\left(\widehat{C}\left(\lambda - T_{\widehat{\Sigma}}\right)^{-1}\right) < 1$. It follows from (10.7.4) that $\lim_{\lambda \to +\infty} r\left(C\left(\lambda - T_\Sigma\right)^{-1}\right) < 1$ and, again by Theorem 2.6.1, $T_\Sigma + C$ is a generator on $L^1(D \times \mathbb{R}^d)$. $\qquad\square$

In particular, we have

Corollary 10.7.1 *Let the cross sections be space homogeneous, i.e.,*

$$\Sigma(x, \zeta) = \Sigma(\zeta), \quad c(x, \zeta, \zeta') = c(\zeta, \zeta'). \tag{10.7.6}$$

If $-\Sigma + C$ *is a generator on* $L^1(\mathbb{R}^d_\zeta)$ *then* $T_\Sigma + C$ *is a generator on* $L^1(D \times \mathbb{R}^d)$.

For space homogeneous *subcritical* cross-sections, the corresponding Kato-Voigt semigroups share the honesty property.

Theorem 10.7.2 *Let the cross sections be space homogeneous and let*

$$\int_{\mathbb{R}^d} c(\zeta', \zeta)\nu(d\zeta') \leq \Sigma(\zeta).$$

If $\overline{-\Sigma + C}$ *is a generator on* $L^1(\mathbb{R}^d_\zeta)$ *then* $\overline{T_\Sigma + C}$ *is a generator on* $L^1(D \times \mathbb{R}^d)$.

Proof By Kato-Voigt perturbation theory, there exists a unique extension of $-\Sigma + C$ generating a minimal substochastic Kato-Voigt semigroup on $L^1(\mathbb{R}^d_\zeta)$ and also a unique extension of $T_\Sigma + C$ generating a minimal substochastic Kato-Voigt semi-group on $L^1(D \times \mathbb{R}^d)$. According to Proposition 2.11.3, the closure $\overline{-\Sigma + C}$ is a generator is equivalent to honesty of the corresponding Kato-Voigt semigroup on $L^1(\mathbb{R}^d_\zeta)$. According to Theorem 2.11.2, honesty of a trajectory on $L^1(\mathbb{R}^d_\zeta)$ emanating from $\psi \in L^1(\mathbb{R}^d_\zeta)$ amounts to

$$\lim_{n \to \infty} \left\| \left(\widehat{C} \left(\lambda + \widehat{\Sigma} \right)^{-1} \right)^n \psi \right\|_{L^1(\mathbb{R}^d_\zeta)} = 0.$$

Then the estimate (10.7.5) together with Theorem 2.11.2 show that any trajectory in $L^1(D \times \mathbb{R}^d)$ emanating from $\psi \in L^1(D \times \mathbb{R}^d)$ is honest so $\overline{T_\Sigma + C}$ is a generator on $L^1(D \times \mathbb{R}^d)$. $\qquad\square$

Remark 10.7.1 We can treat more complex scattering kernels combining "down-shift" and "Bragg" scatterings. The same statements hold also when we replace D by the d-dimensionnal torus $T^d := \mathbb{R}^d/(2\pi\mathbb{Z})^d$.

A similar analysis can be carried out by replacing the transport operator $-\zeta.\nabla_x$ with another generator of a substochastic semigroup acting on the spatial variable (depending on the parameter ζ); for example, a diffusion operator in space with various boundary conditions (see [264]).

10.7.2 Hereditary Mean Ergodicity

We restrict ourselves to the torus T^d and to space homogeneous cross sections. We complement Corollary 10.7.1 by

Theorem 10.7.3 *Let (10.7.6) be satisfied. Suppose that* $-\Sigma + C$ *generates a C_0-semigroup* $\big(\widehat{V}(t)\big)_{t\geq0}$ *on* $L^1(\mathbb{R}^d_\zeta)$. *Let* $(V(t))_{t\geq0}$ *be the C_0-semigroup generated by* $T_\Sigma + C$ *on* $L^1(\mathcal{T}^d \times \mathbb{R}^d)$. *Suppose that* $\big(\widehat{V}(t)\big)_{t\geq0}$ *has an invariant density* $\psi > 0$ *a.e. Then* $(V(t))_{t\geq0}$ *is mean ergodic in* $L^1(\mathcal{T}^d \times \mathbb{R}^d)$.

Proof By Corollary 10.7.1 (and Remark 10.7.1), $T_\Sigma + C$ is a generator in $L^1(\mathcal{T}^d \times \mathbb{R}^d)$. We know that

$$(\lambda - T_\Sigma - C)^{-1}\varphi = (\lambda - T_\Sigma)^{-1}\sum_{j=0}^{\infty}\big(C(\lambda - T_\Sigma)^{-1}\big)^j\varphi.$$

If $\varphi(x, \zeta') = \varphi(\zeta')$ then a simple calculation shows that

$$(\lambda - T_\Sigma - C)^{-1}\varphi = (\lambda + \Sigma)^{-1}\sum_{j=0}^{\infty}\big(C(\lambda + \Sigma)^{-1}\big)^j\varphi$$

$$= (\lambda - (-\Sigma + C))^{-1}\varphi$$

so

$$V(t)\varphi = \widehat{V}(t)\varphi, \quad (\varphi \in L^1(\mathbb{R}^d_\zeta)).$$

Since $\widehat{V}(t)\psi = \psi$ then (up to a normalization in $L^1(\mathcal{T}^d \times \mathbb{R}^d)$) ψ is an invariant density of $(V(t))_{t\geq0}$. Finally, Proposition 2.10.1 ends the proof. $\qquad\square$

10.7.3   Hereditary Sweeping

We restrict ourselves to the torus $\mathcal{T}^d$ and to space-homogeneous cross sections. We easily complement Theorem 10.7.3 with

Theorem 10.7.4 *Let (10.7.6) be satisfied. Suppose that* $-\Sigma + C$ *generates a C_0-semigroup* $\big(\widehat{V}(t)\big)_{t\geq0}$ *on* $L^1(\mathbb{R}^d_\zeta)$. *Let* $(V(t))_{t\geq0}$ *be the C_0-semigroup generated by* $T_\Sigma + C$ *on* $L^1(\mathcal{T}^d \times \mathbb{R}^d)$. *If* $\big(\widehat{V}(t)\big)_{t\geq0}$ *is sweeping from a family of sets* $\Xi \subset \mathbb{R}^d_\zeta$ *then* $(V(t))_{t\geq0}$ *is sweeping from the sets* $\mathcal{T}^d \times \Xi$.

Proof Let $\varphi(.,.) \in L^1(\mathcal{T}^d \times \mathbb{R}^d)$. It is easy to check that

$$\int_{\mathcal{T}^d} V(t)\varphi\, dx = \widehat{V}(t)\psi$$

where

$$\psi(.) = \int_{\mathcal{T}^d} \varphi(x, .)dx \in L^1(\mathbb{R}^d_\zeta).$$

In particular

$$\int_{\mathbb{T}^d \times \Xi} V(t)\varphi \, dx \nu(d\zeta) = \int_{\Xi} \widehat{V}(t)\psi \, \nu(d\zeta) \to 0 \ (t \to +\infty)$$

and we are done. $\square$

10.8 On L^1 Spectral Theory of Neutron Transport

The purpose of this section is to provide a general peripheral spectral analysis for neutron transport C_0-semigroups associated with a suitable class of (abstract) Borel velocity measures. This class includes both the volumic Lebesgue measure and the surface Lebesgue measure on spheres (multigroup models), as well as their combinations. A central challenge lies in addressing the compactness issues that arise in the spectral theory. We develop a systematic functional analytic treatment of these problems. This program was initially developed in [251, 266] for bounded collision operators. Here, we revisit and extend these results to encompass appropriate unbounded collision operators. The main results are presented in Theorems 10.8.3 and 10.8.5. In particular, this theory applies to general space-nonhomogeneous linear Boltzmann equations with power-like hard sphere potentials (see Sect. 10.9 below).

Let $D \subset \mathbb{R}^d$ be an open set with $\mathcal{C}^1$ boundary ∂D. For the time being, D need not be bounded. We endow D with the d-dimensional Lebesgue measure dx. Let $\nu(d\zeta)$ be a general Borel measure on $\mathbb{R}^d$ and $L^1(D \times \mathbb{R}^d)$ with norm

$$\|\varphi\|_{L^1(D \times \mathbb{R}^d)} = \int_{\Omega \times \mathbb{R}^d} |\varphi(x, \zeta)| \, dx \nu(d\zeta).$$

We consider general neutron transport equations

$$\frac{\partial \varphi}{\partial t} + \zeta . \nabla_x \varphi + \Sigma(x, \zeta)\varphi(t, x, \zeta) = \int_{\mathbb{R}^d} c(x, \zeta, \zeta')\varphi(t, x, \zeta')\nu(d\zeta') \qquad (10.8.1)$$

on $L^1(D \times \mathbb{R}^d)$ with the classical nonincoming boundary condition

$$\varphi_{|\Gamma_-} = 0 \qquad (10.8.2)$$

when D is a proper subset of $\mathbb{R}^d$. Of course, we drop (10.8.2) if $D = \mathbb{R}^d$. For the time being Σ is not related to c.

10.8.1 On Σ-Regular Collision Operators

A class of bounded collision operators

$$C : L^1(D \times \mathbb{R}^d) \to L^1(D \times \mathbb{R}^d)$$

(called *regular*)

$$\varphi \to \int_{\mathbb{R}^d} c(x, \zeta, \zeta')\varphi(x, \zeta')\nu(d\zeta') \tag{10.8.3}$$

was considered in previous works [251, 266]. Such collision operators are defined by the fact that the family of operators

$$C_x : \psi \in L^1_\zeta(\mathbb{R}^d) \to \int_{\mathbb{R}^d} c(x, \zeta, \zeta')\psi(\zeta')\nu(d\zeta') \in L^1_\zeta(\mathbb{R}^d) \tag{10.8.4}$$

(indexed by $x \in D$) are *collectively* weakly compact on

$$L^1_\zeta(\mathbb{R}^d) := L^1(\mathbb{R}^d; \nu(d\zeta))$$

in the sense that the set

$$\left\{ C_x\psi; \ \|\psi\|_{L^1_\zeta(\mathbb{R}^d)} \leq 1, \ x \in D \right\}$$

is relatively weakly compact in $L^1_\zeta(\mathbb{R}^d)$. We are going to extend this notion to Σ-bounded collision operators.

Definition 10.8.1 Let $\Sigma : (x, \zeta) \in D \times \mathbb{R}^d \to \Sigma(x, \zeta) \in \mathbb{R}$ be measurable and such that $0 < \Sigma(x, \zeta) < +\infty$ a.e. Let C be a Σ-*bounded* collision operator (10.8.3), i.e., such that $C (\lambda + \Sigma)^{-1} (\lambda > 0)$ is a bounded operator or equivalently

$$\int_{\mathbb{R}^d} \frac{c(x, \zeta, \zeta')}{\lambda + \Sigma(x, \zeta')}\nu(d\zeta) \in L^\infty(D \times \mathbb{R}^d).$$

We say that C is Σ-regular if the class of operators on $L^1_\zeta(\mathbb{R}^d)$

$$\psi \to \int_{\mathbb{R}^d} \frac{c(x, \zeta, \zeta')}{\lambda + \Sigma(x, \zeta')}\psi(\zeta')\nu(d\zeta'), \quad (\lambda > 0) \tag{10.8.5}$$

(indexed by $x \in D$) is *collectively* weakly compact on $L^1_\zeta(\mathbb{R}^d)$. (This definition of Σ-regularity of C is independent of $\lambda > 0$). We say that C is strongly Σ-regular if the class of operators on $L^1_\zeta(\mathbb{R}^d)$

$$\psi \to \int_{\mathbb{R}^d} \frac{c(x, \zeta, \zeta')}{\Sigma(x, \zeta')}\psi(\zeta')\nu(d\zeta')$$

(indexed by $x \in D$) is *collectively* weakly compact on $L^1_\zeta(\mathbb{R}^d)$.

Note that a strongly Σ-regular collision operator C can be viewed as

$$\varphi \to \int_{\mathbb{R}^d} \widetilde{c}(x, \zeta, \zeta') \Sigma(x, \zeta') \varphi(x, \zeta') \nu(d\zeta')$$

where $\varphi \to \int_{\mathbb{R}^d} \widetilde{c}(x, \zeta, \zeta') \varphi(x, \zeta') \nu(d\zeta')$ is a bounded regular collision operator.

Remark 10.8.1 If the cross sections are space homogeneous then the Σ-regularity of C expresses just that $C(\lambda + \Sigma)^{-1}$ $(\lambda > 0)$ is a weakly compact operator on $L^1_\zeta(\mathbb{R}^d)$, i.e., C is Σ-weakly compact on $L^1_\zeta(\mathbb{R}^d)$. The strong Σ-regularity amounts to $C\Sigma^{-1}$ is weakly compact on $L^1_\zeta(\mathbb{R}^d)$.

We recall a key tool.

Proposition 10.8.1 ([207]) *Let C be Σ-regular. Then we can approximate $\widehat{C}$:* $L^1(D \times \mathbb{R}^d) \to L^1(D \times \mathbb{R}^d)$

$$\varphi \to \int_{\mathbb{R}^d} \frac{c(x, \zeta, \zeta')}{\lambda + \Sigma(x, \zeta')} \varphi(x, \zeta') \nu(d\zeta') \ (\lambda > 0)$$

in operator norm by a sequence of operators

$$\widehat{C}_j : \varphi \to \int_{\mathbb{R}^d} k_j(x, \zeta, \zeta') \varphi(x, \zeta') \nu(d\zeta') \tag{10.8.6}$$

such that $k_j(x, \zeta, \zeta') \leq f_j(\zeta)$ where $f_j \in L^1_+(\mathbb{R}^d; \nu(d\zeta))$.

In Proposition 10.8.1 we drop λ when C is strongly Σ-regular.

10.8.2   A Contraction on $L^1(D \times \mathbb{R}^d)$

Let $D \subset \mathbb{R}^d$ be an arbitrary open subset with $\mathcal{C}^1$ boundary ∂D and let $\mathcal{U}_\Sigma = (U_\Sigma(t))_{t \geq 0}$ be the advection semigroup on $L^1(D \times \mathbb{R}^d)$

$$U_\Sigma(t) : \varphi \in L^1(D \times \mathbb{R}^d) \to e^{-\int_0^t \Sigma(x-s\zeta, \zeta)ds} \varphi(x - t\zeta, \zeta) 1_{\{t < \tau(x,\zeta)\}} \tag{10.8.7}$$

with generator T_Σ. The resolvent of the generator is given by

$$(\lambda - T_\Sigma)^{-1} \varphi(x, \zeta) = \int_0^{\tau(x,\zeta)} e^{-\lambda t} e^{-\int_0^t \Sigma(x-s\zeta, \zeta)ds} \varphi(x - t\zeta, \zeta)dt \ \ (\lambda > s(T_\Sigma))$$

where $s(T_\Sigma)$ is "computable"

$$s(T_\Sigma) = -\lim_{t\to+\infty} \inf_{\{(x,\zeta),\ \tau(x,\zeta)>t\}} t^{-1} \int_0^t \Sigma(x - s\zeta, \zeta)ds \qquad (10.8.8)$$

(see [368]). In particular, if Σ is "smooth" then

$$s(T_\Sigma) = \inf_{x\in D} \Sigma(x, 0),$$

(see [178]). Note that according to (10.7.3)(10.7.1)

$$D(T_\Sigma) = \left\{ g \in L^1,\ \zeta.\nabla_x g \in L^1, \Sigma g \in L^1,\ g_{|\Gamma_-} = 0 \right\}.$$

We have:

Lemma 10.8.1 *The operator* $L^1(D \times \mathbb{R}^d) \to L^1(D \times \mathbb{R}^d)$

$$\varphi \to \Sigma(x, \zeta)\left((\lambda - T_\Sigma)^{-1} \varphi\right)\ (\lambda > 0)$$

is a contraction on $L^1(D \times \mathbb{R}^d)$.

Proof Let $\varphi \in L^1_+(D \times \mathbb{R}^d)$ and $g := (\lambda - T_\Sigma)^{-1} \varphi$. Then

$$\lambda g(x, \zeta) + \Sigma(x, \zeta)g(x, \zeta) + \zeta.\nabla_x g(x, \zeta) = \varphi(x, \zeta).$$

By integrating on $D \times \mathbb{R}^d$ we get

$$\lambda \|g\|_{L^1} + \int_{D\times\mathbb{R}^d} \Sigma g + \int_{\partial D\times\mathbb{R}^d} \zeta.n(x)g(x, \zeta)\sigma(dx)\nu(d\zeta) = \|\varphi\|_{L^1}$$

(where $\sigma(dx)$ is the surface-Lebesgue measure on ∂D). Since g vanishes on Γ_- then $\int_{D\times\mathbb{R}^d} \Sigma g \leq \|\varphi\|_{L^1}$. $\qquad\square$

This result is also true if $D = \mathbb{R}^d$. Actually, it is also a consequence of a more general result on absorption C_0-semigroups on L^1 spaces (see Lemma 5.3.3). A useful consequence of Lemma 10.8.1 is that the strong limit

$$\Sigma (0 - T_\Sigma)^{-1} := s \lim_{\lambda\to 0_+} \Sigma (\lambda - T_\Sigma)^{-1}$$

exists in $L^1(D \times \mathbb{R}^d)$ and is a bounded (contraction) operator.

We give a "quick" generation result.

Theorem 10.8.1 *If a collision operator C of the form (10.8.3) is such that*

$$\sup_{(x,\zeta')} \frac{\int_{\mathbb{R}^d} c(x, \zeta, \zeta')\nu(d\zeta)}{\Sigma(x, \zeta')} < 1 \qquad (10.8.9)$$

then

$$T_\Sigma + C : D(T_\Sigma) \subset L^1(D \times \mathbb{R}^d) \to L^1(D \times \mathbb{R}^d)$$

is a generator of a positive semigroup $(V(t))_{t \geq 0}$. In particular, if

$$\Sigma(x, \zeta) = \Sigma_a(x, \zeta) + \Sigma_s(x, \zeta)$$

with $\Sigma_s(x, \zeta') = \int_{\mathbb{R}^d} c(x, \zeta, \zeta') \nu(d\zeta)$ then $(V(t))_{t \geq 0}$ is contractive and (10.8.9) amounts to $\inf \frac{\Sigma_a(x, \zeta')}{\Sigma_s(x, \zeta')} > 0$.

Proof Our assumption amounts to $\left\| C\Sigma^{-1} \right\| < 1$. It suffices to observe that thanks to Lemma 10.8.1

$$\begin{aligned}
\left\| C(\lambda - T_\Sigma)^{-1} \right\| &= \left\| \left(C\Sigma^{-1}\right) \left(\Sigma(\lambda - T_\Sigma)^{-1}\right) \right\| \\
&\leq \left\| C\Sigma^{-1} \right\| \left\| \Sigma(\lambda - T_\Sigma)^{-1} \right\| \\
&\leq \left\| C\Sigma^{-1} \right\| < 1
\end{aligned}$$

and we conclude with Theorem 2.6.1. The contraction of $(V(t))_{t \geq 0}$ follows from the dissipativity of $T_\Sigma + C$. $\qquad\square$

10.8.3 First Compactness Result

To show L^1 compactness results for Neutron Transport, we need suitable assumptions on the velocity measure $\nu(d\zeta)$. We recall a key result from [251].

Theorem 10.8.2 ([251]) *Let C be a bounded regular collision operator. We assume there exists $\alpha > 0$ such that for any $M > 0$ there exists $c_M > 0$ such that*

$$\sup_{e \in S^{d-1}} \nu\{\zeta;\ |\zeta| \leq M,\ |\zeta.e| \leq \varepsilon\} \leq c_M \varepsilon^\alpha. \tag{10.8.10}$$

If D has a finite volume then, for $\lambda > s(T_\Sigma)$, $\left[C(\lambda - T_\Sigma)^{-1}\right]^j$ is weakly compact on $L^1(D \times \mathbb{R}^d)$ for all $j > \frac{d(\alpha+1)}{2\alpha} + 1$.

We are ready to give the first key result.

Theorem 10.8.3 *Let C be a Σ-regular collision operator and let the velocity measure $\nu(d\zeta)$ satisfies (10.8.10). Suppose that D has a finite volume and*

$$\sup_{x \in D} \Sigma(x, \zeta) < +\infty \quad \nu(d\zeta) \text{ a.e.}$$

Then, for $\lambda > s(T_\Sigma)$, $C(\lambda - T_\Sigma)^{-1}$ is power compact in $L^1(D \times \mathbb{R}^d)$. In particular $T_\Sigma + C$ generates a positive C_0-semigroup $(V(t))_{t \geq 0}$.

Proof Without loss of generality, we may assume that Σ is bounded away from zero. By Dunford-Pettis argument, it suffices to show that some power of $C(\lambda - T_\Sigma)^{-1}$ is weakly compact in $L^1(D \times \mathbb{R}^d)$. We note that

$$C(\lambda - T_\Sigma)^{-1} = \left(C\Sigma^{-1}\right)\left(\Sigma(\lambda - T_\Sigma)^{-1}\right).$$

Let $\widehat{C} := C\Sigma^{-1}$. Considering $C(\lambda - T_\Sigma)^{-1}$ as a product of two (unrelated) operators

$$C(\lambda - T_\Sigma)^{-1} = \widehat{C}\left(\Sigma(\lambda - T_\Sigma)^{-1}\right)$$

and noting that $\Sigma(\lambda - T_\Sigma)^{-1}$ is a contraction for $\lambda > 0$ then the estimate

$$\left\|C(\lambda - T_\Sigma)^{-1}\right\| \leq \left\|C\Sigma^{-1}\right\|$$

shows that $C(\lambda - T_\Sigma)^{-1}$ depends continuously on $\widehat{C}$ in operator norm. It follows that any power $\left(C(\lambda - T_\Sigma)^{-1}\right)^j$ depends continuously on $\widehat{C}$ in operator norm. According to Proposition 10.8.1, we can approximate $\widehat{C}$ in operator norm by a suitable sequence $\widehat{C}_j$ (see (10.8.6)). It suffices to show that some power of $\widehat{C}\left(\Sigma(\lambda - T_\Sigma)^{-1}\right)$ is weakly compact when we replace $\widehat{C}$ by $\widehat{C}_j$. Actually, by using a domination argument, it suffices to replace $\widehat{C}$ by

$$\widetilde{C} : \varphi \in L^1(D \times \mathbb{R}^d) \to f(\zeta) \int_{\mathbb{R}^d} \varphi(x, \zeta')\nu(d\zeta')$$

where $f \in L^1_+(\mathbb{R}^d)$. Let

$$\widetilde{\Sigma}(\zeta) := \sup_{x \in D} \Sigma(x, \zeta).$$

Since $\widetilde{\Sigma}$ is finite a.e. then

$$1_{\{\widetilde{\Sigma}(\zeta)\leq m\}} f \to f \text{ in } L^1(\mathbb{R}^d) \ (m \to +\infty)$$

so (since $\widetilde{C}$ depends continuously on $f \in L^1(\mathbb{R}^d)$), without loss of generality, we can replace f by $1_{\{\widetilde{\Sigma}(\zeta)\leq m\}} f$ so that the *support* of f is included in the region $\left\{\zeta; \ \widetilde{\Sigma}(\zeta) \leq m\right\}$. In this case

$$\left(\widetilde{C}(\lambda - T_\Sigma)^{-1}\right)^{j+1} = \left(\widetilde{C}\left(\Sigma(\lambda - T_\Sigma)^{-1}\right)\right)^j \widetilde{C}\left(\Sigma(\lambda - T_\Sigma)^{-1}\right)$$

$$\leq \left(\widetilde{C}\left(\widetilde{\Sigma}(\lambda - T_\Sigma)^{-1}\right)\right)^j \widetilde{C}\left(\Sigma(\lambda - T_\Sigma)^{-1}\right).$$

Since $(\lambda - T_\Sigma)^{-1}$ does not augment the "support with respect to velocities" then

$$\left(\widetilde{C}\left(\widetilde{\Sigma}(\lambda - T_\Sigma)^{-1}\right)\right)^j \widetilde{C}\left(\Sigma(\lambda - T_\Sigma)^{-1}\right)$$

$$\leq m^j \left(\widetilde{C}\left((\lambda - T_\Sigma)^{-1}\right)\right)^j \widetilde{C}\left(\Sigma(\lambda - T_\Sigma)^{-1}\right).$$

According to Theorem 10.8.2 $\left(\widetilde{C}\left((\lambda - T_\Sigma)^{-1}\right)\right)^j$ is weakly compact for j large enough. This ends the proof of the first item. Finally, the generation follows from Theorem 2.6.1. $\qquad\square$

Remark 10.8.2 In a similar way, we can prove Theorem 10.8.3 in $L^1(T^d \times \mathbb{R}^d)$ where T^d is the d-dimensional torus.

The power compactness of $C(\lambda - T_\Sigma)^{-1}$ implies (see Theorem 2.8.1) that

$$\sigma(T_\Sigma + C) \cap \{\operatorname{Re}\lambda > s(T_\Sigma)\}$$

consists (at most) of isolated eigenvalues with finite algebraic multiplicities. A priori the spectrum of the semigroup $V(t)$ in $\left\{\mu \in; \ |\mu| > e^{ts(T)}\right\}$ need not be reduced to isolated eigenvalues because of the lack (a priori) of a spectral mapping theorem [359]. To this end, we need a more complex compactness result relying on a stronger assumption of the velocity measure $\nu(d\zeta)$.

10.8.4 Second Compactness Result

According to Theorem 2.6.1, $(V(t))_{t\geq0}$ is given by a Dyson-Phillips series

$$V(t) = \sum_{j=0}^{\infty} U_j(t)$$

converging in operator norm, where $U_0(t) = U_\Sigma(t)$ and

$$U_{j+1}(t)\varphi = \int_0^t U_j(t-s)CU_\Sigma(s)\varphi ds \ (x \in D(T_\Sigma)). \tag{10.8.11}$$

We know that $U_j(t)$ are bounded operators. Let

$$R_m(t) := \sum_{j=m}^{\infty} U_j(t) \ (m \geq 1)$$

be the remainders of the Dyson-Phillips series.

Lemma 10.8.2 *Let Σ be bounded away from zero. For all $\varphi \in L_+^1(D \times \mathbb{R}^d)$*

$$\left\|U_j(t)\varphi\right\| \leq \left\|\left(C\left(0 - T_\Sigma\right)^{-1}\right)^j \varphi\right\|, \ (j \geq 0). \tag{10.8.12}$$

In particular
$$\left\|U_j(t)\right\| \leq \left\|\left(C\left(0 - T_\Sigma\right)^{-1}\right)^j\right\|, \ (j \geq 0). \tag{10.8.13}$$

Proof We argue by induction. This is true for $j = 0$ since $(U_\Sigma(t))_{t\geq 0}$ is a contraction semigroup. Let (10.8.12) be true for some integer j. Then

$$
\begin{aligned}
\left\| U_{j+1}(t)\varphi \right\| &= \int_0^t \left\| U_j(s)CU_\Sigma(t-s)\varphi \right\| ds \\
&\leq \int_0^t \left\| \left(C\,(0-T_\Sigma)^{-1}\right)^j CU_\Sigma(t-s)\varphi \right\| ds \\
&= \left\| \int_0^t \left(C\,(0-T_\Sigma)^{-1}\right)^j CU_\Sigma(t-s)\varphi ds \right\| \\
&= \left\| \left(C\,(0-T_\Sigma)^{-1}\right)^j \int_0^t CU_\Sigma(t-s)\varphi ds \right\| \\
&\leq \left\| \left(C\,(0-T_\Sigma)^{-1}\right)^j \int_0^{+\infty} CU_\Sigma(s)\varphi ds \right\| \\
&= \left\| \left(C\,(0-T_\Sigma)^{-1}\right)^{j+1} \varphi \right\|
\end{aligned}
$$

and we are done. $\qquad\square$

In particular (10.8.13) implies

$$
\left\| U_j(t) \right\| \leq \left\| \Sigma\,(0-T_\Sigma)^{-1} \right\|^j \left\| C\Sigma^{-1} \right\|^j \quad (j \geq 0). \tag{10.8.14}
$$

We recall a key result from [251].

Theorem 10.8.4 ([251]) *Let C be a bounded regular collision operator. Suppose that D has a finite volume and there exists $\alpha > 0$ such that for all $M > 0$ there exists $c_M > 0$ such that*

$$
\sup_{e \in S^{d-1}} \nu \otimes \nu \left\{ (\zeta, \zeta'); \ |\zeta|, |\zeta'| \leq M, \ |(v - v').e| < \varepsilon \right\} \leq c_M \varepsilon^\alpha. \tag{10.8.15}
$$

Then $R_m(t)$ is weakly compact for all $t \geq 0$ and $m \geq m_0$ where m_0 is the smallest odd integer greater than $\frac{d(\alpha+1)}{\alpha} + 1$.

We note that the assumptions (10.8.10) (10.8.15) on the velocity measure $\nu(d\zeta)$ are satisfied by the Lebesgue measure on $\mathbb{R}^d$ or on spheres of $\mathbb{R}^d$ (multigroup models) or even by the combination of the two. We are ready to show the second key result.

Theorem 10.8.5 *Suppose that C is a Σ-regular collision operator and that D has a finite volume. If the velocity measure $\nu(d\zeta)$ satisfies (10.8.15) then, for m large enough, $R_m(t)$ is weakly compact for all $t \geq 0$. In particular*

$$
\omega_{ess}(V) \leq \omega_{ess}(U_\Sigma). \tag{10.8.16}
$$

Proof Without loss of generality, we may assume that Σ is bounded away from zero. The proof is quite involved and is given in several steps.

Step1. Let us show first that for j large enough, $U_j(t)$ is weakly compact for all $t \geq 0$. Estimate (10.8.14) shows that $U_j(t)$ depends continuously (in operator norm) on $\widehat{C} := C\Sigma^{-1}$. According to Proposition 10.8.1, we can approximate $\widehat{C}$ in operator norm by a suitable sequence $\widehat{C}_j$ (see (10.8.6)). It suffices to replace $\widehat{C}$ by $\widehat{C}_j$. Actually, by using a domination argument, it suffices even to replace $\widehat{C}_j$ by

$$\widetilde{C} : \varphi \in L^1(D \times \mathbb{R}^d) \to f(\zeta) \int_{\mathbb{R}^d} \varphi(x, \zeta')\nu(d\zeta')$$

where $f \in L^1_+(\mathbb{R}^d)$. In this case, $U_j(t)$ depends continuously, in operator norm on $f \in L^1(\mathbb{R}^d)$. Let $\widetilde{\Sigma}(\zeta) := \sup_{x \in D} \Sigma(x, \zeta)$. Since $\widetilde{\Sigma}$ is finite a.e. then

$$1_{\{\widetilde{\Sigma}(\zeta) \leq m\}} f \to f \text{ in } L^1(\mathbb{R}^d) \ (m \to +\infty)$$

so, without loss of generality, we can replace f by $1_{\{\widetilde{\Sigma}(\zeta) \leq m\}} f$. Note that the iteration

$$U_1(t)\varphi = \int_0^t U_\Sigma(t - s)CU_\Sigma(s)\varphi ds$$

is nothing but a time-convolution

$$U_1 = U_\Sigma * [CU_\Sigma]$$

so

$$\begin{aligned} U_2 &= U_1 * [CU_\Sigma] = U_\Sigma * [CU_\Sigma] * [CU_\Sigma] \\ &= U_\Sigma * [CU_\Sigma]^2 \end{aligned}$$

where $[CU_\Sigma]^2 := [CU_\Sigma] * [CU_\Sigma]$. More generally

$$U_j = U_\Sigma * [CU_\Sigma]^j , \ (j \geq 1).$$

We point out that all the U_j's are bounded operators (see the proof of Desch's theorem) and

$$\begin{aligned} U_{j+1} &= \left(U_\Sigma * [CU_\Sigma]^j\right) * [CU_\Sigma] \\ &= \left(U_\Sigma * [(C\Sigma^{-1}) \Sigma U_\Sigma]^j\right) * [CU_\Sigma]. \end{aligned}$$

Note that for all $\psi \in L^1_+(D \times \mathbb{R}^d)$ and all $t \geq 0$ the support of $CU_\Sigma(t)\psi$ is included in $D \times \{\widetilde{\Sigma}(\zeta) \leq m\}$. Since $U_\Sigma(s)$ preserves the "support with respect to velocities" then

$$\left(U_\Sigma * [(C\Sigma^{-1}) \Sigma U_\Sigma]^j\right) * [CU_\Sigma]\psi$$

$$\leq m^j \left(U_\Sigma * \left[\left(C\Sigma^{-1} \right) U_\Sigma \right]^j \right) * [CU_\Sigma] \psi.$$

Step 2. Consider

$$\left(U_\Sigma * \left[\left(C\Sigma^{-1} \right) U_\Sigma \right]^j \right) * [CU_\Sigma]$$

$$= \int_0^t \left(U_\Sigma * \left[\left(C\Sigma^{-1} \right) U_\Sigma \right]^j \right) (s) \left(CU_\Sigma(t - s) \right) ds$$

for a fixed $t > 0$. We observe that

$$\widehat{U}_j(s) := \left(U_\Sigma * \left[\left(C\Sigma^{-1} \right) U_\Sigma \right]^j \right) (s)$$

is nothing but the j-th term of a Dyson-Phillips expansion corresponding to a *bounded* regular collision operator $\widehat{C} := C\Sigma^{-1}$. It follows from Theorem 10.8.4 that for j large enough $\widehat{U}_j(s)$ is compact for all $s \geq 0$. By replacing j by $j + 1$ if necessary, according to Theorem 2.8.3

$$s \geq 0 \to \widehat{U}_j(s)$$

is continuous in operator norm. In particular

$$\left\{ \widehat{U}_j(s); \ s \in [0, t] \right\}$$

are collectively compact. For our purpose, collective weak compactness, (i.e., $\widehat{U}_j(s)$ maps the unit ball of $L^1(D \times \mathbb{R}^d)$ into a weakly compact set independent of $s \in [0, t]$) will suffices.

Step 3. We recall (see e.g. Appendix in [375]) that a set $\Xi \subset L^1(D \times \mathbb{R}^d)$ is relatively weakly compact if and only if for any non-increasing sequence $(A_j)_j$ of measurable subsets of $D \times \mathbb{R}^d$ such that $\cap_j A_j = \varnothing$

$$\sup_{f \in \Xi} \int_{A_j} |f(x, \zeta)| \, dx \nu(d\zeta) \to 0 \, (j \to +\infty).$$

Hence

$$\sup_{s \in [0,t]} \sup_{\|\psi\| \leq 1} \int_{A_j} \left| \left(\widehat{U}_j(s)\psi \right) (x, \zeta) \right| dx \nu(d\zeta) \to 0 \, (j \to +\infty).$$

Step 4. We can formulate this last result as follows: for any $\varepsilon > 0$ there exists j_0 such that

$$\sup_{s \in [0,t]} \int_{A_j} \left| \widehat{U}_j(s)\psi \right| \leq \varepsilon \|\psi\| \ (j \geq j_0).$$

We point out that the bounded operator

$$\int_0^t \widehat{U}_j(s)\,[CU_\Sigma(t-s)]\,ds$$

is nothing but the (unique) bounded extension of

$$\varphi \in D(T) \to \int_0^t \widehat{U}_j(s)\,[CU_\Sigma(t-s)]\,\varphi ds \in L^1(D \times \mathbb{R}^d).$$

With the choice $\varphi \in D(T_\Sigma) \cap L^1_+$

$$\int_{A_j} dx\nu(d\zeta) \int_0^t \widehat{U}_j(s)\,[CU_\Sigma(t-s)]\,\varphi ds$$

$$\leq \varepsilon \int_0^t \|[CU_\Sigma(t-s)]\,\varphi\|\,ds = \varepsilon \int_0^t \|[CU_\Sigma(s)]\,\varphi\|\,ds$$

$$= \varepsilon \left\| \int_0^t CU_\Sigma(s)\varphi \right\| ds = \varepsilon \left\| C \int_0^t U_\Sigma(s)\varphi \right\| ds$$

$$\leq \varepsilon \left\| C\,(0 - T_\Sigma)^{-1}\,\varphi \right\| \leq \varepsilon \left\| C\,(0 - T_\Sigma)^{-1} \right\| \|\varphi\|.$$

This estimate extends to any $\varphi \in L^1_+$ by approximating it in $L^1(D \times \mathbb{R}^d)$ by a sequence $(\varphi_l)_l \subset D(T_\Sigma) \cap L^1_+$ defined by

$$\varphi_l := l \int_0^{l^{-1}} U_\Sigma(s)\varphi ds.$$

Finally the estimate extends to $L^1 = L^1_+ - L^1_+$. Hence

$$\sup_{\|\varphi\|\leq 1} \int_{A_j} \left| \int_0^t \widehat{U}_j(s)\,[CU_\Sigma]\,(t-s)\varphi ds \right|$$

$$\leq \varepsilon \left\| C\,(0 - T_\Sigma)^{-1} \right\|$$

$$\leq \varepsilon \left\| C\Sigma^{-1} \right\| \left\| \Sigma\,(0 - T_\Sigma)^{-1} \right\|$$

$$\leq \varepsilon \left\| C\Sigma^{-1} \right\| \quad (j \geq j_0)$$

which shows that for j large enough

$$\left(U_\Sigma * \left[(C\Sigma^{-1})\,U_\Sigma \right]^j \right) * [CU_\Sigma]$$

is weakly compact and consequently $U_j(t)$ is weakly compact for all $t \geq 0$ if j is large enough.

Step 5. We know by Desch's theorem that the Dyson-Phillips expansion converges in operator norm and therefore some of its remainders is weakly compact. Finally, Theorem 2.8.4 implies $\omega_{ess}(V) \leq \omega_{ess}(U_\Sigma)$. $\qquad\square$

Corollary 10.8.1 *Suppose that C is a Σ-regular collision operator and D has a finite volume. If the velocity measure $\nu(d\zeta)$ satisfies (10.8.15). Then $(V(t))_{t\geq 0}$ has spectral gap provided that $s(T_\Sigma) < s(T_\Sigma + C)$.*

Proof Since $\omega(\mathcal{U}_\Sigma) = s(T_\Sigma)$ then, by Theorem 10.8.5,

$$\omega_{ess}(V) \leq s(T_\Sigma) < s(T_\Sigma + C) = \omega(V)$$

and we are done. $\qquad\qquad\qquad\qquad\qquad\qquad\qquad\qquad\qquad\qquad\qquad\qquad\qquad\square$

Remark 10.8.3 In a similar way, we can prove Theorem 10.8.5 in $L^1(T^d \times \mathbb{R}^d)$ where T^d is the d-dimensional torus.

10.9   Space Nonhomogeneous Linear Boltzmann Equations

In this section, we conduct a peripheral spectral analysis of general linear space nonhomogeneous Boltzmann equations with power-like hard-sphere potentials in various weighted L^1-spaces. We demonstrate how these results follow from the general theory developed in Sect. 10.8. The main findings are presented in Theorem 10.9.1 for non-incoming boundary conditions on domains with finite volume, and in Theorem 10.9.2 for the d-dimensional torus.

We deal with interaction kernel of the form (10.5.1) with angular cut-off (10.5.3) and hard sphere potential (10.5.2) with $\gamma \in (0, d - 2]$. The equation

$$\frac{\partial \varphi}{\partial t} + \zeta.\nabla_x \varphi + \Sigma(\zeta)\varphi(t, x, \zeta) = \int_{\mathbb{R}^d} c(\zeta, \zeta')\varphi(t, x, \zeta')d\zeta'$$

is considered in L^1 weighted spaces for two (velocity) weights

$$m(\zeta) = \exp\left(-a\,|\zeta|^s\right),\; s \in (0, 2) \quad \text{(exponential weight)}$$
$$m(\zeta) = \left(1 + |\zeta|^\beta\right)^{-1},\; \beta > 0 \; \text{(algebraic weight)}$$

and for both non incoming boundary condition and periodic boundary condition.

10.9.1   Non Incoming Boundary Condition

We consider first the boundary condition (10.8.2). Let $\mathcal{U}_\Sigma = (U_\Sigma(t))_{t\geq 0}$ be the advection semigroup on

$$L^1(D \times \mathbb{R}^d) := L^1(D \times \mathbb{R}^d;\; dx \otimes \nu(d\zeta))$$

given in (10.8.7) and let T_Σ be its generator. Here the velocity measure is given by

$$\nu(d\zeta) = m^{-1}(\zeta)d\zeta$$

and the collision operator is

$$C\psi = \int_{\mathbb{R}^d} c(\zeta, \zeta')\psi(\zeta')d\zeta' = \int_{\mathbb{R}^d} \left[c(\zeta, \zeta')m(\zeta)\right]\psi(\zeta')\nu(d\zeta'). \qquad (10.9.1)$$

The main result is

Theorem 10.9.1 *Let D have a finite volume. The interaction kernel is assumed to be of the form (10.5.1) with angular cut-off (10.5.3) and hard sphere potential (10.5.2) such that $\gamma \in (0, d - 2]$. Then*

$$T_\Sigma + C : D(T_\Sigma) \subset L^1(D \times \mathbb{R}^d) \to L^1(D \times \mathbb{R}^d)$$

generates a positive C_0-semigroup $\mathcal{V} = (V(t))_{t\geq 0}$ and $\omega_{ess}(\mathcal{V}) \leq \omega_{ess}(\mathcal{U}_\Sigma)$.

Proof According to Theorem 10.5.1 C is Σ-weakly compact in $L^1(\mathbb{R}^d; m^{-1}(\zeta)d\zeta)$ but, according to Remark 10.8.1, this amounts to saying that C is Σ-regular in the sense of Definition 10.8.1. Thus, according to Theorem 10.8.3, $T_\Sigma + C$ generates a positive C_0-semigroup $\mathcal{V} = (V(t))_{t\geq 0}$ on $L^1(D \times \mathbb{R}^d)$. Then, by Theorem 10.8.5, some remainder terms of the Dyson-Phillips expansion of $(V(t))_{t\geq 0}$ is weakly compact and $\omega_{ess}(\mathcal{V}) \leq \omega_{ess}(\mathcal{U}_\Sigma)$. $\qquad \square$

Remark 10.9.1 The inspection of (10.5.6) (resp. (10.5.7)) shows that C is a bounded operator on $L^1(\mathbb{R}^d; m^{-1}(\zeta)d\zeta)$ if $\gamma \leq s$ (resp. if $\gamma \leq 2$). In this case, $\omega_{ess}(\mathcal{V}) = \omega_{ess}(\mathcal{U}_\Sigma)$, see [245] Theorem 2.10, p. 24.

Remark 10.9.2 (*Conjecture*) We know that $(V(t))_{t\geq 0}$ has spectral gap if and only if

$$\lim_{\lambda \to s(T_\Sigma)} r(C(\lambda - T_\Sigma)^{-1}) > 1. \qquad (10.9.2)$$

We conjecture that (10.9.2) is satisfied if the "size" of D is large enough.

10.9.2   *Periodic Boundary Conditions*

We work now in

$$L^1(\mathcal{T}^d \times \mathbb{R}^d) := L^1(\mathcal{T}^d \times \mathbb{R}^d; \; dx \otimes \nu(d\zeta))$$

where $\mathcal{T}^d$ is the d-dimensional torus and

$$\nu(d\zeta) = m^{-1}(\zeta)d\zeta.$$

We identify any measurable function

$$p : x \in \mathcal{T}^d \to p(x) \in \mathbb{R}$$

to a $[0, 2\pi]^d$-periodic measurable function on $\mathbb{R}^d$ and endow $L^1(\mathcal{T}^d \times \mathbb{R}^d)$ with the norm

$$\|\varphi\|_{L^1(\mathcal{T}^d \times \mathbb{R}^d)} = \int_{[0,2\pi]^d \times \mathbb{R}^d} |\varphi(x, \zeta)| \, m^{-1}(\zeta) dx d\zeta.$$

Let $\mathcal{U}_\Sigma^{(p)} = \left(U_\Sigma^{(p)}(t)\right)_{t \geq 0}$ be the advection semigroup on $L^1(\mathcal{T}^d \times \mathbb{R}^d)$ given by

$$\mathcal{U}_\Sigma^{(p)}(t)\varphi = e^{-t\Sigma(\zeta)}\varphi(x - t\zeta, \zeta)$$

and denote by $T_\Sigma^{(p)}$ its generator. Let C be the collision operator (10.9.1). The main result is

Theorem 10.9.2 *Suppose that the interaction kernel is of the form (10.5.1) with angular cut-off (10.5.3) and hard sphere potential (10.5.2) such that $\gamma \in (0, d - 2]$. Then*

$$T_\Sigma^{(p)} + C : D(T_\Sigma^{(p)}) \subset L^1(\mathcal{T}^d \times \mathbb{R}^d) \to L^1(\mathcal{T}^d \times \mathbb{R}^d)$$

generates a positive C_0-semigroup $\mathcal{V}^{(p)} = \left(V^{(p)}(t)\right)_{t \geq 0}$ and $\omega_{ess}(\mathcal{V}^{(p)}) \leq \omega_{ess}(\mathcal{U}_\Sigma^{(p)})$. In particular $\left(V^{(p)}(t)\right)_{t \geq 0}$ has a spectral gap.

Proof The proof is very similar to that of Theorem 10.9.1. As noted in Remark 10.8.2 and Remark 10.8.3, Theorems 10.8.3 and 10.8.5 are true in $L^1(\mathcal{T}^d \times \mathbb{R}^d)$ if C is Σ-regular. By Theorem 10.5.1, C is Σ-weakly compact in $L^1(\mathbb{R}^d; \, m^{-1}(\zeta)d\zeta)$ and this means (see Remark 10.8.1) that C is Σ-regular. Hence $T_\Sigma^{(p)} + C$ generates a positive C_0-semigroup $\mathcal{V}^{(p)} = \left(V^{(p)}(t)\right)_{t \geq 0}$ on $L^1(\mathcal{T}^d \times \mathbb{R}^d)$ and $\omega_{ess}(\mathcal{V}^{(p)}) \leq \omega_{ess}(\mathcal{U}_\Sigma^{(p)})$.

It remains to show that $\left(V^{(p)}(t)\right)_{t \geq 0}$ has a spectral gap. Since $M(\zeta) = \alpha e^{-\beta|\zeta|^2}$ (for some positive constants α, β) then

$$\int_{\mathbb{R}^d} \Sigma(\zeta) M(\zeta) m^{-1}(\zeta) d\zeta < +\infty$$

(for both weights) and (10.5.5) shows that 0 is an eigenvalue of $-\Sigma + C$ or of $T_\Sigma^{(p)} + C$ since M is space homogeneous. Thus $s(T_\Sigma^{(p)} + C) \geq 0$. Since $\Sigma(\zeta) \geq c_1$ for all ζ (see (10.5.4)) then $s(T_\Sigma^{(p)}) \leq -c_1$ and $s(T_\Sigma^{(p)}) < s(T_\Sigma^{(p)} + C)$. Finally

$$\omega_{ess}(\mathcal{V}^{(p)}) \leq \omega_{ess}(\mathcal{U}_\Sigma^{(p)}) \leq \omega(\mathcal{U}_\Sigma^{(p)}) = s(T_\Sigma^{(p)}) < s(T_\Sigma^{(p)} + C) = \omega(\mathcal{V}^{(p)})$$

ends the proof. $\qquad\square$

In Theorem 10.9.2, we have still a spectral gap if we add a bounded absorption cross section Σ_a with small oscillation, see Theorem 4.4.1.

10.10   On the Leading Eigenvalue of Neutron Transport

In this section, we consider neutron transport operators (10.8.1) with non-incoming boundary conditions (10.8.2) and provide two variational characterizations (Max-inf and Inf-sup) of their leading eigenvalue, which follow from the abstract results established in Sect. 2.5. These are detailed in Theorem 10.10.1 and its dual form in Theorem 10.10.2. Additionally, we present a general existence result for the leading eigenvalue under the assumption that velocities are bounded away from zero (see Theorem 10.10.3). These results were previously obtained with less generality in [245] (Chap. 5) and [256].

10.10.1   Variational Characterizations

Theorem 10.8.3 provides us with power compactness of $C(\lambda - T_\Sigma)^{-1}$ in $L^1(D \times \mathbb{R}^d)$. In particular,

$$T_\Sigma + C : D(T_\Sigma) \to L^1(D \times \mathbb{R}^d)$$

is resolvent positive and we can apply the variational theory of $s(T_\Sigma + C)$ given in Theorem 2.5.2. Let

$$D_{++}(T_\Sigma) = \{\varphi \in D(T_\Sigma), \ \varphi > 0 \text{ a.e.}\}.$$

We have

Theorem 10.10.1 *Let C be a Σ-regular collision operator and let the velocity measure $\nu(d\zeta)$ satisfies (10.8.10). Suppose that D has a finite volume and*

$$\sup_{x \in D} \Sigma(x, \zeta) < +\infty \ \ \nu(d\zeta) \ a.e.$$

Let $C(\lambda - T_\Sigma)^{-1}$ be irreducible. Then $s(T_\Sigma + C) > s(T_\Sigma)$ if and only if there exists some $\psi \in D_{++}(T_\Sigma)$ such that $\inf \frac{T_\Sigma \psi + C\psi}{\psi} > s(T_\Sigma)$. In this case,

$$s(T_\Sigma + C) = \max_{\varphi \in D_{++}(T_\Sigma)} \left(\inf \frac{T_\Sigma \varphi + C\varphi}{\varphi} \right).$$

Similarly, Theorem 10.8.3 and the dual variational characterization of $s(T_\Sigma + C)$ given in Theorem 2.5.4 imply

Theorem 10.10.2 *Let C be a Σ-regular collision operator and let the velocity measure $\nu(d\zeta)$ satisfies (10.8.10). Suppose that D has a finite volume and*

$$\sup_{x \in D} \Sigma(x, \zeta) < +\infty \ \ \nu(d\zeta) \ a.e.$$

Let $C(\lambda - T_\Sigma)^{-1}$ be irreducible. . Then

$$s(T_\Sigma + C) = \inf_{\varphi \in D_{++}(T_\Sigma)} \left(\sup \frac{T_\Sigma \varphi + C\varphi}{\varphi} \right)$$

where $\sup \frac{A\varphi}{\varphi}$ *is the essential supremum of* $\frac{A\varphi}{\varphi}$*.*

10.10.2 Existence Criteria

In this chapter, we have considered that the set of velocities ζ is the whole $\mathbb{R}^d$ endowed with a Borel measure $\nu(d\zeta)$. Actually, the support of $\nu(d\zeta)$ only is relevant since we can identify the spaces

$$L^1(D \times \mathbb{R}^d;\ dx\nu(d\zeta)),\ \ L^1(D \times \Lambda;\ dx\nu(d\zeta))$$

where Λ (a closed subset of $\mathbb{R}^d$) is the support of the measure $\nu(d\zeta)$. In particular, if D is bounded and if

$$0 \notin \Lambda \tag{10.10.1}$$

(i.e., if the velocities under consideration are bounded away from zero) then (10.8.8) shows that

$$s(T_\Sigma) = -\infty.$$

In this case, we have a general existence result, already given in [245] Chap. 5 for bounded collision operators. By using similar arguments, we give here a more general version covering Σ-bounded collision operators.

Theorem 10.10.3 *Suppose that (10.10.1) is satisfied and that D is bounded. Let C be a Σ-regular collision operator and let the velocity measure $\nu(d\zeta)$ satisfies (10.8.10). Suppose that*

$$\sup_{x \in D} \Sigma(x, \zeta) < +\infty \ \ \nu(d\zeta)\ a.e.$$

If some power of $C(\lambda - T_\Sigma)^{-1}$ is positivity improving then $s(T_\Sigma) = -\infty < s(T_\Sigma + C)$ and $s(T_\Sigma + C)$ is an isolated and algebraically simple eigenvalue of $T_\Sigma + C$.

Proof According to Theorem 10.8.3, $C(\lambda - T_\Sigma)^{-1}$ is power compact. Since

$$r\left[(\lambda - T_\Sigma - C)^{-1}\right] = \frac{1}{dit(\lambda, \sigma(T_\Sigma + C))} = \frac{1}{\lambda - s(T_\Sigma + C)}$$

(see [286, Proposition 2.5, p. 67]) then $s(T_\Sigma + C) > -\infty$ if and only if

$$r\left[(\lambda - T_\Sigma - C)^{-1}\right] > 0.$$

According to Proposition 2.5.1

$$(\lambda - T_\Sigma - C)^{-1} = (\lambda - T_\Sigma)^{-1} \sum_{j=0}^{\infty} \left(C(\lambda - T_\Sigma)^{-1}\right)^j \quad (\lambda > s(T_\Sigma + C))$$

so

$$r\left[(\lambda - T_\Sigma - C)^{-1}\right] \geq r\left[(\lambda - T_\Sigma)^{-1}\left(C(\lambda - T_\Sigma)^{-1}\right)^j\right] \quad (\forall j).$$

Note that if $f > 0$ a.e. then so is $(\lambda - T_\Sigma)^{-1} f$. Since some power $\left(C(\lambda - T_\Sigma)^{-1}\right)^j$ is positivity improving and compact then so is $(\lambda - T_\Sigma)^{-1}\left(C(\lambda - T_\Sigma)^{-1}\right)^j$. Hence

$$r\left[(\lambda - T_\Sigma)^{-1}\left(C(\lambda - T_\Sigma)^{-1}\right)^j\right] > 0$$

by Proposition 2.4.2 and finally $r\left[(\lambda - T_\Sigma - C)^{-1}\right] > 0$. $\qquad\square$

Various criteria for irreducibility and for positivity improving are given in [245] Chap. 5. There exists a result by Kingman [186] on superconvexity of spectral radius of positive matrices which has been extended to positive operators on Banach lattices by Kato [182]. An alternative proof of Theorem 10.10.3 consists in showing that $\lim_{\lambda \to -\infty} r(C(\lambda - T_\Sigma)^{-1}) > 1$. Actually, the more precise result $\lim_{\lambda \to -\infty} r(C(\lambda - T_\Sigma)^{-1}) = +\infty$ can be derived from Kato's superconvexity theorem [182], see [245] Theorem 5.12, p. 105.

The situation is more involved when D is bounded and $0 \in \Lambda$. In this case, if Σ is "smooth" then

$$s(T_\Sigma) = - \inf_{x \in D} \Sigma(x, 0)$$

(see e.g. [178]) and $s(T_\Sigma + C) > s(T_\Sigma)$ if and only if

$$\lim_{\lambda \to s(T_\Sigma)} r\left(C(\lambda - T_\Sigma)^{-1}\right) > 1. \tag{10.10.2}$$

Because of the finiteness of $s(T_\Sigma)$, checking (10.10.2) becomes a complicated issue. In particular the "size" of D enters into play and (10.10.2) may be violated if the diameter of D is small enough; this is a well known fact (see [245] Chap. 5 and references therein). In some nice situations, (10.10.2) holds if the "thickness" of D is large enough where the "thickness" of D is measured in terms of the average of the distance function to the boundary ∂D

$$\int_D dis(x, \partial D)dx$$

(see [245, Theorem 6.4, p. 154]) or the average of the exit time function from D

$$\int_{D\times\Lambda} \tau(x,\zeta)dx\nu(d\zeta)$$

(see [245, Corollary 6.1, p. 155]).

Remark 10.10.1 The spectrum of the collision operator C enters into play in the possibility of the condition $s(T_\Sigma + C) > s(T_\Sigma)$. Indeed, if C is a bounded quasi-nilpotent operator (i.e., $r(C) = 0$) then $s(T_\Sigma + C) = s(T_\Sigma)$ regarless of the size of D (see [245, Corollary 5.1, p. 116]); this happens e.g. if $c(x,\zeta,\zeta') = 0$ for $|\zeta| \geq |\zeta'|$. This corresponds to neutron slowing down (i.e., no upscattering) which occurs beyond the energy range of neutron thermalization, see [114, p. 145].

10.11 Critical Neutron Transport Equations

We consider in this section the neutron transport equations (10.8.1) with nonincoming boundary condition (10.8.2) and

$$\begin{cases} \Sigma_s(x,\zeta) = \int_{\mathbb{R}^d} c(x,,\zeta',\zeta)\nu(d\zeta') \\ \Sigma(x,\zeta) = \Sigma_s(x,\zeta) + \Sigma_a(x,\zeta). \end{cases}$$

Let $(U_\Sigma(t))_{t\geq0}$ be the advection semigroup with generator T_Σ and let $(V(t))_{t\geq0}$ be the full substochastic C_0-semigroup generated by an extension of

$$T_\Sigma + C : D(T_\Sigma) \subset L^1(D \times \mathbb{R}^d) \to L^1(D \times \mathbb{R}^d)$$

according to Kato-Voigt perturbation theory. Because of the boundary condition (10.8.2), a priori $(V(t))_{t\geq0}$ is not stochastic. This section is devoted to the *critical* case

$$s(T_\Sigma) = 0. \tag{10.11.1}$$

It follows from (10.8.8) that Σ is not bounded away from zero. We point out that we cannot expect a spectral gap for $(V(t))_{t\geq0}$. This section deals with strong stability of $(V(t))_{t\geq0}$ and with the criticality eigenvalue problem (see below). The main results appear here for the first time and are given in Theorem 10.11.1 and in Theorem 10.11.3) where a key role is played by a compactness result, see Theorem 10.11.2.

10.11.1 Strong Stability

Note that $C(\lambda - T_\Sigma)^{-1}$ $(\lambda > 0)$ is a contraction and then the strong limit

$$C(0 - T_\Sigma)^{-1} := s \lim_{\lambda \to 0} C(\lambda - T_\Sigma)^{-1}$$

exists and is a positive contraction. The main result of this subsection is

Theorem 10.11.1 *Let (10.11.1) be satisfied and*

$$\int_0^{\tau_+(y,\zeta')} \Sigma(y + s\zeta', \zeta')ds > 0 \tag{10.11.2}$$

on a subset of positive $dy\nu(d\zeta')$-measure. Then:

(i) Let $r(C(0 - T_\Sigma)^{-1}) > 0$. If $r(C(0 - T_\Sigma)^{-1})$ is an eigenvalue of $C(0 - T_\Sigma)^{-1}$ with a positive eigenfunction then $r(C(0 - T_\Sigma)^{-1}) < 1$.

(ii) If $r(C(0 - T_\Sigma)^{-1}) < 1$ then

$$\lim_{t \to +\infty} \|V(t)f\| = 0 \ \ (f \in L^1(D \times \mathbb{R}^d)),$$

i.e., $(V(t))_{t \geq 0}$ is strongly stable.

Proof *Step 1.* Let $\varphi \in L^1(D \times \mathbb{R}^d)$. We note that $\left\| C(0 - T_\Sigma)^{-1}\varphi \right\|$ is given by

$$\int_{D \times \mathbb{R}^d} dx\nu(d\zeta) \int_{\mathbb{R}^d} c(x, \zeta, \zeta') \left((0 - T_\Sigma)^{-1} \varphi \right)(x, \zeta')\nu(d\zeta')$$

$$= \int_D dx \int_{\mathbb{R}^d} \Sigma_s(x, \zeta')\nu(d\zeta') \int_0^{\tau(x,\zeta')} e^{-\int_0^t \Sigma(x-s\zeta',\zeta')ds} \varphi(x - t\zeta', \zeta')dt$$

$$= \int_{\mathbb{R}^d} \nu(d\zeta') \int_D \Sigma_s(x, \zeta')dx \int_0^{\tau(x,\zeta')} e^{-\int_0^t \Sigma(x-s\zeta',\zeta')ds} \varphi(x - t\zeta', \zeta')dt$$

$$\leq \int_{\mathbb{R}^d} \nu(d\zeta') \int_D \Sigma(x, \zeta')dx \int_0^{\tau(x,\zeta')} e^{-\int_0^t \Sigma(x-s\zeta',\zeta')ds} \varphi(x - t\zeta', \zeta')dt.$$

Consider first

$$\int_D \Sigma(x, \zeta')dx \int_0^{\tau(x,\zeta')} e^{-\int_0^t \Sigma(x-s\zeta',\zeta')ds} \varphi(x - t\zeta', \zeta')dt$$

$$= \int_D \Sigma(x, \zeta')dx \int_0^{+\infty} 1_{\{t \leq \tau(x,\zeta')\}} e^{-\int_0^t \Sigma(x-s\zeta',\zeta')ds} \varphi(x - t\zeta', \zeta')dt$$

$$= \int_0^{+\infty} dt \int_D \Sigma(x, \zeta') 1_{\{t \leq \tau(x,\zeta')\}} e^{-\int_0^t \Sigma(x-s\zeta',\zeta')ds} \varphi(x - t\zeta', \zeta')dx.$$

The change of variable $y = x - t\zeta'$ gives

$$\int_D \Sigma(x, \zeta') 1_{\{t \leq \tau(x,\zeta')\}} e^{-\int_0^t \Sigma(x-s\zeta',\zeta')ds} \varphi(x - t\zeta', \zeta')dx$$

$$= \int_{D-t\zeta'} \Sigma(y + t\zeta', \zeta') 1_{\{t \leq \tau(y+t\zeta',\zeta')\}} e^{-\int_0^t \Sigma(y+t\zeta'-s\zeta',\zeta')ds} \varphi(y, \zeta')dy.$$

Note that

$$e^{-\int_0^t \Sigma(y+t\zeta'-s\zeta',\zeta')ds} = e^{-\int_0^t \Sigma(y+s\zeta',\zeta')ds}.$$

Since

$$\tau(y + t\zeta', \zeta') = t + \tau(y, \zeta')$$

then $1_{\{t \leq \tau(y+t\zeta',\zeta')\}} = 1$ and

$$\int_D \Sigma(x, \zeta') 1_{\{t \leq \tau(x,\zeta')\}} e^{-\int_0^t \Sigma(x-s\zeta',\zeta')ds} \varphi(x - t\zeta', \zeta')dx$$

$$= \int_D 1_D(y + t\zeta') \Sigma(y + t\zeta', \zeta') \, e^{-\int_0^t \Sigma(y+s\zeta',\zeta')ds} \varphi(y, \zeta')dy.$$

Hence $\left\| C(0 - T)^{-1}\varphi \right\|$ is equal to

$$\int_{D \times \mathbb{R}^d} \left(\int_0^{\tau_+(y,\zeta')} \Sigma(y + t\zeta', \zeta') \, e^{-\int_0^t \Sigma(y+s\zeta',\zeta')ds} \right) \varphi(y, \zeta')dy\nu(d\zeta')$$

where

$$\tau_+(y, \zeta') = \tau(y, -\zeta').$$

Since

$$\frac{d}{dt} e^{-\int_0^t \Sigma(y+s\zeta',\zeta')ds} = -\Sigma(y + t\zeta', \zeta') e^{-\int_0^t \Sigma(y+s\zeta',\zeta')ds}$$

then $\left\| C(0 - T)^{-1}\varphi \right\|$ is equal to

$$\int_{D \times \mathbb{R}^d} \left(1 - e^{-\int_0^{\tau_+(y,\zeta')} \Sigma(y+s\zeta',\zeta')ds} \right) \varphi(y, \zeta')dy\nu(d\zeta').$$

It follows from (10.11.2) that if $\varphi > 0$ then

$$\left\| C(0 - T_\Sigma)^{-1}\varphi \right\| < \|\varphi\|. \tag{10.11.3}$$

If $\alpha := r(C(0 - T_\Sigma)^{-1})$ is an eigenvalue of $C(0 - T_\Sigma)^{-1}$ associated to an eigenfunction $\varphi > 0$ then (10.11.3) shows that $\alpha < 1$.

Step 2. We have a Duhamel identity

$$V(t)f = U_\Sigma(t)f + \int_0^t V(t-s)CU_\Sigma(s)f\,ds, \quad f \in D(T_\Sigma)$$

and for $f \in D_+(T_\Sigma)$

$$\|V(t)f\| = \|U_\Sigma(t)f\| + \int_0^t \|V((t-s)CU_\Sigma(s)f\|\,ds.$$

Since $(V(t))_{t\geq 0}$ is a contraction semigroup then $\lim_{t\to\infty}\|V(t)f\|$ exists and is given by a functional $\chi \in L^\infty(D \times \mathbb{R}^d)$

$$\lim_{t\to\infty}\|V(t)f\| = \langle \chi, f\rangle, \quad (f \in L^1_+(D \times \mathbb{R}^d)).$$

We know that $\lim_{t\to\infty}\|U_\Sigma(t)f\| = 0$. Since $\|V(t-s)CU_\Sigma(s)f\| \leq \|CU_\Sigma(s)f\|$ and

$$\int_0^{+\infty} \|CU_\Sigma(s)f\|\,ds = \left\|\int_0^{+\infty} CU_\Sigma(s)f\,ds\right\|$$
$$= \left\|C(0-T_\Sigma)^{-1}f\right\| < +\infty$$

then

$$\langle \chi, f\rangle = \int_0^{+\infty} \langle \chi, CU_\Sigma(s)f\rangle ds = \langle \chi, \int_0^{+\infty} CU_\Sigma(s)f\,ds\rangle$$
$$= \langle \chi, C(0-T_\Sigma)^{-1}f\rangle = \langle\left(C(0-T_\Sigma)^{-1}\right)^* \chi, f\rangle.$$

Thus
$$\langle \chi, f\rangle = \langle\left(C(0-T_\Sigma)^{-1}\right)^* \chi, f\rangle \qquad (10.11.4)$$

for all $f \in D_+(T_\Sigma)$. Since

$$f_n := n\int_0^{\frac{1}{n}} U_\Sigma(s)f\,ds \to f \quad (n \to \infty)$$

then $D_+(T_\Sigma)$ is dense in $L^1_+(D \times \mathbb{R}^d)$ and then (10.11.4) extends to $L^1_+(D \times \mathbb{R}^d)$. Finally

$$\left(C(0-T_\Sigma)^{-1}\right)^* \chi = \chi$$

since $L^1 = L^1_+ - L^1_+$. This implies that $\chi = 0$ because

$$r\left(\left(C(0 - T_\Sigma)^{-1}\right)^*\right) = r\left(C(0 - T_\Sigma)^{-1}\right) < 1.$$

$\square$

Corollary 10.11.1 *If $C(0 - T_\Sigma)^{-1}$ is power compact then $(V(t))_{t \geq 0}$ is strongly stable.*

Proof If $r\left(C(0 - T_\Sigma)^{-1}\right) = 0$ we are done by Theorem 10.11.1; (such a situation may occur e.g. if $c(x, \zeta, \zeta') = 0$ for $|\zeta| \geq |\zeta'|$; see ([245, Corollary 5.1, p. 116]) and Remark 10.10.1.) Let

$$r\left(C(0 - T_\Sigma)^{-1}\right) > 0.$$

Since $r\left(C(0 - T_\Sigma)^{-1}\right) \in \sigma(C(0 - T_\Sigma)^{-1})$ then $r\left(C(0 - T_\Sigma)^{-1}\right)$ is an eigenvalue and we appeal again to Theorem 10.11.1. $\square$

Of course, the understanding of power compactness of $C(0 - T_\Sigma)^{-1}$ (which is also needed below) is mandatory. Since we deal with the critical case $s(T_\Sigma) = 0$, Σ is not bounded away from zero and we need C to be *strongly Σ-regular*; this condition is stronger than Σ-regularity of C; see Definition 10.8.1. We have

Theorem 10.11.2 *Suppose that D is bounded and that C is strongly Σ-regular. Then $C(0 - T_\Sigma)^{-1}$ is power compact.*

Proof We recall that for $f \geq 0$

$$(\lambda - T_\Sigma)^{-1} f = \int_0^{\tau(x,\zeta)} e^{-\lambda t} e^{-\int_0^t \Sigma(x - s\zeta, \zeta) ds} f(x - t\zeta, \zeta) dt$$

$$\leq \int_0^{\tau(x,\zeta)} f(x - t\zeta, \zeta) dt \quad (\lambda \geq 0)$$

and $\tau_-(x, \zeta) \leq \frac{d}{|\zeta|}$ where d is the diameter of D. By extending trivially f outside D one sees that

$$\int_D dx \int_0^{\tau(x,\zeta)} f(x - t\zeta, \zeta) dt \leq \int_{\mathbb{R}^d} dx \int_0^{\frac{d}{|\zeta|}} f(x - t\zeta, \zeta) dt$$

$$= \frac{d}{|\zeta|} \int_D f(y, \zeta) dy$$

so

$$\int_{D \times \mathbb{R}^d} (\lambda - T_\Sigma)^{-1} f(x, \zeta) |\zeta| \, dx d\zeta \leq d \, \|f\|_{L^1(D \times \mathbb{R}^d)} \quad (\lambda > 0).$$

Note that the strong limit

$$|\zeta|\,(0-T_\Sigma)^{-1} := s\lim_{\lambda\to 0}|\zeta|\,(\lambda-T_\Sigma)^{-1}$$

exists and

$$\left\||\zeta|\,(0-T_\Sigma)^{-1}\right\|_{\mathcal{L}(L^1(D\times\mathbb{R}^d))}\le d.$$

We factorize $C(0-T_\Sigma)^{-1}$ as

$$C(0-T_\Sigma)^{-1} = \left(C\Sigma^{-1}\right)\left(\Sigma(0-T_\Sigma)^{-1}\right).$$

We deal first with weak compactness which can be captured by domination arguments. According to Proposition 10.8.1, $C\Sigma^{-1}$ can be approximated in operator norm by a sequence of collision operators dominated by operators of the form

$$\widehat{C}:\varphi\in L^1(D\times\mathbb{R}^d)\to f(\zeta)\int_{\mathbb{R}^d}\varphi(x,\zeta)\nu(d\zeta)\tag{10.11.5}$$

where $f\in L^1_+(\mathbb{R}^d)$. So it suffices to assume that $C\Sigma^{-1}$ is of the form (10.11.5). By approximation again we may assume that f is continuous with a support included in a spherical shell

$$\left\{\zeta\in\mathbb{R}^d;\ c\le|\zeta|\le c^{-1}\right\}\ (0<c<1).$$

Note that

$$\left(\widehat{C}\left(\Sigma(0-T_\Sigma)^{-1}\right)\right)^j = \left(\widehat{C}\left(\Sigma(0-T_\Sigma)^{-1}\right)\right)^{j-1}\widehat{C}\left(\Sigma(0-T_\Sigma)^{-1}\right).$$

We note that for all $\psi\in L^1(D\times\mathbb{R}^d)$,

$$\begin{cases}\left(\widehat{C}\left(\Sigma(0-T_\Sigma)^{-1}\right)\psi\right)(x,\zeta)=0\\ \quad\text{if } |\zeta|<c \text{ or } |\zeta|>c^{-1}.\end{cases}$$

Since $\Sigma(0-T_\Sigma)^{-1}$ preserve the support "in velocity" then $\left(\widehat{C}\left(\Sigma(0-T_\Sigma)^{-1}\right)\right)^{j-1}$ is actually operating on

$$L^1(D\times\left\{c<|\zeta|<c^{-1}\right\}).\tag{10.11.6}$$

In particular

$$\left(\widehat{C}\left(\Sigma(0-T_\Sigma)^{-1}\right)\right)^{j-1}\le\left(c^{-1}\right)^{j-1}\left(\widehat{C}(0-T_\Sigma)^{-1}\right)^{j-1}.$$

Since we know (see [251]) that $\widehat{C}(0-T_\Sigma)^{-1}$ is power compact (in spaces (10.11.6)) when D is bounded then $\left(\widehat{C}\left(\Sigma(0-T_\Sigma)^{-1}\right)\right)^j$ is weakly compact and consequently $\left(\widehat{C}\left(\Sigma(0-T_\Sigma)^{-1}\right)\right)^{2j}$ is compact by the Dunford-Pettis property. $\qquad\square$

10.11.2   The Criticality Eigenvalue Problem

This subsection continues the previous one. We consider neutron transport equations

$$\frac{\partial\varphi}{\partial t} + \zeta.\nabla_x\varphi + \Sigma(x,\zeta)\varphi(t,x,\zeta) = \int_{\mathbb{R}^d} c(x,\zeta,\zeta')\varphi(t,x,\zeta')\nu(d\zeta')$$
$$+\frac{1}{k}\int_{\mathbb{R}^d} c_f(x,\zeta,\zeta')\varphi(t,x,\zeta')\nu(d\zeta')$$

with nonincoming boundary condition (10.8.2) and with an additional fission operator

$$F : L^1(D \times \mathbb{R}^d) \to L^1(D \times \mathbb{R}^d)$$

given by

$$F\psi = \int_{\mathbb{R}^d} c_f(x,\zeta,\zeta')\psi(x,\zeta')\nu(d\zeta').$$

We restrict ourselves to bounded collision and fission operators. In typical physical examples [114], F is a rank one operator (in velocity) and is space homogeneous. The problem here is twofold:

(1) Look for a value of the parameter $k > 0$ (the so-called *critical eigenvalue*) such that the spectral bound of

$$T_\Sigma + C + \frac{1}{k}F$$

is equal to zero and is an eigenvalue.
(2) Study time asymptotics of $(V(t))_{t\geq 0}$.

This important problem in nuclear reactor theory [114] is well understood in the subcritical case

$$s(T_\Sigma) < 0,$$

(see e.g. [234, 322] ([245, Chap. 5]). As far as we know, the critical case (10.11.1) is completely open and is the object of the following:

Theorem 10.11.3 *Let $s(T_\Sigma) = 0$ and let D be bounded. Suppose that C and F are bounded and strongly Σ-regular collision operators and that $F(0 - T_\Sigma)^{-1}$ is irreducible. Then*
   (i) There exists a unique value $\bar{k} > 0$ such that

$$r\left((C + \frac{1}{\bar{k}}F)(0 - T_\Sigma)^{-1}\right) = 1$$

and

$$s(T_\Sigma + C + \frac{1}{\bar{k}}F) = 0. \tag{10.11.7}$$

(ii) If additionally $|\zeta|^{-1} C \, \Sigma^{-1}$ and $|\zeta|^{-1} F\Sigma^{-1}$ are bounded operators then 0 is an eigenvalue of $T_\Sigma + C + \frac{1}{k}F$ (associated to a nonnegative eigenfunction ψ) and an eigenvalue of the dual operator $\left(T_\Sigma + C + \frac{1}{k}F\right)^{*}$ in $L^\infty(D \times \mathbb{R}^d)$ (associated to a nonnegative eigenfunction φ^{*}).

(iii) Under the normalisation

$$\int \psi(x, \zeta)\varphi^{*}(x, \zeta)dx\nu(d\zeta) = 1$$

$(V(t))_{t \geq 0}$ *extends uniquely to*

$$L^1_{\varphi^{*}} := L^1(D \times \mathbb{R}^d; \ \varphi^{*}(x)dx\nu(d\zeta)) \tag{10.11.8}$$

as a stochastic C_0-semigroup with asymptotically stable invariant density ψ.

Proof (i) Let $\widehat{C}_k := C + \frac{1}{k}F$. The same proof as in Theorem 10.11.2 shows that

$$\widehat{C}_k \, (0 - T_\Sigma)^{-1}$$

is power compact. We consider the function

$$k \in (0, +\infty) \to r\left((C + \frac{1}{k}F) \, (0 - T_\Sigma)^{-1}\right). \tag{10.11.9}$$

By the irreducibility condition and Proposition 2.4.2

$$r\left(F \, (0 - T_\Sigma)^{-1}\right) > 0 \tag{10.11.10}$$

so

$$r\left((C + \frac{1}{k}F) \, (0 - T_\Sigma)^{-1}\right) \geq \frac{1}{k} r\left(F \, (0 - T_\Sigma)^{-1}\right) > 0$$

and consequently $r\left((C + \frac{1}{k}F) \, (0 - T_\Sigma)^{-1}\right)$ is an isolated and algebraically simple eigenvalue of $(C + \frac{1}{k}F) \, (0 - T_\Sigma)^{-1}$.

According to Theorem 10.11.1

$$r\left(C \, (0 - T_\Sigma)^{-1}\right) < 1.$$

By Proposition 2.4.1, the function (10.11.9) is strictly decreasing. By arguing as in Lemma 2.5.2, one sees that the function (10.11.9) is continuous. It follows that

$$\lim_{k \to +\infty} r\left((C + \frac{1}{k}F) \, (0 - T_\Sigma)^{-1}\right) = r\left(C \, (0 - T_\Sigma)^{-1}\right) < 1.$$

Finally

$$r\left((C + \frac{1}{k}F)(0 - T_\Sigma)^{-1}\right) \geq \frac{1}{k}r\left(F(0 - T_\Sigma)^{-1}\right)$$

and (10.11.10) show that

$$\lim_{k\to 0} r\left((C + \frac{1}{k}F)(0 - T_\Sigma)^{-1}\right) = +\infty.$$

Hence there exists a unique $\overline{k} > 0$ such that

$$r\left((C + \frac{1}{\overline{k}}F)(0 - T_\Sigma)^{-1}\right) = 1.$$

By using Proposition 2.4.1

$$r\left((C + \frac{1}{\overline{k}}F)(\lambda - T_\Sigma)^{-1}\right)$$
$$< r\left((C + \frac{1}{\overline{k}}F)(0 - T_\Sigma)^{-1}\right) = 1 \; (\lambda > 0)$$

and consequently $s(T_\Sigma + C + \frac{1}{\overline{k}}F) \leq 0$ by Proposition 2.5.1. Since $s(T_\Sigma + C + \frac{1}{\overline{k}}F) \geq s(T_\Sigma) = 0$ we get (10.11.7).

(ii) There exists a positive eigenfunction $\varphi \in L^1(D \times \mathbb{R}^d)$

$$(C + \frac{1}{\overline{k}}F)(0 - T_\Sigma)^{-1}\varphi = \varphi$$

or

$$(C\Sigma^{-1} + \frac{1}{\overline{k}}F\Sigma^{-1})\Sigma(0 - T_\Sigma)^{-1}\varphi = \varphi.$$

Since $(|\zeta|^{-1}C\Sigma^{-1} + \frac{1}{\overline{k}}|\zeta|^{-1}F\Sigma^{-1})\Sigma(0 - T_\Sigma)^{-1}$ is a bounded operator then $|\zeta|^{-1}\varphi \in L^1(D \times \mathbb{R}^d)$. Thus

$$\psi := (0 - T_\Sigma)^{-1}\varphi = |\zeta|(0 - T_\Sigma)^{-1}\left(|\zeta|^{-1}\varphi\right) \in L^1(D \times \mathbb{R}^d)$$

because $|\zeta|(0 - T_\Sigma)^{-1}$ is also a bounded operator. Since

$$\Sigma\psi = \Sigma(0 - T_\Sigma)^{-1}\varphi \in L^1(D \times \mathbb{R}^d)$$

then ψ belongs to the domain of T_Σ and

$$-T_\Sigma\psi = \varphi = (C + \frac{1}{\overline{k}}F)\psi,$$

i.e., 0 is an eigenvalue of $T_\Sigma + C + \frac{1}{k}F$.

By duality

$$r\left(\left((C + \frac{1}{k}F)\,(0 - T_\Sigma)^{-1}\right)^*\right) = 1$$

is also an eigenvalue of $\left((C + \frac{1}{k}F)\,(0 - T_\Sigma)^{-1}\right)^*$ with eigenvector $\varphi^* \in L^\infty_+(D \times \mathbb{R}^d)$.

$$\left((C + \frac{1}{k}F)\,(0 - T_\Sigma)^{-1}\right)^* \varphi^* = \varphi^*. \qquad (10.11.11)$$

Note that

$$(C + \frac{1}{k}F)\,(0 - T_\Sigma)^{-1} = \left[(C + \frac{1}{k}F)\Sigma^{-1}\right]\left(\Sigma\,(0 - T_\Sigma)^{-1}\right)$$

is a product of two *bounded* operators so

$$\left((C + \frac{1}{k}F)\,(0 - T_\Sigma)^{-1}\right)^* = \left(\Sigma\,(0 - T_\Sigma)^{-1}\right)^*\left[(C + \frac{1}{k}F)\Sigma^{-1}\right]^*.$$

Let us identify $\left(\Sigma\,(0 - T_\Sigma)^{-1}\right)^*$ and $\left[(C + \frac{1}{k}F)\Sigma^{-1}\right]^*$. Note that $\Sigma\,(0 - T_\Sigma)^{-1}$ is the strong limit

$$s\lim_{\lambda \to 0_+} \Sigma\,(\lambda - T_\Sigma)^{-1}$$

and a simple calculation shows that the dual of

$$(\lambda - T_\Sigma)^{-1}\,\varphi(x, \zeta) = \int_0^{\tau(x,\zeta)} e^{-\lambda t} e^{-\int_0^t \Sigma(x - s\zeta, \zeta)ds}\, \varphi(x - t\zeta, \zeta)dt$$

is given by

$$(\lambda - T_\Sigma^*)^{-1}\,g(x, \zeta) = \int_0^{\tau(x,-\zeta)} e^{-\lambda t} e^{-\int_0^t \Sigma(x + s\zeta, \zeta)ds}\, g(x + t\zeta, \zeta)dt.$$

Moreover

$$\left|(\lambda - T_\Sigma^*)^{-1}\,(\Sigma g)\,(x, \zeta)\right|$$

$$= \left|\int_0^{\tau(x,-\zeta)} e^{-\lambda t} e^{-\int_0^t \Sigma(x + s\zeta, \zeta)ds}\, (\Sigma g)\,(x + t\zeta, \zeta)dt\right|$$

$$\leq \|g\|_\infty \int_0^{\tau(x,-\zeta)} e^{-\int_0^t \Sigma(x + s\zeta, \zeta)ds}\, \Sigma(x + t\zeta, \zeta)dt$$

$$= \|g\|_\infty \int_0^{\tau(x,-\zeta)} -\frac{d}{ds} e^{-\int_0^t \Sigma(x+s\zeta,\zeta)ds}\, dt$$

$$= \|g\|_\infty \left[1 - e^{-\int_0^{\tau(x,-\zeta)} \Sigma(x+s\zeta,\zeta)ds} \right] \leq \|g\|_\infty \quad (\lambda > 0)$$

so $\left(0 - T_\Sigma^*\right)^{-1} \Sigma$ exists as a pointwise limit $(\lambda \to 0_+)$ with

$$\left\| \left(0 - T_\Sigma^*\right)^{-1} \Sigma \right\|_{\mathcal{L}(L^\infty)} \leq 1$$

and

$$\left(0 - T_\Sigma^*\right)^{-1} \Sigma = \left(\Sigma \left(0 - T_\Sigma\right)^{-1} \right)^* .$$

We show easily that

$$\left[(C + \frac{1}{k}F)\Sigma^{-1} \right]^* g = \Sigma^{-1}(C^* + \frac{1}{k}F^*)$$

where the kernel of the bounded operator $\Sigma^{-1}(C^* + \frac{1}{k}F^*)$ is given by

$$\Sigma^{-1}(\zeta)c(x, , \zeta', \zeta) + \frac{1}{k}\Sigma^{-1}(\zeta)c_f(x, , \zeta', \zeta).$$

Thus

$$\left((C + \frac{1}{k}F)\,(0 - T_\Sigma)^{-1} \right)^*$$

$$= \left(\left(0 - T_\Sigma^*\right)^{-1} \Sigma \right) \left[\Sigma^{-1}(C^* + \frac{1}{k}F^*) \right].$$

It follows from (10.11.11) that

$$\left(\left(0 - T_\Sigma^*\right)^{-1} \Sigma \right) \left[\Sigma^{-1}(C^* + \frac{1}{k}F^*) \right] \varphi^* = \varphi^*,$$

i.e.,

$$\int_0^{\tau(x,-\zeta)} e^{-\int_0^t \Sigma(x+s\zeta,\zeta)ds} \left((C^* + \frac{1}{k}F^*)\varphi^* \right)(x + t\zeta, \zeta)dt = \varphi^*(x, \zeta).$$

Let

$$\varphi_\lambda^*(x, \zeta) := \int_0^{\tau(x,-\zeta)} e^{-\lambda t} e^{-\int_0^t \Sigma(x+s\zeta,\zeta)ds} \left((C^* + \frac{1}{k}F^*)\varphi^* \right)(x + t\zeta, \zeta)dt$$

$$= \left(\lambda - T_\Sigma^*\right)^{-1}\left(C^* + \frac{1}{k}F^*\right)\varphi^*.$$

Then $\varphi_\lambda^* \in D(T_\Sigma^*)$ and $\varphi_\lambda^* \to \varphi^*$ $(\lambda \to 0_+)$ monotonically and pointwisely and therefore in the weak star topology of $L^\infty(D \times \mathbb{R}^d)$. Since

$$\left(\lambda - T_\Sigma^*\right)\varphi_\lambda^* = \left(C^* + \frac{1}{k}F^*\right)\varphi^*$$

and $\lambda\left\|\varphi_\lambda^*\right\|_\infty \to 0$ $(\lambda \to 0)$ then

$$- T_\Sigma^*\varphi_\lambda^* \to \left(C^* + \frac{1}{k}F^*\right)\varphi^* \quad (\lambda \to 0_+)$$

in $L^\infty(D \times \mathbb{R}^d)$. Hence for all $\varphi \in D(T_\Sigma)$

$$\langle -T_\Sigma^*\varphi_\lambda^*, \varphi\rangle \to \langle (C^* + \frac{1}{k}F^*)\varphi^*, \varphi\rangle,$$

i.e.,

$$\langle \varphi^*, T_\Sigma\varphi\rangle = \langle -(C^* + \frac{1}{k}F^*)\varphi^*, \varphi\rangle \quad \forall \varphi \in D(T_\Sigma).$$

This shows that $\varphi^* \in D(T_\Sigma^*)$ and

$$T_\Sigma^*\varphi^* = -(C^* + \frac{1}{k}F^*)\varphi^*,$$

i.e.,

$$T_\Sigma^*\varphi^* + (C^* + \frac{1}{k}F^*)\varphi^* = 0.$$

Hence

$$(T_\Sigma + C + \frac{1}{k}F)^*\varphi^* = 0$$

and

$$V^*(t)\varphi^* = \varphi^* \quad (t \geq 0).$$

(iii) Note that $L^1(D \times \mathbb{R}^d)$ is imbedded continuously (and densely) into the weighted space (10.11.8) with norm

$$\|f\|_{\varphi^*} = \int_{D \times \mathbb{R}^d} |f(x, \zeta)|\, \varphi^*(x, \zeta)\, dx\nu(d\zeta).$$

Let $\langle ., .\rangle$ be the duality pairing between $L^1(D \times \mathbb{R}^d)$ and $L^\infty(D \times \mathbb{R}^d)$. Then, for all $f \in L_+^1(D \times \mathbb{R}^d)$

$$\|V(t)f\|_{\varphi^*} = \int_{D \times \mathbb{R}^d} V(t)f(x, \zeta)\varphi^*(x, \zeta)dx\nu(d\zeta)$$
$$= \langle V(t)f, \varphi^* \rangle = \langle f, V^*(t)\varphi^* \rangle$$
$$= \langle f, \varphi^* \rangle = \|f\|_{\varphi^*}$$

and $(V(t))_{t \geq 0}$ extends uniquely to $L^1_{\varphi^*}$ as a stochastic C_0-semigroup. Finally Theorem 2.10.3 ends the proof. $\qquad\qquad\square$

10.12 Neutron Transport on the Whole Space

To deal with spectral theory, compactness results were obtained in Sect. 10.8 for spatial domain D with finite volume. In this section, we explore neutron transport equations on $L^1(\mathbb{R}^d_x \times \mathbb{R}^d_\zeta)$

$$\frac{\partial \varphi}{\partial t} + \zeta.\nabla_x\varphi + \Sigma(x, \zeta)\varphi(t, x, \zeta) = \int_{\mathbb{R}^d} c(x, \zeta, \zeta')\varphi(t, x, \zeta')\nu(d\zeta'), \quad (10.12.1)$$

i.e., when $D = \mathbb{R}^d_x$. This model appears in the so-called *multiple scattering* problem where the cross sections $\Sigma(x, \zeta')$ and $c(x, , \zeta, \zeta')$ have compact support in space, more precisely

$$\Sigma(x, \zeta') = c(x, , \zeta, \zeta') = 0 \ (x \notin D) \qquad\qquad (10.12.2)$$

where $D \subset \mathbb{R}^d$ is a bounded open set, (see e.g. [159] ([178, Chap. 13]) [165]). The difference with the usual *reactor* problem is that the dynamics takes place in the whole space even if the cross sections are compactly supported in space while the reactor problem is confined in a bounded space region $D \subset \mathbb{R}^d_x$ with non incoming boundary condition.

In this section, we explore this equation when the cross sections are not compactly supported in space and develop a new and general construction in two (opposite) directions:

In the first direction, we replace (10.12.2) by the assumption that the cross sections vanish at infinity in a suitable sense (see (10.12.7) below). In this case

$$s(T_\Sigma) = 0,$$

so a spectral gap is excluded. We extend most of the compactness results obtained for bounded geometries (see Theorems 10.12.2 and 10.12.3). Despite these results, unlike the case on the torus, the existence of an invariant density is still an open question; one aspect of the difficulty (but it is not the only one) is the fact that

$$\int_0^{+\infty} \Sigma(y + s\zeta, \zeta)ds = +\infty \ \text{a.e.} \qquad\qquad (10.12.3)$$

is a necessary condition for the existence of an invariant density while the construction relies on the assumption that

$$\sup_{\zeta} \Sigma(y, \zeta) \to 0 \text{ as } y \to \infty. \tag{10.12.4}$$

In any case, this excludes the existence of an invariant density for compactly supported (in space) cross sections.

The second direction does not rely on compactness results and deals with *scattering theory*. The cross sections are not necessarily compactly supported in space and a priori we do not assume (10.12.4). Actually, in this case, the relevant hypotheses must relate to the mean free path

$$h(x, \zeta) := \frac{\Sigma(x, \zeta)}{|\zeta|}$$

rather than $\Sigma(x, \zeta)$. The subject of scattering theory has a rich tradition in the context of Transport Theory which goes back to [158, 337] (see [18, 75, 85, 118, 127, 128, 243, 309, 329, 363]); see also the monograph [119] for more information. A general result on the existence of wave operators is given in Theorem 10.12.4.

10.12.1 Preliminaries

Let $(U_\Sigma(t))_{t\geq0}$ be the advection C_0-semigroup on $L^1(\mathbb{R}_x^d \times \mathbb{R}_\zeta^d)$ defined by

$$U_\Sigma(t)\varphi = e^{-\int_0^t \Sigma(x-s\zeta,\zeta)ds} \varphi(x - t\zeta, \zeta)$$

where

$$\begin{cases} \Sigma(x, \zeta) = \Sigma_s(x, \zeta) + \Sigma_a(x, \zeta) \\ \Sigma_s(x, \zeta) = \int_{\mathbb{R}^d} c(x, \zeta', \zeta)\nu(d\zeta'). \end{cases}$$

We denote by T_Σ the generator of $(U_\Sigma(t))_{t\geq0}$. We recall that the spectral bound of T_Σ is given by

$$s(T_\Sigma) = -\lim_{t\to+\infty} \inf_{(x,\zeta)} t^{-1} \int_0^t \Sigma(x - s\zeta, \zeta)ds, \tag{10.12.5}$$

(see (10.8.8)). In connection with the general theory in Chap. 8, note that here the flow π_t on $\mathbb{R}_x^d \times \mathbb{R}_\zeta^d$ (endowed with the measure $dx\nu(d\zeta)$) is given here by

$$\pi_t(x, \zeta) = \begin{cases} (x + t\zeta, \zeta) \text{ if } \zeta \neq 0 \\ (x, 0) \text{ if } \zeta = 0. \end{cases}$$

and is measure-preserving. Note also that

$$\pi_t(x, \zeta) \to \infty \text{ as } (x, \zeta) \to \infty$$

and the results in Chap. 8 give:

Theorem 10.12.1 *(i) $s(T_\Sigma) < 0$ if and only if there exist $c_i > 0$ $(i = 1, 2)$ such that*

$$\int_0^{c_1} \Sigma(x - s\zeta, \zeta)ds \geq c_2 \text{ a.e.}$$

(ii) If Σ is continuous on some characteristic curve and vanishes identically on it then $s(T_\Sigma) = 0$.
  (iii) If $\lim_{|x|+|\zeta|\to\infty} \Sigma(x, \zeta) = 0$ then $s(T_\Sigma) = 0$.
  (iv) Let Σ be continuous. If $\liminf_{|x|+|\zeta|\to\infty} \Sigma(x, \zeta) > 0$ and if there exists no characteristic curve on which Σ vanishes identically then $s(T_\Sigma) < 0$.

Notice that $(U_\Sigma(t))_{t\geq 0}$ extends to a C_0-group $(U_\Sigma(t))_{t\in\mathbb{R}}$ where

$$U_\Sigma(-t)\varphi = e^{\int_0^t \Sigma(x+s\zeta,\zeta)ds}\varphi(x + t\zeta, \zeta) \ (t > 0)$$

provided that

$$\sup_{(x,\zeta)} \int_0^t \Sigma(x + s\zeta, \zeta)ds < +\infty \text{ for } t \text{ small enough} \tag{10.12.6}$$

(which insures the strong continuity of $(U_\Sigma(t))_{t\in\mathbb{R}}$ at 0). We denote by

$$C : D(C) \subset L^1(\mathbb{R}_x^d \times \mathbb{R}_\zeta^d) \to L^1(\mathbb{R}_x^d \times \mathbb{R}_\zeta^d)$$

the collision operator defined by

$$C\varphi = \int_{\mathbb{R}^d} c(x, \zeta, \zeta')\varphi(x, \zeta')\nu(d\zeta')$$

with maximal domain $D(C)$ given by

$$\left\{\varphi \in L^1(\mathbb{R}_x^d \times \mathbb{R}_\zeta^d); \ C|\varphi| \in L^1(\mathbb{R}_x^d \times \mathbb{R}_\zeta^d)\right\} = \left\{\varphi \in L^1(\mathbb{R}_x^d \times \mathbb{R}_\zeta^d); \ \Sigma_s\varphi \in L^1(\mathbb{R}_x^d \times \mathbb{R}_\zeta^d)\right\}.$$

It follows from the subcriticality condition

$$\int_{\mathbb{R}^d} c(x, \zeta', \zeta)\nu(d\zeta') \leq \Sigma(x, \zeta)$$

and the general Kato-Voigt perturbation theory that there exists a unique extension

$$G \supset T_\Sigma + C : D(T_\Sigma) \to L^1(\mathbb{R}^d_x \times \mathbb{R}^d_\zeta)$$

which generates a minimal substochastic C_0-semigroup $(V(t))_{t\geq 0}$. We are going to explore two "opposite" situations.

10.12.2 Compactness Results

In this section, we restrict ourselves to bounded collision operators C

$$\sup_{(x,\zeta)} \Sigma_s(x,\zeta) = \sup_{(x,\zeta)} \int_{\mathbb{R}^d} c(x,\zeta',\zeta)\nu(d\zeta') < +\infty,$$

in which case $G = T_\Sigma + C$. We assume also that C is regular, i.e., the family of operators on $L^1(\mathbb{R}^d_\zeta)$

$$C_x : \varphi \in L^1(\mathbb{R}^d_\zeta) \to \int_{\mathbb{R}^d} c(x,\zeta,\zeta')\varphi(\zeta')\nu(d\zeta') \in L^1(\mathbb{R}^d_\zeta)$$

(indexed by $x \in \mathbb{R}^d$) is collectively weakly compact. We assume additionally that

$$\lim_{|x|\to\infty} \|C_x\|_{\mathcal{L}(L^1(\mathbb{R}^d_\zeta))} = 0. \tag{10.12.7}$$

For conservative models (i.e., $\Sigma = \Sigma_s$), we have

$$\|C_x\|_{\mathcal{L}(L^1(\mathbb{R}^d_\zeta))} = \sup_{\zeta\in\mathbb{R}^d} \Sigma_s(x,\zeta)$$

and the condition (10.12.7) amounts to

$$\sup_{\zeta\in\mathbb{R}^d} \Sigma_s(x,\zeta) \to 0 \ (|x| \to \infty). \tag{10.12.8}$$

It follows from (10.12.5) and (10.12.8) that

$$s(T_\Sigma) = 0$$

for conservative models, i.e., if $\Sigma_a = 0$. The first result is

Theorem 10.12.2 *Suppose that C is a bounded regular collision operator satisfying (10.12.7). Let the velocity measure $\nu(d\zeta)$ satisfy (10.8.10). Then $C(\lambda - T_\Sigma)^{-1}$ $(\lambda > 0)$ is power compact in $L^1(\mathbb{R}^d_x \times \mathbb{R}^d_\zeta)$.*

Proof By assumption

$$1_{\{|x|\leq c\}}C \to C \ (c \to \infty) \tag{10.12.9}$$

in operator norm. (In this proof and in some subsequent proofs, in (10.12.9), c is a positive real number and should not be confused with the scattering kernel $c(.,.,.)$). Since $\left(C(\lambda - T)^{-1}\right)^j$ ($\lambda > 0$) depends continuously (in operator norm) on C, it suffices to give a proof for $1_{\{|x| \leq c\}}C$ in place of C. In this case, the proof follows the same lines as in bounded domains with non incoming boundary conditions (see [251]); we omit the details. $\qquad\square$

The second result is

Theorem 10.12.3 *Suppose that C is a bounded regular collision operator satisfying (10.12.7). Let the velocity measure $\nu(d\zeta)$ satisfy (10.8.15). Then, for m large enough, $R_m(t)$ is weakly compact for all $t \geq 0$. In particular $\omega_{ess}(V) = \omega_{ess}(U_\Sigma)$.*

Proof Since $1_{\{|x| \leq c\}}C \to C$ ($c \to \infty$) in operator norm and $U_j(t)$ depends continuously (in operator norm) on C, it suffices to give a proof for $1_{\{|x| \leq c\}}C$ in place of C. In this case, the proof is the same as in bounded domains with non incoming boundary conditions (see [251]); we omit the details. $\qquad\square$

For conservative models (i.e., $\Sigma = \Sigma_s$), $(V(t))_{t \geq 0}$ is a stochastic C_0-semigroup on $L^1(\mathbb{R}_x^d \times \mathbb{R}_\zeta^d)$ so $\omega(V) = 0$. Since $\omega_{ess}(U_\Sigma) = \omega(U_\Sigma) = s(T_\Sigma)$ then $\omega_{ess}(V) = \omega(V) = 0$ and consequently $(V(t))_{t \geq 0}$ cannot exhibit a spectral gap.

10.12.3   Open Problem

Despite the power compactness of $C(\lambda - T_\Sigma)^{-1}$ ($\lambda > 0$) (see Theorem 10.12.2), the existence of an invariant density is unclear. Indeed, for conservative models (i.e., $\Sigma = \Sigma_s$), $(V(t))_{t \geq 0}$ is a stochastic C_0-semigroup on $L^1(\mathbb{R}_x^d \times \mathbb{R}_\zeta^d)$ and $\omega_{ess}(V) = \omega(V) = 0$ so we are faced with technical problems similar to those arising in critical conservative equations on the d-dimensional torus [279]; actually, due to unboundedness of spatial domain, here the problem is much more involved. We note that $C(0 - T_\Sigma)^{-1}$ is stochastic. If $C(0 - T_\Sigma)^{-1}$ is power compact in $L^1(\mathbb{R}_x^d \times \mathbb{R}_\zeta^d)$ (or at least quasi compact) then its spectral radius is equal to one and is an eigenvalue so

$$C(0 - T_\Sigma)^{-1}u = u$$

for a positive $u \in L^1(\mathbb{R}_x^d \times \mathbb{R}_\zeta^d)$ and, by Theorem 3.3.1, we have an alternative according to $v := (0 - T_\Sigma)^{-1}u$ belongs to L^1 or not.

In the first case, (after normalization) v is an invariant density and is asymptotically stable (see Theorem 2.10.3) because $(V(t))_{t \geq 0}$ is partially integral (see Theorem 10.12.3) since $V(t)$ dominates the weakly compact remainder $R_m(t)$. In the second case, $(V(t))_{t \geq 0}$ is sweeping from the sets of integrability of v.

We have *not* been able to show that $C(0 - T_\Sigma)^{-1}$ is power compact in $L^1(\mathbb{R}_x^d \times \mathbb{R}_\zeta^d)$ and we wonder whether this is merely a technical difficulty or if there exists a deeper reason preventing such a result. The analysis of the existence of an invariant density in the whole space is an interesting open problem. We end this subsection with two remarks:

(i) The infiniteness condition (10.12.3) excludes collision frequencies Σ with compact support in space. Indeed, if there exists $c > 0$ such that $\Sigma(x, \zeta) = 0$ for $|x| > c$ then

$$\int_0^{+\infty} \Sigma(y + s\zeta, \zeta)ds = \int_0^{\frac{c+|y|}{|\zeta|}} \Sigma(y + s\zeta, \zeta)ds \leq \|\Sigma\|_\infty \frac{c + |y|}{|\zeta|}.$$

(ii) The infiniteness condition (10.12.3) is compatible with (10.12.4). For instance, if $\Sigma(y, \zeta) = \alpha(y)\beta(\zeta)$ with $\beta \in L^\infty$ and

$$\alpha(y) = \frac{1}{1 + |y|^\delta} \quad (0 < \delta \leq 1)$$

then both (10.12.4) and (10.12.3) are satisfied.

10.12.4 *Scattering Theory for Neutron Transport*

In this subsection, we explore another regime for Neutron Transport related to scattering theory given in Chap. 8. We write the collision operator C in the form

$$C\varphi = \int_{\mathbb{R}^d} c(x, \zeta, \zeta')\varphi(x, \zeta')\nu(d\zeta') = \int_{\mathbb{R}^d} k(x, \zeta, \zeta')\Sigma(x, \zeta')\varphi(x, \zeta')\nu(d\zeta')$$

where

$$k(x, \zeta, \zeta') := \frac{c(x, \zeta, \zeta')}{\Sigma(x, \zeta')}.$$

Then $C = K\Sigma$ where

$$K : L^1(\mathbb{R}^d_x \times \mathbb{R}^d_\zeta) \rightarrow L^1(\mathbb{R}^d_x \times \mathbb{R}^d_\zeta)$$

is a substochastic operator given by

$$K\varphi = \int_{\mathbb{R}^d} k(x, \zeta, \zeta')\varphi(x, \zeta')\nu(d\zeta').$$

We can write (10.12.1) in the more general form

$$\frac{\partial \varphi}{\partial t} + \zeta.\nabla_x\varphi + \Sigma\varphi = K\Sigma\varphi$$

where K is just an abstract substochatic operator on $L^1(\mathbb{R}_x^d \times \mathbb{R}_\zeta^d)$ which need not be an integral operator. To apply the theory in Chap. 8, we have just to observe that the flow π_t on $\mathbb{R}_x^d \times \mathbb{R}_\zeta^d$ (endowed with the measure $dx\nu(d\zeta)$) is given here by

$$\pi_t(x, \zeta) = (x + t\zeta, \zeta)$$

and is measure-preserving. The key condition (8.5.1) becomes here

$$\sup_{(y,\zeta)} \int_{-\infty}^{+\infty} \Sigma(y + s\zeta, \zeta)ds < +\infty \tag{10.12.10}$$

and the advection C_0-semigroup $(U_\Sigma(t))_{t\geq 0}$ extends a bounded positive C_0-group $(U_\Sigma(t))_{t\in\mathbb{R}}$ on $L^1(\mathbb{R}_x^d \times \mathbb{R}_\zeta^d)$. Even if $(U_\Sigma(t))_{t\in\mathbb{R}}$ is a positive group on $L^1(\mathbb{R}_x^d \times \mathbb{R}_\zeta^d)$, $(V(t))_{t\geq 0}$ need not be a group. Moreover, even if $(V(t))_{t\geq 0}$ extends to a group (e.g. if C is bounded), $V(t)$ need not be positive for $t < 0$; in some sense, $(V(t))_{t\in\mathbb{R}}$ "distinguishes" between the past and the future; this is reminiscent of the Boltzmann equation of the kinetic theory of gases where this distinction is manifested in Boltzmann's H-theorem [80]. Note that (10.12.10) is opposite to (10.12.3). Finally, we can express Theorem 8.5.1 as

Theorem 10.12.4 *Let K be either a stochastic operator or a substochastic integral operator. If (10.12.10) is satisfied then the wave operators*

$$\begin{cases} \Omega^+ := s\lim_{t\to+\infty} V(t)U_\Sigma(-t) \\ \Omega^- := s\lim_{t\to+\infty} U_\Sigma(-t)V(t) \end{cases}$$

exist in $L^1(\mathbb{R}_x^d \times \mathbb{R}_\zeta^d)$. In particular

$$\left\| V(t)\varphi - U_\Sigma(t)\Omega^-\varphi \right\| \to 0 \ (t \to +\infty) \ (\varphi \in L^1(\mathbb{R}_x^d \times \mathbb{R}_\zeta^d)),$$

i.e., any trajectory $(V(t)\varphi)_{t\geq 0}$ of the full dynamics (with initial data φ) behaves asymptotically (as $t \to +\infty$) as a free trajectory $(U_\Sigma(t)\psi)_{t\geq 0}$ (with initial data $\psi = \Omega^-\varphi$).

We recover the locally decaying property of $(V(t))_{t\geq 0}$ which appeared in early days of scattering theory of Neutron Transport [363].

Corollary 10.12.1 *Let the conditions in Theorem 10.12.4 be satisfied. For any $\varphi \in L^1(\mathbb{R}_x^d \times \mathbb{R}_\zeta^d)$*

$$\int_{\Xi\times\mathbb{R}_\zeta^d} |(V(t)\varphi)(x, \zeta)| \, dx\nu(d\zeta) \to 0 \ (t \to +\infty)$$

for any bounded subset $\Xi \subset \mathbb{R}_x^d$.

Proof According to Corollary 8.5.1, it suffices to show that $(U_\Sigma(t))_{t\geq 0}$ has this property. Since $(U_\Sigma(t))_{t\geq 0}$ is bounded, it suffices to choose φ in a dense subspace $\mathcal{S}$ of $L^1(\mathbb{R}^d_x \times \mathbb{R}^d_\zeta)$. Let

$$\mathcal{S} := \left\{ \varphi \in L^1(\mathbb{R}^d_x \times \mathbb{R}^d_\zeta);\ \exists \varepsilon > 0,\ \varphi(x, \zeta) = 0 \text{ if } |\zeta| \leq \varepsilon \text{ or } |x| \geq \varepsilon^{-1} \right\}.$$

Then for any $\varphi \in \mathcal{S}$ and any bounded subset $\Xi \subset \mathbb{R}^d_x$

$$\int_{\Xi \times \mathbb{R}^d_\zeta} |(U_\Sigma(t)\varphi)(x, \zeta)|\, dx \nu(d\zeta)$$

$$\leq \int_\Xi dy \int_{|\zeta|>\varepsilon} |\varphi(y - t\zeta, \zeta)|\, \nu(d\zeta) = 0$$

for t large enough because $|y - t\zeta| > \varepsilon^{-1}$ for all $y \in \Xi$ and all $|\zeta| > \varepsilon$ if t large enough. $\qquad\square$

In nuclear reactor theory (up to notations) the collision frequency $\Sigma(x, \zeta)$ is given by $|\zeta|\, h(x, \zeta)$ where $h(x, \zeta)$ is the so-called mean free path (see [114]) so

$$\int_{-\infty}^{+\infty} \Sigma(y + s\zeta, \zeta)ds = \int_{-\infty}^{+\infty} h(y + s\omega, \zeta)ds \quad (\omega = |\zeta|^{-1}\zeta).$$

Note that (10.12.10) is of course satisfied if the mean free path is bounded and has a compact support in space. Actually, we have the following much more general result.

Proposition 10.12.1 *Let $\Sigma(x, \zeta) = |\zeta|\, h(x, \zeta)$ where $h(x, \zeta)$ is the mean free path. Suppose that*

$$h(x, \zeta) \leq p(|x|, \zeta)$$

where $p : [0, +\infty) \times \mathbb{R}^d_\zeta$ is such that $s \geq 0 \to p(s, \zeta)$ is non increasing and

$$\sup_{\zeta \in \mathbb{R}^d} \int_0^{+\infty} p(s, \zeta)ds < +\infty.$$

Then

$$\sup_{(y,\zeta)} \int_{-\infty}^{+\infty} \Sigma(y + s\zeta, \zeta)ds \leq 6 \sup_{\zeta \in \mathbb{R}^d} \int_0^{+\infty} p(s, \zeta)ds.$$

In particular (10.12.10) is satisfied.

Proof Since

$$\int_{-\infty}^{+\infty} h(y + s\omega, \zeta)ds \leq \int_{-\infty}^{+\infty} p(|y + s\omega|, \zeta)ds \quad (\omega = |\zeta|^{-1}\zeta)$$

it suffices to check the boundedeness of

$$\int_{-\infty}^{+\infty} p(|y + s\omega|, \varsigma)ds.$$

We have

$$|y + s\omega| = \sqrt{|y|^2 + s^2 + 2s\,(y, \omega)}$$
$$= \sqrt{|y|^2 + s^2 + 2s\,|y|\,(\widehat{y}, \omega)}$$

where $\widehat{y} = |y|^{-1}\,y$. We introduce the set

$$\Lambda_{(\widehat{y},\omega)} = \{s \in \mathbb{R}; s\,(\widehat{y}, \omega) \geq 0\}.$$

On the set $\Lambda_{(\widehat{y},\omega)}$, we have $|y + s\omega| \geq |s|$. Then

$$\int_{-\infty}^{+\infty} p(|y + s\omega|, \varsigma)ds$$
$$= \int_{\Lambda_{(\widehat{y},\omega)}} p(|y + s\omega|, \varsigma)ds$$
$$+ \int_{\Lambda^c_{(\widehat{y},\omega)}} p(|y + s\omega|, \varsigma)ds$$

where

$$\int_{\Lambda_{(\widehat{y},\omega)}} p(|y + s\omega|, \varsigma)ds \leq \int_{-\infty}^{+\infty} p(|s|, \varsigma)ds$$
$$-2 \int_0^{+\infty} p(s, \varsigma)ds.$$

On the set $\Lambda^c_{(\widehat{y},\omega)}$ we have $s\,(\widehat{y}, \omega) = -|s|\,|(\widehat{y}, \omega)|$ so

$$|y|^2 + s^2 + 2s\,|y|\,(\widehat{y}, \omega)$$
$$= |y|^2 + s^2 - 2\,|s|\,|y| + 2\,|s|\,|y| - 2\,|s|\,|y|\,|(\widehat{y}, \omega)|$$
$$= (|y| - |s|)^2 + 2\,|s|\,|y|\,(1 - |(\widehat{y}, \omega)|)$$
$$\geq (|y| - |s|)^2$$

and

$$\int_{\Lambda^c_{(\widehat{y},\omega)}} p(|y + s\omega|, \varsigma)ds \leq \int_{-\infty}^{+\infty} p(||y| - |s||, \varsigma)ds = 2 \int_0^{+\infty} p(||y| - s|, \varsigma)ds.$$

The last integral decomposes into two parts

$$\int_0^{|y|} p(|y| - s, \zeta)ds = \int_0^{|y|} p(s, \zeta)ds \le \int_0^{+\infty} p(s, \zeta)ds$$

and

$$\int_{|y|}^{+\infty} p(s - |y|, \zeta)ds = \int_0^{+\infty} p(s, \zeta)ds.$$

Hence

$$\int_{-\infty}^{+\infty} p(|y + s\omega|, \zeta)ds \le 6 \int_0^{+\infty} p(s, \zeta)ds$$

which ends the proof. $\square$

Note that Proposition 10.12.1 applies e.g. if there exists $C > 0$ such that

$$h(y, \zeta) \le \frac{C}{1 + |y|^\delta} \quad (\delta > 1).$$

10.13 On Kinetic Theory of Particle Swarms

In this section, we consider general equations of the form

$$\frac{\partial f}{\partial t} + a(\zeta).\nabla_\zeta f + \Sigma(\zeta)f(t, \zeta) = \int_{\mathbb{R}^d} k(\zeta, \zeta')\Sigma(\zeta')f(t, \zeta')d\zeta' \qquad (10.13.1)$$

on $L^1(\mathbb{R}^d) := L^1(\mathbb{R}^d_\zeta; d\zeta)$ where

$$\int_{\mathbb{R}^d} k(\zeta', \zeta)d\zeta' = 1$$

and $a \in C^1(\mathbb{R}^d, \mathbb{R}^d)$ with bounded partial derivatives. Suppose that a is divergence-free

$$\operatorname{div} a = 0.$$

If we introduce the stochastic operator

$$K : \varphi \to \int_{\mathbb{R}^d} k(\zeta, \zeta')\varphi(\zeta')d\zeta'$$

then we can write (10.13.1) in the abstract form

$$\frac{\partial f}{\partial t} + a(\zeta).\nabla_\zeta f + \Sigma(\zeta)f(t,\zeta) = K(\Sigma f) \qquad (10.13.2)$$

where K is just a stochastic operator and (a priori) need not be an integral operator. We point out that most of the literature (see [18, 19, 127–129, 308]) is concerned with constant accelerations

$$a(\zeta) = a \in \mathbb{R}^d,$$

see however [28, 29] for non constant ones.

Here, we consider general smooth divergence-free vector fields and show how this broad class of equations fits within the theory of perturbed Frobenius–Perron semigroups developed in Chap. 8. We also explore in greater detail the special, and particularly important, cases of constant and Lorentz-type vector fields. To our knowledge, the peripheral spectral analysis and scattering theory for such general kinetic models are new.

The advection equation

$$\frac{\partial f}{\partial t} + a(\zeta).\nabla_\zeta f + \Sigma(\zeta)f(t,\zeta) = 0 \qquad (10.13.3)$$

can be solved explicitly by the method of characteristics. Indeed, the characteristic curves

$$\frac{d\widehat{\zeta}}{dt} = a(\widehat{\zeta}(t)), \quad \widehat{\zeta}(0) = \zeta \in \mathbb{R}^d \ (t \in \mathbb{R}) \qquad (10.13.4)$$

are such that $h(t) := f(t, \widehat{\zeta}(t))$ satisfies

$$h'(t) = \frac{\partial f}{\partial t}(t, \widehat{\zeta}(t)) + a(\widehat{\zeta}(t)).\nabla_\zeta f(t, \widehat{\zeta}(t))$$

so that (10.13.3) amounts $h'(t) + \Sigma(\widehat{\zeta}(t))h(t) = 0$ and

$$f(t, \widehat{\zeta}(t)) = f_0(\zeta)e^{-\int_0^t \Sigma(\widehat{\zeta}(\tau))d\tau}.$$

If we introduce the flow

$$\pi_t : \zeta \in \mathbb{R}^d \to \widehat{\zeta}(t) \in \mathbb{R}^d \ (t \in \mathbb{R})$$

where $\widehat{\zeta}(.)$ is the solution to (10.13.4) then

$$f(t, \pi_t(\zeta)) = f_0(\zeta)e^{-\int_0^t \Sigma(\pi_\tau(\zeta))d\tau}.$$

Since $\pi_t : \mathbb{R}^d \to \mathbb{R}^d$ is a C^1 difféomorphism then, by replacing ζ by $\pi_{-t}(\zeta)$, we get

$$f(t, \zeta) = f_0(\pi_{-t}(\zeta))e^{-\int_0^t \Sigma(\pi_\tau(\pi_{-t}(\zeta)))d\tau}$$
$$= f_0(\pi_{-t}(\zeta))e^{-\int_0^t \Sigma(\pi_{\tau-t}(\zeta))d\tau}$$
$$= f_0(\pi_{-t}(\zeta))e^{-\int_0^t \Sigma(\pi_{-(t-\tau)}(\zeta))d\tau}$$
$$= f_0(\pi_{-t}(\zeta))e^{-\int_0^t \Sigma(\pi_{-\tau}(\zeta))d\tau}.$$

By assuming that

$$\lim_{t \to 0_+} \int_0^t \Sigma((\pi_{-\tau}(\zeta))d\tau = 0 \text{ a.e.}$$

we define the absorption semigroup $(U_\Sigma(t))_{t \geq 0}$ on $L^1(\mathbb{R}^d; d\zeta)$

$$U_\Sigma(t)f = f(\pi_{-t}(\zeta))e^{-\int_0^t \Sigma(\pi_{-\tau}(\zeta))d\tau}$$

and denote by T_Σ its generator. We assume in this section that

$$D(T) \cap D(\Sigma) \text{ is a core of } T \tag{10.13.5}$$

where T is the generator of the Frobenius-Perron semigroup $(U(t))_{t \geq 0}$

$$U(t)f = f(\pi_{-t}(\zeta)) \tag{10.13.6}$$

corresponding to $\Sigma = 0$. A sufficient condition for (10.13.5) is provided by

Proposition 10.13.1 *If $\Sigma \in L^1_{loc}(\mathbb{R}^d; d\zeta)$ then (10.13.5) is satisfied.*

Proof Note that $C_c^1(\mathbb{R}^d)$ (the C^1-functions with compact support) is included in $D(T) \cap D(\Sigma)$. Hence it suffices to show that $C_c^1(\mathbb{R}^d)$ is a core of T. According to ([97, Theorem 1.9]) it suffices to show that $(U(t))_{t \geq 0}$ leaves invariant $C_c^1(\mathbb{R}^d)$. By standard results on differential equations $\zeta \to \pi_{-t}(\zeta)$ is C^1 and then so is

$$\zeta \to f(\pi_{-t}(\zeta)) \tag{10.13.7}$$

if f is C^1. In addition, if f is compactly supported then so is (10.13.7) since π_t is a C^1 difféomorphism. This ends the proof. $\qquad\qquad\square$

According to Lemma 5.3.2, (10.13.5) implies that Σ is U-regular and

$$\begin{cases} T_\Sigma = T - \Sigma \\ D(T_\Sigma) = D(T) \cap D(\Sigma). \end{cases}$$

Since the Frobenius-Perron semigroup is stochastic then $\int T\varphi = 0$ ($\varphi \in D(T)$) and consequently

$$\int (T_\Sigma \varphi + C\varphi) = \int T\varphi - \int \Sigma\varphi + \int C\varphi = 0$$

where $C = K\Sigma$ and there exists a unique extension

$$G \supset T_\Sigma + C$$

which generates a minimal substochastic C_0-semigroup $(V(t))_{t\geq 0}$. It follows that the time asymptotics in the theory of particle swarms are encompassed by the general framework developed in Chap. 8, allowing us to state the main results without detailed proofs. The question of whether $s(T_\Sigma) = 0$ or $s(T_\Sigma) < 0$ is addressed in Theorem 10.13.1. When K is weakly compact, the existence of a spectral gap for the semigroup $(V(t))_{t\geq 0}$ in the case $s(T_\Sigma) < 0$ is examined in Theorem 10.13.3. Meanwhile, the presence of an invariant density or the sweeping behavior of $(V(t))_{t\geq 0}$ when $s(T_\Sigma) = 0$ is analyzed in Theorem 10.13.4. For a general (abstract) substochastic operator K, a scattering theory for $(V(t))_{t\geq 0}$ is developed in Theorem 10.13.6. Cases involving constant or Lorentz-type vector fields are also considered.

10.13.1 On Spectral Bound of Advection Semigroups

According to the general theory given in Chap. 8 the spectral bound of T_Σ is given by

$$s(T_\Sigma) = -\lim_{t\to+\infty} \inf_{\zeta\in\mathbb{R}^d} t^{-1} \int_0^t \Sigma(\pi_s(\zeta))ds$$

where π_t is the flow defined above. We note that preimages of compact sets by π_t are compact sets because $\pi_t : \mathbb{R}^d \to \mathbb{R}^d$ is a C^1 difféomorphism so

$$|\pi_t(\zeta)| \to \infty \text{ as } |\zeta| \to \infty.$$

As a consequence of Theorems 8.3.2 and 8.3.3, we have

Theorem 10.13.1 *Suppose that Σ is continuous. Then:*

(i) If there exists some characteristic curve $\{\widehat{\zeta}(s, \zeta);\ s \geq 0\}$ on which Σ vanishes identically then $s(T_\Sigma) = 0$.

(ii) If $\lim_{x\to\infty} \Sigma(x) = 0$ then $s(T_\Sigma) = 0$.

(iii) If there exists a compact set outside which $\Sigma(.)$ is bounded away from zero and if there is no characteristic curve $\{\widehat{\zeta}(s, \zeta);\ s \geq 0\}$ on which Σ vanishes identically then $s(T_\Sigma) < 0$.

10.13.2 Spectral Theory and Asymptotics for Particle Swarms

The general theory given in Chap. 8 implies the following results:

Theorem 10.13.2 *If K is quasi-compact then $(V(t))_{t\geq 0}$ is stochastic and genuinely honest (i.e., $G = T_\Sigma + C$).*

Theorem 10.13.3 *If K be weakly compact then $(V(t))_{t\geq 0}$ and $(U_\Sigma(t))_{t\geq 0}$ share the same essential spectrum; in particular $\omega_{ess}(V) = \omega_{ess}(U_\Sigma)$. If additionaly $s(T_\Sigma) < 0$ then $(V(t))_{t\geq 0}$ has a spectral gap, i.e., $\omega_{ess}(V) < \omega(V) = 0$.*

Theorem 10.13.4 *Let K be weakly compact and $s(T_\Sigma) = 0$. Let $C\,(0 - T_\Sigma)^{-1}$ be irreducible. Then 1 is an eigenvalue of $C\,(0 - T_\Sigma)^{-1}$ associated to a positive eigenfunction φ. Moreover:*
  (i) If $\psi := (0 - T_\Sigma)^{-1}\,\varphi \in L^1\left(\mathbb{R}^d; d\zeta\right)$ then $\psi \in D(T_\Sigma)$ and (after normalization) is a unique invariant density of $(V(t))_{t\geq 0}$ and is asymptotically stable.
  (ii) If $\psi := (0 - T_\Sigma)^{-1}\,\varphi \notin L^1\left(\mathbb{R}^d; d\zeta\right)$ then $(V(t))_{t\geq 0}$ is sweeping from the sets

$$\Xi_\varepsilon := \{\Sigma \geq \varepsilon\} \quad (\varepsilon > 0).$$

As in Chap. 8, if $(V(t))_{t\geq 0}$ is irreducible and has an invariant density then $\int_{-\infty}^{+\infty} \Sigma(\widehat{\zeta}(s, \zeta))ds = +\infty$ a.e.

10.13.3 Scattering Theory for Particle Swarms

In this subsection, K is a general (abstract) substochastic operator and need not be an integral operator. Similarly, following the results in Chap. 8 we can state

Theorem 10.13.5 *If the condition*

$$\sup_{\zeta \in \mathbb{R}^d} \int_{-\infty}^{+\infty} \Sigma(\widehat{\zeta}(s, \zeta))ds < +\infty \tag{10.13.8}$$

is satisfied then $(U_\Sigma(t))_{t\geq 0}$ extends to a positive bounded C_0-group $(U_\Sigma(t))_{t\in\mathbb{R}}$ on $L^1(\mathbb{R}^d; d\zeta)$.

Theorem 10.13.6 *Let K be either a general stochastic operator or a substochastic integral operator. If the finite "optical length" condition (10.13.8) is satisfied then the waves operators*

$$\begin{cases} \Omega^+ := s\lim_{t\to+\infty} V(t)U_\Sigma(-t) \\ \Omega^- := s\lim_{t\to+\infty} U_\Sigma(-t)V(t) \end{cases}$$

exist on $L^1(\mathbb{R}^d; d\zeta)$. In particular

$$\|V(t)\varphi - U_\Sigma(t)\Omega^-\varphi\| \to 0 \ (t \to +\infty) \ \ (\varphi \in L^1(\mathbb{R}^d; d\zeta)),$$

i.e., any trajectory $(V(t)\varphi)_{t\geq 0}$ of the full dynamics (with initial data φ) behaves asymptotically (as $t \to +\infty$) as a free trajectory $(U_\Sigma(t)\psi)_{t\geq 0}$ (with initial data $\psi = \Omega^-\varphi$).

10.13.4 Constant Vector Fields

In this subsection, we discuss briefly, without formal statements, the case of constant vector fields

$$a(\zeta) = a \in \mathbb{R}^d$$

considered in the literature, (see e.g. [18, 127, 128, 308]). In this case, $\pi_t(\zeta) = \zeta + ta$ and

$$U_\Sigma(t)f = f(\zeta - ta)e^{-\int_0^t \Sigma((\zeta - sa)ds}$$

so (10.13.8) amounts to

$$\sup_{\zeta \in \mathbb{R}^d} \int_{-\infty}^{+\infty} \Sigma(\zeta + sa)ds < +\infty. \tag{10.13.9}$$

Arguing as in the proof of Proposition 10.12.1, if there exists $p : \mathbb{R}_+ \to \mathbb{R}_+$ such that

$$\begin{cases} \Sigma(\zeta) \leq p(|\zeta|) \\ s \geq 0 \to p(s) \text{ is nonincreasing} \\ \int_0^{+\infty} p(s)ds < +\infty \end{cases}$$

then

$$\sup_{\zeta \in \mathbb{R}^d} \int_{-\infty}^{+\infty} \Sigma(\zeta + sa)ds \leq 6\,|a|^{-1} \int_0^{+\infty} p(s)ds$$

so (10.13.9) is satisfied and the regime of scattering theory described in Theorem 8.5.1 occurs. On the contrary, if

$$\int_{-\infty}^{+\infty} \Sigma(\zeta + sa)ds = +\infty \text{ a.e.}$$

(e.g. if there exist $c > 0$ and $0 < \delta \leq 1$ such that $\Sigma(\zeta) \geq \frac{c}{1+|\zeta|^\delta}$) and if K is weakly compact then several possibilities may occur:

(1) If $s(T_\Sigma) < 0$ then $(V(t))_{t\geq 0}$ has a spectral gap.
(2) If $s(T_\Sigma) = 0$ then there exists $\varphi > 0$ such that

$$C(0 - T_\Sigma)^{-1}\varphi = \varphi$$

and two subcases may occur:

(i) If $\psi := (0 - T_\Sigma)^{-1}\varphi \in L^1\left(\mathbb{R}^d\right)$ then (after mormalization) ψ is an invariant density and $(V(t))_{t\geq 0}$ is asymptotically stable.

(ii) If $\psi := (0 - T_\Sigma)^{-1}\varphi \notin L^1\left(\mathbb{R}^d\right)$ then $(V(t))_{t\geq 0}$ is sweeping from the sets

$$\Lambda_\varepsilon := \left\{\zeta \in \mathbb{R}^d;\ \Sigma(\zeta) \geq \varepsilon\right\}\ \ (\varepsilon > 0).$$

In other words, in the case (ii) any trajectory $(V(t)\varphi)_{t\geq 0}$ "concentrates" (as $t \to +\infty$) in the regions where Σ gets small. For instance, if Σ is locally bounded from below and tends to zero at infinity then $(V(t))_{t\geq 0}$ is sweeping from the bounded subsets of $\mathbb{R}^d$. However, we point out that the concentration phenomenon occuring in the case (ii) has nothing to do with the sweeping occuring in the context of scattering theory, i.e., under (10.13.9); indeed, in this case $s(T_\Sigma) = 0$ and $(U_\Sigma(t))_{t\geq 0}$ is sweeping from the bounded subsets $\Xi \subset \mathbb{R}^d$ since

$$\int_\Xi |U_\Sigma(t)f| \leq \int_\Xi |f(\zeta - ta)|\,d\zeta$$

$$= \int_{\Xi - ta} \left|f(\zeta')\right|\,d\zeta' \to 0\ (t \to +\infty)$$

and, according to Remark 8.5.1, $(V(t))_{t\geq 0}$ is also sweeping from the bounded subsets $\Xi \subset \mathbb{R}^d$.

We note also that according to Theorem 8.3.3 if Σ is continuous then:

(i) If there exists some half line $\{\zeta + sa;\ s \geq 0\}$ on which Σ vanishes identically then $s(T_\Sigma) = 0$.

(ii) If $\lim_{x\to\infty} \Sigma(x) = 0$ then $s(T_\Sigma) = 0$.

(iii) If there exists a compact set outside which $\Sigma(.)$ is bounded away from zero and if there is no half line $\{\zeta + sa;\ s \geq 0\}$ on which Σ vanishes identically then $s(T_\Sigma) < 0$.

10.13.5   *Vector Fields of Lorentz Type*

This subsection is devoted to Lorentz fields

$$a(\zeta) = a + b \times \zeta \tag{10.13.10}$$

where $a \in \mathbb{R}^3$ (the electric field) and $b \in \mathbb{R}^3$ (the magnetic field) are given constants. Note that

$$\zeta \to b \times \zeta = \begin{pmatrix} b_2\zeta_3 - b_3\zeta_2 \\ b_3\zeta_1 - b_1\zeta_3 \\ b_1\zeta_2 - b_2\zeta_1 \end{pmatrix}$$

is divergence free. Note also that (10.13.10) covers the case $a(\zeta) = a$ (considered [18, 19, 127, 128, 308]) and the case $a(\zeta) = b \times \zeta$ which, as far as we know, has not been considered up to now. Suppose that

$$a.b = 0$$

so there exists $e \in \mathbb{R}^3$ such that $a = -b \times e$. In particular, e is a stationary point (i.e., $a(e) = 0$). Actually all the points in

$$C_s := \{e + \lambda b; \ \lambda \in \mathbb{R}\}$$

are stationary points. By using the change of function $\widetilde{\zeta}(t) := \widehat{\zeta}(t) - e$, the characteristic equation

$$\frac{d\widehat{\zeta}}{dt} = a + b \times \widehat{\zeta}(t), \ \ \widehat{\zeta}(0) = \zeta \notin C_s$$

becomes

$$\frac{d\widetilde{\zeta}(t)}{dt} = b \times \widetilde{\zeta}(t), \widetilde{\zeta}(0) = \zeta - e.$$

Then

$$\frac{d}{dt} \left|\widetilde{\zeta}(t)\right|^2 = 2\widetilde{\zeta}(t).\frac{d\widetilde{\zeta}(t)}{dt} = 2\widetilde{\zeta}(t).\left(b \times \widetilde{\zeta}(t)\right) = 0$$

gives

$$\left|\widetilde{\zeta}(t)\right| = \left|\widetilde{\zeta}(0)\right| = |\zeta - e| \ \ (t \in \mathbb{R}) \tag{10.13.11}$$

while

$$\frac{d\left(\widetilde{\zeta}(t).b\right)}{dt} = \frac{d\widetilde{\zeta}(t)}{dt}.b = \left(b \times \widetilde{\zeta}(t)\right).b = 0$$

gives

$$\widetilde{\zeta}(t).b = \widetilde{\zeta}(0).b \ \ (t \in \mathbb{R}). \tag{10.13.12}$$

Hence the orbit $\left\{\widetilde{\zeta}(t); \ t \in \mathbb{R}\right\}$ is included in the intersection of the affine plane

$$\left\{v \in \mathbb{R}^3, \ v.b = (\zeta - e).b\right\}$$

(orthogonal to b) and the sphere

$$\left\{v \in \mathbb{R}^3, \ |v| = |\zeta - e|\right\},$$

i.e., in a circle $\widetilde{C}_\zeta$. The center of this circle is given by

$$\left(\widetilde{\zeta}(0).\frac{b}{|b|}\right)\frac{b}{|b|} = |b|^{-2}\left((\zeta - e).b\right)b$$

and its radius by

$$\sqrt{|\zeta - e|^2 - |b|^{-2}\left((\zeta - e).b\right)^2}.$$

Hence $\widetilde{\zeta}(.)$ is periodic. Similarly, the orbit $\left\{\widehat{\zeta}(t) = e + \widetilde{\zeta}(t);\ t \in \mathbb{R}\right\}$ is included in the circle

$$\widehat{C}_\zeta := \widetilde{C}_\zeta + e$$

and the characteristic curves $\widehat{\zeta}(.)$ are periodic. This implies that the condition

$$\int_{-\infty}^{+\infty} \Sigma(\pi_s(\zeta))ds < +\infty \ \text{a.e.}$$

is never satisfied (unless Σ is identically zero). Hence the regime of scattering theory is no longer possible in the case of Lorentz forces. According to (10.13.11) (10.13.12)

$$\left|\frac{d\widetilde{\zeta}(t)}{dt}\right|^2 = \left|b \times \widetilde{\zeta}(t)\right|^2 = |b|^2\left|\widetilde{\zeta}(t)\right|^2 - \left(b.\widetilde{\zeta}(t)\right)^2$$

$$= |b|^2\left|\widetilde{\zeta}(0)\right|^2 - \left(b.\widetilde{\zeta}(0)\right)^2,$$

i.e., the scalar velocity is time independent

$$\sqrt{|b|^2\,|\zeta - e|^2 - (b.\,(\zeta - e))^2}.$$

It follows that the period T_ζ of each characteristic curve C_ζ is equal to

$$T_\zeta = \frac{2\pi\sqrt{|\zeta - e|^2 - |b|^{-2}\left((\zeta - e).b\right)^2}}{\sqrt{|b|^2\,|\zeta - e|^2 - (b.\,(\zeta - e))^2}}$$

$$= \frac{1}{|b|}\frac{2\pi\sqrt{|\zeta - e|^2 - \left((\zeta - e).\overline{b}\right)^2}}{\sqrt{|\zeta - e|^2 - \left(\overline{b}.\,(\zeta - e)\right)^2}} = \frac{2\pi}{|b|}$$

$(\widehat{b} := |b|^{-1}b)$ so $T_\zeta\ (\zeta \notin C_s)$ is ζ-independent. This allows us to "compute" the spectral bound

$$s(T_\Sigma) = -\lim_{t \to +\infty}\inf_{\zeta \in \mathbb{R}^3} t^{-1}\int_0^t \Sigma(\pi_s(\zeta)))ds.$$

Lemma 10.13.1 *Suppose that Σ is bounded and continuous at $\zeta \in C_s$. Then*

$$s(T_\Sigma) = -\inf_{\zeta \in \mathbb{R}^3} \frac{|b|}{2\pi} \int_0^{\frac{2\pi}{|b|}} \Sigma(\pi_s(\zeta)))ds.$$

Proof Let $\zeta \in \mathbb{R}^3$ ($\zeta \notin C_s$) be fixed. For any $t > 0$

$$t = [t]\frac{2\pi}{|b|} + r(t) \ \ (r(t) < \frac{2\pi}{|b|})$$

with $[t] \in \mathbb{N}$ and

$$\int_0^t \Sigma(\pi_s(\zeta)))ds = \int_0^{[t]\frac{2\pi}{|b|}} \Sigma(\pi_s(\zeta)))ds + \int_{[t]\frac{2\pi}{|b|}}^t \Sigma(\pi_s(\zeta)))ds$$

$$= [t]\int_0^{\frac{2\pi}{|b|}} \Sigma(\pi_s(\zeta)))ds + \int_0^{r(t)} \Sigma(\pi_s(\zeta)))ds.$$

Hence

$$t^{-1}\int_0^t \Sigma(\pi_s(\zeta)))ds = t^{-1}[t]\int_0^{\frac{2\pi}{|b|}} \Sigma(\pi_s(\zeta)))ds + t^{-1}\int_0^{r(t)} \Sigma(\pi_s(\zeta)))ds$$

and

$$\inf_{\zeta \notin C_s} t^{-1}\int_0^t \Sigma(\pi_s(\zeta)))ds$$

$$= \inf_{\zeta \notin C_s} \left(t^{-1}[t]\int_0^{\frac{2\pi}{|b|}} \Sigma(\pi_s(\zeta)))ds + t^{-1}\int_0^{r(t)} \Sigma(\pi_s(\zeta)))ds \right).$$

Since Σ is bounded then

$$t^{-1}\int_0^{r(t)} \Sigma(\pi_s(\zeta)))ds \le \|\Sigma\|_\infty \frac{2\pi}{t\,|b|} \to 0 \ (t \to +\infty)$$

uniformly in $\zeta \in \mathbb{R}^3$ so (using the fact that $t^{-1}[t] \to \frac{|b|}{2\pi}$)

$$\lim_{t \to +\infty} \inf_{\zeta \notin C_s} t^{-1}\int_0^t \Sigma(\pi_s(\zeta)))ds$$

$$= \lim_{t \to +\infty} \inf_{\zeta \notin C_s} \left(t^{-1}[t]\int_0^{\frac{2\pi}{|b|}} \Sigma(\pi_s(\zeta)))ds \right)$$

$$= \lim_{t \to +\infty} t^{-1}[t] \left(\inf_{\zeta \notin C_s} \int_0^{\frac{2\pi}{|b|}} \Sigma(\pi_s(\zeta)))ds \right)$$

$$= \inf_{\zeta \notin \mathcal{C}_s} \frac{|b|}{2\pi} \int_0^{\frac{2\pi}{|b|}} \Sigma(\pi_s(\zeta)))ds.$$

Since $\pi_s(v) = v$ $(v \in \mathcal{C}_s)$ then

$$\lim_{t \to +\infty} t^{-1} \inf_{\zeta \in \mathbb{R}^3} \int_0^t \Sigma(\pi_s(\zeta)))ds$$

$$= \lim_{t \to +\infty} \min \left\{ \inf_{\zeta \notin \mathcal{C}_s} t^{-1} \int_0^t \Sigma(\pi_s(\zeta)))ds, \ \inf_{v \in \mathcal{C}_s} \Sigma(v) \right\}$$

$$= \min \left\{ \inf_{\zeta \notin \mathcal{C}_s} \frac{|b|}{2\pi} \int_0^{\frac{2\pi}{|b|}} \Sigma(\pi_s(\zeta)))ds, \ \inf_{v \in \mathcal{C}_s} \Sigma(v) \right\}.$$

Since $\Sigma(.)$ is continuous at $v \in \mathcal{C}_s$ then

$$\lim_{\zeta \to v} \frac{|b|}{2\pi} \int_0^{\frac{2\pi}{|b|}} \Sigma(\pi_s(\zeta))ds = \Sigma(v) \ (v \in \mathcal{C}_s)$$

so

$$s(T_\Sigma) = - \inf_{\zeta \in \mathbb{R}^3} \frac{|b|}{2\pi} \int_0^{\frac{2\pi}{|b|}} \Sigma(\pi_s(\zeta))ds.$$

$\square$

The question of whether $s(T_\Sigma) = 0$ or $s(T_\Sigma) < 0$ is a key issue. The answer follows directly from Theorem 8.3.3:

Proposition 10.13.2 *Suppose that Σ is continuous. Then:*
(i) If Σ vanishes on some circle $\widehat{C_{\widehat{\zeta}}}$ $(\zeta \notin \mathcal{C}_s)$ or at some point $\zeta \in \mathcal{C}_s$ then $s(T_\Sigma) = 0$.
(ii) If $\lim_{|\zeta| \to \infty} \Sigma(\zeta) = 0$ then $s(T_\Sigma) = 0$.
(iii) If $\liminf_{|\zeta| \to \infty} \Sigma(\zeta) > 0$ and $\Sigma(\zeta) > 0$ $(\zeta \in \mathcal{C}_s)$ and if there exists no circle $\widehat{C_{\widehat{\zeta}}}$ $(\zeta \notin \mathcal{C}_s)$ on which Σ vanishes identically then $s(T_\Sigma) < 0$.

Then Theorem 8.4.2 gives

Theorem 10.13.7 *Suppose that K is weakly compact. If $s(T_\Sigma) < 0$ then $(V(t))_{t \geq 0}$ has a spectral gap (i.e., $\omega_{ess}(V) < \omega(V) = 0$).*

Finally Theorem 8.4.3 gives

Theorem 10.13.8 *Suppose that K is weakly compact. Let $s(T_\Sigma) = 0$ and let $C(0 - T_\Sigma)^{-1}$ be irreducible. Then 1 is an eigenvalue of $C(0 - T_\Sigma)^{-1}$ associated to a positive eigenfunction φ. Moreover:*
(1) If $\psi := (0 - T_\Sigma)^{-1} \varphi \in L^1(\mathbb{R}^3; d\zeta)$ then $\psi \in D(T_\Sigma)$ and (after normalization) is a unique invariant density of $(V(t))_{t \geq 0}$ and is asymptotically stable.

(2) If $\psi := (0 - T_\Sigma)^{-1} \varphi \notin L^1 \left(\mathbb{R}^3; d\zeta\right)$ then $(V(t))_{t \geq 0}$ is sweeping from the sets

$$\Xi_\varepsilon := \left\{ \zeta \in \mathbb{R}^3;\ \Sigma(\zeta) \geq \varepsilon \right\} \quad (\varepsilon > 0).$$

10.14 Neutron Transport with Partly Elastic Collisions

In the 1970s, Larsen and Zweifel [192] introduced, for neutron transport operators
with vacuum boundary conditions, collision operators of notable complexity where
the standard inelastic operator is complemented by *"downshift"* and *"Bragg"* (elastic) collision operators. Further developments were presented later in [324]. In this
section, we revisit the results of [324] with greater depth, focusing on conservative partly elastic neutron transport semigroups in the context of the d-dimensional
torus. Specifically, we omit the "downshift" component of the collision operator and
consider conservative models on the torus involving both inelastic and elastic collision operators. This new construction extends in several directions the general theory
developed in [266], which was devoted to inelastic scattering.[1] A key difficulty arises
from the presence of elastic scattering, which induces a loss of compactness in the
velocity variable: more precisely, the collision operator C is not weakly compact
in velocity (i.e., C is not regular), and therefore the analysis falls outside the scope
of [266]. It is also known that curves of essential spectrum appear in this context [192].
The goal of this section is to show how compactness results can be partially recovered due to the presence of the inelastic component. However, the construction is
substantially more involved than in the absence of elastic scattering; in particular, a
crucial step relies on a harmonic analysis result for measures from [310].

We consider general partly elastic neutron transport equations on the torus

$$\frac{\partial \varphi}{\partial t} + \rho\omega.\frac{\partial \varphi}{\partial x} + \sigma(x, \rho, \omega)\varphi(x, \rho, \omega, t) = C\varphi \tag{10.14.1}$$

where the collision operator

$$C := C_i + C_e$$

combines an inelastic collision operator C_i and an elastic one C_e where

$$C_i\varphi(x, \rho, \omega) = \int_a^b \int_{S^{d-1}} c_i(x, \rho, \omega, \rho', \omega')\varphi(x, \rho', \omega', t)d\rho'd\omega'$$

and

$$C_e\varphi(x, \rho, \omega) = \int_{S^{d-1}} c_e(x, \rho, \omega, \omega')\varphi(x, \rho, \omega', t)d\omega'.$$

---

[1] This construction was partly presented at the 2016 conference [270] held in Bedlewo (Poland)
and remains unpublished.

To avoid too much additional technicalities, we restrict ourselves to bounded collision operators.

Here $(x, \rho) \in \mathcal{T}^d \times [a, b]$ where

$$\mathcal{T}^d := \mathbb{R}^d / (2\pi\mathbb{Z})^d$$

where $0 < a < b < +\infty$ and $\omega \in S^{d-1}$. The assumption that $a > 0$ (i.e., the velocities are bounded away from zero) is not essential and is made just to avoid cumbersone technical details. Of course (10.14.1) is supplemented by an initial condition

$$\varphi(x, \rho, \omega, 0) = \varphi_0(x, \rho, \omega).$$

All the cross sections σ, c_i, c_e are measurable and nonnegative. We work in the space

$$L^1 := L^1(\mathcal{T}^d \times [a, b] \times S^{d-1})$$

endowed with Lebesgue measure $dx\,d\rho\,d\omega$ on $\mathcal{T}^d \times [a, b] \times S^{d-1}$ and identify any measurable function

$$p : x \in \mathcal{T}^d \to p(x) \in \mathbb{R}$$

to a $[0, 2\pi]^d$-periodic measurable function on $\mathbb{R}^d$. We define a weighted translation C_0-semigroup $(U(t))_{t \geq 0}$ of contractions on L^1 by

$$U(t)\varphi \to e^{-\int_0^t \sigma(x - s\rho\omega, \rho, \omega)ds} \varphi(x - t\rho\omega, \rho, \omega) \ \ (t \geq 0)$$

and denote by T its generator; there is a small change of notations here: we denote by σ the cross section instead of Σ and note $U(t)$ (resp. T) in place of $U_\Sigma(t)$ (resp. T_Σ). We recall that the type of $(U(t))_{t \geq 0}$ or equivalently the spectral bound of its generator T is given by

$$s(T) = -\lim_{t \to +\infty} \inf_{(x,\rho,\omega) \in \mathcal{T}^d \times [a,b] \times S^{d-1}} t^{-1} \int_0^t \sigma(x + s\rho\omega, \rho, \omega)ds \qquad (10.14.2)$$

and $s(T) < 0$ if and only if there exist two positive constants C_1 and C_2 such that

$$\int_0^{C_1} \sigma(x + s\rho\omega, \rho, \omega)ds \geq C_2 \ \forall(x, \rho, \omega), \qquad (10.14.3)$$

(see [266]). We denote by $(U^e(t))_{t \geq 0}$ the C_0-semigroup generated by $T + C_e$ and by $(V(t))_{t \geq 0}$ the C_0-semigroup generated by $T + C_e + C_i$. Besides numerous preliminary results, key (quite involved) weak compactness results, given in Lemma 10.14.2, are the core of a stability of essential type

$$\omega_{ess}(V) = \omega_{ess}(U^e),$$

(see Theorem 10.14.1) and finally of a spectral gap result (see Theorem 10.14.2) when $s(T) < 0$. The construction of this section is organized as follows:

10.14.1   *Mono-energetic Operators*

For any fixed $\rho \in [a, b]$ we define the *monoenergetic* C_0-semigroup $\left(U_\rho(t)\right)_{t \geq 0}$ of contractions on the space

$$L^1(T^d \times S^{d-1})$$

by

$$U_\rho(t)\varphi \rightarrow e^{-\int_0^t \sigma(x - s\rho\omega, \rho, \omega)ds}\,\varphi(x - t\rho\omega, \omega)\ \ (t \geq 0)$$

and denote by T_ρ its generator. Arguing as in [266], one can prove that the type of $\left(U_\rho(t)\right)_{t \geq 0}$ or equivalently the spectral bound of its generator T_ρ is given by

$$s(T_\rho) = -\lim_{t \to +\infty}\ \inf_{(x,\omega)\in T^d \times S^{d-1}}\ t^{-1}\int_0^t \sigma(x + s\rho\omega, \rho, \omega)ds. \qquad (10.14.4)$$

Lemma 10.14.1   *We always have*

$$s(T) \geq \sup_{\rho \in [a,b]} s(T_\rho).$$

If the limit (10.14.4) is uniform in $\rho \in [a, b]$ then

$$s(T) = \sup_{\rho \in [a,b]} s(T_\rho).$$

Proof Let us show first that

$$s(T_\rho) \in \sigma(T)\ \ \forall \rho \in [a, b]$$

and use the abreviation

$$\eta_\rho := s(T_\rho).$$

Fix some $\overline{\rho} \in [a, b]$. The fact that $\eta_{\overline{\rho}}$ belongs to the *boundary* of the spectrum of $T_{\overline{\rho}}$ implies that $\eta_{\overline{\rho}}$ is an approximate eigenvalue or equivalently, for a fixed $\lambda > 0$,

$$\frac{1}{\lambda - \eta_{\overline{\rho}}}$$

is an approximate eigenvalue of $(\lambda - T_{\overline{\rho}})^{-1}$, i.e., there exists a sequence $(f_j)_j \subset L^1(T^d \times S^{d-1})$ such that

$$\|f_j\|_{L^1(\mathcal{T}^d \times S^{d-1})} = 1$$

and

$$\left\| (\lambda - T_{\overline{\rho}})^{-1} f_j - \frac{1}{\lambda - \eta_{\overline{\rho}}} f_j \right\|_{L^1(\mathcal{T}^d \times S^{d-1})} \to 0,$$

i.e.,

$$\int_{\mathcal{T}^d \times S^{d-1}} \left| \int_0^{+\infty} e^{-\lambda t} e^{-\int_0^t \sigma(x - s\overline{\rho}\omega, \overline{\rho}, \omega)ds} f_j(x - t\overline{\rho}\omega, \omega)dt - \frac{1}{\lambda - \eta_{\overline{\rho}}} f_j(x, \omega) \right| dx d\omega \to 0.$$

We build a normalized sequence $(\psi_j)_j \subset L^1(\mathcal{T}^d \times [a, b] \times S^{d-1})$ by

$$\psi_j(x, \rho, \omega) = \begin{cases} \frac{1}{2\varepsilon_j} f_j(x, \omega) & \text{if } \rho \in \left[\overline{\rho} - \varepsilon_j, \overline{\rho} + \varepsilon_j\right] \\ 0 & \text{otherwise} \end{cases}$$

where $\varepsilon_j \to 0$ as $j \to +\infty$; in the sequel we write ε instead of ε_j for the simplicity of notations. Then

$$\left\| (\lambda - T)^{-1} \psi_j - \frac{1}{\lambda - \eta_{\overline{\rho}}} \psi_j \right\|_{L^1(\mathcal{T}^d \times [a,b] \times S^{d-1})}$$

is equal to

$$\int_{\mathcal{T}^d \times [a,b] \times S^{d-1}} \left| \int_0^{+\infty} e^{-\lambda t} e^{-\int_0^t \sigma(x - s\rho\omega, \rho, \omega)ds} \psi_j(x - t\rho\omega, \rho, \omega)dt - \frac{1}{\lambda - \eta_{\overline{\rho}}} \psi_j(x, \rho, \omega) \right| dx d\omega$$

$$= \frac{1}{2\varepsilon} \int_{\overline{\rho}-\varepsilon}^{\overline{\rho}+\varepsilon} d\rho \int_{\mathcal{T}^d \times S^{d-1}} \left| \int_0^{+\infty} e^{-\lambda t} e^{-\int_0^t \sigma(x - s\overline{\rho}\omega, \rho, \omega)ds} f_j(x - t\overline{\rho}\omega, \omega)dt - \frac{1}{\lambda - \eta_{\overline{\rho}}} f_j(x, \omega) \right| dx d\omega.$$

We use the decomposition

$$\int_0^{+\infty} e^{-\lambda t} e^{-\int_0^t \sigma(x - s\rho\omega, \rho, \omega)ds} f_j(x - t\overline{\rho}\omega, \omega)dt$$

$$= \int_0^{+\infty} e^{-\lambda t} \left(e^{-\int_0^t \sigma(x - s\rho\omega, \rho, \omega)ds} - e^{-\int_0^t \sigma(x - s\overline{\rho}\omega, \overline{\rho}, \omega)ds} \right) f_j(x - t\overline{\rho}\omega, \omega)dt$$

$$+ \int_0^{+\infty} e^{-\lambda t} e^{-\int_0^t \sigma(x - s\overline{\rho}\omega, \overline{\rho}, \omega)ds} f_j(x - t\overline{\rho}\omega, \omega)dt$$

and note that for any $c > 0$

$$\int_{\mathcal{T}^d \times S^{d-1}} \int_0^{+\infty} e^{-\lambda t} \left| e^{-\int_0^t \sigma(x - s\rho\omega, \rho, \omega)ds} - e^{-\int_0^t \sigma(x - s\overline{\rho}\omega, \overline{\rho}, \omega)ds} \right| |f_j(x - t\overline{\rho}\omega, \omega)| \, dt dx d\omega$$

$$= \int_{\mathcal{T}^d \times S^{d-1}} \int_0^c e^{-\lambda t} \left| e^{-\int_0^t \sigma(x - s\rho\omega, \rho, \omega)ds} - e^{-\int_0^t \sigma(x - s\overline{\rho}\omega, \overline{\rho}, \omega)ds} \right| |f_j(x - t\overline{\rho}\omega, \omega)| \, dt dx d\omega$$

$$+ \int_{T^d \times S^{d-1}} \int_c^{+\infty} e^{-\lambda t} \left| e^{-\int_0^t \sigma(x - s\rho\omega, \rho, \omega)ds} - e^{-\int_0^t \sigma(x - s\overline{\rho}\omega, \overline{\rho}, \omega)ds} \right| |f_j(x - t\overline{\rho}\omega, \omega)| \, dt\, dx\, d\omega$$

where

$$\int_{T^d \times S^{d-1}} \int_c^{+\infty} e^{-\frac{\lambda}{2} t} \left| e^{-\int_0^t \sigma(x - s\rho\omega, \rho, \omega)ds} - e^{-\int_0^t \sigma(x - s\overline{\rho}\omega, \overline{\rho}, \omega)ds} \right| e^{-\frac{\lambda}{2} t} |f_j(x - t\overline{\rho}\omega, \omega)| \, dt\, dx\, d\omega$$

$$\leq 2 e^{-\frac{\lambda}{2} c} \int_{T^d \times S^{d-1}} \int_c^{+\infty} e^{-\frac{\lambda}{2} t} |f_j(x - t\overline{\rho}\omega, \omega)| \, dt\, dx\, d\omega$$

$$\leq \frac{4 e^{-\frac{\lambda}{2} c}}{\lambda} \|f_j\|_{L^1(T^d \times S^{d-1})} = \frac{4 e^{-\frac{\lambda}{2} c}}{\lambda}$$

so

$$\frac{1}{2\varepsilon} \int_{\overline{\rho}-\varepsilon}^{\overline{\rho}+\varepsilon} d\rho \int_{T^d \times S^{d-1}} \int_c^{+\infty} e^{-\lambda t} \left| e^{-\int_0^t \sigma(x - s\rho\omega, \rho, \omega)ds} - e^{-\int_0^t \sigma(x - s\overline{\rho}\omega, \overline{\rho}, \omega)ds} \right| |f_j(x - t\overline{\rho}\omega, \omega)| \, dt\, dx\, d\omega$$

$$\leq \frac{4 e^{-\frac{\lambda}{2} c}}{\lambda}$$

is as small as we want (uniformly in j) if c is large enough. On the other hand

$$\int_{T^d \times S^{d-1}} \int_0^c e^{-\lambda t} \left| e^{-\int_0^t \sigma(x - s\rho\omega, \rho, \omega)ds} - e^{-\int_0^t \sigma(x - s\overline{\rho}\omega, \overline{\rho}, \omega)ds} \right| |f_j(x - t\overline{\rho}\omega, \omega)| \, dt\, dx\, d\omega$$

$$\leq \sup_{\substack{\rho \in [\overline{\rho}-\varepsilon, \overline{\rho}+\varepsilon] \\ s \leq t \leq c \\ \omega \in S^{d-1}}} \left| e^{-\int_0^t \sigma(x - s\rho\omega, \rho, \omega)ds} - e^{-\int_0^t \sigma(x - s\overline{\rho}\omega, \overline{\rho}, \omega)ds} \right| \frac{\|f_j\|_{L^1(T^d \times S^{d-1})}}{\lambda}$$

$$= \frac{1}{\lambda} \sup_{\substack{\rho \in [\overline{\rho}-\varepsilon, \overline{\rho}+\varepsilon] \\ s \leq t \leq c \\ \omega \in S^{d-1}}} \left| e^{-\int_0^t \sigma(x - s\rho\omega, \rho, \omega)ds} - e^{-\int_0^t \sigma(x - s\overline{\rho}\omega, \overline{\rho}, \omega)ds} \right|$$

so

$$\frac{1}{2\varepsilon} \int_{\overline{\rho}-\varepsilon}^{\overline{\rho}+\varepsilon} d\rho \int_{T^d \times S^{d-1}} \left| \int_0^c e^{-\lambda t} \left(e^{-\int_0^t \sigma(x - s\rho\omega, \rho, \omega)ds} - e^{-\int_0^t \sigma(x - s\overline{\rho}\omega, \overline{\rho}, \omega)ds} \right) f_j(x - t\overline{\rho}\omega, \omega)dt \right| dx\, d\omega$$

$$\leq \frac{1}{\lambda} \sup_{\substack{\rho \in [\overline{\rho}-\varepsilon, \overline{\rho}+\varepsilon] \\ s \leq t \leq c \\ \omega \in S^{d-1}}} \left| e^{-\int_0^t \sigma(x - s\rho\omega, \rho, \omega)ds} - e^{-\int_0^t \sigma(x - s\overline{\rho}\omega, \overline{\rho}, \omega)ds} \right|$$

is as small as we want (uniformly in j) if ε is small enough. We observe that

$$\frac{1}{2\varepsilon} \int_{\overline{\rho}-\varepsilon}^{\overline{\rho}+\varepsilon} d\rho \int_{T^d \times S^{d-1}} \left| \int_0^{+\infty} e^{-\lambda t} e^{-\int_0^t \sigma(x - s\overline{\rho}\omega, \overline{\rho}, \omega)ds} f_j(x - t\overline{\rho}\omega, \omega)dt - \frac{1}{\lambda - \eta_{\overline{\rho}}} f_j(x, \omega) \right| dx\, d\omega$$

$$= \int_{T^d \times S^{d-1}} \left| \int_0^{+\infty} e^{-\lambda t} e^{-\int_0^t \sigma(x - s\overline{\rho}\omega, \overline{\rho}, \omega)ds} f_j(x - t\overline{\rho}\omega, \omega)dt - \frac{1}{\lambda - \eta_{\overline{\rho}}} f_j(x, \omega) \right| dx\, d\omega$$

$$= \left\| (\lambda - T_{\overline{\rho}})^{-1} f_j - \frac{1}{\lambda - \eta_{\overline{\rho}}} f_j \right\|_{L^1(T^d \times S^{d-1})} \to 0 \text{ as } j \to +\infty$$

so, by fixing c large enough, one sees that

$$\left\| (\lambda - T)^{-1} \psi_j - \frac{1}{\lambda - \eta_{\bar{\rho}}} \psi_j \right\|_{L^1(\mathcal{T}^d \times [a,b] \times S^{d-1})} \to 0 \text{ as } j \to +\infty$$

and $\frac{1}{\lambda - \eta_{\bar{\rho}}}$ is an approximate eigenvalue of $(\lambda - T)^{-1}$ or equivalently $\eta_{\bar{\rho}}$ is an approximate eigenvalue of T. Thus $s(T_\rho) \in \sigma(T) \quad \forall \rho \in [a, b]$ and the closedness of the spectrum show that $\sup_{\rho \in [a,b]} s(T_\rho) \in \sigma(T)$. This shows our first claim

$$\sup_{\rho \in [a,b]} s(T_\rho) \leq s(T).$$

Let now

$$\lambda > \gamma := \sup_{\rho \in [a,b]} s(T_\rho)$$

be arbitrary and

$$c := \frac{1}{2} (\lambda - \gamma).$$

Let us show that $\lambda \notin \sigma(T) \quad \forall \lambda > \gamma$. This will imply that $s(T) \leq \gamma$ and ends the proof. Consider the equation in $L^1(\mathcal{T}^d \times [a, b] \times S^{d-1})$

$$\lambda \varphi - T \varphi = g.$$

For each $\rho \in [a, b]$, we define the functions of $L^1(\mathcal{T}^d \times S^{d-1})$

$$g_\rho(x, \omega) := g(x, \rho, \omega), \quad \varphi_\rho(x, \omega) := \varphi(x, \rho, \omega).$$

Since $\lambda \notin \sigma(T_\rho)$ for all $\rho \in [a, b]$ then

$$\varphi_\rho = (\lambda - T_\rho)^{-1} g_\rho$$

and

$$\|\varphi_\rho\|_{L^1(\mathcal{T}^d \times S^{d-1})}$$
$$\leq \left\| (\lambda - T_\rho)^{-1} \right\|_{\mathcal{L}(L^1(\mathcal{T}^d \times S^{d-1}))} \|\varphi_\rho\|_{L^1(\mathcal{T}^d \times S^{d-1})}.$$

It suffices to show that

$$\sup_{\rho \in [a,b]} \left\| (\lambda - T_\rho)^{-1} \right\|_{\mathcal{L}(L^1(\mathcal{T}^d \times S^{d-1}))} < +\infty \quad \forall \lambda > \sup_{\rho \in [a,b]} s(T_\rho) \qquad (10.14.5)$$

since

$$\|\varphi\|_{L^1(\mathcal{T}^d \times [a,b] \times S^{d-1})}$$

$$= \int_a^b \|\varphi_\rho\|_{L^1(\mathcal{T}^d \times S^{d-1})} \, d\rho$$

$$\leq \left(\sup_{\rho \in [a,b]} \|(\lambda - T_\rho)^{-1}\|_{\mathcal{L}(L^1(\mathcal{T}^d \times S^{d-1}))} \right) \int_a^b \|g_\rho\|_{L^1(\mathcal{T}^d \times S^{d-1})} \, d\rho$$

$$= \left(\sup_{\rho \in [a,b]} \|(\lambda - T_\rho)^{-1}\|_{\mathcal{L}(L^1(\mathcal{T}^d \times S^{d-1}))} \right) \|g\|_{L^1(\mathcal{T}^d \times [a,b] \times S^{d-1})} \cdot$$

Consider the strong integral

$$(\lambda - T_\rho)^{-1} = \int_0^{+\infty} e^{-\lambda t} U_\rho(t) dt.$$

Since

$$\eta(U_\rho) = \lim_{t \to +\infty} \ln \|U_\rho(t)\|$$

$$= - \lim_{t \to +\infty} \left(\inf t^{-1} \int_0^t \sigma(x + s\rho\omega, \rho, \omega) ds \right)$$

$$= \lim_{t \to +\infty} \left(\sup_{(x,\omega) \in \mathcal{T}^d \times S^{d-1}} -t^{-1} \int_0^t \sigma(x + s\rho\omega, \rho, \omega) ds \right)$$

uniformly in $\rho \in [a, b]$ then there exists $\bar{t} > 0$ such that for $t \geq \bar{t}$

$$\sup_{(x,\omega) \in \mathcal{T}^d \times S^{d-1}} -t^{-1} \int_0^t \sigma(x + s\rho\omega, \rho, \omega) ds \leq \eta(U_\rho) + \frac{c}{2} \; \forall \rho,$$

i.e.,

$$- t^{-1} \int_0^t \sigma(x + s\rho\omega, \rho, \omega) ds \leq \eta(U_\rho) + \frac{c}{2} \; \forall (x, \rho, \omega).$$

Then

$$\left| U_\rho(t)\varphi \right| = e^{-\int_0^t \sigma(x - s\rho\omega, \rho, \omega) ds} \, |\varphi(x - t\rho\omega, \omega)|$$

$$\leq e^{\left(\eta(U_\rho) + \frac{c}{2}\right)t} \, |\varphi(x - t\rho\omega, \omega)|$$

and

$$e^{-\lambda t} \left| U_\rho(t)\varphi \right| \leq e^{-\lambda t} e^{\left(\eta(U_\rho) + \frac{c}{2}\right)t} \, |\varphi(x - t\rho\omega, \omega)|$$

$$= e^{-\left(\lambda - \eta(U_\rho) - \frac{c}{2}\right)t} \, |\varphi(x - t\rho\omega, \omega)|$$

$$\leq e^{-\frac{c}{2}t} \, |\varphi(x - t\rho\omega, \omega)| \quad \forall (x, \rho, \omega)$$

so

$$e^{-\lambda t}\left\|U_\rho(t)\right\| \le e^{-\frac{c}{2}t} \quad \forall t \ge \bar{t}.$$

Since $\left\|U_\rho(t)\right\| \le 1 \;\forall t \ge 0$ then

$$e^{-\lambda t}\left\|U_\rho(t)\right\| \le \begin{array}{l} 1 \text{ if } t \in \left[0,\bar{t}\right] \\ e^{-\frac{c}{2}t} \;\; \forall t \ge \bar{t} \end{array}$$

and consequently

$$\left\|(\lambda - T_\rho)^{-1}\right\| \le \int_0^{+\infty} e^{-\lambda t}\left\|U_\rho(t)\right\| dt$$
$$\le \int_0^1 e^{-\lambda t} dt + \int_1^{+\infty} e^{-\frac{c}{2}t} dt \;\; \forall \rho$$

which shows (10.14.5). $\square$

10.14.2 Weak Compactness Results

Consider the restriction to $L^1(S^{n-1})$ of the elastic collision operator, i.e.,

$$L^1(S^{d-1}) \ni \varphi \to \int_{S^{n-1}} c_e(x,\rho,\omega,\omega')\varphi(\omega')d\omega' \in L^1(S^{d-1}).$$

This is a family of operators

$$C_e(x,\rho) : L^1(S^{d-1}) \to L^1(S^{d-1})$$

indexed by $(x,\rho) \in \mathcal{T}^d \times [a,b]$. Similarly, the restriction to $L^1([a,b] \times S^{d-1})$ of the inelastic collision operator is a family of operators

$$C_i(x) : L^1([a,b] \times S^{d-1}) \to L^1([a,b] \times S^{d-1})$$

(indexed by $x \in \mathcal{T}^d$) where

$$(C_i(x)\varphi)(\rho,\omega) = \int_a^b \int_{S^{d-1}} c_i(x,\rho,\omega,\rho',\omega')\varphi(\rho',\omega')d\rho'd\omega'.$$

Definition 10.14.1 We say that C_e is a regular elastic collision operator if the family

$$\left\{C_e(x,\rho);\; (x,\rho) \in \mathcal{T}^d \times [a,b]\right\}$$

is collectively weakly compact on $L^1(S^{d-1})$. This means that the image of the unit ball of $L^1(S^{d-1})$ by $C_e(x, \rho)$ is included in a weakly compact set independent of $(x, \rho) \in T^d \times [a, b]$. (For example, an elastic collision operator with bounded kernel is regular.)

Definition 10.14.2 We say that C_i is a regular inelastic collision operator if the family

$$\{C_i(x); \ x \in T^d\}$$

is collectively weakly compact on $L^1([a, b] \times S^{d-1})$. This means that the image of the unit ball of $L^1([a, b] \times S^{d-1})$ by $C_i(x)$ is included in a weakly compact set independent of $x \in T^d$. (For example, an inelastic collision operator with bounded kernel is regular.)

Proposition 10.14.1 *Let C_e and C_i be regular. Then there exists a sequence of elastic collision operators $(C_e^j)_j$ converging in operator norm to C_e and with kernels $c_e^j(x, \rho, \omega, \omega')$ such that*

$$c_e^j(x, \rho, \omega, \omega') \le f_e^j(\omega) \qquad (10.14.6)$$

where $f_e^j \in L^1(S^{d-1})$. Similarly, there exists a sequence of inelastic collision operators $(C_i^j)_j$ converging in operator norm to C_i whose kernels $c_i^j(x, \rho, \omega, \rho', \omega')$ are such that

$$c_i^j(x, \rho, \omega, \rho', \omega') \le f_i^j(\rho, \omega) \qquad (10.14.7)$$

where $f_i^j \in L^1([a, b] \times S^{d-1})$.

Proof It is a simple application of Proposition 10.8.1.   □

We give now some key weak compactness results.

Lemma 10.14.2 *Suppose that C_e and C_i are regular. Then:*
(i) For all $\rho \in [a, b]$

$$C_e^\rho(\lambda - T_\rho)^{-1} C_e^\rho \ \ (\lambda > 0)$$

is a weakly compact operator on $L^1(T^d \times S^{d-1})$.
(ii) $C_i U^e(t) C_i \ (t > 0)$ is a weakly compact operator on $L^1(T^d \times [a, b] \times S^{d-1})$.

Proof By approximation (Proposition 10.14.1) it suffices to deal with collision operators with kernels satisfying (10.14.6) (10.14.7). By a domination argument, we may even replace these kernels respectively by f_e^j and f_i^j. By approximation again, we may suppose that f_e^j and f_i^j are continuous with compact supports. We end up with the case

$$c_e(x, \rho, \omega, \omega') = f_e(\omega), \quad c_i(x, \rho, \omega, \rho', \omega') = f_i(\rho, \omega)$$

where f_e and f_i are continuous with compact supports. Since

$$U(t)\varphi \to e^{-\int_0^t \sigma(x - s\rho\omega, \rho, \omega)ds} \varphi(x - t\rho\omega, \rho, \omega) \le \varphi(x - t\rho\omega, \omega) \ \ (t \ge 0)$$

for nonnegative φ, we may assume that $\sigma(.) = 0$ and finally we may replace f_e and f_i by some positive constants.

(i) Let $\rho \in [a, b]$ be fixed. If $\varphi \in L^1(T^d \times S^{d-1})$ then, up to a multiplicative constant, $C_e^\rho \varphi$ is equal to

$$\int_{S^{n-1}} \varphi(x, \omega')d\omega' = \tilde{\varphi}(x)$$

so it suffices to show that

$$L^1(T^d) \ni \psi \to \int_{S^{d-1}} \int_0^{+\infty} e^{-\lambda t}\varphi(x - t\rho\omega)d\omega dt \in L^1(T^d)$$

is weakly compact. On the other hand

$$\int_{S^{d-1}} \int_0^{+\infty} e^{-\lambda t}\varphi(x - t\rho\omega)d\omega dt$$
$$= \frac{1}{\rho} \int_{S^{d-1}} \int_0^{+\infty} e^{-\frac{\lambda}{\rho}\tau}\varphi(x - \tau\omega)d\omega d\tau$$

and the change $y = x - \tau\omega$ shows that

$$dy = \tau^{d-1}d\tau d\omega = |x - y|^{d-1} d\tau d\omega$$

and

$$\int_{S^{d-1}} \int_0^{+\infty} e^{-\lambda t}\varphi(x - t\rho\omega, \rho)d\omega dt$$
$$= \frac{1}{\rho} \int_{\mathbb{R}^d} \frac{e^{-\frac{\lambda}{\rho}|x-y|}}{|x - y|^{d-1}}\varphi(y).$$

Note that this is a convolution in the whole space $\mathbb{R}^d$ and φ is $[0, 2\pi]^d$-periodic and

$$x \in [0, 2\pi]^d .$$

Kolmogorov's criterion of compactness on $L^1([0, 2\pi]^d)$ shows that this operator is compact on $L^1(T^d)$ and ends the proof.

(ii) Let us show first that $C_i U(t)C_i$ $(t > 0)$ is weakly compact on $L^1(T^d \times [a, b] \times S^{d-1})$. Up to a multiplicative constant, $C_i\varphi$ is equal to

$$\int_a^b \int_{S^{d-1}} \varphi(x, \rho', \omega')d\rho'd\omega' = \tilde{\varphi}(x)$$

so it suffices to show that

$$L^1(T^d) \ni \psi \to \int_a^b \int_{S^{d-1}} \psi(x - t\rho\omega)d\rho d\omega \in L^1(T^d)$$

is weakly compact. Note that

$$\int_a^b \int_{S^{d-1}} \psi(x - t\rho\omega)d\rho d\omega = \frac{1}{t}\int_{ta}^{tb} \int_{S^{d-1}} \psi(x - \tau\omega)d\tau d\omega$$

so the change $y = x - \tau\omega$ and

$$dy = \tau^{d-1}d\tau d\omega = |x - y|^{d-1}\, d\tau d\omega$$

show that

$$\int_a^b \int_{S^{d-1}} \psi(x - t\rho\omega)d\rho d\omega = \int_{ta \le |x-y| \le tb} \psi(y)\frac{dy}{|x - y|^{d-1}}$$

where

$$x \in [0, 2\pi]^d .$$

Note that

$$y \in \Xi_x := \{y;\, ta \le |x - y| \le tb\}$$

and

$$\Xi := \cup_{x \in [0,2\pi]^d}\, \Xi_x \text{ is bounded.}$$

Hence this operator is dominated by

$$\psi \in L^1(\Xi) \to \int_\Xi \frac{\psi(y)}{|x - y|^{d-1}}dy \in L^1([0, 2\pi]^d)$$

which can be shown to be compact by Kolmogorov's criterion of compactness.
We recall that

$$U^e(t) = U(t) + \sum_{k=1}^\infty U_k(t)$$

where (see [245, Lemma 2.1, p. 14])

$$U_k(.) = [U(.)C_e]^k * U(.)$$

and

$$[U(.)C_e]^k = [U(.)C_e] * [U(.)C_e] * \dots * [U(.)C_e]$$

(k times) where

$$f * g = \int_0^t f(t-s)g(s)ds$$

for any strongly continuous $f : \mathbb{R}_+ \ni t \to f(t) \in \mathcal{L}(X)$ and $g : \mathbb{R}_+ \ni t \to g(t) \in \mathcal{L}(X)$. It follows that

$$C_i U^e(t) C_i = C_i U(t) C_i + \sum_{k=1}^{\infty} C_i U_k(t) C_i$$

where the series converges in operator norm. We already know that $C_i U(t) C_i$ ($t > 0$) is a weakly compact operator. Consider

$$C_i U_k(t) C_i \varphi = C_i \left[[U(.)C_e]^k * U(.) \right] C_i \varphi, \quad (k \geq 1).$$

Since

$$C_i \varphi = \int_a^b \int_{S^{d-1}} \varphi(x, \rho', \omega') d\rho' d\omega' = \widetilde{\varphi}(.) \in L^1(\mathcal{T}^d)$$

it suffices to show that the restriction of $C_i \left[[U(.)C_e]^k * U(.) \right]$ to $L^1(\mathcal{T}^d)$ is a weakly compact operator. Note that $[U(.)C_e]^k * U(.)$ is a purely elastic operator (in the sense that ρ is just a parameter) and identifies to

$$\left[U_\rho(.)C_e^\rho \right]^k * U_\rho(.).$$

Since $C_e \psi = \int_{S^{d-1}} \psi(x, \rho', \omega') d\omega'$ and

$$C_i \psi(x, \rho, \omega) = \int_a^b \int_{S^{d-1}} \psi(x, \rho', \omega') d\rho' d\omega'$$
$$= \int_a^b d\rho' \left(\int_{S^{d-1}} \psi(x, \rho', \omega') d\omega' \right) = \int_a^b (C_e \psi) \, d\rho'$$

then $C_i \left[[U(.)C_e]^k * U(.) \right]$ identifies to a strong integral with respect to the parameter $\rho \in [a, b]$ of

$$C_e^\rho \left[U_\rho(.)C_e^\rho \right]^k * U_\rho(.).$$

Thanks to Theorem 2.9.1, it suffices to show (for a fixed $\rho > 0$) that the restriction of

$$C_e^\rho \left[U_\rho(.)C_e^\rho \right]^k * U_\rho(.)$$

to $L^1(\mathcal{T}^d)$ is weakly compact. Notice that

$$C_e^\rho \left[U_\rho(.)C_e^\rho \right]^k * U_\rho(.) = \left[C_e^\rho U_\rho(.) \right]^{k+1} \quad (k \geq 1).$$

Since

$$\left[C_e^\rho U_\rho(.)\right]^{k+1} = \left[C_e^\rho U_\rho(.)\right]^{k-1} * \left[C_e^\rho U_\rho(.)C_e^\rho\right] * U_\rho(.)$$

then, by appealing again to Theorem 2.9.1 (now it is a strong integral in time), it suffices to show that the restriction of

$$\left[C_e^\rho U_\rho(.)C_e^\rho\right] * U_\rho(.)$$

to $L^1(T^d)$ is weakly compact. This last operator is nothing but

$$L^1(T^d) \ni \varphi \to \int_{S^{d-1}} \left(\int_0^t U_\rho(t-s)C_e^\rho U_\rho(s)\varphi ds\right) d\omega \in L^1(T^d).$$

Since

$$\int_\varepsilon^{t-\varepsilon} U_\rho(t-s)C_e^\rho U_\rho(s)ds \to \int_0^t U_\rho(t-s)C_e^\rho U_\rho(s)ds \text{ as } \varepsilon \to 0$$

in operator norm then it suffices to deal with

$$L^1(T^d) \ni \varphi \to \int_\varepsilon^{t-\varepsilon} \left[\int_{S^{d-1}} \left(U_\rho(t-s)C_e^\rho U_\rho(s)\varphi\right) d\omega\right] ds \in L^1(T^d).$$

The explicit action of this operator on $\varphi \in L^1(T^d)$ is

$$\int_\varepsilon^{t-\varepsilon} \left[\int_{S^{d-1}} \int_{S^{d-1}} \varphi(x - (t-s)\rho\omega - s\rho\omega')d\omega'd\omega\right] ds.$$

By appealing again to Theorem 2.9.1, it suffices to show (for each $0 < s < t$) that

$$O_{s,t,\rho} : L^1(T^d) \to L^1(T^d)$$

defined by

$$L^1(T^d) \ni \varphi \to \int_{S^{d-1}} \int_{S^{d-1}} \varphi(x - (t-s)\rho\omega - s\rho\omega')d\omega'd\omega \in L^1(T^d)$$

is a weakly compact operator.

Consider now the (rotationally invariant) Radon measures with compact supports

$$\mu_{s,\rho} : C_c(\mathbb{R}^d) \ni f \to \int_{S^{d-1}} f(s\rho\omega')d\omega'.$$

One sees that

$$\mu_{s,\rho} * \mu_{(t-s),\rho} : f \to \int_{S^{d-1} \times S^{d-1}} f((t-s)\rho\omega + s\rho\omega')d\omega'd\omega$$

so that

$$\int_{S^{d-1}} \int_{S^{d-1}} f(x - (t-s)\rho\omega - s\rho\omega')d\omega'd\omega$$

is the convolution of f with the measure $\mu_{s,\rho} * \mu_{(t-s),\rho}$. On the other hand, by ([310, Theorem 2.6]) the compactly supported measure

$$\mu_{s,\rho} * \mu_{(t-s),\rho}$$

is absolutely continuous with respect to Lebesgue measure on $\mathbb{R}^d$, i.e., there exists $h \in L^1(\mathbb{R}^d)$ (with a compact support) such that

$$\int_{S^{d-1} \times S^{d-1}} f(x - (t-s)\rho\omega - s\rho\omega')d\omega'd\omega = \int_{\mathbb{R}^d} f(y)h(x-y)dy.$$

By using again Kolmogorov's criterion of compactness, one can show that

$$O_{s,t,\rho} : L^1(\mathcal{T}^d) \to L^1(\mathcal{T}^d)$$

is a compact operator. $\square$

10.14.3   Spectral Bound of Elastic Generators

For each $\rho \in [a, b]$ define

$$C_e^\rho : L^1(\mathcal{T}^d \times S^{d-1}) \to L^1(\mathcal{T}^d \times S^{d-1})$$

$$C_e^\rho \varphi = \int_{S^{d-1}} k_e(x, \rho, \omega, \omega')\varphi(x, \omega')d\omega'$$

and

$$T_\rho + C_e^\rho : D(T_\rho) \subset L^1(\mathcal{T}^d \times S^{d-1}) \to L^1(\mathcal{T}^d \times S^{d-1}).$$

According to Lemma 10.14.2 $\left[(\lambda - T_\rho)^{-1}C_e^\rho\right]^2$ is weakly compact on $L^1(\mathcal{T}^d \times S^{d-1})$ and then $\left[(\lambda - T_\rho)^{-1}C_e^\rho\right]^4$ is compact on $L^1(\mathcal{T}^d \times S^{d-1})$. It follows from Theorem 2.8.1 that the set

$$\sigma(T_\rho + C_e^\rho) \cap \left\{\operatorname{Re}\lambda > s(T_\rho)\right\}$$

consists at most of isolated eigenvalues. This set is either empty, in which case

$$s(T_\rho + C_e^\rho) = s(T_\rho),$$

or not empty, in which case

$$s(T_\rho + C_e^\rho) > s(T_\rho),$$

and $s(T_\rho + C_e^\rho)$ is an isolated leading eigenvalue of $T_\rho + C_e^\rho$.

Lemma 10.14.3 *Suppose that C_e and C_i are regular. We have*

$$\sup_{\rho \in [a,b]} s(T_\rho + C_e^\rho) \le s(T + C_e).$$

If

$$[a, b] \ni \rho \to (\lambda - T_\rho)^{-1} C_e^\rho \in \mathcal{L}(L^1(\mathcal{T}^d \times S^{d-1})) \tag{10.14.8}$$

is continuous in operator norm then

$$s(T + C_e) = \sup_{\rho \in [a,b]} s(T_\rho + C_e^\rho).$$

Proof By performing similar calculations as in the proof of Lemma 10.14.1 one sees that

$$s(T_\rho + C_e^\rho) \in \sigma(T + C_e) \ \ \forall \rho \in [a, b]$$

and then by a closedness argument

$$\sup_{\rho \in [a,b]} s(T_\rho + C_e^\rho) \in \sigma(T + C_e).$$

Let us show that any $\lambda > \sup_{\rho \in [a,b]} s(T_\rho + C_e^\rho)$ belongs to the resolvent set of $T + C_e$. To this end, consider the equation in $L^1(\mathcal{T}^d \times (a, b) \times S^{d-1})$

$$\lambda \varphi - T \varphi - C_e \varphi = g.$$

As previously, we first solve in $L^1(\mathcal{T}^d \times S^{d-1})$ the equation

$$\lambda \varphi_\rho - T_\rho \varphi - C_e^\rho \varphi_\rho = g_\rho$$

by

$$\varphi_\rho = (\lambda - T_\rho - C_e^\rho)^{-1} g_\rho$$

since $\lambda > \sup_{\rho \in [a,b]} s(T_\rho + C_e^\rho)$. Since

$$\left\| \varphi_\rho \right\|_{L^1(\mathcal{T}^d \times S^{d-1})}$$
$$\le \left\| (\lambda - T_\rho - C_e^\rho)^{-1} \right\|_{\mathcal{L}(L^1(\mathcal{T}^d \times S^{d-1}))} \left\| g_\rho \right\|_{L^1(\mathcal{T}^d \times S^{d-1})}$$

then as previously it suffices to show that

$$\sup_{\rho\in[a,b]} \left\|(\lambda - T_\rho - C_e^\rho)^{-1}\right\|_{\mathcal{L}(L^1(T^d\times S^{d-1}))} < +\infty.$$

This amounts to showing that if χ lives in a bounded subset of $L^1(T^d \times S^{d-1})$ then the norm of the solution $\psi \in L^1(T^d \times S^{d-1})$ of

$$\lambda\psi - T_\rho\psi - C_e^\rho\psi = \chi$$

is uniformly bounded in $\rho \in [a, b]$. Let us argue by contradiction. Thus there exists a sequence $(\rho_j)_j \subset [a, b]$ such that $\rho_j \to \bar\rho$ and a bounded sequence $(\chi_j)_j \subset L^1(T^d \times S^{d-1})$ such that

$$\left\|\psi_j\right\|_{L^1(T^d\times S^{d-1})} \to +\infty$$

where

$$\lambda\psi_j - T_{\rho_j}\psi_j - C_e^{\rho_j}\psi_j = \chi_j.$$

Note that

$$\psi_j - (\lambda - T_{\rho_j})^{-1}C_e^{\rho_j}\psi_j = (\lambda - T_{\rho_j})^{-1}\chi_j$$

and

$$\frac{\psi_j}{\|\psi_j\|} - (\lambda - T_{\rho_j})^{-1}C_e^{\rho_j}\frac{\psi_j}{\|\psi_j\|} = \frac{(\lambda - T_{\rho_j})^{-1}\chi_j}{\|\psi_j\|} \to 0$$

in norm. Thus

$$\widehat{\psi_j} - (\lambda - T_{\rho_j})^{-1}C_e^{\rho_j}\widehat{\psi_j} \to 0 \qquad (10.14.9)$$

where $\widehat{\psi_j} = \frac{\psi_j}{\|\psi_j\|}$ has a unit norm. We note that

$$(\lambda - T_{\rho_j})^{-1}C_e^{\rho_j} \to (\lambda - T_{\bar\rho})^{-1}C_e^{\bar\rho}$$

in operator norm and $\left((\lambda - T_\rho)^{-1}C_e^\rho\right)^4$ is compact so the family of operators

$$\left\{\left((\lambda - T_{\rho_j})^{-1}C_e^{\rho_j}\right)^4, \ j \in \mathbb{N}\right\}$$

is collectively compact and consequently

$$\left((\lambda - T_{\rho_j})^{-1}C_e^{\rho_j}\right)^3\widehat{\psi_j} - \left((\lambda - T_{\rho_j})^{-1}C_e^{\rho_j}\right)^4\widehat{\psi_j} \to 0$$

shows that $\left\{\left((\lambda - T_{\rho_j})^{-1}C_e^{\rho_j}\right)^3\widehat{\psi_j}\right\}_j$ is relatively compact in $L^1(T^d \times S^{d-1})$. Similarly, it follows that $\left\{\left((\lambda - T_{\rho_j})^{-1}C_e^{\rho_j}\right)^2\widehat{\psi_j}\right\}_j$ is relatively compact in $L^1(T^d \times$

S^{d-1}) and then so is the sequence $\left\{(\lambda - T_{\rho_j})^{-1} C_e^{\rho_j} \widehat{\psi_j}\right\}_j$. Then (10.14.9) shows that $\left\{\widehat{\psi_j}\right\}_j$ is relatively compact in $L^1(T^d \times S^{d-1})$ so that a subsequence converges in norm to some normalized function $\psi \in L^1(T^d \times S^{d-1})$. By passing to the limit in (10.14.9) we get

$$(\lambda - T_{\bar{\rho}})^{-1} C_e^{\bar{\rho}} \psi = \psi, \quad \|\psi\| = 1.$$

This shows that λ is an eigenvalue of $T_{\bar{\rho}} + C_e^{\bar{\rho}}$ and contradicts the fact that $\lambda > s(T_\rho + C_e^\rho) \ \forall \rho \in [a, b]$. $\qquad \square$

A simple calculation shows that

$$(\lambda - T_\rho)^{-1}\varphi = \frac{1}{\rho}\int_0^{+\infty} e^{-\frac{\lambda}{\rho}\tau} e^{-\frac{1}{\rho}\left(\lambda\tau + \int_0^\tau \sigma(x - s'\omega, \rho, \omega)ds'\right)} \varphi(x - \tau\omega, \omega)d\tau$$

and

$$[a, b] \ni \rho \rightarrow (\lambda - T_\rho)^{-1} \in \mathcal{L}(L^1(T^d \times S^{d-1}))$$

is continuous in operator norm if $\sigma(x, \rho, \omega)$ depends continuously in ρ (uniformly in (x, ω)). Thus (10.14.8) is satisfied e.g. if $k_e(x, \rho, \omega, \omega')$ depends continuously in ρ (uniformly in (x, ω, ω')).

10.14.4   The Conservativity

Lemma 10.14.4 *If*

$$\int_{(a,b)\times S^{d-1}} c_i(x, \rho, \omega, \rho', \omega')d\rho d\omega + \int_{S^{d-1}} c_e(x, \rho', \omega, \omega')d\omega = \sigma(x, \rho', \omega')$$

then $(V(t))_{t\geq 0}$ *is a stochastic* C_0*-semigroup.*

Proof Let

$$\widehat{c_i}(x, \rho', \omega') := \int_{(a,b)\times S^{d-1}} c_i(x, \rho, \omega, \rho', \omega')d\rho d\omega$$

and

$$\widehat{c_e}(x, \rho', \omega') := \int_{S^{d-1}} c_e(x, \rho', \omega, \omega')d\omega.$$

Let $\varphi \in D(T) \cap L^1_+(T^d \times (a, b) \times S^{d-1})$. Then

$$\int_{T^d \times (a,b) \times S^{d-1}} T\varphi$$

$$= -\int_{T^d \times (a,b) \times S^{d-1}} \sigma(x, \rho, \omega)\varphi(x, \rho, \omega)\,dx\,d\rho\,d\omega.$$

On the other hand

$$\int_{(a,b) \times S^{d-1}} C_e\varphi\,d\rho\,d\omega$$

$$= \int_{(a,b) \times S^{d-1}} \left[\int_{S^{d-1}} c_e(x, \rho, \omega, \omega')\varphi(x, \rho, \omega')\,d\omega' \right] d\rho\,d\omega$$

$$= \int_{(a,b) \times S^{d-1}} \left[\int_{S^{d-1}} c_e(x, \rho, \omega, \omega')\,d\omega \right] \varphi(x, \rho, \omega')\,d\rho\,d\omega'$$

$$= \int_{(a,b) \times S^{d-1}} \left[\int_{S^{d-1}} c_e(x, \rho', \omega, \omega')\,d\omega \right] \varphi(x, \rho', \omega')\,d\rho'\,d\omega'$$

and

$$\int_a^b \int_{S^{d-1}} c_i(x, \rho, \omega, \rho', \omega')\varphi(x, \rho', \omega', t)\,dv'\,d\omega'$$

is given by

$$\int_{(a,b) \times S^{d-1}} C_i\varphi\,d\rho\,d\omega$$

$$= \int_{(a,b) \times S^{d-1}} \left[\int_a^b \int_{S^{d-1}} c_i(x, \rho, \omega, \rho', \omega')\varphi(x, \rho', \omega')\,dv'\,d\omega' \right] d\rho\,d\omega$$

$$= \int_a^b \int_{S^{d-1}} \left[\int_{(a,b) \times S^{d-1}} c_i(x, \rho, \omega, \rho', \omega')\,d\rho\,d\omega \right] \varphi(x, \rho', \omega')\,dv'\,d\omega'.$$

Hence

$$\int_{(a,b) \times S^{d-1}} C_e\varphi\,d\rho\,d\omega + \int_{(a,b) \times S^{d-1}} C_i\varphi\,d\rho\,d\omega$$

is equal to

$$\int_a^b \int_{S^{d-1}} \left[\widehat{k_i}(x, \rho', \omega') + \widehat{k_e}(x, \rho', \omega') \right] \varphi(x, \rho', \omega')\,d\rho'\,d\omega'.$$

Finally

$$\int_{T^d \times (a,b) \times S^{d-1}} T\varphi + K_e\varphi + K_i\varphi,$$

i.e.,

$$\int_{T^d} \left[\int_a^b \int_{S^{d-1}} \left(\widehat{c_i}(x, \rho', \omega') + \widehat{c_e}(x, \rho', \omega') - \sigma(x, \rho', \omega') \right) \varphi(x, \rho', \omega') d\rho' d\omega' \right] dx$$

is equal to 0. This ends the proof. $\square$

10.14.5 Essential Type of Partly Elastic Semigroups

We give here a key (spectral) stability result inspired by Sbihi's paper [324] (and also [245, Chap. 2]).

Theorem 10.14.1 *Suppose that C_e and C_i are regular. Then $(U^e(t))_{t \geq 0}$ and $(V(t))_{t \geq 0}$ have the same essential type.*

Proof We recall hat

$$V(t) = U^e(t) + \sum_{k=1}^{\infty} U_k(t)$$

where (see [245, Lemma 2.1, p. 14])

$$U_k(.) = \left[U^e(.)C_i \right]^k * U^e(.)$$

and $[U^e(.)C_i]^k$ is given by

$$\left[U^e(.)C_i \right]^k = \left[U^e(.)C_i \right] * \left[U^e(.)C_i \right] * \ldots * \left[U^e(.)C_i \right]$$

(k times). Since $C_i U^e(.)C_i$ is weakly compact on $L^1(T^d \times [a, b] \times S^{d-1})$ (Lemma 10.14.2 (ii)) then so is

$$\begin{aligned}
\left[U^e(.)C_i \right]^2 &= \left[U^e(.)C_i \right] * \left[U^e(.)C_i \right] \\
&= \left[U^e(.) \right] * \left[C_i U^e(.)C_i \right]
\end{aligned}$$

(Theorem 2.9.1) and this is equivalent to weak compactness of the second order remainder $\sum_{k=2}^{\infty} U_k(t)$, (see [245, Theorem 2.6, p. 16]). Finally, by Theorem 2.8.2, $(U^e(t))_{t \geq 0}$ and $(V(t))_{t \geq 0}$ have the same essential type. $\square$

10.14.6 Existence of a Spectral Gap

We are ready to show a key result of this section.

Theorem 10.14.2 *Suppose that C_e and C_i are regular and that (10.14.8) is satisfied. Let*

$$c_e(x, \rho', \omega, \omega') > 0 \ \ a.e. \tag{10.14.10}$$

and let the conservativity condition

$$\sigma(x, \rho', \omega') = \int_{(a,b) \times S^{d-1}} c_i(x, \rho, \omega, \rho', \omega') d\rho d\omega + \int_{S^{d-1}} c_e(x, \rho', \omega, \omega') d\omega$$

be satisfied. Suppose that $c_i(x, \rho, \omega, \rho', \omega')$ is not identically zero. If $s(T) < 0$ (or equivalently if (10.14.3) is satisfied) then $(V(t))_{t \geq 0}$ is a stochastic semigroup with a spectral gap.

Proof We recall first that

$$s(T_\rho) \leq s(T) \ \ \forall \rho \in [a, b].$$

According to Lemma 10.14.4 $(V(t))_{t \geq 0}$ is a stochastic semigroup so that its type (or equivalently the spectral bound of its generator) is equal to zero

$$s(T + C_e + C_i) = 0.$$

According to Lemma 10.14.2 some power of $(\lambda - T_\rho)^{-1} C_e^\rho$ is compact in $L^1(T^d \times S^{d-1})$ so (see Theorem 2.8.1)

$$\sigma(T_\rho + C_e^\rho) \cap \{ \operatorname{Re} \lambda > s(T_\rho) \}$$

consists at most of isolated eigenvalues so that either this set is empty in which case

$$s(T_\rho + C_e^\rho) = s(T_\rho) \leq s(T) < 0$$

or this set is not empty and then

$$s(T_\rho + C_e^\rho) > s(T_\rho)$$

and $s(T_\rho + C_e^\rho)$ is an isolated leading eigenvalue of $T_\rho + C_e^\rho$ associated to a nonnegative eigenfuntion φ

$$T_\rho \varphi + C_e^\rho \varphi = s(T_\rho + C_e^\rho) \varphi.$$

It follows from (10.14.10) that $\varphi(x, \omega) > 0$ a.e. Since $c_i(x, \rho, \omega, \rho', \omega')$ is not identically zero then

$$- \sigma(x, \rho, \omega) + \int_{S^{d-1}} c_e(x, \rho, \omega', \omega)d\omega < 0$$

on a set of positive measure. By integration,

$$\int_{T^d \times S^{d-1}} \left[-\sigma(x, \rho, \omega) + \int_{S^{d-1}} c_e(x, \rho, \omega', \omega)d\omega \right] \varphi(x, \omega)dxd\omega$$
$$= s(T_\rho + C_e^\rho) \int_{T^d \times S^{d-1}} \varphi(x, \omega)dxd\omega$$

shows that

$$s(T_\rho + C_e^\rho) \int_{T^d \times S^{d-1}} \varphi(x, \omega)dxd\omega < 0$$

or equivalently $s(T_\rho + C_e^\rho) < 0$. Hence

$$s(T_\rho + C_e^\rho) < 0 \; \forall \rho \in [a, b]. \tag{10.14.11}$$

Let us show that

$$\sup_{\rho \in [a,b]} s(T_\rho + C_e^\rho) < 0.$$

Arguing by contradiction, assume that there exists a sequence $(\rho_k)_k \subset [a, b]$ such that $\rho_k \to \overline{\rho}$ and

$$s(T_{\rho_k} + C_e^{\rho_k}) \to 0.$$

In particular, for k large enough

$$s(T_{\rho_k} + C_e^{\rho_k}) > s(T) \geq s(T_\rho) \; \forall \rho \in [a, b]$$

and then $s(T_{\rho_k} + C_e^{\rho_k})$ is an isolated leading eigenvalue of $T_{\rho_k} + C_e^{\rho_k}$ associated to a nonnegative eigenfuntion φ_k

$$T_{\rho_k} \varphi_k + C_e^{\rho_k} \varphi_k = s_k \varphi_k$$

where

$$s_k := s(T_{\rho_k} + C_e^{\rho_k}) \to 0.$$

Thus

$$\varphi_k = (s_k - T_{\rho_k})^{-1} C_e^{\rho_k} \varphi_k.$$

By using compactness arguments as in the proof of Lemma 10.14.3 one can pass to the limit and get

$$\varphi = (0 - T_{\overline{\rho}})^{-1} C_e^{\overline{\rho}} \varphi$$

for some nontrivial $\varphi \geq 0$. This shows that

$$s(T_{\bar{p}} + C_e^{\bar{\rho}}) = 0$$

which contradicts (10.14.11). Thus, by Lemma 10.14.3

$$s(T + C_e) < 0.$$

Since

$$s(T + C_e + C_i) = \omega(V) = 0$$

and $(U^e(t))_{t \geq 0}$ and $(V(t))_{t \geq 0}$ have the same essential type (see Theorem 10.14.1) then

$$\omega_{ess}(V) = \omega_{ess}(U^e) \leq \omega(U^e) = s(T + C_e) < s(T + C_e + C_i) = \omega(V)$$

and we are done. $\qquad\square$

10.14.7 *Absence of Spectral Gaps and Strong Convergence*

Note that if (10.14.3) is not satisfied then

$$s(T + C_e) = s(T + C_e + C_i) = 0$$

and we have no spectral gap. The second key result is:

Theorem 10.14.3 *Suppose that (10.14.3) is not satisfied and that $(V(t))_{t \geq 0}$ is irreducible. If $(V(t))_{t \geq 0}$ has an invariant density then it is asymptotically stable.*

Proof We have seen in the proof of Theorem 10.14.1 that

$$R_2(t) = \sum_{j=2}^{+\infty} U_j(t)$$

is a weakly compact operator and consequently $R_2(t)$ is an integral operator (see [115]. Finally $\left(e^{t(T+C_e+C_i)}\right)_{t \geq 0}$ is partially integral and we conclude with Theorem 2.10.3. We could also conclude with a different argument; indeed, the weak compactness of $R_2(t)$ implies the compactness of $R_5(t)$ (see [245, Corollary 2.1, p.16]) and consequently

$$\mathbb{R}_+ \ni t \to R_6(t)$$

is continuous in operator norm (see [245, Theorem 2.7, p. 18 and Corollary 2.2, p. 19]). Finally Theorem 2.10.8 ends the proof. $\qquad\square$

Note that under the symmetry assumptions

$$c_i(x, \rho, \omega, \rho', \omega') = c_i(x, \rho', \omega', \rho, \omega), \quad c_e(x, \rho', \omega, \omega') = c_e(x, \rho', \omega', \omega)$$

$$(10.14.12)$$

0 is an eigenvalue of $T + C_e + C_i$ associated to the constant eigenfunction 1 and then $(V(t))_{t \geq 0}$ is asymptotically stable.

10.15   Neutron Transport with Non Local Boundary Operators

In the previous sections, we considered neutron transport semigroups with classical vacuum boundary condition or on the d-dimensional torus. In this section, we explore them for more complex boundary conditions. Let $D \subset \mathbb{R}^d$ $(d \geq 2)$ be a bounded open subset with a smooth (say $\mathcal{C}^1$) boundary ∂D. We denote by $n(x)$ the outward unit norm at $x \in \partial D$. Let

$$\begin{cases} \Gamma_+ = \left\{ (x, \zeta) \in \partial D \times \mathbb{R}^d, \ \zeta.n(x) > 0 \right\} \\ \Gamma_- = \left\{ (x, \zeta) \in \partial D \times \mathbb{R}^d, \ \zeta.n(x) < 0 \right\}. \end{cases}$$

Let $d\zeta$ be the usual Lebesgue measure on $\mathbb{R}^d$ and let

$$\begin{cases} L^1(\Gamma_+) := L^1(\Gamma_+; \ d\mu_+) \\ L^1(\Gamma_-) := L^1(\Gamma_-; \ d\mu_-) \end{cases}$$

endowed with the measures

$$\begin{cases} d\mu_+ = \zeta.n(x)d\zeta d\sigma(x) \\ d\mu_- = |\zeta.n(x)| \, d\zeta d\sigma(x) \end{cases}$$

where $d\sigma(x)$ is the Lebesgue surface measure on ∂D. Note that

$$\|f\|_{L^1(\Gamma_+)} = \int_{\Gamma_+} |f(x, \zeta)| \, \zeta.n(x)d\zeta d\sigma(x)$$

$$= \int_{\partial D} \left(\int_{\Gamma_+(x)} |f(x, \zeta)| \, \zeta.n(x)d\zeta \right) d\sigma(x)$$

where $\Gamma_+(x)$ is the half space of velocities (indexed by $x \in \partial D$)

$$\Gamma_+(x) := \left\{ \zeta \in \mathbb{R}^d; \ \zeta.n(x) > 0 \right\}$$

endowed with the measure $\zeta.n(x)d\zeta$. So we may view $f \in L^1(\Gamma_+)$ as a field

$$x \in \partial D \to f(x,.) \in L^1(\Gamma_+(x))$$

where

$$\|f\|_{L^1(\Gamma_+)} = \int_{\partial D} \|f(x,.)\|_{L^1(\Gamma_+(x))} \, d\sigma(x).$$

Similarly, if $\Gamma_-(x)$ is the half space of velocities (indexed by $x \in \partial D$)

$$\Gamma_-(x) := \left\{ \zeta \in \mathbb{R}^d; \ \zeta.n(x) < 0 \right\}$$

then any $f \in L^1(\Gamma_-)$ is identified to a field

$$x \in \partial D \to f(x,.) \in L^1(\Gamma_-(x))$$

and

$$\|f\|_{L^1(\Gamma_-)} = \int_{\partial D} \|f(x,.)\|_{L^1(\Gamma_-(x))} \, d\sigma(x). \tag{10.15.1}$$

There exists an important spectral literature on neutron transport equations with a boundary operator in one space dimension

$$d = 1.$$

Without claiming to be complete, we refer e.g. to [68, 87, 198, 208, 340] and to the numerous references therein. As far as we know, apart from bounce back boundary condition [197], the passage to *higher* space dimensions is open because of the complexity of the analysis of compactness problems behind spectral theory for non zero boundary operators. We provide here the first sytematic peripheral spectral analysis of neutron transport semigroups, in multi-dimensional space geometry, with *stochastic diffuse* boundary operators, (see the details below).

In a completely different direction, we point out the existence of a general result (in higher space dimensions) of strong convergence to equilibrium for linear conservative Boltzmann equations (even with external forces) with stochastic boundary operators provided that a detailed balance holds inside the domain D and also at the boundary ∂D (relative to the boundary operator) [304]; in this case, the invariant density (i.e., the equilibrium) is given for free and the strong convergence relies on relative entropy techniques. Our goal here is to give various spectral results with time asymptotics in higher space dimensions without appealing to any detailed balance condition.

We recall that there exists a trace theory on $\Gamma_\pm$ for the Sobolev space

$$W^{(1)} = \left\{ \varphi \in L^1(D \times \mathbb{R}^d), \ \zeta.\nabla_x f \in L^1(D \times \mathbb{R}^d) \right\}$$

(see [82, 83, 365]; see also [23, 24, 41] and references therein for more general vector fields). In particular, [365] contains a thorough analysis of multidimensional collisionless transport equation

$$\frac{\partial f}{\partial t} + \zeta.\nabla_x f = 0$$

in

$$L^1(D \times \mathbb{R}^d) := L^1\left(D \times \mathbb{R}^d; dx\nu(d\zeta)\right)$$

(for general velocity measure $\nu(d\zeta)$) with (sub)stochastic boundary operators

$$\mathcal{B} : L^1(\Gamma_+) \to L^1(\Gamma_-)$$

relating the incoming and outgoing fluxes f_{Γ_-} and f_{Γ_+} (the traces of f on Γ_- and Γ_+), i.e.,

$$f_{\Gamma_-} = \mathcal{B}(f_{\Gamma_+}).$$

Let
$$\begin{cases} T_\mathcal{B} : \left\{\varphi \in W^{(1)}, \; f_{\Gamma_-} = \mathcal{B}(f_{\Gamma_+})\right\} \to L^1(D \times \mathbb{R}^d) \\ \qquad\quad T_\mathcal{B} f = -\zeta.\nabla_x f. \end{cases} \qquad (10.15.2)$$

We point out that if $d \geq 2$ (and $\|\mathcal{B}\| = 1$) then $T_\mathcal{B}$ need not be closed and even its closure need not be a generator. However, there exists an extension A of $T_\mathcal{B}$

$$A \supset T_\mathcal{B}$$

which generates a minimal substochastic C_0-semigroup $(U_\mathcal{B}(t))_{t \geq 0}$ on $L^1(D \times \mathbb{R}^d)$.
We recall that if

$$\tau(x, \zeta) := \inf \{s > 0, \; x - s\zeta \notin D\}$$

then $M_0 : L^1(\Gamma_-) \to L^1(\Gamma_+)$ defined by

$$M_0 u(x, \zeta) = u(x - \tau(x, \zeta)\zeta, \zeta) \quad (x, \zeta) \in \Gamma_+$$

is a stochastic operator [23]. A honesty theory for $(U_\mathcal{B}(t))_{t \geq 0}$, in the spirit of honesty for additive perturbations (see Remark 2.11.2) is given in [259] for stochastic boundary operators $\mathcal{B}$ (see also [21]). We do not present this theory here but just mention that if there exists $u \in L^1(\Gamma_+)$, $u > 0$ a.e. such that

$$M_0 \mathcal{B} u \leq u$$

then $\overline{T_\mathcal{B}}$ generates a stochastic C_0-semigroup $(U_\mathcal{B}(t))_{t \geq 0}$, (see [259] Theorem 21). In particular, since

$$M_0 \mathcal{B} : L^1(\Gamma_+) \to L^1(\Gamma_+)$$

is stochastic then its spectral radius $r(M_0\mathcal{B})$ is equal to one and $\overline{T_\mathcal{B}}$ generates a stochastic C_0-semigroup provided that 1 is an eigenvalue of $M_0\mathcal{B}$. This strongly suggests that the spectral nature of $r(M_0\mathcal{B})$ should be a key point.

More recent developments on collisionless transport equations with stochastic boundary operators $\mathcal{B}$ are given in [212]. In particular if there exists $u \in L^1(\Gamma_+)$ such that $M_0 \mathcal{B} u = u$ and

$$\int_{\{\zeta, \zeta.n(x)>0\}} \frac{u(x, \zeta)}{|\zeta|} \, \zeta.n(x)d\zeta < +\infty$$

then $(U_\mathcal{B}(t))_{t \geq 0}$ has an invariant density which is asymptotically stable [212]. Much more precise results are given for boundary operators which are convex combinations of a stochastic diffuse part

$$\mathcal{D} : L^1(\Gamma_+) \to L^1(\Gamma_-)$$

given by

$$\mathcal{D}u(x, \zeta) = \int_{\{\zeta', \, \zeta'.n(x)>0\}} k(x, \zeta, \zeta')u(x, \zeta') \, \zeta'.n(x)d\zeta', \quad (x, \zeta) \in \Gamma_-$$

and a stochastic (non diffuse) deterministic reflexion operator $\mathcal{R}$

$$\mathcal{B} = \beta(.)\mathcal{D} + (1 - \beta(.))\mathcal{R}$$

where

$$0 \leq \beta(x) \leq 1 \ \ (x \in \partial D).$$

The main assumption on the diffuse part $\mathcal{D}$ is that for each $x \in \partial D$ the integral operator with respect velocities

$$u \in L^1(\Gamma_+(x)) \to \int_{\Gamma_+(x)} k(x, \zeta, \zeta')u(\zeta') \, \zeta'.n(x)d\zeta' \in L^1(\Gamma_-(x))$$

is weakly compact and the weak compactness of this family of operators (indexed by $x \in \partial D$) is collective in a suitable sense, see [212]. Such boundary operators are called *regular* diffusive operators by analogy with regular (additive) collision operators. A key result is the weak compactness of

$$\mathcal{D}M_0\mathcal{D} : L^1(\Gamma_+) \to L^1(\Gamma_-)$$

for any diffusive operator $\mathcal{D}$ (see [212, Theorem 5.1]).

Remark 10.15.1 We note that if $\mathcal{D}_1$ and $\mathcal{D}_2$ are two different *regular* diffusive operators then

$$\mathcal{D}_1 M_0 \mathcal{D}_2 : L^1(\Gamma_+) \to L^1(\Gamma_-)$$

(dominated by $(\mathcal{D}_1 + \mathcal{D}_2)M_0(\mathcal{D}_1 + \mathcal{D}_2)$) is also weakly compact.

One shows that

$$r_{ess}(M_0\mathcal{B}) < 1$$

(in particular 1 is an eigenvalue of $M_0\mathcal{B}$) provided that

$$\inf \beta > 1 + \beta_\infty - \sqrt{1 + \beta_\infty^2} \tag{10.15.3}$$

where $\beta_\infty = \sup \beta$, (see [212, Theorem 5.6]). (The condition (10.15.3) is of course satisfied e.g. if β is a positive constant.) In this case, $T_\mathcal{B}$ itself is the generator of $(U_\mathcal{B}(t))_{t\geq 0}$ and its resolvent is given by

$$(\lambda - T_\mathcal{B})^{-1} f = (\lambda - T_0)^{-1} f + \sum_{j=0}^{+\infty} \Xi_\lambda \mathcal{B} \, (M_\lambda \mathcal{B})^j \, G_\lambda \, (\lambda > 0) \tag{10.15.4}$$

where $(\lambda - T_0)^{-1}$ is the (free) resolvent corresponding to $\mathcal{B} = 0$

$$(\lambda - T_0)^{-1} f = \int_0^{\tau(x,\zeta)} e^{-\lambda t} f(x - t\zeta, \zeta)dt,$$

$M_\lambda : L^1(\Gamma_-) \to L^1(\Gamma_+)$ is given by

$$M_\lambda u(x, \zeta) = u(x - \tau(x, \zeta)\zeta, \zeta)e^{-\lambda\tau(x,\zeta)} \quad (x, \zeta) \in \Gamma_+$$

$G_\lambda : L^1(D \times \mathbb{R}^d) \to L^1(\Gamma_+)$ is given by

$$G_\lambda f(x, \zeta) = \int_0^{\tau(x,v)} e^{-\lambda t} f(x - t\zeta, \zeta)dt, \quad (x, \zeta) \in \Gamma_+$$

and $\Xi_\lambda : L^1(\Gamma_-) \to L^1(D \times \mathbb{R}^d)$ by

$$\Xi_\lambda f(x, \zeta) = e^{-\lambda\tau(x,\zeta)} f(x - \tau(x, \zeta)\zeta, \zeta).$$

We point out that the series

$$\sum_{j=0}^{+\infty} \Xi_\lambda \mathcal{B} \, (M_\lambda \mathcal{B})^j \, G_\lambda$$

converges in operator norm because $r_\sigma(M_\lambda\mathcal{B}) < 1 \;\; (\lambda > 0)$; we refer to [212] for the details. If the boundary operator $\mathcal{B}$ reduces to its diffuse part $\mathcal{D}$ (i.e., $\mathcal{B} = \mathcal{D}$) and if the set of velocities ζ is not all $\mathbb{R}^d$ but is bounded away from zero

$$\{\zeta \in \mathbb{R}^d; \; |\zeta| \geq v_{\min} > 0 \} \tag{10.15.5}$$

then $(U_\mathcal{D}(t))_{t\geq 0}$ is eventually compact, i.e., $U_\mathcal{D}(t)$ is compact for t large enough [214].

In this section, we consider the full linear Boltzmann equation

$$\frac{\partial f}{\partial t} + \zeta.\nabla_x f + \Sigma(x,\zeta)\varphi(t,x,\zeta) = \int_{\mathbb{R}^d} c(x,\zeta,\zeta')\varphi(t,x,\zeta')d\zeta'$$

with purely diffuse stochastic boundary operator $(\mathcal{B} = \mathcal{D})$

$$f_{\Gamma_-} = \mathcal{D}(f_{\Gamma_+})$$

(i.e., $\beta = 1$) where

$$\Sigma(x,\zeta) = \int_{\mathbb{R}^d} c(x,\zeta',\zeta)d\zeta'.$$

Some of the results are true even when $\mathcal{B} \neq \mathcal{D}$. For the sake of simplicity, we restrict ourselves to

$$\Sigma(.,.) \in L^\infty(D \times \mathbb{R}^d);$$

in this case,

$$C : f \in L^1(D \times \mathbb{R}^d) \to \int_{\mathbb{R}^d} c(x,\zeta,\zeta')f(x,\zeta')d\zeta'$$

is a bounded operator on $L^1(D \times \mathbb{R}^d)$. Let $\left(U_\mathcal{B}^\Sigma(t)\right)_{t\geq 0}$ be the C_0-semigroup on $L^1(D \times \mathbb{R}^d)$ generated by

$$\begin{cases} T_\mathcal{B}^\Sigma = T_\mathcal{B} - \Sigma \\ D(T_\mathcal{B}^\Sigma) = D(T_\mathcal{B}) \end{cases}$$

and let $(V_\mathcal{B}(t))_{t\geq 0}$ be the C_0-semigroup on $L^1(D \times \mathbb{R}^d)$ generated by

$$\begin{cases} A_\mathcal{B} := T_\mathcal{B} - \Sigma + C \\ D(A_\mathcal{B}) = D(T_\mathcal{B}). \end{cases}$$

Since

$$\int_{D\times\mathbb{R}^d} (-\Sigma f + Cf) = 0$$

and $\int_{D\times\mathbb{R}^d} T_\mathcal{B} f = 0$ for $f \in D(T_\mathcal{B})$ $((U_\mathcal{B}(t))_{t\geq 0}$ is stochastic) then $(V_\mathcal{B}(t))_{t\geq 0}$ is a stochastic C_0-semigroup. A simple calculation shows that the resolvent of $T_\mathcal{B}^\Sigma$ is given by

$$\left(\lambda - T_\mathcal{B}^\Sigma\right)^{-1} f = \left(\lambda - T_0^\Sigma\right)^{-1} f + \sum_{j=0}^{+\infty} \Xi_\lambda^\Sigma \mathcal{B} \left(M_\lambda^\Sigma \mathcal{B}\right)^j G_\lambda^\Sigma \quad (\lambda > 0) \quad (10.15.6)$$

where

$$\left(\lambda - T_0^{\Sigma}\right)^{-1} f = \int_0^{\tau(x,\zeta)} e^{-\lambda t} e^{-\int_0^t \Sigma(x-s\zeta,\zeta)ds} f(x - t\zeta, \zeta)dt ,$$

$M_\lambda^{\Sigma} : L^1(\Gamma_-) \to L^1(\Gamma_+)$ is given by

$$M_\lambda^{\Sigma} u(x, \zeta) = u(x - \tau(x, \zeta)\zeta, \zeta) e^{-\int_0^{\tau(x,\zeta)} \Sigma(x-s\zeta,\zeta)ds} e^{-\lambda \tau(x,\zeta)} \quad (x, \zeta) \in \Gamma_+$$

$G_\lambda^{\Sigma} : L^1(D \times \mathbb{R}^d) \to L^1(\Gamma_+)$ is given by

$$G_\lambda^{\Sigma} f(x, \zeta) = \int_0^{\tau(x,\zeta)} e^{-\lambda t} e^{-\int_0^t \Sigma(x-s\zeta,\zeta)ds} f(x - t\zeta, \zeta)dt, \quad (x, \zeta) \in \Gamma_+.$$

and $\Xi_\lambda^{\Sigma} : L^1(\Gamma_-) \to L^1(D \times \mathbb{R}^d)$ by

$$\Xi_\lambda^{\Sigma} f(x, \zeta) = e^{-\lambda \tau(x,\zeta)} e^{-\int_0^{\tau(x,\zeta)} \Sigma(x-s\zeta,\zeta)ds} f(x - \tau(x, \zeta)\zeta, \zeta).$$

Note that $\left(U_0^{\Sigma}(t)\right)_{t \geq 0}$ (resp. T_0^{Σ}) were denoted previously $(U_\Sigma(t))_{t \geq 0}$ (resp. T_Σ). We recall a useful formula

$$\int_{D \times \mathbb{R}^d} h(x, \zeta)dxd\zeta = \int_{\Gamma_-} d\zeta d\sigma(z) \, |n(z).\zeta| \int_0^{\tau_+(z,\zeta)} h(z + s\zeta, \zeta)ds, \quad (10.15.7)$$

(for nonnegative measurable functions h) where

$$\tau_+(y, \zeta) := \inf \{s > 0, \ y + s\zeta \notin D\} = \tau(y, -\zeta).$$

(see [23]). A natural parameter in our construction is the spectral bound $s(T_0^{\Sigma})$ or equivalently the type of the advection semigroup $\left(U_0^{\Sigma}(t)\right)_{t \geq 0}$. We recall (see (10.8.8)) that

$$s(T_0^{\Sigma}) = - \lim_{t \to +\infty} \inf_{\{(x,\zeta), \ \tau(x,\zeta)>t\}} t^{-1} \int_0^t \Sigma(x - s\zeta, \zeta)ds \leq 0.$$

One of the main key preliminary results is the hereditary subcriticality

$$s(T_0^{\Sigma}) < 0 \Rightarrow s(T_{\mathcal{B}}^{\Sigma}) < 0$$

(see Lemma 10.15.1). In the case

$$s(T_0^{\Sigma}) < 0,$$

we show that $\left(\lambda - T_{\mathcal{D}}^{\Sigma}\right)^{-1} C \ (\lambda > s(T_{\mathcal{D}}^{\Sigma}))$ is power compact in $L^1(D \times \mathbb{R}^d)$ (see Lemma 10.15.3). This implies that 0 is an (isolated leading) eigenvalue of $T_{\mathcal{B}}^{\Sigma}$ and the stochastic semigroup $(V_{\mathcal{D}}(t))_{t \geq 0}$ has an invariant density and is asymptotically stable (see Theorem 10.15.1); we conjecture that $(V_{\mathcal{B}}(t))_{t\geq 0}$ has spectral gap but we have not proved it. The critical case

$$s(T_0^{\Sigma}) = 0$$

is much more involved and requires additional assumptions on the boundary operator $\mathcal{D}$ and the collision operator C. Besides numerous technical preliminary results, we show that the strong limit

$$\left(0 - T_{\mathcal{D}}^{\Sigma}\right)^{-1} C := s \lim_{\lambda \to 0} \left(\lambda - T_{\mathcal{D}}^{\Sigma}\right)^{-1} C$$

exists and is power compact in $L^1(D \times \mathbb{R}^d)$ (see Lemma 10.15.6) from which we deduce that $(V_{\mathcal{D}}(t))_{t\geq 0}$ has an invariant density and is asymptotically stable (see Theorem 10.15.2).

Remark 10.15.2 We point out that the spectral bound $s(T_0^{\Sigma})$ (or equivalently the type of the corresponding advection semigroup $\left(U_0^{\Sigma}(t)\right)_{t\geq 0}$) is the same in all L^p spaces [367]. Moreover, since the spectral bound is stable by duality then we have also

$$s(T_0^{\Sigma}) = - \lim_{t \to +\infty} \inf_{\{(x,\zeta),\, \tau_+(x,\zeta)>t\}} t^{-1} \int_0^t \Sigma(x + s\zeta, \zeta)ds.$$

This section is organized as follows

10.15.1 Hereditary Subcriticality

We start with a key result where $\mathcal{B}$ need not be completely diffuse.

Lemma 10.15.1 *Let k be the kernel of $\mathcal{D}$. Suppose that $k(x, \zeta, \zeta') > 0$ a.e.[2] If $s(T_0^{\Sigma}) < 0$ then $s(T_{\mathcal{B}}^{\Sigma}) < 0$.*

Proof We already know that $s(T_{\mathcal{B}}^{\Sigma}) \leq 0$ since $\left(U_{\mathcal{B}}^{\Sigma}(t)\right)_{t\geq 0}$ is substochastic. It suffices to show that 0 is not in the spectrum of $T_{\mathcal{B}}^{\Sigma}$. By assumption, 0 is not in the spectrum of T_0^{Σ}.

Step 1. Let us check that

$$h(x, \zeta) := \Xi_0^{\Sigma} f(x, \zeta) = e^{-\int_0^{\tau(x,\zeta)} \Sigma(x-s'\zeta,\zeta)ds'} f(x - \tau(x, \zeta)\zeta, \zeta)$$

belongs to $L^1(D \times \mathbb{R}^d)$. It suffices to consider $f \in L_+^1(\Gamma_-)$. According to (10.15.7) $\int_{D\times\mathbb{R}^d} h(x, \zeta)dxd\zeta$ is given by

---

[2] This assumption could be relaxed to some extent.

$$\int_{\Gamma_-} d\zeta d\sigma(z)\, |n(z).\zeta| \int_0^{\tau_+(z,\zeta)} h(z+s\zeta,\zeta)ds$$

$$= \int_{\Gamma_-} d\zeta d\sigma(z)\, |n(z).\zeta| \times$$

$$\int_0^{\tau_+(z,\zeta)} e^{-\int_0^{\tau(z+s\zeta,\zeta)} \Sigma(z+s\zeta-s'\zeta,\zeta)ds'} f(z+s\zeta-\tau(z+s\zeta,\zeta)\zeta,\zeta)ds$$

Notice that if $(z,\zeta) \in \Gamma_-$ then

$$z + s\zeta - \tau(z+s\zeta,\zeta)\zeta = z$$

so

$$f(z + s\zeta - \tau(z+s\zeta,\zeta)\zeta,\zeta) = f(z,\zeta)$$

and $\int_{D\times\mathbb{R}^d} h(x,\zeta)dxd\zeta$ is equal to

$$\int_{\Gamma_-} d\zeta d\sigma(z)\, |n(z).\zeta| \times \int_0^{\tau_+(z,\zeta)} e^{-\int_0^s \Sigma(z+(s-s')\zeta,\zeta)ds'} f(z,\zeta)ds$$

$$= \int_{\Gamma_-} d\zeta d\sigma(x)\, |n(z).\zeta|\, f(z,\zeta) \int_0^{\tau_+(z,\zeta)} e^{-\int_0^s \Sigma(z+s'\zeta,\zeta)ds'} ds$$

It suffices to check that the function

$$\int_0^{\tau_+(z,\zeta)} e^{-\int_0^s \Sigma(z+s'\zeta,\zeta)ds'} ds$$

is bounded. Consider this function in the regions

$$\{\tau_+(z,\zeta) \le c\} \ \text{ and } \ \{\tau_+(z,\zeta) > c\}$$

where c will be fixed later. In the first region

$$\int_0^{\tau_+(z,\zeta)} e^{-\int_0^s \Sigma(z+s'\zeta,\zeta)ds'} ds \le c.$$

To deal with $\{\tau_+(z,\zeta) > c\}$ we note that

$$\int_0^s \Sigma(z+s'\zeta,\zeta)ds' = s\left(s^{-1}\int_0^s \Sigma(z+s'\zeta,\zeta)ds'\right)$$

and (see Remark 10.15.2)

$$\lim_{s\to+\infty} \inf_{\{(x,\zeta),\ \tau_+(x,\zeta)>s\}} s^{-1}\int_0^s \Sigma(x+s'\zeta,\zeta)ds' = -s(T_0^\Sigma) > 0$$

so there exists $c > 0$ large enough such that

$$\inf_{\{(x,\zeta),\ \tau_+(x,\zeta)>s\}} s^{-1} \int_0^s \Sigma(x + s'\zeta, \zeta)ds' \geq -\frac{s(T_0^\Sigma)}{2} > 0 \ \forall s \geq c.$$

Hence

$$s^{-1} \int_0^s \Sigma(z - s'\zeta, \zeta)ds' \geq -\frac{s(T_0^\Sigma)}{2} > 0 \quad (s \geq c)$$

in the region $\{(z, \zeta),\ \tau_+(z, \zeta) > s\}$ and

$$\int_0^{\tau_+(z,\zeta)} e^{-\int_0^s \Sigma(z+s'\zeta,\zeta)ds'} ds$$

$$= \int_0^c e^{-\int_0^s \Sigma(z+s'\zeta,\zeta)ds'} ds + \int_c^{\tau_+(z,\zeta)} e^{-\int_0^s \Sigma(z+s'\zeta,\zeta)ds'} ds$$

$$\leq c + \int_c^{\tau_+(z,\zeta)} e^{-(-\frac{s(T_0^\Sigma)}{2})s} ds \leq c + \int_c^{+\infty} e^{-(-\frac{s(T_0^\Sigma)}{2})s} ds$$

ends the proof of step 1.

Step 2. $r\left(M_0^\Sigma \mathcal{B}\right) < 1$.

We know (see [212]) that

$$M_0\mathcal{B} : L^1(\Gamma_+) \rightarrow L^1(\Gamma_+)$$

is a stochastic operator and

$$M_0^\Sigma \mathcal{B} \leq M_0\mathcal{B}. \tag{10.15.8}$$

Note that $\mathcal{B} = \mathcal{D} + \mathcal{R}$. Under the condition (10.15.3) we know (by [212]) that

$$r_{ess}(M_0\mathcal{B}) < r(M_0\mathcal{B}) = 1.$$

By Theorem 2.4.1

$$r_{ess}\left(M_0^\Sigma \mathcal{B}\right) \leq r_{ess}(M_0\mathcal{B})$$

so

$$r_{ess}\left(M_0^\Sigma \mathcal{B}\right) < 1.$$

If $r\left(M_0^\Sigma \mathcal{B}\right) = 1$ then $r\left(M_0^\Sigma \mathcal{B}\right)$ is Riesz point of $M_0^\Sigma \mathcal{B}$ and consequently, by Proposition 2.4.1,

$$r\left(M_0^\Sigma \mathcal{B}\right) < r(M_0\mathcal{B}) = 1$$

(because $M_0\mathcal{B} \geq M_0\mathcal{D}$ is irreducible) and we get a contradiction; therefore

$$r\left(M_0^\Sigma \mathcal{B}\right) < 1. \tag{10.15.9}$$

(We point out that (10.15.9) does *not* rely on $s(T_0^\Sigma) < 0$, i.e., it is true even if $s(T_0^\Sigma) = 0$.)

Step 3. Since $r\left(M_0^\Sigma \mathcal{B}\right) < 1$ then (10.15.6) extends to $\lambda = 0$, i.e., 0 is not in the spectrum of $T_\mathcal{B}^\Sigma$ and

$$\left(0 - T_\mathcal{B}^\Sigma\right)^{-1} f = \left(0 - T_0^\Sigma\right)^{-1} f + \sum_{j=0}^{+\infty} \Xi_0^\Sigma \mathcal{B} \left(M_0^\Sigma \mathcal{B}\right)^j G_0^\Sigma f.$$

$\square$

10.15.2   The Subcritical Case $s(T_0^\Sigma) < 0$

In this subsection, we need to restrict ourselves to purely diffuse boundary operators. A general class of diffuse boundary operators, called regular, was introduced in [212]. Roughly speaking these operators (which are local in $x \in \partial D$) are weakly compact in velocity ζ where the weak compactness is collective (in a suitable sense) in $x \in \partial D$. These boundary operators enjoy a nice approximation property (see [212, Lemma 3.7]):

There exists a sequence of diffuse boundary operators $\mathcal{D}_n$ converging to $\mathcal{D}$ in operator norm such that each $\mathcal{D}_n$ is dominated by a diffuse boundary operator of the form

$$u \in L^1(\Gamma_+) \to q(\zeta) \int_{\{\zeta',\ \zeta'.n(x)>0\}} u(x, \zeta') \, \zeta'.n(x) d\zeta' \qquad (10.15.10)$$

where $q \geq 0$ is a continuous function with compact support in a spherical shell

$$\{c_1 \leq |\zeta| \leq c_2\} \ (c_i > 0).$$

We need two key preliminary results.

Lemma 10.15.2 *Suppose that ∂D is of class $C^{1,\alpha}$ for some $\alpha > 0$. Let the boundary operator be diffuse (i.e., $\mathcal{B} = \mathcal{D}$) and regular and let the collision operator C be regular. If $s(T_0^\Sigma) < 0$ then*

$$\sum_{j=0}^{+\infty} \Xi_0^\Sigma \mathcal{D} \left(M_0^\Sigma \mathcal{D}\right)^j G_0^\Sigma C$$

is weakly compact on $L^1(D \times \mathbb{R}^d)$.

Proof The series above converges in operator norm so it suffices to show that the operators

$$\Xi_0^\Sigma \mathcal{D} \left(M_0^\Sigma \mathcal{D}\right)^j G_0^\Sigma C \ (j \geq 0)$$

are weakly compact. To deal with $j = 1$ note that

$$\Xi_0^\Sigma \mathcal{D} M_0^\Sigma \mathcal{D} G_0^\Sigma C$$

is weakly compact because $\mathcal{D} M_0^\Sigma \mathcal{D} \le \mathcal{D} M_0 \mathcal{D}$ and we know that $\mathcal{D} M_0 \mathcal{D}$ is weakly compact. It follows that the weak compactness property is true for all $j \ge 1$. It suffices to consider the case $j = 0$

$$\Xi_0^\Sigma \mathcal{D} G_0^\Sigma C$$

or simply

$$\mathcal{D} G_0^\Sigma C : L^1(D \times \mathbb{R}^d) \to L^1(\Gamma_-).$$

By approximation (in operator norm) we can restrict ourselves to a boundary operator of the form (10.15.10) and to a collision operator

$$C : f \in L^1(D \times \mathbb{R}^d) \to \widehat{c}(\zeta) \int_{\mathbb{R}^d} f(x, \zeta') d\zeta'$$

where $\widehat{c} \ge 0$ is continuous function with compact support in a spherical shell

$$\{c_1 \le |\zeta| \le c_2\} \ (c_i > 0).$$

In this case, by factorization, it suffices to deal with

$$\varphi \in L^1(D) \to q(\zeta) \int_{\{\zeta', \ \zeta'.n(x)>0\}} \left(\widehat{c}(\zeta') \int_0^{\tau(x,\zeta')} \varphi(x - t\zeta') dt \right) \zeta'.n(x) d\zeta'.$$

Note that $\tau(x, \zeta') \le \frac{\widehat{d}}{|\zeta|} \le \frac{\widehat{d}}{c_1}$ where $\widehat{d}$ is the diameter of D. By extending φ by zero outside D the operator is dominated by

$$\varphi \in L^1(D) \to q(\zeta) \int_{\mathbb{R}^d} \left(\widehat{c}(\zeta') \int_0^{\frac{\widehat{d}}{c_1}} \varphi(x - t\zeta') dt \right) |\zeta'.n(x)| d\zeta'.$$

Now the change $x - t\zeta' = y$ gives

$$\int_{\mathbb{R}^d} \left(\widehat{c}(\zeta') \int_0^{\frac{\widehat{d}}{c_1}} \varphi(x - t\zeta') dt \right) |\zeta'.n(x)| d\zeta'$$

$$= \int_0^{\frac{\widehat{d}}{c_1}} \left[\int_{\mathbb{R}^d} \widehat{c}(\frac{x - y}{t}) \varphi(y) \ |(x - y).n(x)| \frac{dy}{t^{d+1}} \right] dt$$

$$= \int_D \left(\int_0^{\frac{\widehat{d}}{c_1}} \widehat{c}(\frac{x - y}{t}) \frac{dt}{t^{d+1}} \right) |x - y| \, \varphi(y) \left| \left(\frac{x - y}{|x - y|} \right).n(x) \right| dy.$$

Since $|\zeta| \leq c_2$ then $\frac{|x-y|}{t} \leq c_2$ implies

$$\int_0^{\frac{\widehat{d}}{c_1}} \widehat{c}(\frac{x-y}{t}) \frac{dt}{t^{d+1}} \leq \int_{\frac{|x-y|}{c_2}}^{\frac{\widehat{d}}{c_1}} \widehat{c}(\frac{x-y}{t}) \frac{dt}{t^{d+1}}$$

$$\leq d \, \|\widehat{c}\|_\infty \left[\frac{1}{t^d}\right]_{\frac{\widehat{d}}{c_1}}^{\frac{|x-y|}{c_2}}$$

$$\leq \frac{c}{|x-y|^d}$$

for a positive constant c. Hence

$$\int_{\mathbb{R}^d} \left(\widehat{c}(\zeta') \int_0^{\frac{\widehat{d}}{c_1}} \varphi(x - t\zeta')dt\right) |\zeta'.n(x)| \, d\zeta'$$

$$\leq c \int_D \frac{\varphi(y)}{|x-y|^{d-1}} \left|\left(\frac{x-y}{|x-y|}\right).n(x)\right| dy.$$

It suffices to show that

$$B : \varphi \in L^1(D) \to \int_D \frac{\varphi(y)}{|x-y|^{d-1}} \left|\left(\frac{x-y}{|x-y|}\right).n(x)\right| dy \in L^1(\partial D)$$

is weakly compact. To this end, we note first that the assumption ∂D is $C^{1,\alpha}$ implies (see [214, Lemma 3.2.]) that

$$\left(\frac{x-y}{|x-y|}\right).n(x) = O(|x-y|^\alpha)$$

so the kernel of B is

$$b(x, y) := \frac{1}{|x-y|^{d-1}} \left|\left(\frac{x-y}{|x-y|}\right).n(x)\right| = O(\frac{1}{|x-y|^{d-\alpha-1}}).$$

Let us approximate B by B_ε with kernel

$$b_\varepsilon(x, y) := \begin{cases} b(x, y) \text{ if } |x-y| \geq \varepsilon \\ 0 \text{ otherwise.} \end{cases}$$

Then

$$\|B - B_\varepsilon\|_{\mathcal{L}(L^1(D), L^1(\partial D))} = \sup_{y \in D} \int_{\partial D} (b(x, y) - b_\varepsilon(x, y)) \, d\sigma(x)$$

and there exists a constant $c' > 0$ such that

$$\sup_{y\in D} \int_{\partial D} (b(x, y) - b_\varepsilon(x, y))\, d\sigma(x)$$

$$\leq c' \sup_{y\in D} \int_{\partial D\cap\{|x-y|<\varepsilon\}} \frac{1}{|x - y|^{d-1-\alpha}} d\sigma(x).$$

Standard estimates, similar to those in ([108, Lemma 2.2. p. 119]), show that

$$\sup_{y\in D} \int_{\partial D\cap\{|x-y|<\varepsilon\}} \frac{1}{|x - y|^{d-1-\alpha}} d\sigma(x) \to 0 \ (\varepsilon \to 0).$$

Since b_ε is bounded then B_ε is weakly compact and then so is B.

To end the proof of the Lemma, note that $q(\zeta)$ is bounded and compactly supported so

$$\psi \in L^1(\partial D) \to q\psi \in L^1(\Gamma_-)$$

is a bounded operator and finally $\mathcal{D}G_0^\Sigma C : L^1(D \times \mathbb{R}^d) \to L^1(\Gamma_-)$ is weakly compact. $\qquad\square$

The second key preliminary result is

Lemma 10.15.3 *Suppose that ∂D be $C^{1,\alpha}$ for some $\alpha > 0$. Let the boundary operator be diffuse (i.e., $\mathcal{B} = \mathcal{D}$) and regular and let the collision operator C be regular. Then $\left(\lambda - T_{\mathcal{B}}^\Sigma\right)^{-1} C \ (\lambda > s(T_{\mathcal{B}}^\Sigma))$ is power compact in $L^1(D \times \mathbb{R}^d)$.*

Proof Thanks to Lemma 10.15.2

$$\mathcal{W} := \left(\lambda - T_{\mathcal{B}}^\Sigma\right)^{-1} C - \left(\lambda - T_0^\Sigma\right)^{-1} C$$

is weakly compact. On the other hand, it is known (see [251] for more general velocity measures than the Lebesgue measure) that there exists an integer N such that $\left(\left(\lambda - T_0^\Sigma\right)^{-1} C\right)^N$ is compact in $L^1(D \times \mathbb{R}^d)$. Then the product

$$\left(\left(\lambda - T_{\mathcal{B}}^\Sigma\right)^{-1} C\right)^N = \left(\left(\lambda - T_0^\Sigma\right)^{-1} C + \mathcal{W}\right)^N$$

is the sum of $\left(\left(\lambda - T_{\mathcal{B}}^\Sigma\right)^{-1} C\right)^N$ and other (product) terms which are weakly compact because the weakly compact operator $\mathcal{W}$ appears (as a factor) at least once. Hence $\left(\left(\lambda - T_0^\Sigma\right)^{-1} C + \mathcal{W}\right)^N$ is weakly compact and $\left(\left(\lambda - T_0^\Sigma\right)^{-1} C + \mathcal{W}\right)^{2N}$ is compact. $\qquad\square$

We are now ready to state the main result.

Theorem 10.15.1 *Let ∂D be $\mathcal{C}^{1,\alpha}$ for some $\alpha > 0$. Suppose that $s(T_0^{\Sigma}) < 0$. Let the boundary operator be diffuse (i.e., $\mathcal{B} = \mathcal{D}$) and regular and let the collision operator C be regular. Then the stochastic semigroup $(V_{\mathcal{B}}(t))_{t \geq 0}$ has an invariant density. If $(V_{\mathcal{B}}(t))_{t \geq 0}$ is irreducible then its invariant density is unique and is asymptotically stable.*

Proof We know that $\left(\lambda - T_{\mathcal{B}}^{\Sigma}\right)^{-1} C$ is power compact in $L^1(D \times \mathbb{R}^d)$. According to Theorem 2.8.1, $T_{\mathcal{B}}^{\Sigma}$ and $T_{\mathcal{B}}^{\Sigma} + C$ have the same unbounded component of the essential resolvent set. In particular

$$\sigma(T_{\mathcal{B}}^{\Sigma} + C) \cap \left\{ \operatorname{Re} \lambda > s(T_{\mathcal{B}}^{\Sigma}) \right\}$$

consists (at most) of isolated eigenvalues with finite algebraic multiplicities. According to Lemma 10.15.1 $s(T_{\mathcal{B}}^{\Sigma}) < 0$. Since $(V_{\mathcal{B}}(t))_{t \geq 0}$ is stochastic then $s(T_{\mathcal{B}}^{\Sigma} + C) = 0$. Hence 0 is an isolated eigenvalue with finite algebraic multiplicity. Thus $(V_{\mathcal{B}}(t))_{t \geq 0}$ has an invariant density. Note that $V_{\mathcal{B}}(t) \geq U_{\mathcal{B}}^{\Sigma}(t)$. We know that for $\Sigma = 0$, $\left(U_{\mathcal{B}}^0(t)\right)_{t \geq 0}$ is partially integral (see [212, Lemma 7.3]). Since Σ is just a shift then $\left(U_{\mathcal{B}}^{\Sigma}(t)\right)_{t \geq 0}$ is partially integral. Finally $(V_{\mathcal{B}}(t))_{t \geq 0}$ is also partially integral and Theorem 2.10.3 ends the proof. $\qquad\square$

Note that $V_{\mathcal{B}}(t) \geq V_0(t)$ so $(V_{\mathcal{B}}(t))_{t \geq 0}$ is irreducible if $(V_0(t))_{t \geq 0}$ is. We refer e.g. to [245, Chap. 5] for various irreducibility criteria for $(V_0(t))_{t \geq 0}$.

10.15.3 The Critical Case $s(T_0^{\Sigma}) = 0$

The analysis of the critical case is much more involved and depends on additional assumptions. We need to extend the series

$$\sum_{j=0}^{+\infty} \Xi_{\lambda}^{\Sigma} \mathcal{D} \left(M_{\lambda}^{\Sigma} \mathcal{D} \right)^j G_{\lambda}^{\Sigma} f \quad (\lambda > 0)$$

to $\lambda = 0$ and to explore its properties. Several preliminary results are necessary. We start with

Lemma 10.15.4 $f \in L^1(\Gamma_-) \to \Xi_0^{\Sigma}\left(|\zeta| f\right) \in L^1(D \times \mathbb{R}^d)$ *is a bounded operator.*

Proof Note that $\Xi_{\lambda}^{\Sigma} : L^1(\Gamma_-) \to L^1(D \times \mathbb{R}^d)$

$$\Xi_{\lambda}^{\Sigma} f(x, \zeta) = e^{-\lambda \tau(x, \zeta)} e^{-\int_0^{\tau(x, \zeta)} \Sigma(x - s\zeta, \zeta) ds} f(x - \tau(x, \zeta)\zeta, \zeta),$$

is a bounded operator if $\lambda > 0$. Note that for $f \geq 0$

$$\Xi_{\lambda}^{\Sigma} f(x, \zeta) \leq f(x - \tau(x, \zeta)\zeta, \zeta).$$

Let

$$h(x, \zeta) := |\zeta| \, f(x - \tau(x, \zeta)\zeta, \zeta).$$

According to (10.15.7) $\int_{D \times \mathbb{R}^d} h(x, \zeta)dxd\zeta$ is given by

$$\int_{D \times \mathbb{R}^d} h(x, \zeta)dxd\zeta = \int_{\Gamma_-} d\zeta d\sigma(z) \, |n(z).\zeta| \int_0^{\tau_+(z,\zeta)} h(z + s\zeta, \zeta)ds$$

$$= \int_{\Gamma_-} d\zeta d\sigma(z) \, |n(z).\zeta| \times$$

$$\int_0^{\tau_+(z,\zeta)} |\zeta| \, f(z + s\zeta - \tau(z + s\zeta, \zeta)\zeta, \zeta)ds$$

$$= \int_{\Gamma_-} d\zeta d\sigma(z) \, |n(z).\zeta| \times \int_0^{\tau_+(z,\zeta)} |\zeta| \, f(z, \zeta)ds$$

$$= \int_{\Gamma_-} d\zeta d\sigma(z) \, |n(z).\zeta| \, |\zeta| \, f(z, \zeta) \int_0^{\tau_+(z,\zeta)} ds$$

$$\leq \widehat{d} \int_{\Gamma_-} d\zeta d\sigma(z) \, |n(z).\zeta| \, f(z, \zeta)$$

where the estimate $\tau_+(z, \zeta) \leq \frac{\widehat{d}}{|\zeta|}$ is used in the last step. This shows that the operator

$$f \in L^1(\Gamma_-) \rightarrow \Xi_\lambda \left(|\zeta| \, f\right) \in L^1(D \times \mathbb{R}^d)$$

is a bounded operator for all $\lambda \geq 0$, i.e., $\Xi_\lambda \left(|\zeta| \, f\right)$ extends to $\lambda = 0$. $\square$

The second preliminary result is

Lemma 10.15.5 *Let the diffuse boundary operator $\mathcal{D} : L^1(\Gamma_+) \rightarrow L^1(\Gamma_-)$ be such that*

$$\varphi \in L^1(\Gamma_+) \rightarrow \frac{1}{|\zeta|}\mathcal{D}\varphi \in L^1(\Gamma_-)$$

is a regular boundary operator. Then for any regular collision operator C the strong limit

$$\sum_{j=0}^{+\infty} \Xi_0^\Sigma \mathcal{D} \left(M_0^\Sigma \mathcal{D}\right)^j G_0^\Sigma C := s \lim_{\lambda \to 0} \sum_{j=0}^{+\infty} \Xi_\lambda^\Sigma \mathcal{D} \left(M_\lambda^\Sigma \mathcal{D}\right)^j G_\lambda^\Sigma C \qquad (10.15.11)$$

exists and is weakly compact on $L^1(D \times \mathbb{R}^d)$.

Proof *Step 1.* According to (10.15.9) we have $r(M_0^\Sigma \mathcal{D}) < 1$ (for $s(T_0^\Sigma) \leq 0$). We know also that $G_\lambda^\Sigma \leq G_0^\Sigma \leq G_0^0$ and G_0^0 is stochastic. This is true also when $s(T_0^\Sigma) = 0$. Thus the series $\sum_{j=0}^{+\infty} \left(M_0^\Sigma \mathcal{D}\right)^j G_0^\Sigma C$ converges in operator norm. Finally $\Xi_0^\Sigma \mathcal{D}$ factorizes as

$$\Xi_0^\Sigma \mathcal{D} = \left(\Xi_0^\Sigma \, |\zeta|\right) \left(\frac{1}{|\zeta|}\mathcal{D}\right)$$

which is a bounded operator. Hence the strong limit (10.15.11) exists.

Step 2. We note that for $j \geq 1$

$$\Xi_0^\Sigma \mathcal{D} \left(M_0^\Sigma \mathcal{D}\right)^j G_0^\Sigma C$$
$$= \Xi_0^\Sigma \mathcal{D} \left(M_0^\Sigma \mathcal{D}\right) \left(M_0^\Sigma \mathcal{D}\right)^{j-1} G_0^\Sigma C$$
$$= \left(\Xi_0^\Sigma \, |\zeta|\right) \left(|\zeta|^{-1} \mathcal{D} M_0^\Sigma \mathcal{D}\right) \left(M_0^\Sigma \mathcal{D}\right)^{j-1} G_0^\Sigma C.$$

Since

$$\left(|\zeta|^{-1} \mathcal{D}\right) M_0^\Sigma \mathcal{D} \leq \left(|\zeta|^{-1} \mathcal{D}\right) M_0 \mathcal{D}$$

and $\left(|\zeta|^{-1} \mathcal{D}\right) M_0 \mathcal{D}$ is weakly compact (see Remark 10.15.1) then $\Xi_0^\Sigma \mathcal{D} \left(M_0^\Sigma \mathcal{D}\right)^j G_0^\Sigma C$ is weakly compact for all $j \geq 1$. To deal with $j = 0$, i.e., with the weak compactness of

$$\Xi_0^\Sigma \mathcal{D} G_0^\Sigma C = \left(\Xi_0^\Sigma \, |\zeta|\right) \left(|\zeta|^{-1} \mathcal{D}\right) G_0^\Sigma C$$

it suffices to show the weak compactness of

$$\left(|\zeta|^{-1} \mathcal{D}\right) G_0^\Sigma C : L^1(D \times \mathbb{R}^d) \to L^1(\Gamma_-);$$

the proof is identical to that of $\mathcal{D} G_0^\Sigma C$ in the subcritical case (see Lemma 10.15.2). $\qquad\qquad\qquad\qquad\qquad\qquad\qquad\qquad\qquad\qquad\qquad\square$

The third preliminary result is

Lemma 10.15.6 *Let the diffuse boundary operator $\mathcal{D} : L^1(\Gamma_+) \to L^1(\Gamma_-)$ be such that*

$$\varphi \in L^1(\Gamma_+) \to \frac{1}{|\zeta|}\mathcal{D}\varphi \in L^1(\Gamma_-)$$

is a regular boundary operator. Then for any regular collision operator C the strong limit

$$\left(0 - T_\mathcal{D}^\Sigma\right)^{-1} C := s \lim_{\lambda \to 0} \left(\lambda - T_\mathcal{D}^\Sigma\right)^{-1} C$$

exists if and only if the strong limit

$$\left(0 - T_0^\Sigma\right)^{-1} C := s \lim_{\lambda \to 0} \left(\lambda - T_0^\Sigma\right)^{-1} C$$

exists. In particular both limits exist if $|\zeta|^{-1} C$ is bounded. If additionally, $|\zeta|^{-1} C \, |\zeta|$ is strongly Σ-regular then both limits are power compact in $L^1(D \times \mathbb{R}^d)$.

Proof Note that

$$\left(\lambda - T_{\mathcal{B}}^{\Sigma}\right)^{-1} C = \left(\lambda - T_0^{\Sigma}\right)^{-1} C + \sum_{j=0}^{+\infty} \Xi_{\lambda}^{\Sigma} \mathcal{B} \left(M_{\lambda}^{\Sigma} \mathcal{B}\right)^j G_{\lambda}^{\Sigma} C \quad (\lambda > 0).$$

According to Lemma 10.15.5, the strong limit

$$\left(0 - T_{\mathcal{B}}^{\Sigma}\right)^{-1} C - \left(0 - T_0^{\Sigma}\right)^{-1} C := s \lim_{\lambda \to 0} \left(\left(\lambda - T_{\mathcal{B}}^{\Sigma}\right)^{-1} C - \left(\lambda - T_0^{\Sigma}\right)^{-1} C\right)$$

exists and is weakly compact. If $|\zeta|^{-1} C$ is bounded then the strong limit

$$\left(0 - T_0^{\Sigma}\right)^{-1} C = \left(\left(0 - T_0^{\Sigma}\right)^{-1} |\zeta|\right) \left(|\zeta|^{-1} C\right)$$

exists because $\left(0 - T_0^{\Sigma}\right)^{-1} |\zeta|$ is a bounded operator. In addition, $\left(0 - T_{\mathcal{D}}^{\Sigma}\right)^{-1} C$ is power compact if and only if $\left(0 - T_0^{\Sigma}\right)^{-1} C$ is. On the other hand, by Theorem 10.11.2, $C(0 - T_{\Sigma})^{-1}$ is power compact if C is strongly Σ-regular, i.e., if $C\Sigma^{-1}$ is regular. We note that $\left(0 - T_0^{\Sigma}\right)^{-1}$ *commutes* with multiplication operators by space homogeneous functions (in particular by $|\zeta|^{-1}$) so

$$\left(\left(0 - T_0^{\Sigma}\right)^{-1} C\right)^{j+1}$$

$$= \left(\left(0 - T_0^{\Sigma}\right)^{-1} |\zeta|\right) \left(\left(|\zeta|^{-1} C |\zeta|\right) \left(0 - T_0^{\Sigma}\right)^{-1}\right)^j \left(|\zeta|^{-1} C\right)$$

is compact for j large enough since $\left(\left(|\zeta|^{-1} C |\zeta|\right) \left(0 - T_0^{\Sigma}\right)^{-1}\right)^j$ is (see Theorem 10.11.2 with $|\zeta|^{-1} C |\zeta|$ in place of C). This ends the proof. $\qquad\square$

The last preliminary result is

Lemma 10.15.7 *Suppose that the diffuse boundary operator* $\mathcal{D} : L^1(\Gamma_+) \to L^1(\Gamma_-)$ *is such that*

$$\varphi \in L^1(\Gamma_+) \to \frac{1}{|\zeta|} \mathcal{D}\varphi \in L^1(\Gamma_-)$$

is a regular boundary operator. Then the C_0-semigroup $\left(U_{\mathcal{B}}^{\Sigma}(t)\right)_{t \geq 0}$ *is strongly stable (i.e., converges strongly to zero as $t \to +\infty$).*

Proof It is easy to see that $\left(U_0^{\Sigma}(t)\right)_{t \geq 0}$ is strongly stable. This is equivalent to saying that $\left(U_0^{\Sigma}(t)\right)_{t \geq 0}$ is mean ergodic with zero ergodic projection because $\left(U_0^{\Sigma}(t)\right)_{t \geq 0}$ is contractive and then for any $f \in L_+^1(D \times \mathbb{R}^d)$

$$t \to \int_{D \times \mathbb{R}^d} U_0^{\Sigma}(t) f$$

is nonincreasing and then has a limit which must coincide with the ergodic limit

$$\lim_{t\to+\infty} \frac{1}{t} \int_0^t \left(\int_{D\times\mathbb{R}^d} U_0^{\Sigma}(t) f \right) ds.$$

According to Theorem 2.10.1, this is equivalent to

$$\lim_{\lambda\to+0_+} \lambda(\lambda - T_0^{\Sigma})^{-1} f = 0 \ (f \in L^1(D \times \mathbb{R}^d)).$$

Since (10.15.6) gives

$$\lambda \left(\lambda - T_{\mathcal{B}}^{\Sigma}\right)^{-1} f - \lambda \left(\lambda - T_0^{\Sigma}\right)^{-1} f = \lambda \sum_{j=0}^{+\infty} \Xi_{\lambda}^{\Sigma} \mathcal{B} \left(M_{\lambda}^{\Sigma} \mathcal{B}\right)^j G_{\lambda}^{\Sigma} \quad (\lambda > 0)$$

then

$$\lim_{\lambda\to+0_+} \lambda \left(\lambda - T_{\mathcal{B}}^{\Sigma}\right)^{-1} f = 0 \ (f \in L^1(D \times \mathbb{R}^d))$$

because $r\left(M_{\lambda}^{\Sigma}\mathcal{B}\right) \le r\left(M_0^{\Sigma}\mathcal{B}\right) < 1 \ (\lambda \ge 0)$, see (10.15.9). By Theorem 2.10.1 again $\left(U_{\mathcal{B}}^{\Sigma}(t)\right)_{t\ge 0}$ is mean ergodic with zero ergodic projection or equivalently $\left(U_{\mathcal{B}}^{\Sigma}(t)\right)_{t\ge 0}$ is strongly stable. $\qquad\square$

We are ready to give the main statement of this subsection.

Theorem 10.15.2 *Let $s(T_0^{\Sigma}) = 0$. Let ∂D be $C^{1,\alpha}$ for some $\alpha > 0$ and let $\frac{1}{|\zeta|}D$ be a regular diffuse boundary operator. Suppose that $\frac{1}{|\zeta|}C\frac{1}{\Sigma}$ is bounded, $C\frac{1}{\Sigma}$ is regular and that $|\zeta|^{-1} C |\zeta|$ is strongly Σ-regular. Then the stochastic semigroup $(V_{\mathcal{B}}(t))_{t\ge 0}$ has an invariant density. If $(V_{\mathcal{B}}(t))_{t\ge 0}$ is irreducible then its invariant density is unique and $(V_{\mathcal{B}}(t))_{t\ge 0}$ is asymptotically stable.*

Proof Step 1. $C\left(0 - T_{\mathcal{D}}^{\Sigma}\right)^{-1}$ is a stochastic operator. Note first that $\Sigma\left(0 - T_{\mathcal{D}}^{\Sigma}\right)^{-1}$ is stochastic. Indeed, let

$$f := \lim_{\lambda\to 0} \Sigma \left(\lambda - T_{\mathcal{D}}^{\Sigma}\right)^{-1} \varphi,$$

i.e., $f = \lim_{\lambda\to 0} \Sigma\psi_{\lambda}$ where

$$\psi_{\lambda} = \left(\lambda - T_{\mathcal{D}}^{\Sigma}\right)^{-1} \varphi$$

so

$$\lambda\psi_{\lambda} - T_{\mathcal{D}}^{\Sigma}\psi_{\lambda} = \varphi$$

and

$$\lambda\int \psi_{\lambda} + \int \Sigma\psi_{\lambda} = \int \varphi.$$

By Lemma 10.15.7 $\lambda\left(\lambda - T_{\mathcal{D}}^{\Sigma}\right)^{-1}\varphi \to 0$ so $\int f = \int \varphi$. It follows that $C\left(0 - T_{\mathcal{D}}^{\Sigma}\right)^{-1}$ is also stochastic since

$$\int C\left(0 - T_{\mathcal{D}}^{\Sigma}\right)^{-1}\varphi = \int \Sigma\left(0 - T_{\mathcal{D}}^{\Sigma}\right)^{-1}\varphi = \int \varphi.$$

Step 2. $C\left(0 - T_{\mathcal{D}}^{\Sigma}\right)^{-1}$ is power compact:

$$\left(C\left(0 - T_{\mathcal{D}}^{\Sigma}\right)^{-1}\right)^{j+2}$$
$$= C\left(\left(0 - T_{\mathcal{D}}^{\Sigma}\right)^{-1}C\right)^{j}\left(0 - T_{\mathcal{D}}^{\Sigma}\right)^{-1}C\left(0 - T_{\mathcal{D}}^{\Sigma}\right)^{-1}$$
$$= C\left(\left(0 - T_{\mathcal{D}}^{\Sigma}\right)^{-1}C\right)^{j}\left[\left(\left(0 - T_{\mathcal{D}}^{\Sigma}\right)^{-1}\right)\left(C\frac{1}{\Sigma}\right)\right]\Sigma\left(0 - T_{\mathcal{D}}^{\Sigma}\right)^{-1}$$

shows that $C\left(0 - T_{\mathcal{D}}^{\Sigma}\right)^{-1}$ is power compact because the strong limit $\left(\left(0 - T_{\mathcal{D}}^{\Sigma}\right)^{-1}\right)$ $\left(C\frac{1}{\Sigma}\right)$ exists (see Lemma 10.15.6 with $C\frac{1}{\Sigma}$ in place of C) and $\Sigma\left(0 - T_{\mathcal{D}}^{\Sigma}\right)^{-1}$ is a contraction. Hence there exists non trivial $\varphi \in L_{+}^{1}$ such that

$$C\left(0 - T_{\mathcal{D}}^{\Sigma}\right)^{-1}\varphi = \varphi,$$

i.e.,

$$\lim_{\lambda \to 0} C\left(\lambda - T_{\mathcal{D}}^{\Sigma}\right)^{-1}\varphi = \varphi.$$

Since $\left(0 - T_{\mathcal{D}}^{\Sigma}\right)^{-1}C\left(0 - T_{\mathcal{D}}^{\Sigma}\right)^{-1}$ is bounded then

$$\psi := \lim_{\lambda \to 0}\left(\lambda - T_{\mathcal{D}}^{\Sigma}\right)^{-1}\varphi \in L^{1}$$

exists and $C\psi = \varphi$. In particular $\psi \neq 0$. Note that

$$\psi_{\lambda} := \left(\lambda - T_{\mathcal{D}}^{\Sigma}\right)^{-1}\varphi \to \psi$$

in L^{1} and
$$\lambda\psi_{\lambda} - T_{\mathcal{D}}^{\Sigma}\psi_{\lambda} = \varphi$$

implies
$$T_{\mathcal{D}}^{\Sigma}\psi_{\lambda} \to -\varphi = -C\psi.$$

By closedness of $T_{\mathcal{D}}^{\Sigma}$, $\psi \in D(T_{\mathcal{D}}^{\Sigma})$ and

$$T_{\mathcal{D}}^{\Sigma}\psi + C\psi = 0.$$

Finally, after normalization, ψ is an invariant density and, as in the previous subsection, Theorem 2.10.3 ends the proof. $\square$

10.16   On L^1 Essential Spectrum For Neutron Transport

In all the preceding sections, we focused primarily on the (peripheral) discrete spectra of neutron transport operators. In this section, we turn our attention to the complementary part of the spectrum. The analysis of this component relies on fundamentally different tools and assumptions. We present a systematic construction for vacuum boundary conditions.[3] Before entering into the subject, some preliminaries are necessary.

Let $D \subset \mathbb{R}^d$ be an open, not necessarily bounded, subset and let $\nu(d\zeta)$ be an arbitrary positive sigma finite Borel measure on $\mathbb{R}^d$ with support Λ. Let $(U_\Sigma(t))$ be the absorption C_0-semigroup on $L^p(D \times \Lambda)$ $(1 \le p < +\infty)$

$$U_\Sigma(t) : f \to e^{-\int_0^t \Sigma(x-s\zeta,\zeta)ds} f(x - t\zeta, \zeta)\chi(t < \tau(x, \zeta))$$

where

$$\tau(x, \zeta) := \inf \{t > 0; x - t\zeta \notin D\}$$

is the "exit time" function from D. We denote by T_Σ the generator of $(U_\Sigma(t))$. We recall that

$$s(T_\Sigma) = - \lim_{t \to +\infty} \inf_{\{(x,\zeta),\ \tau(x,\zeta)>t\}} t^{-1} \int_0^t \Sigma(x - s\zeta, \zeta)ds. \tag{10.16.1}$$

We assume in all this section that

$$\tau(x, \zeta) < +\infty \text{ a.e.} \tag{10.16.2}$$

Note that this assumption does not exclude D from being unbounded or even having infinite volume. The spectra $\sigma(T_\Sigma)$ and $\sigma(T_\Sigma + C)$ (C is a collision operator) were investigated (mostly for bounded $D \subset \mathbb{R}^d$) for $d = 1$ in the seminal paper by Lehner and Wing [204] and since then in numerous papers for more or less general models, see [178] and references therein. The first systematic analysis of $\sigma(T_\Sigma)$ and $\sigma(U_\Sigma(t))$ (for general semiflows) is given in [367] where it is proved that

$$\sigma(T_\Sigma) = \{\lambda;\ \mathrm{Re}\,\lambda \le s(T_\Sigma)\}, \quad \sigma(U_\Sigma(t)) = \{\mu;\ |\mu| \le e^{s(T_\Sigma)t}\}. \tag{10.16.3}$$

Other systematic approaches were given in [210, 255] with the same conclusions. In the previous sections, we investigated in depth the so-called asymptotic spectra

---

[3] The results appeared in a preprint [247] but were never published.

$$\sigma(T_\Sigma + C) \cap \{\mathrm{Re}\,\lambda > s(T_\Sigma)\} \ \text{and}\ \sigma(e^{t(T_\Sigma+C)}) \cap \left\{\mu;\ |\mu| > e^{s(T_\Sigma)t}\right\}$$

in L^1 setting by means of suitable weak compactness tools, (see [252] for the case $p > 1$). This section is devoted to the analysis of their complements

$$\sigma(T_\Sigma + C) \cap \{\mathrm{Re}\,\lambda \le s(T_\Sigma)\} \ \text{and}\ \sigma(e^{t(T_\Sigma+C)}) \cap \left\{\mu;\ |\mu| \le e^{s(T_\Sigma)t}\right\}$$

which are rather related to essential spectra. We recall that there exist at least five (non equivalent) definitions of essential spectra, $\sigma_{ek}(O)$ ($1 \le k \le 5$), for a closed and densely defined operators O in a complex Banach space X (see e.g. [116, Chap. IX]). Although these definitions may differ, they all yield the same essential spectral radius when O is a bounded operator. Perhaps the most common is

$$\sigma_{e3}(O) = \{\lambda \in \mathbb{C};\ \lambda - O \notin F(X)\}$$

where $F(X)$ denotes the set of Fredholm operators on X. If O is bounded, $\sigma_{e3}(O)$ coincides with the notion of essential spectrum given in Sect. 2.7 in the Calkin algebra $C(X) := \frac{\mathcal{L}(X)}{\mathcal{K}(X)}$ since any Fredholm operator is invertible modulo a compact operator. Another concept of essential spectra is

$$\sigma_{e2}(O) = \{\lambda \in \mathbb{C};\ \lambda - O \notin F_+(X)\}$$

(see [116, Chap. IX], for the definition of semi-Fredholm operators in $F_+(X)$). We recall that $\sigma_{e2}(O) \subset \sigma_{e3}(O)$ and that $\lambda \in \sigma_{e2}(O)$ if and only if there exists a singular sequence corresponding to λ, i.e., a sequence $(x_n)_n \subset D(O)$ containing no norm convergent subsequence in X such that

$$\|x_n\| = 1 \text{ and } \|Ox_n - \lambda x_n\| \to 0,$$

(see [116, Theorem 1.3, p. 415]). It turns out that the concept of σ_{e2} is well suited to transport operators. Indeed, (10.16.3) are nothing but $\sigma_{e2}(T_\Sigma)$ and $\sigma_{e2}(U_\Sigma(t))$, see [210, 255, 367].

We recall that if two closed and densely defined operators O_i ($i = 1, 2$) are such that $(\lambda - O_1)^{-1} - (\lambda - O_2)^{-1}$ is compact for some $\lambda \in \rho(O_1) \cap \rho(O_2)$ then $\sigma_{ek}(O_1) = \sigma_{ek}(O_2)$ ($1 \le k \le 4$), see [116, Theorem 2.4, p. 420].

In $L^p(D \times \Lambda)$ ($1 < p < +\infty$), any bounded regular collision operator C is T_Σ-compact provided that the volume of D is finite and the hyperplanes of $\mathbb{R}^d$ have zero $\nu(d\zeta)$-measure, see [252]. It follows that

$$(\lambda - T_\Sigma - C)^{-1} - (\lambda - T_\Sigma)^{-1} \text{ is compact in } L^p \ (1 < p < +\infty)$$

and consequently (see [116, Theorem 2.3, p. 419 and Theorem 2.4, p. 420])

$$\sigma_{e2}(T_\Sigma + C) = \sigma_{e2}(T_\Sigma) = \{\lambda;\ \mathrm{Re}\,\lambda \le s(T_\Sigma)\}. \tag{10.16.4}$$

In addition, if the affine hyperplanes of $\mathbb{R}^d$ have zero $\nu(d\zeta)$-measure then

$$e^{t(T_\Sigma+C)} - e^{tT_\Sigma} \text{ is compact in } L^p \ (1 < p < +\infty)$$

(see [252]) and therefore

$$\sigma_{e2}(e^{t(T_\Sigma+C)}) = \sigma_{e2}(e^{tT_\Sigma}) = \left\{\mu; \ |\mu| \leq e^{s(T_\Sigma)t}\right\}. \tag{10.16.5}$$

The situation is much more intricate in $L^1(D \times \Lambda)$ because of the negative results

$$(\lambda - T_\Sigma - C)^{-1} - (\lambda - T_\Sigma)^{-1} \text{ and } e^{t(T_\Sigma+C)} - e^{tT_\Sigma} \ (t > 0) \tag{10.16.6}$$

are never weakly compact in $L^1(D \times \Lambda)$ if $d \geq 3$ even if they are *Dunford-Pettis* operators [249], i.e., send weakly compact sets into compact ones. However, the differences (10.16.6) are weakly compact if $d = 1$ (see [251]) while the case $d = 2$ is open. After several preliminary results, we find again the known result

$$\begin{cases} \sigma(T_\Sigma) = \sigma_{e2}(T_\Sigma) = \{\lambda; \ \text{Re}\,\lambda \leq s(T_\Sigma)\} \\ \sigma(U_\Sigma(t)) = \sigma_{e2}(U_\Sigma(t)) = \left\{\mu; \ |\mu| \leq e^{s(T_\Sigma)t}\right\} \end{cases}$$

(see Theorem 10.16.1). What is new here is the proof, in which the singular sequences underlying these results are explicitly constructed and precisely described. The second result is that the spectral inclusions

$$\begin{cases} \{\lambda; \ \text{Re}\,\lambda \leq s(T_\Sigma)\} \subset \sigma_{e2}(T_\Sigma + C) \\ \left\{\mu; \ |\mu| \leq e^{s(T_\Sigma)t}\right\} \subset \sigma_{e2}(e^{t(T_\Sigma+C)}) \end{cases}$$

are always true under quite general assumptions (see Theorems 10.16.2 and 10.16.3). Unlike the analysis of asymptotic spectra

$$\begin{cases} \sigma(T_\Sigma + C) \cap \{\lambda; \ \text{Re}\,\lambda > s(T_\Sigma)\} \\ \sigma(e^{t(T_\Sigma+C)}) \cap \left\{\mu; \ |\mu| > e^{s(T_\Sigma)t}\right\} \end{cases}$$

considered in the previous sections (where the velocity measure $\nu(d\zeta)$ must satisfy suitable assumptions), the spectral results given here are true for more general Borel measures $\nu(d\zeta)$, see Remark 10.16.2. More details are given in the following subsections.

10.16.1   *Essential Spectra of Collisionless Operators*

We deal with T_Σ and $e^{s(T_\Sigma)t}$ in $L^1(D \times \Lambda)$. Even if their analysis in all $L^p(D \times \Lambda)$ is given e.g. in [367], we provide here a more precise construction in the case $p = 1$ in terms of singular sequences whose supports are finely analyzed. We define, for each $c > 0$, the closed set

$$E_c = \{(x, \zeta); \ \tau(x, -\zeta) \le c\}$$

and the space $L^1(E_c)$ identified with the closed subspace of $L^1(D \times \Lambda)$ consisting of (class of) functions vanishing a.e. outside of E_c. We start with a simple observation:

Lemma 10.16.1 *The subspaces $L^1(E_c)$ $(c > 0)$ are invariant under the action of $\{U_\Sigma(t); t \ge 0\}$.*

Proof Let $f \in L^1(E_c)$, i.e.,

$$f(x, \zeta) = 0 \ \ a.e \ \text{on} \ \{(x, \zeta); \ \tau(x, -\zeta) > c\}.$$

Let

$$g = U_\Sigma(t)f = e^{-\int_0^t \Sigma(x-s\zeta, \zeta)ds} f(x - t\zeta, \zeta)\chi(t < \tau(x, \zeta)).$$

Let us show that $g \in L^1(E_c)$, i.e.,

$$g(x, \zeta) = 0 \ \ \text{a.e on} \ \{(x, \zeta); \ \tau(x, -\zeta) > c\}.$$

We note that

$$\tau(x - t\zeta, -\zeta) = \tau(x, -\zeta) + t$$

so that, on the set $\{(x, \zeta); \ \tau(x, -\zeta) > c\}$,

$$\tau(x - t\zeta, -\zeta) > c,$$

it follows that $f(x - t\zeta, \zeta) = 0$ and $g(x, \zeta) = 0$. $\qquad\qquad\square$

Thus, for each $c > 0$, we can define the induced semigroup $\left(U_\Sigma^c(t)\right)_{t \ge 0}$, i.e., $U_\Sigma^c(t) = U_\Sigma(t)_{|L^1(E_c)}$, and denote by T_Σ^c its generator.

Lemma 10.16.2 *For each $c > 0$, the induced semigroup $\left(U_\Sigma^c(t)\right)_{t \ge 0}$ is nilpotent; more precisely, $U_\Sigma^c(t)$ vanishes for $t \ge c$.*

Proof Let $f \in L^1(E_c)$. Then

$$\|U_\Sigma^c(t)f\| = \int_{D \times \Lambda} e^{-\int_0^t \Sigma(x-s\zeta, \zeta)ds} |f(x - t\zeta, \zeta)| \chi(t < \tau(x, \zeta))dx\nu(d\zeta).$$

Observe that

$$x \in D \ \text{and} \ t < \tau(x, \zeta)$$

are equivalent to

$$y(= x - t\zeta) \in D \ \text{and} \ \tau(y, -\zeta) > t$$

so that, a change of variables gives

$$\left\| U_\Sigma^c(t) f \right\| = \int_{D\times\Lambda} e^{-\int_0^t \Sigma(y+s\zeta,\zeta)ds} \, |f(y,\zeta)| \, \chi(\tau(y,-\zeta) > t) dy\nu(d\zeta). \quad (10.16.7)$$

Since $f(y,\zeta) = 0$ $a.e.$ on $\{(y,\zeta); \ \tau(y,-\zeta) > c\}$ then

$$\left\| U_\Sigma^c(t) f \right\| = 0 \text{ if } t \geq c$$

and the proof is complete. $\qquad\qquad\qquad\qquad\qquad\qquad\qquad\qquad\qquad\qquad\qquad\qquad\square$

It follows that the resolvent of T_c is entire

$$(\lambda - T_\Sigma^c)^{-1} = \int_0^c e^{-\lambda t} U_\Sigma^c(t) dt \quad (\lambda \in \mathbb{C}).$$

The computation of the norm of $(\lambda - T_\Sigma^c)^{-1}$ for real λ is given in the following

Lemma 10.16.3 *Let* $\lambda \in \mathbb{R}$*. Then*

$$\left\| (\lambda - T_\Sigma^c)^{-1} \right\| = \sup_{\{\tau(y,-\zeta)\leq c\}} \int_0^{\tau(y,-\zeta)} e^{-\int_0^t (\lambda+\Sigma(y+s\zeta,\zeta))ds} dt$$

where sup denotes the essential supremum.

Proof We note that $(\lambda - T_\Sigma^c)^{-1}$ is a positive operator for real λ. Let $f \in L_+^1(E_c)$. By Fubini theorem

$$\left\| (\lambda - T_\Sigma^c)^{-1} f \right\| = \int_0^c e^{-\lambda t} \left\| U_c(t) f \right\| dt.$$

Hence (10.16.7) yields to

$$\left\| (\lambda - T_\Sigma^c)^{-1} f \right\| - \int_0^c e^{-\lambda t} dt \int_{D\times\Lambda} e^{-\int_0^t \Sigma(y+s\zeta,\zeta)ds} f(y,\zeta)\chi(\tau(y,-\zeta) > t) dy\nu(d\zeta)$$

$$= \int_{E_c} f(y,\zeta) \left[\int_0^{\tau(y,-\zeta)} e^{-\int_0^t (\lambda+\Sigma(y+s\zeta,\zeta))ds} dt \right] dy\nu(d\zeta).$$

Now

$$\left\| (\lambda - T_\Sigma^c)^{-1} \right\| = \sup_{\|f\|_{L_+^1(E_c)}=1} \left\| (\lambda - T_\Sigma^c)^{-1} f \right\| = \sup_{(y,\zeta)\in E_c} \int_0^{\tau(y,-\zeta)} e^{-\int_0^t (\lambda+\Sigma(y+s\zeta,\zeta))ds} dt$$

$$= \sup_{\{\tau(y,-\zeta)\leq c\}} \int_0^{\tau(y,-\zeta)} e^{-\int_0^t (\lambda+\Sigma(y+s\zeta,\zeta))ds} dt$$

ends the proof. $\qquad\qquad\qquad\qquad\qquad\qquad\qquad\qquad\qquad\qquad\qquad\qquad\qquad\square$

Lemma 10.16.4 *Let $\lambda < s(T_\Sigma)$. Then there exist positive constants $a > 0$ and $p_0 > 0$ such that*

$$\sup_{\{p \leq \tau(y,-\zeta)\}} \int_0^{\tau(y,-\zeta)} e^{-\int_0^t (\lambda+\Sigma(y+s\zeta,\zeta))ds}\, dt \geq p e^{ap} \quad \forall p \geq p_0.$$

Proof If $\tau(y, -\zeta) \geq p$ then

$$\int_0^{\tau(y,-\zeta)} e^{-\int_0^t (\lambda+\Sigma(y+s\zeta,\zeta))ds}\, dt \geq \int_0^p e^{-\int_0^t (\lambda+\Sigma(y+s\zeta,\zeta))ds}\, dt$$
$$\geq \int_0^p e^{-\int_0^p (\lambda+\Sigma(y+s\zeta,\zeta))ds}\, dt$$
$$= p e^{-\int_0^p (\lambda+\Sigma(y+s\zeta,\zeta))ds}.$$

It follows that

$$\sup_{\{\tau(y,-\zeta)\geq p\}} \int_0^{\tau(y,-\zeta)} e^{-\int_0^t (\lambda+\Sigma(y+s\zeta,\zeta))ds}\, dt \geq p e^{-\inf_{\{p\leq\tau(x,-\zeta)\}} \int_0^p (\lambda+\Sigma(y+s\zeta,\zeta))ds}.$$

On the other hand $\lambda < s(T_\Sigma)$ amounts to

$$\lambda + \lim_{p\to+\infty} \inf_{\{p\leq\tau(x,-\zeta)\}} p^{-1} \int_0^p \Sigma(x+s\zeta,\zeta)ds < 0.$$

Hence there exist $a > 0$ and $p_0 > 0$ such that

$$\lambda + \inf_{\{p\leq\tau(x,-\zeta)\}} p^{-1} \int_0^p \Sigma(x+s\zeta,\zeta)ds < -a \quad \text{for } p \geq p_0,$$

i.e.,

$$\inf_{\{p\leq\tau(x,-\zeta)\}} \left\{ \lambda p + \int_0^p \Sigma(x+s\zeta,\zeta)ds \right\} < -ap \quad \text{for } p \geq p_0$$

or

$$\inf_{\{p\leq\tau(x,-\zeta)\}} \int_0^p (\lambda + \Sigma(x+s\zeta,\zeta))ds < -ap \quad \text{for } p \geq p_0$$

and consequently

$$\sup_{\{\tau(y,-\zeta)\geq p\}} \int_0^{\tau(y,-\zeta)} e^{-\int_0^t (\lambda+\Sigma(y+s\zeta,\zeta))ds}\, dt \geq p e^{ap} \quad \text{for } p \geq p_0.$$

$\square$

Corollary 10.16.1 *Let (10.16.2) be satisfied and let $\lambda < s(T_\Sigma)$. Then*

$$\left\| (\lambda - T_\Sigma^c)^{-1} \right\|_{\mathcal{L}(L^1(E_c))} \to \infty \text{ as } c \to \infty.$$

Proof According to Lemma 10.16.3

$$\left\| (\lambda - T_\Sigma^c)^{-1} \right\|_{\mathcal{L}(L^1(E_c))} = \sup_{\{\tau(y,-\zeta)\le c\}} \int_0^{\tau(y,-\zeta)} e^{-\int_0^t (\lambda + \Sigma(y+s\zeta,\zeta))ds} dt.$$

Hence

$$\lim_{c\to\infty} \left\| (\lambda - T_\Sigma^c)^{-1} \right\|_{\mathcal{L}(L^1(E_c))} = \sup_{\{\tau(y,-\zeta)<\infty\}} \int_0^{\tau(y,-\zeta)} e^{-\int_0^t (\lambda + \Sigma(y+s\zeta,\zeta))ds} dt$$

$$= \sup_{\{\tau(y,-\zeta)\le\infty\}} \int_0^{\tau(y,-\zeta)} e^{-\int_0^t (\lambda + \Sigma(y+s\zeta,\zeta))ds} dt.$$

It follows that

$$\lim_{c\to\infty} \left\| (\lambda - T_\Sigma^c)^{-1} \right\|_{\mathcal{L}(L^1(E_c))} \ge \sup_{\{\tau(y,-\zeta)\ge p\}} \int_0^{\tau(y,-\zeta)} e^{-\int_0^t (\lambda + \Sigma(y+s\zeta,\zeta))ds} dt \ \ \forall p$$

and we conclude by Lemma 10.16.4. $\qquad\qquad\qquad\qquad\qquad\qquad\qquad\qquad\square$

Lemma 10.16.5 *Let (10.16.2) be satisfied. Let $\lambda < s(T_\Sigma)$. There exist a sequence $\{\tau_n\}_n \subset \mathbb{R}_+$ such that $\tau_n < n$, $\tau_n \to \infty$ and a normalized sequence $\{\varphi_n\} \subset L^1_+(D \times \Lambda)$ such that $supp\varphi_n \subset \{\tau_n \le \tau(y, -\zeta) \le n\}$ and*

$$\left\| (\lambda - T_\Sigma^n)^{-1} \varphi_n \right\| \ge \frac{1}{2} \left\| (\lambda - T_\Sigma^n)^{-1} \right\|_{\mathcal{L}(L^1(E_n))}.$$

Proof Let $\alpha_n = \left\| (\lambda - T_\Sigma^n)^{-1} \right\|_{\mathcal{L}(L^1(E_n))}$. We know by Corollary 10.16.1 that $\alpha_n \to \infty$ as $n \to \infty$. By Lemma 10.16.3,

$$\alpha_n = \sup_{\{\tau(y,-\zeta)\le n\}} \int_0^{\tau(y,-\zeta)} e^{-\int_0^t (\lambda + \Sigma(y+s\zeta,\zeta))ds} dt.$$

Let

$$\Xi_n = \left\{ (y, \zeta); \ \tau(y, -\zeta) \le n, \ \int_0^{\tau(y,-\zeta)} e^{-\int_0^t (\lambda + \Sigma(y+s\zeta,\zeta))ds} dt \ge \frac{\alpha_n}{2} \right\}.$$

This set has a positive $dy\nu(d\zeta)$-measure. We note that on Ξ_n

$$\int_0^{\tau(y,-\zeta)} e^{-(\lambda+\widehat{\sigma})t} dt \ge \frac{\alpha_n}{2}$$

where $\widehat{\sigma} = \inf \Sigma(.,.)$. Moreover, formula (10.16.1) shows that $\widehat{\sigma} \leq -s(T_\Sigma)$ so

$$\lambda + \widehat{\sigma} < 0.$$

There exists a unique $\tau_n > 0$ defined by

$$\int_0^{T_n} e^{-(\lambda+\widehat{\sigma})t} \, dt = \frac{\alpha_n}{2} \tag{10.16.8}$$

and $\tau_n \to \infty$ as $n \to \infty$. Moreover (10.16.8) shows that

$$\Xi_n \subset \{(y, \zeta); \ \tau_n \leq \tau(y, -\zeta) \leq n\}.$$

Now, for all $f \in L^1_+(E_n)$ with support on Ξ_n

$$\left\| (\lambda - T_\Sigma^n)^{-1} f \right\| = \int_{E_n} f(y, \zeta) \left[\int_0^{\tau(y,-\zeta)} e^{-\int_0^t (\lambda + \Sigma(y+s\zeta, \zeta))ds} \, dt \right] dy \nu(d\zeta)$$

$$\geq \frac{\alpha_n}{2} \, \|f\|_{L^1(\Xi_n)}.$$

It suffices to choose a normalized element $\varphi_n \in L^1_+(\Xi_n)$. $\qquad\square$

We are now ready to determine the essential spectrum of absorption semigroups and their generators.

Theorem 10.16.1 *Let (10.16.2) be satisfied. Then*
(i) $\sigma(T_\Sigma) = \sigma_{e2}(T_\Sigma) = \{\lambda; \ \text{Re}\,\lambda \leq s(T_\Sigma)\}$
(ii) $\sigma(U_\Sigma(t)) = \sigma_{e2}(U_\Sigma(t)) = \left\{\mu; \ |\mu| \leq e^{s(T_\Sigma)t}\right\}$.

Proof (i) We already know that $\sigma(T_\Sigma) \subset \{\lambda; \ \text{Re}\,\lambda \leq s(T_\Sigma)\}$. Let $\lambda < s(T_\Sigma)$. According to Lemma 10.16.5, there exists $\varphi_n \in L^1_+(E_n)$ such that $\|\varphi_n\| = 1$ and

$$\left\| (\lambda - T_\Sigma^n)^{-1} \varphi_n \right\| \geq \frac{\alpha_n}{2} \to \infty.$$

Let

$$\psi_n = \frac{(\lambda - T_\Sigma^n)^{-1} \varphi_n}{\left\| (\lambda - T_\Sigma^n)^{-1} \varphi_n \right\|}.$$

Then $\|\psi_n\| = 1$ and

$$\|\lambda\psi_n - T_\Sigma \psi_n\| = \frac{\|\varphi_n\|}{\left\| (\lambda - T_\Sigma^n)^{-1} \varphi_n \right\|}$$

$$= \frac{1}{\left\| (\lambda - T_\Sigma^n)^{-1} \varphi_n \right\|} \to 0. \tag{10.16.9}$$

Following ([367, Proposition A1]), we define, for each $\eta \in R$, the isometric isomorphism

$$M_\eta : f \in L^1(\Omega \times \Lambda) \to e^{-i\eta\alpha(.,.)} f(.,.)$$

where $\alpha(x, \zeta) = \frac{x.\zeta}{|\zeta|^2}$. It is easy to verify that

$$M_\eta^{-1} U_\Sigma(t) M_\eta = e^{i\eta t} U_\Sigma(t)$$

and that

$$M_\eta^{-1} T_\Sigma M_\eta = T_\Sigma + i\eta I.$$

We can write (10.16.9) as

$$\left\| \lambda\psi_n - M_\eta^{-1} T_\Sigma M_\eta \psi_n - i\eta\psi_n \right\| \to 0$$

or

$$\left\| \lambda M_\eta \psi_n - T_\Sigma M_\eta \psi_n - i\eta M_\eta \psi_n \right\| \to 0,$$

i.e.,

$$\left\| \lambda\widehat{\psi}_n - T_\Sigma \widehat{\psi}_n - i\eta\widehat{\psi}_n \right\| \to 0$$

where $\widehat{\psi}_n = M_\eta \psi_n$. We first observe that the approximate eigenfunction $\widehat{\psi}_n = M_\eta \psi_n$ has the same support as ψ_n because $\left|\widehat{\psi}_n\right| = \psi_n$. On the other hand, since

$$\operatorname{supp}\varphi_n \subset \{\tau_n \leq \tau(y, -\zeta) \leq n\} \subset \{\tau_n \leq \tau(y, -\zeta)\} \tag{10.16.10}$$

and

$$\cap_n \{\tau_n \leq \tau(y, -\zeta)\} = \{\tau(y, -\zeta) = +\infty\}$$

has zero measure then $\left(\widehat{\psi}_n\right)_n$ has no convergent in norm subsequence and $\lambda - i\eta \in \sigma_{e2}(T_\Sigma)$. This ends the proof of (i) since $\sigma_{e2}(T_\Sigma)$ is closed.

(ii) follows easily by a standard argument from semigroup theory. $\qquad\square$

We note that Theorem 10.16.1 holds with no assumption on the Borel velocity measure $\nu(d\zeta)$. In particular, the results are true for discrete (collisionless) velocity models where $\nu(d\zeta)$ consists of a (finite or infinite) sum of Dirac measures.

10.16.2 *Essential Spectrum with Collision Operators*

We consider now the full neutron transport operator

$$T_\Sigma + C : D(T_\Sigma) \subset L^1(D \times \Lambda) \to L^1(D \times \Lambda)$$

with a bounded collision operator

$$C : f \in L^1(D \times \Lambda) \to \int_\Lambda c(x, \zeta, \zeta') f(x, \zeta') \nu(d\zeta').$$

This amounts to

$$\widehat{c}(.,.) := \int_\Lambda c(., \zeta, .) \nu(d\zeta) \in L^\infty(D \times \Lambda)$$

and

$$\|C\|_{\mathcal{L}(L^1)} = \|\widehat{c}(.,.)\|_{L^\infty(D \times \Lambda)}.$$

We denote by $(V(t))_{t \geq 0}$ the C_0-semigroup generated by $T_\Sigma + C$. We obtain two main results. The first one is:

Theorem 10.16.2 *Let (10.16.2) be satisfied. If*

$$\sup_{\{\tau(y,\zeta) + \tau(y,-\zeta) \geq c\}} \widehat{c}(.,.) \to 0 \text{ as } c \to \infty \tag{10.16.11}$$

then we have the following inclusions
(i) $\{\lambda;\ \mathrm{Re}\,\lambda \leq s(T_\Sigma)\} \subset \sigma_{e2}(T_\Sigma + C)$
(ii) $\left\{\mu;\ |\mu| \leq e^{s(T_\Sigma)t}\right\} \subset \sigma_{ap}(V(t))$.

Proof Let $\lambda < s(T_\Sigma)$ and $\eta \in \mathbb{R}$. It is shown in Theorem 10.16.1 that

$$\left\| \lambda \widehat{\psi}_n - T_\Sigma \widehat{\psi}_n - i\eta \widehat{\psi}_n \right\| \to 0$$

where $\widehat{\psi}_n = M_\eta \psi_n$ and $\widehat{\psi}_n \in D(T_\Sigma)$, $|\widehat{\psi}_n| = \psi_n$ and $\|\widehat{\psi}_n\| = 1$. We note that

$$\begin{aligned}
\|C\widehat{\psi}_n\| &= \int \int \left| \int c(x, \zeta, \zeta') \widehat{\psi}_n(x, \zeta') \nu(d\zeta') \right| dx \nu(d\zeta) \\
&\leq \int \int \int |c(x, \zeta, \zeta')| \, |\widehat{\psi}_n|(x, \zeta') \nu(d\zeta') dx \nu(d\zeta) \\
&= \int \int \int |c(x, \zeta, \zeta')| \, \psi_n(x, \zeta') \nu(d\zeta') dx \nu(d\zeta) \\
&= \int \int \widehat{c}(x, \zeta') \psi_n(x, \zeta') dx \nu(\zeta').
\end{aligned}$$

Since

$$\mathrm{supp}\,\psi_n \subset \{\tau(y, \zeta) + \tau(y, -\zeta) \geq \tau_n\}$$

(see (10.16.10)) then

$$\|C\widehat{\psi}_n\| \leq \int_{\{\tau(x,\zeta)+\tau(x,-\zeta)\geq \tau_n\}} \widehat{c}(x,\zeta')\psi_n(x,\zeta')dx\nu(d\zeta')$$

$$\leq \sup_{\{\tau(x,\zeta)+\tau(x,-\zeta)\geq \tau_n\}} \widehat{c}(.,.) \times \|\psi_n\|$$

$$= \sup_{\{\tau(x,\zeta)+\tau(x,-\zeta)\geq \tau_n\}} \widehat{c}(.,.) \to 0 \text{ as } n \to \infty.$$

Hence

$$\left\| \lambda\widehat{\psi}_n - i\eta\widehat{\psi}_n - (T_\Sigma\widehat{\psi}_n + C\widehat{\psi}_n) \right\| \to 0.$$

Since

$$\cap_n \subset \{\tau(y,\zeta) + \tau(y,-\zeta) \geq \tau_n\} = \{\tau(y,\zeta) + \tau(y,-\zeta) = \infty\}$$

has zero measure then $\{\widehat{\psi}_n\}$ has no norm convergent subsequence. $\qquad \square$

We point out that the condition (10.16.11) "amounts" to

$$\widehat{c}(x,\zeta') = 0 \text{ on the null set } \left\{\tau(x,\zeta') + \tau(x,-\zeta') = \infty\right\};$$

in particular, if D is bounded, it amounts to $\widehat{c}(x,0) = 0$. The following results avoid the condition (10.16.11) by adapting and extending a classical argument, (see e.g. [283]). To this end, we introduce the following assumption on $\nu(d\zeta)$

$$\int_\Pi e^{iz\cdot\zeta}\nu(d\zeta) \to 0 \text{ as } |z| \to \infty, (\Pi \subset \mathbb{R}^d \text{ bounded}). \tag{10.16.12}$$

Remark 10.16.1 We note that (10.16.12) is satisfied if $\nu(d\zeta)$ is absolutely continuous with respect to the Lebesgue measure on $\mathbb{R}^d$ or the surface Lebesgue measures on spheres about the origin (multigroup models). Indeed, the Fourier transform of the surface Lebesgue measure on a sphere vanishes at infinity (see e.g. [206]) and consequently so is the case for any measure absolutely continuous with respect to it (see e.g. [217]). We note also that (10.16.12) is stronger than the assumption, used previously, that the hyperplanes have zero $\nu(d\zeta)$-measure (see [245, Corollary 4.2 p. 70]). The analysis of (10.16.12) in full generality is a much more complex subject [217].

The second main result is:

Theorem 10.16.3 *Let (10.16.2) (10.16.12) be satisfied. Then we have the following inclusions*

(i) $\{\lambda; \text{ Re } \lambda \leq s(T_\Sigma)\} \subset \sigma_{e2}(T_\Sigma + C)$
(ii) $\left\{\mu; |\mu| \leq e^{s(T_\Sigma)t}\right\} \subset \sigma_{e2}(V(t)).$

Proof The trick consists in modifying suitably the singular sequence $\{\widehat{\psi}_n\}$ in order to introduce oscillations with respect to velocities. Let $\lambda < s(T_\Sigma)$ and $\eta \in \mathbb{R}$. It is shown in Theorem 10.16.1 that

$$\left\| \lambda\widehat{\psi}_n - T_\Sigma\widehat{\psi}_n - i\eta\widehat{\psi}_n \right\| \to 0 \ \text{ as } n \to \infty.$$

where $\widehat{\psi}_n = M_\eta\psi_n$ and $\widehat{\psi}_n \in D(T_\Sigma)$, $\left|\widehat{\psi}_n\right| = \psi_n$ and $\left\|\widehat{\psi}_n\right\| = 1$. We replace $\widehat{\psi}_n(x, \zeta)$ by

$$\overline{\psi}_{n,m}(x, \zeta) = \widehat{\psi}_n(x, \zeta)e^{im.\zeta}$$

where $m \in \mathbb{Z}^N$ is a multiindex. We note that $\overline{\psi}_{n,m}$ has the same support as $\widehat{\psi}_n$. It is clear that

$$\left\| \lambda\overline{\psi}_{n,m} - T_\Sigma\overline{\psi}_{n,m} - i\eta\overline{\psi}_{n,m} \right\| = \left\| \lambda\widehat{\psi}_n - T_\Sigma\widehat{\psi}_n - i\eta\widehat{\psi}_n \right\| \to 0 \ \text{ as } n \to \infty$$

uniformly in $m \in \mathbb{Z}^N$. On the other hand

$$\left\| C\overline{\psi}_{n,m} \right\| = \int \int \left| \int c(x, \zeta, \zeta')\overline{\psi}_{n,m}(x, \zeta')\nu(d\zeta') \right| dx\nu(d\zeta)$$

$$= \int \int \left| \int c(x, \zeta, \zeta')\widehat{\psi}_n(x, \zeta')e^{im.\zeta'}\nu(d\zeta') \right| dx\nu(d\zeta).$$

Assumption (10.16.12) implies the Riemann-Lebesgue property

$$\int \varphi(v')e^{iz.\zeta'}\nu(d\zeta') \to 0 \text{ as } |z| \to \infty \ (\varphi \in L^1(\mathbb{R}^d; \nu(d\zeta)). \tag{10.16.13}$$

Indeed, (10.16.13) which is clearly true for simple functions φ, remains true for $\varphi \in L^1(\mathbb{R}^d)$ by a density argument. We note that

$$\int c(x, \zeta, \zeta') \left|\widehat{\psi}_n(x, \zeta')\right| \nu(d\zeta') = \int c(x, \zeta, \zeta')\psi_n(x, \zeta')d\nu(\zeta') < \infty \ \ a.e.$$

because

$$\int \int dx\nu(d\zeta) \left[\int c(x, \zeta, \zeta')\psi_n(x, \zeta')\nu(d\zeta') \right] = \int \int \widehat{c}(x, \zeta')\psi_n(x, \zeta')dx\nu(d\zeta') < \infty.$$

Hence, for almost all $(x, \zeta) \in D \times \Lambda$ and for each n

$$\int k(x, \zeta, \zeta')\widehat{\psi}_n(x, \zeta')e^{im.\zeta'}\nu(d\zeta') \to 0 \text{ as } |m| \to \infty$$

by the Riemann-Lebesgue property. Moreover,

$$\left| \int c(x, \varsigma, \varsigma') \widehat{\overline{\psi}}_n(x, \varsigma') e^{im \cdot \varsigma'} \nu(d\varsigma') \right| \leq \int c(x, \varsigma, \varsigma') \psi_n(x, \varsigma') \nu(d\varsigma') \in L^1(D \times \Lambda)$$

so, by the dominated convergence theorem,

$$\forall n \quad \int \int \left| \int c(x, \varsigma, \varsigma') \widehat{\overline{\psi}}_n(x, \varsigma') e^{im \cdot \varsigma'} \nu(d\varsigma') \right| dx d\nu(\varsigma) \to 0 \text{ as } |m| \to \infty.$$

In particular, for each n there exists $m_n \in Z^N$ such that

$$\int \int \left| \int c(x, \varsigma, \varsigma') \widehat{\overline{\psi}}_n(x, \varsigma') e^{im_n \cdot \varsigma'} \nu(d\varsigma') \right| dx \nu(d\varsigma) \leq \frac{1}{n}.$$

Thus $\left\| C \overline{\psi}_{n,m_n} \right\| \leq \frac{1}{n}$ and consequently

$$\left\| \lambda \overline{\psi}_{n,m_n} - i\eta \overline{\psi}_{n,m_n} - (T_\Sigma \overline{\psi}_{n,m_n} + C \overline{\psi}_{n,m_n}) \right\|$$
$$\leq \left\| \lambda \overline{\psi}_{n,m_n} - i\eta \overline{\psi}_{n,m_n} - T_\Sigma \overline{\psi}_{n,m_n} \right\| + \left\| C \overline{\psi}_{n,m_n} \right\|$$
$$\leq \left\| \lambda \widehat{\psi}_n - T_\Sigma \widehat{\psi}_n - i\eta \widehat{\psi}_n \right\| + \frac{1}{n} \to 0 \text{ as } n \to \infty.$$

We have obtained a singular sequence $\left\{ \overline{\psi}_{n,m_n} \right\}$ and we can end the proof as previously. $\qquad\qquad\square$

Remark 10.16.2 Without additional assumptions, a priori Theorems 10.16.2 and 10.16.3 do not exhaust the essential spectra of $T_\Sigma + C$ and $V(t)$. Indeed, the spectral properties of $T_\Sigma + C$ (resp. $V(t)$) in the half space $\{\text{Re } \lambda > s(T_\Sigma)\}$ (resp. in $\{\mu; \ |\mu| > e^{s(T_\Sigma)t}\}$) rely on various technical assumptions (in particular on regularity of collisions operators [251, 252]) while we have assumed nothing about the collision operator C except that it is a kernel operator with respect to velocities; in particular, C need not be a regular collision operator.

10.17   Diffusion Models in Nuclear Reactor Theory

We end the chapter on Transport Theory by the analysis of neutron *diffusion* equations with fissions occuring in nuclear reactor theory. In this section, we use the notations of variables of the Physics literature where v is the velocity, $\omega = \frac{v}{|v|} \in S^2$ is the direction (or angle) of the velocity and $E = \frac{1}{2} m |v|^2$ is the kinetic energy of the neutron with mass m. To avoid confusion with the matrix of diffusion, the spatial domain is denoted by Ω. The diffusion equation for the "flux" of neutrons reads

$$\frac{1}{|v|}\frac{d\varphi}{dt}(t, x, \omega, E) - \mathrm{div}_x\left(D(x)\nabla_x\varphi(t, x, \omega, E)\right) + \widetilde{\Sigma}(x, \omega, E)\varphi(t, x, \omega, E)$$

$$= \int_\alpha^\beta \int_{S^2} \widetilde{c}(x, \omega, E, \omega', E')\varphi(t, x, \omega', E')dE'd\omega'$$

$$+ \int_\alpha^\beta \int_{S^2} \widetilde{c}_f(x, \omega, E, \omega', E')\varphi(t, x, \omega', E')dE'd\omega'$$

where $\nabla_x\varphi$ is gradient with respect to the space variable $x \in \Omega \subset \mathbb{R}^3$, $D(x)$ is a bounded, symmetric and coercive matrix of diffusion, div_x is a divergence in space variable and $E \in (\alpha, \beta)$ with

$$0 < \alpha < \beta < +\infty.$$

If the cross sections $\widetilde{\Sigma}$, $\widetilde{c}$ and $\widetilde{c}_f$ do not depend on the direction $\omega \in S^2$ of velocities then by averaging the equation in ω we get a simpler equation for

$$\psi(t, x, E) := \int_{S^2} \varphi(t, x, \omega', E')d\omega'$$

($d\omega'$ is the Lebesgue surface measure on S^2) which is the one used in practice in nuclear reactor theory, (see e.g. [94, Chap. I, Part A, Sect. 5]). In this section, we ignore this simplification and deal with non isotropic equations which can be written

$$\frac{d\varphi}{dt}(t, x, \omega, E) - |v|\,\mathrm{div}_x\left(D(x)\nabla_x\varphi(t, x, \omega, E)\right) + \Sigma(x, \omega, E)\varphi(t, x, \omega, E)$$

$$= \int_\alpha^\beta \int_{S^2} c(x, \omega, E, \omega', E')\varphi(t, x, \omega', E')dE'd\omega'$$

$$+ \int_\alpha^\beta \int_{S^2} c_f(x, \omega, E, \omega', E')\varphi(t, x, \omega', E')dE'd\omega'$$

where

$$\begin{cases} \Sigma(x, \omega, E) = |v|\,\widetilde{\Sigma}(x, \omega, E) \\ c(x, \omega, E, \omega', E') = |v|\,\widetilde{c}(x, \omega, E, \omega', E') \\ |v|\,c_f(x, \omega, E, \omega', E') \end{cases}$$

(and $|v| = \sqrt{\frac{2E}{m}}$). The functional space we consider is

$$L^1\left(\Omega \times S^2 \times (\alpha, \beta)\right)$$

with norm

$$\|\psi\| = \int_{\Omega \times S^2 \times (\alpha, \beta)} |\psi(x, \omega, E)|\, dx d\omega dE.$$

To "restore" the usual notation of velocities, let

$$\zeta = (\omega, E) \in \Lambda := S^2 \times (\alpha, \beta)$$

where Λ is endowed with the measure

$$\nu(d\zeta) := d\omega dE$$

and write the above equation in $L^1(\Omega \times \Lambda)$ as

$$\frac{d\varphi}{dt} - \zeta_2 \operatorname{div}_x (D(x)\nabla_x \varphi(t, x, \zeta)) + \Sigma(x, \zeta)\varphi(t, x, \zeta)$$
$$= \int_\Lambda c(x, \zeta, \zeta')\varphi(t, x, \zeta')\nu(d\zeta') + \int_\Lambda c_f(x, \zeta, \zeta')\varphi(t, x, \zeta')\nu(d\zeta')$$
$$= C\varphi + C_f\varphi$$

where

$$\zeta_2 := \sqrt{\frac{2E}{m}}$$

and C (resp. C_f) is a collision (resp. a fission) operator. We observe that in place of the usual transport operator $\zeta.\nabla_x$, we have now $\zeta_2 \operatorname{div}_x (D(x)\nabla_x)$ which is nothing but an elliptic operator with respect to the space variable $x \in \Omega$ depending on a *parameter* ζ_2. The spatial domain Ω is assumed to be a bounded open set with smooth boundary.

Several boundary conditions with respect to the space variable are possible: the Neumann, Dirichlet, Robin or periodic boundary conditions. Actually, the Robin boundary condition

$$\varphi(t, x, \zeta) + \lambda(x)\frac{\partial \varphi}{\partial \nu}(t, x, \zeta) = 0 \ (x \in \partial\Omega)$$

seems to be the most physical one (see e.g. [94, Chap. I, Part A, Sect. 5]) where the function $\lambda(.) \geq 0$ is the so-called extrapolation length. In this section, we consider all of them simultaneously.

As far as we know, spectral analysis of C_0-semigroups governing such equations has never been considered even for simple examples; see however some multigroup versions [150, 342]. We provide a general construction which appears here for the first time. However, contrary to what one might expect, their peripheral spectral analysis is considerably less technical than that of neutron transport semigroups.

We recall that for all the boundary conditions above $\operatorname{div}_x (D(x)\nabla_x)$ generates a holomorphic and compact irreducible contraction C_0-semigroup $(U(t))_{t\geq 0}$ in all $L^p(\Omega)$ $(1 \leq p < +\infty)$ see e.g. [10]. We extend $(U(t))_{t\geq 0}$ to $L^1(\Omega \times \Lambda)$ as a holomorphic positive contraction semigroup as follows: Indeed, the equation in $L^1(\Omega \times S^2 \times [\alpha, \beta])$

$$\lambda\varphi(x,\zeta) - \zeta_2 \operatorname{div}_x\left(D(x)\nabla_x\varphi(x,\zeta)\right) = g(x,\zeta) \quad (\lambda > 0)$$

(for $g \in L^1_+$) is solved for a fixed ζ by

$$\frac{\lambda}{\zeta_2}\varphi(x,\zeta) - \operatorname{div}_x\left(D(x)\nabla_x\varphi(x,\zeta)\right) = \frac{1}{\zeta_2}g(x,\zeta),$$

i.e.,

$$\varphi(.,\zeta) = \left(\frac{\lambda}{\zeta_2} - \operatorname{div}_x\left(D(x)\nabla_x\right)\right)^{-1}\left(\frac{1}{\zeta_2}g(.,\zeta)\right) \tag{10.17.1}$$

so

$$|\varphi(.,\zeta)| \le \left(\frac{\lambda}{\zeta_2} - \operatorname{div}_x\left(D(x)\nabla_x\right)\right)^{-1}\left(\frac{1}{\zeta_2}|g(x,\zeta)|\right)$$

so

$$\int_\Omega |\varphi(x,\zeta)|\,dx \le \frac{\zeta_2}{\lambda}\int_\Omega \frac{1}{\zeta_2}|g(x,\zeta)|\,dx = \frac{1}{\lambda}\int_\Omega |g(x,\zeta)|\,dx$$

and consequently

$$\|\varphi\|_{L^1(\Omega\times\Lambda)} \le \frac{1}{\lambda}\|g\|_{L^1(\Omega\times\Lambda)}.$$

We denote by G the operator $\operatorname{div}_x\left(D(x)\nabla_x\right)$ in $L^1(\Omega)$

$$G := \operatorname{div}_x\left(D(x)\nabla_x\right)$$

(with one of the boundary conditions above). Then the domain of

$$G^\zeta := \zeta_2 G := \zeta_2 \operatorname{div}_x\left(D(x)\nabla_x\right)$$

in $L^1(\Omega \times \Lambda)$ is nothing but $L^1(\Lambda,\ D(G))$. We denote by $\left(U^\zeta(t)\right)_{t\ge 0}$ the positive contraction C_0-semigroup generated by G^ζ. Since $(U(t))_{t\ge 0}$ is holomorphic in $L^1(\Omega)$ then this amounts to the existence of $c > 0$ and $r > 0$ such that

$$\left\|(\lambda - G)^{-1}\right\|_{\mathcal{L}(L^1(\Omega))} \le \frac{c}{|\lambda|} \quad (\operatorname{Re}\lambda > 0, |\lambda| \ge r)$$

(see [286, A-II-Theorem 1.14]) and for complex λ (10.17.1) gives

$$\int_\Omega |\varphi(x,\zeta)|\,dx \le \frac{c\zeta_2}{|\lambda|}\int_\Omega \frac{1}{\zeta_2}|g(x,\zeta)|\,dx = \frac{c}{|\lambda|}\int_\Omega |g(x,\zeta)|\,dx$$

and finally

$$\|\varphi\|_{L^1(\Omega\times\Lambda)} \le \frac{c}{|\lambda|}\|g\|_{L^1(\Omega\times\Lambda)} \quad (\operatorname{Re}\lambda > 0, |\lambda| \ge r),$$

i.e.,

$$\left\| (\lambda - G^\varsigma)^{-1} \right\|_{\mathcal{L}(L^1(\Omega))} \le \frac{c}{|\lambda|} \quad (\operatorname{Re}\lambda > 0, \; |\lambda| \ge r)$$

which shows (see again [286, A-II-Theorem 1.14]) the holomorphy of $\left(U^\varsigma(t)\right)_{t\ge 0}$ in $L^1(\Omega \times \Lambda)$.

Let $\left(U^\varsigma_\Sigma(t)\right)_{t\ge 0}$ be the corresponding absorption semigroup in $L^1(\Omega \times \Lambda)$ with generator G^ς_Σ. We note that $\left(U^\varsigma_\Sigma(t)\right)_{t\ge 0}$ is holomorphic since $\left(U^\varsigma(t)\right)_{t\ge 0}$ is [169]. In this section, we suppose that $L^1(\Lambda, \; D(G)) \cap D(\Sigma)$ is a core for $L^1(\Lambda, \; D(G))$, we do not deal with this additional technical problem which is certainly connected to a smoothness of the diffusion matrix $D(.)$; this is so e.g. if G is just the Laplacian $\triangle$ (in space variable) and $\Sigma \in L^1_{loc}(\Omega \times \Lambda)$ since $C^\infty_c(\Omega \times \Lambda)$ (the restrictions to $\Omega \times \Lambda$ of $C^\infty_c(\Omega \times \mathbb{R}^n)$) is a core of $L^1(\Lambda; D(\triangle))$ and is included in $D(\Sigma)$. In this case (see Lemma 5.3.2 and Remark 5.3.1))

$$\begin{cases} G^\varsigma_\Sigma = G^\varsigma - \Sigma \\ D(G^\varsigma_\Sigma) = L^1(\Lambda; D(G)). \end{cases} \tag{10.17.2}$$

Let

$$\widehat{C} := C + C_f.$$

The first technical result is that $\left(\widehat{C}(\lambda - G^\varsigma_\Sigma)^{-1}\right)^2$ is weakly compact on $L^1(\Omega \times \Lambda)$, (see Lemma 10.17.1) which implies that

$$G^\varsigma_\Sigma + \widehat{C} : D(G^\varsigma_\Sigma) \subset L^1(\Omega \times \Lambda) \to L^1(\Omega \times \Lambda) \tag{10.17.3}$$

generates an analytic positive C_0-semigroup $(V(t))_{t\ge 0}$ on $L^1(\Omega \times \Lambda)$ such that $\omega_{ess}(V) \le \omega_{ess}(U^\varsigma_\Sigma)$, (see Theorem 10.17.1). Under conservativity assumptions and Neumann or periodic boundary conditions, $(V(t))_{t\ge 0}$ has spectral gap, (see Theorem 10.17.2).

According to the general theory (see Lemma 5.3.3)

$$\Sigma(\lambda - G^\varsigma_\Sigma)^{-1} \text{ is a contraction in } L^1(\Omega \times \Lambda) \quad (\lambda > 0).$$

We start with a key preliminary result.

Lemma 10.17.1 *Let C and C_f be Σ-regular operators and let*

$$\widehat{\Sigma}(\varsigma) := \sup_{x\in\Omega} \Sigma(x, \varsigma) < +\infty \; a.e. \tag{10.17.4}$$

Let $\widehat{C} := C + C_f$. Then $\left(\widehat{C}(\lambda - G^\varsigma_\Sigma)^{-1}\right)^2$ is weakly compact on $L^1(\Omega \times \Lambda)$.

Proof Let $\lambda > 0$ and $\psi \in L^1_+(\Omega \times \Lambda)$. Let $\widehat{c}(x, \zeta, \zeta')$ be the kernel of $\widehat{C}$

$$
\widehat{C}(\lambda - \Delta_\Sigma)^{-1}\psi = \int_\Lambda \widehat{c}(x, \zeta, \zeta')((\lambda - G^\zeta_\Sigma)^{-1}\psi)(x, \zeta')\nu(d\zeta')
$$

$$
= \int_\Lambda \frac{\widehat{c}(x, \zeta, \zeta')}{\Sigma(x, \zeta')}\Sigma(x, \zeta')((\lambda - G^\zeta_\Sigma)^{-1}\psi)(x, \zeta')\nu(d\zeta')
$$

can be viewed as a product of the contraction $\Sigma(\lambda - G^\zeta_\Sigma)^{-1}$ and the bounded operator

$$
\psi \to \int_\Lambda \frac{\widehat{c}(x, \zeta, \zeta')}{\Sigma(x, \zeta')}\psi(x, \zeta')\nu(d\zeta').
$$

By approximation (see Proposition 10.8.1) and using a domination argument, without loss of generality, we may assume that

$$
\frac{\widehat{c}(x, \zeta, \zeta')}{\Sigma(x, \zeta')} = f(\zeta)
$$

where $f \in L^1_+(\Lambda)$. Since

$$
f 1_{\{\widehat{\Sigma} \le j\}} \to f \text{ in } L^1(\Lambda) \ (j \to +\infty)
$$

then, by a density argument, we may assume that the support of f is included in a set $\{\widehat{\Sigma} \le j\}$ for some integer j. In this case,

$$
\widehat{C}(\lambda - G^\zeta_\Sigma)^{-1}\psi \in L^1(\Omega \times \Lambda_j)
$$

where $\Lambda_j := \{\widehat{\Sigma} \le j\}$ and

$$
\left(\widehat{C}(\lambda - G^\zeta_\Sigma)^{-1}\right)^2 \psi = \widehat{C}(\lambda - G^\zeta_\Sigma)^{-1}\varphi
$$

where

$$
\varphi \in L^1(\Omega \times \Lambda_j).
$$

Note that

$$
\zeta_2 = \sqrt{\frac{2E}{m}} \in \left(\sqrt{\frac{2\alpha}{m}}, \sqrt{\frac{2\beta}{m}}\right)
$$

so

$$
(\lambda - G^\zeta_\Sigma)^{-1}\psi \le (\lambda - G^\zeta)^{-1}\psi
$$
$$
= (\lambda - \zeta_2 G)^{-1}\psi
$$

$$=\zeta_2^{-1}(\lambda\zeta_2^{-1}-G)^{-1}\psi$$

$$\leq\left(\frac{2\alpha}{m}\right)^{-\frac{1}{2}}(\lambda\left(\frac{2\beta}{m}\right)^{-\frac{1}{2}}-G)^{-1}\psi$$

$$=c_1(\lambda c_2-G)^{-1}\psi$$

where we have used the monotonicity of the resolvent in $\frac{\lambda}{\zeta_2}$ and

$$c_1=\left(\frac{2\alpha}{m}\right)^{-\frac{1}{2}},\ c_2=\left(\frac{2\beta}{m}\right)^{-\frac{1}{2}}.$$

Hence

$$\left(\widehat{C}(\lambda-G_\Sigma^\zeta)^{-1}\right)^2\psi$$

$$=f(\zeta)\int_\Lambda\widehat{\Sigma}(\zeta')((\lambda-G_\Sigma^\zeta)^{-1}\varphi)(x,\zeta')\nu(d\zeta')$$

$$\leq c_1 f(\zeta)\int_\Lambda\widehat{\Sigma}(\zeta')((\lambda c_2-G)^{-1}\varphi)(x,\zeta')\nu(d\zeta')$$

$$=c_1 f(\zeta)(\lambda c_2-G)^{-1}\left(\int_\Lambda\widehat{\Sigma}(\zeta')\varphi(.,\zeta')\nu(d\zeta')\right)$$

$$\leq jc_1 f(\zeta)(\lambda c_2-G)^{-1}\left(\int_\Lambda\varphi(.,\zeta')\nu(d\zeta')\right)$$

where we used the key facts that $(\lambda c_2-G)^{-1}$ *commutes* with the multiplication operator by $\widehat{\Sigma}(\zeta)$ and $\varphi\in L^1(\Omega\times\Lambda_j)$. Finally, up to a mutilplicative factor, $\left(\widehat{C}(\lambda-G_\Sigma^\zeta)^{-1}\right)^2$ is dominated by a product of several operators namely, $\widehat{C}(\lambda-G_\Sigma^\zeta)^{-1}$, the bounded averaging operator

$$h\in L^1(\Omega\times\Lambda)\to\widetilde{h}=\int_\Lambda h(.,\zeta')\nu(d\zeta')\in L^1(\Omega),$$

the compact operator

$$(\lambda c_2-G)^{-1}:L^1(\Omega)\to L^1(\Omega)$$

and the bounded operator

$$\varphi\in L^1(\Omega)\to\varphi f\in L^1(\Omega\times\Lambda).$$

This ends the proof. $\qquad\qquad\square$

We are ready to state the main spectral result.

Theorem 10.17.1 *Let C and C_f be Σ-regular operators and let (10.17.4) be satisfied. Then*

$$G_{\Sigma}^{\zeta} + \widehat{C} : D(G_{\Sigma}^{\zeta}) \subset L^1(\Omega \times \Lambda) \to L^1(\Omega \times \Lambda) \tag{10.17.5}$$

generates an analytic positive C_0-semigroup $(V(t))_{t \geq 0}$ on $L^1(\Omega \times \Lambda)$. Moreover $\omega_{ess}(V) \leq \omega_{ess}(U_{\Sigma}^{\zeta})$.

Proof It is a direct consequence of Lemma 10.17.1 and Theorem 2.6.5. $\square$

We complement Theorem 10.17.1 by a spectral gap result.

Theorem 10.17.2 *Let C and C_f be Σ-regular operators and let (10.17.4) be satisfied. We assume additionally that $\beta < +\infty$ and that*

$$\underline{\Sigma}(x) := \inf_{\zeta \in \Lambda} \Sigma(x, \zeta) \text{ is not identically zero}$$

and

$$\Psi(x, \zeta) := \int_{\Lambda} c(x, \zeta', \zeta)\nu(d\zeta') + \int_{\Lambda} c_f(x, \zeta', \zeta)\nu(d\zeta') - \Sigma(x, \zeta) \geq 0.$$

Then, for both Neumann and periodic boundary condition, $(V(t))_{t \geq 0}$ has spectral gap in $L^1(\Omega \times \Lambda)$.

Proof We note that

$$G_{\Sigma}^{\zeta} = G^{\zeta} - \Sigma = \zeta_2 G - \Sigma = \zeta_2 \left(G - \zeta_2^{-1}\Sigma\right)$$

is a family of resolvent compact operators on $L^1(\Omega)$ indexed by $\zeta \in \Lambda$. The point spectrum of G_{Σ}^{ζ} (in $L^1(\Omega)$) is equal to ζ_2 times the point spectrum of $G - \zeta_2^{-1}\Sigma$ in particular the spectral bound of G_{Σ}^{ζ} (in $L^1(\Omega)$) is equal to ζ_2 times the spectral bound of $G - \zeta_2^{-1}\Sigma$. Since $\zeta_2^{-1} \in \left(c_2^{-1}, c_1^{-1}\right)$ then the spectral bound of $G - \zeta_2^{-1}\Sigma$ is less than or equal to the spectral bound (in $L^1(\Omega)$) of

$$G - c_2^{-1}\underline{\Sigma} = G_{c_2^{-1}\underline{\Sigma}}.$$

Let us show that this spectral bound is < 0. We observe that

$$(\lambda - G_{c_2^{-1}\underline{\Sigma}})^{-1} \leq (\lambda - G)^{-1}$$

and

$$(\lambda - G_{c_2^{-1}\underline{\Sigma}})^{-1} \neq (\lambda - G)^{-1}$$

so, by using Proposition 2.4.1,

$$r\left((\lambda - G_{c_2^{-1}\underline{\Sigma}})^{-1}\right) < r\left((\lambda - G)^{-1}\right)$$

or equivalently

$$s(G_{\beta^{-1}\underline{\Sigma}}) < s(G) = 0$$

because $(U(t))_{t \geq 0}$ is stochastic in $L^1(\Omega)$ for Neumann or periodic boundary conditions. It follows that the spectral bound of $G_{\underline{\Sigma}}^{\zeta}$ in $L^1(\Omega \times \Lambda)$ is < 0 so

$$\omega(U_{\underline{\Sigma}}^{\zeta}) = s(G_{\underline{\Sigma}}^{\zeta}) < 0$$

and by Theorem 10.17.1

$$\omega_{ess}(V) \leq \omega_{ess}(U_{\underline{\Sigma}}^{\zeta}) \leq \omega(U_{\underline{\Sigma}}^{\zeta}) < 0.$$

Let us show that $s(G_{\underline{\Sigma}}^{\zeta} + \widehat{C}) \geq 0$. Indeed, if $s(G_{\underline{\Sigma}}^{\zeta} + \widehat{C}) < 0$ then for $\lambda < 0$ and small enough $\left(\lambda - G_{\underline{\Sigma}}^{\zeta} - \widehat{C}\right)^{-1}$ exists and is positive so for non trivial $\varphi \geq 0$

$$\psi = \left(\lambda - G_{\underline{\Sigma}}^{\zeta} - \widehat{C}\right)^{-1} \varphi \geq 0$$

is not trivial and, by integrating

$$\lambda \psi - G_{\underline{\Sigma}}^{\zeta} \psi - \widehat{C}\psi = \varphi,$$

we get

$$\lambda \int_{\Omega \times \Lambda} \psi - \int_{\Omega \times \Lambda} \Psi(x, \zeta)\psi = \int_{\Omega \times \Lambda} \varphi$$

which implies that $\lambda > 0$. Finally $s(G_{\underline{\Sigma}}^{\zeta} + \widehat{C}) \geq 0$ and

$$\omega_{ess}(V) < 0 \leq s(G_{\underline{\Sigma}}^{\zeta} + \widehat{C}) = \omega(V),$$

this ends the proof. $\qquad\qquad\qquad\qquad\qquad\qquad\qquad\qquad\qquad\square$

Remark 10.17.1 (Open question) The existence of a spectral gap for Dirichlet or Robin boundary condition is much more involved and is left open. In fact it relies on the (abstract) condition $\lim_{\lambda \to s(G_{\underline{\Sigma}}^{\zeta})} r(\widehat{C}(\lambda - G_{\underline{\Sigma}}^{\zeta})^{-1} > 1$. As for neutron transport semigroups (see Remark 10.9.2), this condition should probably be satisfied at least if the "size" of Ω is large enough or if $\int_{\Lambda} c(x, \zeta', \zeta)\nu(d\zeta') = \Sigma(x, \zeta)$ and the fission operator C_f is "sufficiently strong".

10.18 Comments

The origins of spectral theory for Neutron Transport can be traced back to the seminal work of Lehner and Wing [204], who studied a model case in slab geometry. It became apparent early on that, in general, one could not hope for more than an understanding of the time asymptotics of neutron transport equations [174, 359]; even highly idealized (and physically unrealistic) rod models with only two velocities offer limited additional insight [378]. Nevertheless, certain one-dimensional transport operators were successfully analyzed through the celebrated Case's method of eigen-distribution expansions [79]. This method was later revisited and placed on firm mathematical foundations via two distinct functional analytic approaches [151, 191], giving rise to a rich and elegant spectral theory for abstract kinetic models, as developed in the monographs [144, 178]. The literature on the spectral theory of Neutron Transport, produced by physicists, engineers, and mathematicians alike, is vast and cannot be fully surveyed here. A representative sample of references up to 1997 can be found in [245], while more recent developments up to 2013 are covered in [268]; selected references are also listed below. In bounded spatial domains, neutron transport semigroups are eventually compact, that is, they become compact for sufficiently large times, provided the set of velocities is bounded away from zero. This general result was established early on by Jörgens [174]. In contrast, when arbitrarily small velocities are allowed, a half-plane of essential spectrum may appear for the generators of neutron transport semigroups. In this more delicate setting, two fundamental functional analytic contributions to the understanding of the (asymptotic) discrete spectrum of neutron transport semigroups and their generators are due to I. Vidav. He introduced into transport theory the analytic Fredholm alternative (also known as the Gohberg–Shmul'yan alternative) for power-compact operators, which provides a framework for analyzing the asymptotic discrete spectrum of the generator. I. Vidav also applied the Krein–Rutman theorem to study the existence and properties of the leading eigenvalue [358]. In addressing the time asymptotics of neutron transport semigroups, I. Vidav observed that knowledge of the asymptotic spectrum of the generator alone is insufficient, due to the possible failure of a spectral mapping theorem in this context. To overcome this, he emphasized the necessity of establishing the compactness of a suitable remainder term in a Dyson–Phillips expansion [359]. I. Vidav's foundational ideas were subsequently developed and extended in a substantial body of work, including [223, 242, 331, 362, 368, 373, 375] ([245, Chap. 2]) [71, 72, 287]. While physical cross sections arising from free gas models were already incorporated in the 1960s into spectral analyses conducted in weighted L^2-spaces, the first spectral analysis of neutron transport generators in (weighted) L^1-spaces using such physically relevant cross sections is due to Suhadolc [343]. Subsequent advances in the spectral theory of neutron transport semigroups in L^1-spaces, including the works [146, 238, 240, 344, 368], did not, however, accommodate these physically accurate cross sections until the contribution of Weis [375], which successfully addressed them. Building on earlier ideas already present in [344], the collision operators introduced in [375] were shown to be collectively (with respect to the spatial variable x) weakly compact in the velocity variable ζ. This insight motivated the introduction, in [207], of the general notion of regular bounded collision operators in L^1-spaces, a concept that proved to be well suited to the spectral analysis of neutron transport semigroups in L^1-spaces for a wide class of velocity measures $\nu(d\zeta)$, as developed in [251][266]. The treatment of neutron transport semigroups with unbounded (but Σ-bounded) collision operators and general velocity measures $\nu(d\zeta)$, as developed in Sect. 10.8, became possible following the introduction of the concept of Σ-regular collision operators. A concrete application of this framework is presented in Sect. 10.9, where we address general space-nonhomogeneous linear Boltzmann equations with power-like hard sphere potentials. This analysis relies on a refined understanding of space-homogeneous models in various weighted L^1-spaces, as reported in [211] and detailed in Sect. 10.5. It is worth noting that, in these weighted spaces, the associated collision operators are not always bounded, see Remark 10.5.3.

Many open questions (or comments) are worth mentioning:

(1) In Sect. 10.7, we can treat more complex scattering kernels combining "downshift" and "Bragg" scatterings. The same statements hold also when we replace D by the d-dimensional torus $T^d := \mathbb{R}^d/(2\pi\mathbb{Z})^d$.

(2) In connection with Sect. 10.7.2: The existence of an invariant density for space-homogeneous equations relies on suitable Σ-weak compactness of C on $L^1(\mathbb{R}^d_\zeta)$ only. A priori, this is compatible with arbitrary velocity measure $\nu(d\zeta)$. Thus, the mean ergodicty of $(V(t))_{t\geq0}$ (on the torus) may hold for arbitrary velocity measures, e.g. for combinations of Dirac masses (i.e., for discrete velocity models). However, the strong convergence of $(V(t))_{t\geq0}$ to the ergodic projection relies on compactness results on $L^1(T^d \times \mathbb{R}^d)$ which depend on quite specific conditions on the velocity measure $\nu(d\zeta)$, see [266] and Sect. 10.8. More generally, the non locality (in velocity) of jump operators (which is encoded in the velocity measure $\nu(d\zeta)$) plays a fundamental role in Transport Theory, in particular in compactness problems behind spectral theory. Among the earlier works, see [2, 65, 104, 146, 174, 238, 334, 341, 343, 344, 354, 368, 375]; for more recent works, see [152, 251, 252, 266, 279, 324, 325]. There exists also a rich literature which goes back to [142] on fractional Sobolev regularity (in space) induced by velocity averages which is used in non linear Kinetic Theory, see e.g. [303].

(3) Theorems 10.8.3 and 10.8.5 admit similar versions on the multidimensional torus.

(4) The L^p spectral theory of Neutron Transport ($1 < p < +\infty$) is given by different approaches in [252, 267, 325] for bounded regular collision operators, i.e., the operators (10.8.4) on $L^p_\zeta(\mathbb{R}^d) := L^p(\mathbb{R}^d; \nu(d\zeta))$ (indexed by $x \in \partial D$) are collectively compact on $L^p_\zeta(\mathbb{R}^d)$. The main result is: If the affine hyperplanes of $\mathbb{R}^d$ have zero $\nu(d\zeta)$-measure and if D has a finite volume then $R_1(t) = V(t) - U(t)$ is compact in $L^p(D \times \mathbb{R}^d; dx\nu(d\zeta))$. In particular $(V(t))_{t\geq0}$ and $(U(t))_{t\geq0}$ have the same essential spectrum; this property is stronger than the stability of the essential type. We wonder whether this result

remains true for (unbounded) Σ-regular collision operators (i.e., the operators (10.8.5) on $L^p_\zeta(\mathbb{R}^d) := L^p(\mathbb{R}^d; \nu(d\zeta))$ (indexed by $x \in \partial D$) are collectively compact on $L^p_\zeta(\mathbb{R}^d)$)? We refer to Theorem 14.5.1 for an alternative approach relying on L^1 predual analysis. We point out that if $d \geq 3$, $R_1(t) = V(t) - U(t)$ is never weakly compact in L^1 but is a Dunford-Pettis operator, i.e., maps weakly compact sets into compact sets; for $d = 1$, $R_1(t)$ is weakly compact while the case $d = 2$ is open, see [249]. This peculiarity of the dimension $d = 2$ also appears in a different context in some inverse problems for Neutron Transport (see [245, Chap. 11]). We wonder what the underlying reason for this dimension-specific behavior might be.

(5) In Theorem 10.8.5 (in L^1 spaces) the assumption (10.8.15) is technically important. (Note that we have an equality $\omega_{ess}(V) = \omega_{ess}(U_\Sigma)$ for bounded regular collision operators [251].) We wonder whether the spectral estimate (10.8.16) remains true if we replace (10.8.15) by the more general assumption that the affine hyperplanes of $\mathbb{R}^d$ have zero $\nu(d\zeta)$-measure? This should be more in the spirit of the L^p theory ($1 < p < +\infty$) [252, 267, 325].

(6) We note that neutron transport equations with delayed neutrons are governed by a system. The determination of essential type of the semigroup governing this system (for bounded collision operators) is considered in ([245, Chap. 4, Sect. 5]) in L^p spaces ($p > 1$) for a general class of velocity measures $\nu(d\zeta)$. The result relies on the compactness of a third order remainder term $R_3(t)$ of a Dyson-Phillips expansion (see [245, Theorem 4.15, p. 89]). The case $p = 1$ is open even for Lebesgue velocity measure $d\zeta$. We conjecture that in this case, $R_4(t)$ should be compact in L^1 spaces and we should capture the essential type as for $p > 1$.

(7) Homogenization results for spectral problems of Neutron Transport with (space) locally periodic collision frequencies and bounded regular collision operators are given in [250]; (see also [172] for non linear versions). We conjecture that the results extend to (unbounded) Σ-regular collision operators.

(8) See the open question in Remark 10.9.2

(9) In connection with Sect. 10.10.1: We wonder whether we can derive "reasonable" estimates of the leading neutron transport eigenvalue from its variational (Max-inf and Inf-sup) characterizations? (We refer to [256] for some rough estimates in this direction). At present, we do not know if these variational characterizations hold significant numerical value. It is worth noting that these characterizations also apply to space homogeneous kinetic equations. Additionally, a variational characterization of the critical eigenvalue exists, as shown in [209].

(10) In connection with existence criteria given in Sect. 10.10.2: As pointed out in Remark 10.10.1, the spectral properties of the collision operator C are essential in assessing whether the spectral gap condition condition $s(T_\Sigma + C) > s(T_\Sigma)$ can be satisfied.: if C is a bounded quasi-nilpotent operator (i.e., $r(C) = 0$) then $s(T_\Sigma + C) = s(T_\Sigma)$ regardless of the size of the spatial domain (see [245, Corollary 5.1, p. 116]). This situation arises, for instance, when the kernel

$c(x, \zeta, \zeta') = 0$ for $|\zeta| \geq |\zeta'|$; which corresponds to neutron slowing down (i.e., the absence of upscattering), a regime that typically occurs beyond the energy range of neutron thermalization, see [114, p. 145]. See also ([245, Theorem 6.6, p. 158]) for another spectral role of the collision operator C.

(11) In connection with the critical case (10.11.1):

(i) In Theorem 10.11.3, is $(V(t))_{t \geq 0}$ uniformly bounded in $L^1(D \times \mathbb{R}^d)$ when $k = \overline{k}$? If so then we can appeal to Theorem 2.10.4 to assert that $(V(t))_{t \geq 0}$ is asymptotically stable in $L^1(D \times \mathbb{R}^d)$ which is a stronger result than the mere asymptotic stability in $L^1_{\varphi^*}$ since φ^* is not bounded away from zero.

(ii) The diffusion approximation of Neutron Transport relies on the assumption that the collision frequency is bounded away from zero (in particular it relies on a subcriticality condition $s(T_\Sigma) < 0$), see e.g. [42, 246]. What is the correct theory of diffusion approximation under the critical assumption $s(T_\Sigma) = 0$? See [43] for a special case.

(iii) The theory of persisting chain reactions relies on the assumption that the collision frequency is bounded away from zero [50, 171, 298, 299]. How to develop this theory under the critical assumption (10.11.1)?

(12) A spectral analysis of neutron transport semigroups (with compactly supported cross sections) in exterior domains with specular reflections on an obstacle is given in [341]. By using suitable density arguments, it is possible to extend this theory to non compactly supported cross sections, in the spirit of the analysis above.

(13) See the open problem in Sect. 10.12.3 about the existence of an invariant density in the whole space.

(14) In connection with Sect. 10.12.4: A scattering theory for neutron transport C_0-semigroups in exterior domains with specular reflections on an obstacle is given in [85, 341]. It is possible to improve it in the spirit of the analysis of Sect. 10.12.4.

(15) In connection with Sect. 10.14.7 and Theorem 10.14.3: If (10.14.3) is not satisfied, the question about how to decide whether 0 is an eigenvalue of $T + C_e + C_i$ is left open. We expect that the construction given in [279] for inelastic scattering could be adapted to deal with an additional elastic collision operator. Note that if the symmetry conditions (10.14.12) are satisfied then 0 is an eigenvalue of $T + C_e + C_i$ associated to the constant eigenfunction 1. Note also that if the cross sections are space homogeneous then it suffices to ignore the transport operator $\rho \omega . \frac{\partial}{\partial x}$ and to show (in the space $L^1([a, b] \times S^{d-1})$) that 0 is an eigenvalue of $-\sigma + C_e + C_i$ and this is answered by Theorem 10.6.3. The treatment of full neutron transport semigroups with partly elastic collisions on the torus taking into account also of "downshift" scattering is an interesting open problem.

(16) Neutron Transport with stochastic boundary operators is also rich in open questions. We conjecture that Theorem 10.15.1) is not optimal and actually $(V_{\mathcal{B}}(t))_{t \geq 0}$ has a spectral gap.

(17) The fact that the boundary operator is purely diffusive plays a key role in the construction. The extension to partly diffusive boundary operator $B = D + R$ (with suitable control of the norm of R) is an interesting open problem.

(18) Let $s(T_0^\Sigma) = 0$. What happens if some of the conditions in Theorem 10.15.2 are not satisfied? What kind of sweeping phenomenon may occur ?

(19) A general Tauberian approach to (algebraic) rates of convergence to equilibrium is given in [213] for the collisionless case (i.e., $\Sigma = 0$ and $C = 0$). Its extension to the full model with conservative boundary operators is an interesting open problem. (A Tauberian approach, in the same spirit, is given in [215] for transport equations on the torus with conservative scattering).

(20) Both Theorems 10.16.2 and 10.16.3 in Sect. 10.16 are devoted to the description of L^1-essential spectra of neutron transport semigroups (for vacuum boundary condition) under the general assumption of finite exit time function

$$\tau(x, \zeta) < +\infty \text{ a.e.}$$

which is trivially true for bounded spatial domains but can be satisfied even by suitable spatial domains with infinite volume. The first theorem, under Condition (10.16.11), holds for arbitrary Borel velocity measure $\nu(d\zeta)$ while Theorem 10.16.2 (independent from (10.16.11)) needs Assumption (10.16.12) on $\nu(d\zeta)$. Note that Theorem 10.16.2 allows $\nu(d\zeta)$ to be e.g. a sum of Dirac masses (discrete velocity models). We point out that the (weak) compactness results at the core of the analysis of the asymptotic spectrum (i.e., the discrete part of the spectrum) are no longer true for discrete velocity models and needs both the spatial domain to have a finite volume and the condition (10.8.15) on $\nu(d\zeta)$. These remarks highlight that the analysis of the essential spectrum and that of the asymptotic (i.e., discrete) spectrum rely on fundamentally different types of assumptions on the velocity measure. See also Remark 10.16.1.

(21) It is possible to develop a peripheral spectral theory for very general one-dimensional hyperbolic systems, which in particular encompasses neutron transport equations with discrete velocities and boundary operators, see [5].

(22) See Remarks 10.16.1 and 10.16.2 for various comments and open questions on the velocity measure $\nu(d\zeta)$.

(23) In Sect. 10.17 devoted to neutron diffusion models, the assumption that $\nu(d\zeta) = d\omega dE$ plays no role in the construction; for instance one could instead take $\nu(d\zeta)$ to be a finite sum of Lebesgue measures supported on spheres centered at the origin, thereby covering multigroup diffusion models and complementing the existing literature [150, 342]. The existence of a spectral gap for Dirichlet or Robin boundary condition is open, see Remark 10.17.1.

(24) See Sect. 14.6 for further comments and references on certain classes of neutron transport operators that allow the analysis to go beyond peripheral spectral theory.

Chapter 11
Mathematical Population Dynamics

11.1 Chapter Aims

This chapter focuses on the spectral analysis and long-time asymptotics of several classes of structured population models. We develop a spectral gap theory for a class of integro-differential models of diffusive type, subject to generalized Wentzell-Robin boundary conditions. We also address growth-fragmentation equations (conservative or with mass loss) in different L^1-settings, under various assumptions on the growth and fragmentation rates; this leads to several theories, including one describing a scattering regime (runaway phenomenon). Additionally, we explore growth equations without fragmentation and with a nonlocal McKendrick-von Foerster boundary condition.

11.2 Chapter Overview

In this chapter, we investigate the peripheral spectral analysis and long-time asymptotics of several classes of structured population models. Our aim is to provide a broad overview of recent results in this area, with a particular focus on introducing key mathematical tools relevant to their analysis. In addition to surveying existing literature, we also present several new contributions. Each class of models is treated in a separate section. In Sect. 11.3, we develop a spectral gap theory for a class of integro-differential models of *diffusive* type, subject to generalized Wentzell–Robin boundary conditions. For a detailed discussion of the biological relevance of these models, we refer the reader to [124]. The remaining sections are devoted to growth–fragmentation equations, which arise in the modeling of various physical and biological systems involving the concentration of aggregates that undergo both growth and fragmentation. Notable applications include phytoplankton dynamics, prion propagation, and related phenomena. Over the past two decades, this topic has

M. Mokhtar-Kharroubi, *Peripheral Spectra of Perturbed Positive Semigroups*, Lecture Notes in Mathematics 2388, https://doi.org/10.1007/978-3-032-11173-9_11

attracted significant attention; we refer to [131] for a comprehensive overview of the diverse contexts in which these models are employed. Section 11.4 is devoted to the spectral gap analysis of *growth–fragmentation* equations with local boundary conditions, considered in the so-called "finite mass and aggregate number" space. Under suitable assumptions on the growth rate, the analysis relies crucially on the presence of the fragmentation operator and on the unbounded behavior of the fragmentation rate $a(.)$ at infinity. In Sect. 11.5, we demonstrate how a *runaway* phenomenon arises in the "finite mass" space under an alternative assumption on the growth rate. In Sect. 11.6, we study growth equations without fragmentation, including a death term $a(.)$, posed in the "finite mass and aggregate number" space with nonlocal boundary conditions. We present two spectral gap theories: one that relies on the unboundedness of the death term $a(.)$ at infinity, and another that holds independently of this condition. More information is given in the introductions of the different sections. Some remarks, comments or open questions are relegated to the end of this chapter in the section "Comments".

11.3   Diffusion with Wentzell-Robin Boundary Condition

A model of structured populations with generalized Wentzell-Robin boundary conditions

$$\begin{cases} u_t(s,t) + (\gamma(s)u(s,t))_s = (d(s)u_s(s,t))_s - \mu(s)u(s,t) + \int_0^m \beta(s,y)u(y,t)dy \\ \quad [(d(s)u_s(s,t))_s]_{s=0} - b_0 u_s(0,t) + c_0 u(0,t) = 0, \\ \quad [(d(s)u_s(s,t))_s]_{s=m} + b_m u_s(m,t) + c_m u(m,t) = 0 \end{cases}$$

with $m < +\infty$ and

$$b_0 - \gamma(0) > 0, \quad b_m + \gamma(m) > 0$$

(and initial conditions) was considered first in [124]. This work was extended in different directions in [275]; in particular when $m = +\infty$ where we have to drop the boundary condition at m. We refer the reader to the introductions of these two papers and to the references therein for more motivation and references on such structured population models. From a mathematical point of view, the main issues are well-posedness of this evolution system and the understanding of its time asymptotics. Further extensions of [275] are given in [278].

To acquaint the reader with the mathematical techniques applied in these recent studies, we recall in this section some results from [278]. Keeping the notations in [275], the general assumptions are

$$\gamma, d \in W^{1,\infty}(0,m), \ \mu \in L^1_{loc}(0,m) \tag{11.3.1}$$

$$\mu, \gamma' \text{ and } s \mapsto \beta(s,.) \text{ are continuous at } s = 0 \text{ and } s = m, \tag{11.3.2}$$

$$b_0, b_m > 0, \ c_0, c_m \geq 0, \ \gamma, \mu \geq 0 \text{ and } d(s) \geq d_0 > 0 \, a.e. \ s \in [0, m] \qquad (11.3.3)$$

$$\int_0^m \int_0^m \beta(s, y) \, ds \, dy < +\infty, \quad \int_0^m \beta_0(y) \, dy < +\infty, \quad \int_0^m \beta_m(y) \, dy < +\infty$$
$$(11.3.4)$$

where $\beta_0(y) = \beta(0, y)$ and $\beta_m(y) = \beta(m, y)$.

Unlike [124, 275] where the reproduction operator

$$\beta : u \to \int_0^m \beta(., y) u(y) \, dy \qquad (11.3.5)$$

is a bounded operator, β need not be bounded on $L^1(0, m)$. As in [124, 275], the Cauchy problem above is written in the following matrix form

$$\begin{cases} U'(t) = \mathcal{A} U(t) = (A + K) U(t), \\ U(0) = (u^0, u_0^0, u_m^0) \in \mathcal{X}, \end{cases}$$

where

$$A \begin{pmatrix} u \\ u_0 \\ u_m \end{pmatrix} = \begin{pmatrix} (du')' - (\gamma u)' - \mu u \\ (b_0 - \gamma(0)) u'(0) - \rho_0 u_0 \\ -(b_m + \gamma(m)) u'(m) - \rho_m u_m \end{pmatrix}$$

$$K \begin{pmatrix} u \\ u_0 \\ u_m \end{pmatrix} = \begin{pmatrix} \int_0^m \beta(., y) u(y) \, dy \\ \int_0^m \beta_0(y) u(y) \, dy \\ \int_0^m \beta_m(y) u(y) \, dy \end{pmatrix}$$

in the space

$$\mathcal{X} = (L^1(0, m) \times \mathbb{R}^2, \|.\|_{\mathcal{X}})$$

endowed with the norm

$$\|(u, u_0, u_m)\|_{\mathcal{X}} = \|u\|_{L^1(0, m)} + c_1 |u_0| + c_2 |u_m|$$

where

$$c_1 = \frac{d(0)}{b_0 - \gamma(0)}, \quad c_2 = \frac{d(m)}{b_m + \gamma(m)}.$$

The domain of A is

$$D(A) = \{(u, u_0, u_m) \in W^{2,1}(0, m) \times \mathbb{R}^2 : u(0) = u_0, u(m) = u_m\}$$

and A is shown to be the generator of a positive C_0-semigroup $(T(t))_{t \geq 0}$, (see [124] or [275]). Moreover, by using the Hopf maximum principle, $(T(t))_{t \geq 0}$ is shown to be irreducible, or equivalently the resolvent of its generator is positivity improving, see ([275] Theorem 2.4 and Proposition 2). We show that $\mathcal{A} := A + K$ generates a positive C_0-semigroup $(U(t))_{t\geq 0}$ (see Theorem 11.3.1) and the stability of essential type $\omega_{ess}(T) = \omega_{ess}(U)$ (see Theorem 11.3.2); both results rely on weak compactness arguments. We show also a spectral gap result $\omega_{ess}(U) < \omega(U)$ (see Theorem 11.3.3); the proof relies on strict comparison of spectral radius of positive operators.

11.3.1   Generation Theorem

The first key result is:

Theorem 11.3.1 *Let (11.3.1) (11.3.2) (11.3.3) (11.3.4) be satisfied. Then K is A-bounded and*

$$\mathcal{A} := A + K$$

with $D(\mathcal{A}) = D(A)$ generates a positive C_0-semigroup $(U(t))_{t\geq 0}$ on $\mathcal{X}$.

Proof (i) We observe that the (positive) reproduction operator decomposes as

$$\int_0^m \beta(x,y)u(y)dy = -\int_0^m \left(\int_0^y \beta(x,s)ds\right)u'(y)dy + \left(\int_0^m \beta(x,s)ds\right)u(m)$$

so

$$K\begin{pmatrix} u \\ u_0 \\ u_m \end{pmatrix} := \begin{pmatrix} -\int_0^m \left(\int_0^y \beta(x,s)ds\right)u'(y)dy + \widehat{\beta}(x)u(m) \\ \int_0^m \beta_0(y)u(y)dy \\ \int_0^m \beta_m(y)u(y)dy \end{pmatrix}$$

where

$$\widehat{\beta}(x) := \int_0^m \beta(x,s)ds.$$

Note that $W^{1,1}(0,m)$ imbedds continuously into $C([0,m])$ (see e.g. [73] Chap. 8) so

$$u \in W^{1,1}(0,m) \to u(m) \in \mathbb{R}$$

$$u \in W^{1,1}(0,m) \to \int_0^m \beta_0(y)u(y)dy \in \mathbb{R}$$

and

$$u \in W^{1,1}(0,m) \to \int_0^m \beta_m(y)u(y)dy \in \mathbb{R}$$

are continuous linear functionals on $W^{1,1}(0,m)$. Furthermore, the positive operator

$$\varphi \in L^1(0,m) \to \int_0^m \left(\int_0^y \beta(x,s)ds\right)\varphi(y)dy \in L^1(0,m)$$

is weakly compact since it is dominated by the rank-one operator

$$\varphi \in L^1(0, m) \to \left(\int_0^m \varphi(y)dy \right) \widehat{\beta}(x) \in L^1(0, m)$$

so

$$(u, u_0, u_m) \in W^{1,1}(0, m) \times \mathbb{R}^2 \to \begin{pmatrix} \int_0^m \beta(x, y)u(y)dy \\ \int_0^m \beta_0(y)u(y)dy \\ \int_0^m \beta_m(y)u(y)dy \end{pmatrix}$$

is weakly compact. Thus

$$K : D(A) \to \mathcal{X}$$

is weakly compact, i.e., K is A-weakly compact, and

$$\mathcal{A} := A + K : D(A) \to \mathcal{X}$$

is a generator of a positive semigroup $(U(t))_{t \geq 0}$ by virtue of Theorem 2.6.3. $\square$

The generation result given in Theorem 11.3.1 is also true, with the same arguments, when $m = +\infty$ where we drop the boundary condition at infinity, see [278]. Since $(T(t))_{t \geq 0}$ is irreducible (see [275]) and $T(t) \leq U(t)$ then $(U(t))_{t \geq 0}$ is irreducible regardless of K.

11.3.2   Stability of Essential Spectra

The second key result is:

Theorem 11.3.2 *Let (11.3.1) (11.3.2) (11.3.3) (11.3.4) be satisfied. Then $U(t) - T(t)$ is a weakly compact operator and therefore $(U(t))_{t \geq 0}$ and $(T(t))_{t \geq 0}$ have the same essential spectrum and consequently the same essential type.*

Proof Let $\beta^k(., .)$ be defined on $[0, m]^2$, bounded and such that

$$\beta^k(x, s) \leq \beta(x, s)$$

$$\beta^k \to \beta \text{ in } L^1\left[(0, m)^2\right] \ (k \to +\infty).$$

Let β_0^k be defined on $[0, m]$, bounded and such that

$$\beta_0^k \leq \beta_0 \text{ and } \beta_0^k \to \beta_0 \text{ in } L^1(0, m) \ (k \to +\infty).$$

Similarly, let β_m^k be defined on $[0, m]$, bounded and such that

$$\beta_m^k \leq \beta_m \text{ and } \beta_m^k \to \beta_m \text{ in } L^1(0, m) \ (k \to +\infty).$$

We introduce the bounded operator

$$K^k \begin{pmatrix} u \\ u_0 \\ u_m \end{pmatrix} := \begin{pmatrix} \int_0^m \beta^k(.,y)u(y)dy \\ \int_0^m \beta_0^k(y)u(y)dy \\ \int_0^m \beta_m^k(y)u(y)dy. \end{pmatrix} \tag{11.3.6}$$

Since

$$L^1(0,m) \ni u \to \int_0^m \beta^k(.,y)u(y)dy \in L^1(0,m)$$

is weakly compact and

$$L^1(0,m) \ni u \to \begin{pmatrix} \int_0^m \beta_0^k(y)u(y)dy \\ \int_0^m \beta_m^k(y)u(y)dy \end{pmatrix} \in \mathbb{R}^2$$

is compact then K^k is a weakly compact operator and consequently

$$\int_0^t U(t-s)K^kT(s)ds \ \ (0 < t < +\infty) \text{ is weakly compact} \tag{11.3.7}$$

(see Theorem 2.9.1). Let

$$\varphi \in D(A) \cap \mathcal{X}_+.$$

Note that $T(s)\varphi \in D(A)$ so $KT(s)\varphi$ is meaningfull since K is A-bounded. There exists $C \geq 1$ and $\lambda \in \mathbb{R}$ such that $\|U(t)\| \leq Ce^{\lambda t}$. Then

$$\left\| \int_0^t U(t-s)KT(s)\varphi ds - \int_0^t U(t-s)K^kT(s)\varphi ds \right\|$$

$$= \left\| \int_0^t U(t-s)\left(K - K^k\right)T(s)\varphi ds \right\|$$

$$\leq \int_0^t \|U(t-s)\| \left\|\left(K - K^k\right)T(s)\varphi\right\| ds$$

$$\leq C \int_0^t e^{\lambda(t-s)} \left\|\left(K - K^k\right)T(s)\varphi\right\| ds$$

$$= Ce^{\lambda t} \int_0^t \left\|\left(K - K^k\right)e^{-\lambda s}T(s)\varphi\right\| ds.$$

Since $K - K^k$ is positive then, by the additivity of the L^1 norm on the positive cone,

$$\int_0^t \left\| \left(K - K^k \right) e^{-\lambda s} T(s)\varphi \right\| ds = \left\| \int_0^t \left(K - K^k \right) e^{-\lambda s} T(s)\varphi ds \right\|$$

$$= \left\| \left(K - K^k \right) \int_0^t e^{-\lambda s} T(s)\varphi ds \right\|$$

$$\leq \left\| \left(K - K^k \right) \int_0^{+\infty} e^{-\lambda s} T(s)\varphi ds \right\|$$

$$= \left\| \left(K - K^k \right) \left(\lambda - A \right)^{-1} \varphi \right\| .$$

Since $D(A) \cap \mathcal{X}_+$ is dense in $\mathcal{X}_+$ and $(\lambda - A)^{-1} : \mathcal{X} \to L^\infty(0, m) \times \mathbb{R}^2$ is continuous then

$$\left\| \left(\int_0^t U(t - s) K T(s) ds \right) \varphi - \left(\int_0^t U(t - s) K^k T(s) ds \right) \varphi \right\|$$

$$\leq C e^{\lambda t} \left\| \left(K - K^k \right) \left(\lambda - A \right)^{-1} \varphi \right\| \quad (\varphi \in \mathcal{X}_+).$$

We note that for any constant $c > 0$

$$\begin{pmatrix} \int_0^m \beta^k(., y) u(y) dy \\ \int_0^m \beta_0^k(y) u(y) dy \\ \int_0^m \beta_m^k(y) u(y) dy \end{pmatrix} \to \begin{pmatrix} \int_0^m \beta(., y) u(y) dy \\ \int_0^m \beta_0(y) u(y) dy \\ \int_0^m \beta_m(y) u(y) dy \end{pmatrix} \quad (k \to +\infty)$$

in $\mathcal{X} = L^1(0, m) \times \mathbb{R}^2$ uniformly in $\|u\|_{L^\infty(0,m)} \leq c$. Since

$$(\lambda - A)^{-1} : \mathcal{X} \to L^\infty(0, m) \times \mathbb{R}^2$$

is continuous then

$$\sup_{\varphi \in \mathcal{X}_+, \ \|\varphi\| \leq 1} \left\| \left(K - K^k \right) \left(\lambda - A \right)^{-1} \varphi \right\| \to 0 \quad (k \to +\infty)$$

and consequently

$$\sup_{\varphi \in \mathcal{X}_+, \ \|\varphi\| \leq 1} \left\| \left(\int_0^t U(t - s) K T(s) ds \right) \varphi - \left(\int_0^t U(t - s) K^k T(s) ds \right) \varphi \right\| \to 0 \ (k \to +\infty).$$

Since

$$\int_0^t U(t - s) K T(s) ds - \int_0^t U(t - s) K^k T(s) ds$$

is a positive operator then

$$\int_0^t U(t - s) K^k T(s) ds \to \int_0^t U(t - s) K T(s) ds \ (k \to +\infty)$$

in operator norm. Then (11.3.7) implies that $\int_0^t U(t-s)KT(s)ds$ is weakly compact. Thus, for any $t > 0$, $U(t) - T(t)$ is a weakly compact operator and therefore $(U(t))_{t\geq 0}$ and $(T(t))_{t\geq 0}$ have the same essential spectrum, (see [181]). $\square$

Under an additional assumption, Theorem 11.3.2 is also true for $m = +\infty$, see [278].

11.3.3   Spectral Gap

We complement Theorem 11.3.2 by

Theorem 11.3.3 *Let (11.3.1) (11.3.2) (11.3.3) (11.3.4) be satisfied. Then $(U(t))_{t\geq 0}$ has a spectral gap.*

Proof Note first that $(\lambda - A)^{-1}$ is compact (since the imbedding of $D(A)$ into $\mathcal{X}$ is compact) and positivity improving and therefore irreducible so

$$r\left[(\lambda - A)^{-1}\right] > 0, \quad (\lambda > s(A)),$$

(see Proposition 2.4.2). On the other hand

$$r\left[(\lambda - A)^{-1}\right] = (\lambda - s(A))^{-1}$$

(see [286]) whence $s(A) > -\infty$ and

$$s(\mathcal{A}) \geq s(A) > -\infty.$$

We have seen in Theorem 11.3.2 that

$$\omega_{ess}(U) = \omega_{ess}(T).$$

We note that the irreducible compact operators $(\lambda - \mathcal{A})^{-1}$ and $(\lambda - A)^{-1}$ are such that
$$(\lambda - A)^{-1} \leq (\lambda - \mathcal{A})^{-1}.$$

Since $K \neq 0$ then $(\lambda - A)^{-1} \neq (\lambda - \mathcal{A})^{-1}$ whence

$$r\left[(\lambda - A)^{-1}\right] < r\left[(\lambda - \mathcal{A})^{-1}\right],$$

by Proposition 2.4.1. On the other hand $r\left[(\lambda - A)^{-1}\right] = (\lambda - s(A))^{-1}$ and

$$r\left[(\lambda - \mathcal{A})^{-1}\right] = (\lambda - s(\mathcal{A}))^{-1}$$

so $s(A) < s(\mathcal{A})$. Finally

$$\omega_{ess}(U) = \omega_{ess}(T) \le \omega(T) = s(A) < s(\mathcal{A}) = \omega(U)$$

and we are done. $\qquad\qquad\qquad\qquad\qquad\qquad\qquad\qquad\qquad\qquad\qquad\qquad\square$

Under an additional assumption, Theorem 11.3.3 extends to the case $m = +\infty$, see [278].

11.4 Growth-Fragmentation Equations

In this section, we deal with well-posedness and spectral gap analysis of growth-fragmentation equations

$$\frac{\partial}{\partial t}u(x,t) + \frac{\partial}{\partial x}(r(x)u(x,t)) + a(x)u(x,t) \qquad (11.4.1)$$
$$= \int_{x}^{+\infty} a(y)b(x,y)u(y,t)dy, \quad u(x,0) = u_0(x), \quad x,t > 0$$

with measurable nonnegative fragmentation kernel $b(.,.)$ and positive growth rate $r(.)^1$ satisfying the general structural assumptions

$$\int_0^y xb(x,y)dx = y\left(1 - \eta(y)\right), \quad 0 \le \eta(y) \le 1 \ (y \ge 0)$$

$$r(.) \in C\left(0, +\infty\right), \ \int_0^\infty \frac{1}{r(\tau)}d\tau = +\infty;$$

while the total fragmentation rate $a(.)$ is nonnegative and

$$a \in L^1_{loc}(0, +\infty).$$

Among the physical examples of growth rates we can find in the literature, note for instance the typical ones

$$r(x) = 1 \ \text{ or } r(x) = x, \quad (x > 0).$$

The equations above arise in the modeling of various physical or biological phenomena involving concentration of agregates which experience both growth and fragmentation. Typical biological examples are provided by phytoplankton dynamics [16, 32] or by prions dynamics [76, 121]; we refer to [131] and references therein for a lot of contexts where these equations arise; (see also [39, 40, 54–56, 235, 280] for more recent developments).

---

[1] In this monograph, $r(O)$ denotes in general the spectral radius of an operator O. This does not be confused with the growth rate function $r(.)$ used in this section.

The unknown $u(x,t)$ represents the concentration at time t of "agregates" with mass $x > 0$ while $b(x,y)$ $(x < y)$ describes the distribution of mass x agregates, called daughter agregates, spawned by the fragmentation of a mass y agregates. The local *mass conservation* in the fragmentation process corresponds to

$$\frac{1}{y} \int_0^y xb(x,y)dx = 1, \tag{11.4.2}$$

i.e., to $\eta(.) = 0$. In this case, we say that the kernel $b(.,.)$ is conservative. Most studies in the literature deal with fragmentation kernels that are conservative. On the other hand

$$\eta(.) \neq 0 \tag{11.4.3}$$

signifies that *mass loss* occurs during the fragmentation process, i.e.,

$$\frac{1}{y} \int_0^y xb(x,y)dx \leq 1$$

where

$$\eta(y) = 1 - \frac{1}{y} \int_0^y xb(x,y)dx$$

quantifies this mass loss ([33] Chap. 9). Some physical situations exhibit mass loss, see e.g. [164]. We point out that for a given concentration $u(.,.)$,

$$\int_0^{+\infty} u(x,t)xdx \ \text{ and } \ \int_0^{+\infty} u(x,t)dx$$

are respectively the total mass and the total number of agregates at time $t \geq 0$. Three natural L^1 functional spaces are of particular interest

$$\begin{cases} X_1 := L^1(\mathbb{R}_+; \ xdx), \\ X_0 := L^1(\mathbb{R}_+; \ dx), \\ X_{0,1} := L^1(\mathbb{R}_+; \ (1+x)\,dx), \end{cases}$$

where X_1 is the "finite mass" space, X_0 is the "finite agregates number" space and $X_{0,1}$ is the "finite mass and agregates number" space. A systematic construction in these functional spaces about well-posedness in the sense of semigroup theory and spectral gap results is given in [280].

As in the previous section, to acquaint the reader with some mathematical tools used in [280], we recall the main steps of the construction, mostly without proofs, in the "finite mass and agregates number" space $X_{0,1}$ under the assumption

$$\int_0^1 \frac{1}{r(\tau)}d\tau < +\infty, \quad \int_1^\infty \frac{1}{r(\tau)}d\tau = +\infty. \tag{11.4.4}$$

The main results consist in two key a priori estimates (Lemmas 11.4.1 and 11.4.2), a generation result (Theorem 11.4.2), the resolvent compactness of the generator (Lemma 11.4.3) and a spectral gap result (Theorem 11.4.3).

11.4.1   Preliminary Results

The method of characteristics shows that the partial differential equation

$$\frac{\partial}{\partial t}u(x,t) + \frac{\partial}{\partial x}[r(x)u(x,t)] = 0, \quad (x, t > 0)$$

with initial condition $u(x, 0) = f(x)$ and boundary condition

$$\lim_{y \to 0} r(y)u(y, t) = 0 \quad (t > 0)$$

has a unique solution given by

$$u(y, t) = \begin{cases} \frac{r(X(y,t))f(X(y,t))}{r(y)} & \text{if } \int_0^y \frac{1}{r(\tau)}d\tau > t \\ = 0 \text{ if } \int_0^y \frac{1}{r(\tau)}d\tau < t \end{cases}$$

where $X(y, t) > 0 \ (t > 0)$ is defined by

$$\int_{X(y,t)}^y \frac{1}{r(\tau)}d\tau = t, \quad \left(\text{if } \int_0^y \frac{1}{r(\tau)}d\tau > t\right). \tag{11.4.5}$$

Theorem 11.4.1 ([280] Theorem 2.9) *Let (11.4.4) be satisfied and let $X(y, t)$ be defined by (11.4.5). Then*

$$U_0(t)f := \chi_{\left\{\int_0^y \frac{1}{r(\tau)}d\tau > t\right\}} \frac{r(X(y,t))f(X(y,t))}{r(y)}$$

defines a positive C_0-semigroup on $X_{0,1}$ if and only if

$$[0, +\infty) \ni t \to \sup_{x>0} \frac{1 + y(x,t)}{1 + x}$$

is locally bounded where $y(x, t)$ is defined by

$$\int_x^{y(x,t)} \frac{1}{r(\tau)}d\tau = t \ (x > 0).$$

This occurs if

$$C_r := \sup_{x>0} \frac{r(x)}{1+x} < +\infty; \tag{11.4.6}$$

in this case, $\frac{1+y(x,t)}{x+1} \leq e^{C_r t}$ $(x > 0)$.

Let T_0 be the generator of $(U_0(t))_{t\geq 0}$. Then its resolvent is given by

$$\left((\lambda - T_0)^{-1} f\right)(y) = \frac{1}{r(y)} \int_0^y e^{-\int_x^y \frac{\lambda}{r(s)} ds} f(x) dx, \quad \mathrm{Re}\,\lambda > s(T_0)$$

where $s(T_0)$ is the spectral bound of T_0, (see [280] Proposition 6). Note that

$$L^1(\mathbb{R}_+; (1+x)\,dx) \subset L^1(\mathbb{R}_+; dx)$$

and

$$T_0 f = -\frac{\partial}{\partial x}\left(r(x) f(x)\right)$$

with domain

$$D(T_0) = \left\{ f \in X_{0,1}; \ \frac{\partial (rf)}{\partial y} \in X_{0,1}, \ \lim_{y\to 0} r(y) f(y) = 0 \right\}$$

where $\frac{\partial(rf)}{\partial y}$ is the derivative (in the sens of distributions on $(0, +\infty)$) of the function $rf \in L^1(\mathbb{R}_+; dx)$. Note that $rf \in W^{1,1}(\mathbb{R}_+)$ so that $\lim_{y\to 0} r(y) f(y)$ exists.

Lemma 11.4.1   (see [280] Lemma 2.10) *The resolvent enjoys a pointwise estimate*

$$\left|(\lambda - T_0)^{-1} f\right|(y) \leq \frac{1}{(1+y)r(y)} \|f\|_{X_{0,1}} \quad (\lambda > C_r). \tag{11.4.7}$$

Similarly,

$$\frac{\partial}{\partial t} u(x,t) + \frac{\partial}{\partial x}[r(x)u(x,t)] + a(x)u(x,t) = 0, \ u(x,0) = f(x)$$

with boundary condition $\lim_{y\to 0} r(y)u(y,t) = 0$ $(t > 0)$ gives rise to a positive C_0-semigroup $\left(U_0^{(a)}(t)\right)_{t\geq 0}$ on $X_{0,1}$ given by

$$U_0^{(a)}(t)f := e^{-\int_{X(y,t)}^y \frac{a(p)}{r(p)} dp} f(X(y,t)) \frac{\partial X(y,t)}{\partial y} = e^{-\int_{X(y,t)}^y \frac{a(p)}{r(p)} dp} U_0(t)f.$$

The resolvent of its generator is given by

$$\left((\lambda - T_0^{(a)})^{-1} f\right)(y) = \frac{1}{r(y)} \int_0^y e^{-\int_x^y \frac{\lambda+a(\tau)}{r(\tau)} d\tau} f(x) dx \quad (\mathrm{Re}\,\lambda > s(T_0^{(a)}))$$

$(s(T_0^{(a)})$ is the spectral bound of $T_0^{(a)})$ and

$$D(T_0^{(a)}) = \{f \in D(T_0);\ af \in X_{0,1}\},\ T_0^{(a)}f = T_0 f - af.$$

We note that $(\lambda - T_0^{(a)})^{-1}$ is dominated by $(\lambda - T_0)^{-1}$ and consequently inherites the pointwise estimate (11.4.7)

$$\left|(\lambda - T_0^{(a)})^{-1}f\right|(y) \le \frac{1}{(1+y)r(y)} \|f\|_{X_{0,1}}\ (\lambda > C_r),\ (f \in X_{0,1}) \qquad (11.4.8)$$

A second key estimate of $(\lambda - T_0^{(a)})^{-1}$ is given by

Lemma 11.4.2 ([280] Lemma 2.11) *Let (11.4.6) be satisfied. Then for all $\lambda > C_r$*

$$\int_0^{+\infty} \left|\left((\lambda - T_0^{(a)})^{-1}f\right)(y)\right| a(y)y\,dy \le \int_0^{+\infty} |(f(y)|\,y\,dy,\ \left(f \in X_{0,1}\right).$$
$$(11.4.9)$$

We introduce now the fragmentation operator

$$B : \varphi \in D(T_0^{(a)}) \to \int_x^{+\infty} a(y)b(x, y)\varphi(y)dy. \qquad (11.4.10)$$

The main generation result relies on a key estimate and on Theorem 2.6.1.

Theorem 11.4.2 ([280] Theorem 2.12) *Let (11.4.6) be satisfied. Suppose that* $\sup_{y>0} \frac{n(y)}{1+y} < +\infty$ *where* $n(y) := \int_0^y b(x, y)dx$. *Then the fragmentation operator (11.4.10) is* $T_0^{(a)}$*-bounded in* $X_{0,1}$ *and*

$$\lim_{\lambda \to +\infty} \left\| B(\lambda - T_0^{(a)})^{-1} \right\|_{\mathcal{L}(X_{0,1})} \le \lim \sup_{a(y) \to +\infty} \frac{y(1 - \eta(y)) + n(y)}{1 + y}.$$

In particular, $T_0^{(a)} + B : D(T_0^{(a)}) \subset X_{0,1} \to X_{0,1}$ *generates a positive semigroup* $(V(t))_{t \ge 0}$ *in* $X_{0,1}$ *if*

$$\begin{cases} \liminf_{a(y) \to +\infty} \eta(y) > 0 \\ \limsup_{a(y) \to +\infty} \frac{n(y)}{1+y} = 0. \end{cases} \qquad (11.4.11)$$

One observes that the assumption of mass loss is needed only in neighborhoods of points where the fragmentation rate $a(.)$ diverges. Due to (11.4.4), the existence of a spectral gap depends on a (weak) compactness result which relies on the "singularity" of the fragmentation rate $a(.)$ at infinity. Indeed, let

$$\Omega_c = \{x > 0;\ a(x) < c\}\quad (c > 0)$$

be the sublevel sets the fragmentation rate $a(.)$. We say that the sublevel sets of a are thin at infinity relatively to r (thin at infinity for short) if

$$\int_1^\infty \frac{1_{\Omega_c}(\tau)}{r(\tau)} d\tau < +\infty \quad (c > 0). \tag{11.4.12}$$

In particular, if $\lim_{y \to +\infty} a(y) = +\infty$ then the sublevel sets of a are automatically thin at infinity.

Lemma 11.4.3 ([280] Theorem 2.13) *Let (11.4.6) (11.4.12) be satisfied. Then $T_0^{(a)}$ is resolvent compact on $X_{0,1}$.*

Proof The proof is based on the estimates (11.4.8) (11.4.9). Indeed, the pointwise estimate (11.4.8) insures that $(\lambda - T_0^{(a)})^{-1}$ is "locally weakly compact" on $[0, +\infty)$ while (11.4.9) and the fact that the sublevel sets of a are thin at infinity capture weak compactness at infinity; (see [280] for the details). $\square$

It follows from Lemma 11.4.3 that $T_0^{(a)} + B : D(T_0^{(a)}) \subset X_{0,1} \to X_{0,1}$ is also resolvent compact in $X_{0,1}$ since

$$(\lambda - T_0^{(a)} - B)^{-1} = (\lambda - T_0^{(a)})^{-1} \sum_{j=0}^{+\infty} (B(\lambda - T_0^{(a)})^{-1})^j$$

for λ large enough. One can show that if the support of $a(.)$ is not bounded then $(V(t))_{t \geqslant 0}$ is irreducible in $X_{0,1}$, (see [280] Lemma 2.14).

11.4.2   Spectral Gap for Growth-Fragmentation Semigroups

We are ready to give the key idea for a proof of the main statement.

Theorem 11.4.3 *Let (11.4.6) (11.4.11) (11.4.12) be satisfied and let the support of $a(.)$ be unbounded. Then $(V(t))_{t \geqslant 0}$ has spectral gap.*

Proof The idea of the proof consists in introducing a (bounded) weakly compact operator $\widehat{B}$ on $X_{0,1}$ such that

$$\widehat{B}\varphi \leq B\varphi, \quad (\varphi \in D_+(T_0^{(a)})).$$

Then the same construction shows that $T_0^{(a)} + (B - \widehat{B})$ is resolvent compact and generates a positive semigroup $(\widehat{V}(t))_{t \geqslant 0}$. Note that

$$s(T_0^{(a)} + (B - \widehat{B})) \geq s(T_0^{(a)}).$$

Since

$$V(t) - \widehat{V}(t) = \int_0^t \widehat{V}(t - s)\widehat{B}V(s)ds$$

is weakly compact (see Theorem 2.9.1) then

$$\omega_{ess}(V) = \omega_{ess}(\widehat{V}) \le s\left(T_0^{(a)} + (B - \widehat{B})\right).$$

On the other hand the inequality between compact operators

$$\left(\lambda - T_0^{(a)} - (B - \widehat{B})\right)^{-1} \le \left(\lambda - T_0^{(a)} - B\right)^{-1}$$

(it is not an equality) implies

$$r\left(\left(\lambda - T_0^{(a)} - (B - \widehat{B})\right)^{-1}\right) < r\left(\left(\lambda - T_0^{(a)} - B\right)^{-1}\right)$$

(see Proposition 2.4.1) or equivalently

$$s\left(T_0^{(a)} + (B - \widehat{B})\right) < s\left(T_0^{(a)} + B\right).$$

Finally $\omega_{ess}(V) < s\left(T_0^{(a)} + B\right) = \omega(V)$. $\square$

We can build a spectral theory similar to that of Sect. 11.4 in the "finite mass" space $X_1 := L^1(\mathbb{R}_+;\ xdx)$ when

$$\int_0^1 \frac{1}{r(\tau)}d\tau = +\infty, \quad \int_1^\infty \frac{1}{r(\tau)}d\tau = +\infty. \tag{11.4.13}$$

The only change is that we have to assume that the sublevel sets of a are thin (relatively to r) at both zero and infinity in the sense that

$$\int_0^{+\infty} \frac{1_{\Omega_c}(\tau)}{r(\tau)}d\tau < +\infty \ \ (c > 0), \tag{11.4.14}$$

(for instance $a(.)$ tends to infinity at both 0 and $+\infty$). In this case, the previous assumption (11.4.11) simplifies to

$$\lim_{a(y)\to+\infty} \inf\ \eta(y) > 0, \tag{11.4.15}$$

see [280].

11.5   A Runaway Phenomenon for Growth Fragmentation Equations

In this section, we explore new problems. We consider growth fragmentation equations in the "finite mass" space X_1 under (11.4.13) and drop the assumption that the sublevel sets of a are thin relatively to r; more precisely, we suppose that

$$\int_0^{+\infty} \frac{a(y)}{r(y)} dy < +\infty; \tag{11.5.1}$$

roughly speaking, instead of assuming that $a(y)$ tends to infinity as $y \to 0$ and $y \to +\infty$, we now assume that $a(y)$ tends to zero at both ends. We drop also the mass loss assumption, i.e., we suppose (11.4.2). We show here that this setting falls within the framework of *scattering theory*. To the best of our knowledge, the results presented in this section appear here for the first time. To enable fully explicit calculations, we consider a specific choice of growth rate

$$\begin{cases} r(x) = \alpha x \ \ (x > 0) \\ \quad \text{with } \alpha > 0. \end{cases} \tag{11.5.2}$$

In this case, the growth equation (with $a(.) = 0$) is governed by a C_0-group $(U_0(t))_{t \in \mathbb{R}}$ (with generator T_0) on X_1 given by

$$(U_0(t)f)(y) = f(X(y,t))\frac{\partial X(y,t)}{\partial y}, \ \ (t \in \mathbb{R})$$

where $X(y,t) > 0$ is defined uniquely by

$$\int_{X(y,t)}^{y} \frac{1}{r(\tau)} d\tau = t, \ \ (t \in \mathbb{R})$$

where $X(y,t) \le y$ if $t \ge 0$ and $X(y,t) > y$ if $t < 0$. Since $r(\tau) = \alpha\tau$ then $\ln \frac{y}{X(y,t)} = \alpha t$ or

$$X(y,t) = ye^{-\alpha t}.$$

Similarly

$$\int_y^{Y(y,t)} \frac{1}{r(\tau)} d\tau = t, \ \ (t \in \mathbb{R})$$

shows that $Y(y,-t) = X(y,t)$, $(t \in \mathbb{R})$. The change of variable $X(y,t) = x$ shows that

$$\|U_0(t)f\|_{X_1} = \int_0^{+\infty} |f(X(y,t))| \frac{\partial X(y,t)}{\partial y} ydy$$

$$= e^{\alpha t} \int_0^{+\infty} |f(x)|\, x\, dx = e^{\alpha t}\, \|f\|_{X_1};$$

in particular $\left(e^{-\alpha t} U_0(t)\right)_{t\in\mathbb{R}}$ is a stochastic C_0-group on X_1. If we introduce a fragmentation rate

$$a(.) \in L^1_{loc}(0, +\infty)$$

satisfying (11.5.1) then we get a positive (absorption) weighted C_0-group $\left(U_0^{(a)}(t)\right)_{t\in\mathbb{R}}$ on X_1 given by

$$U_0^{(a)}(t)f = e^{-\int_{X(y,t)}^y \frac{a(p)}{r(p)}dp} U_0(t)f, \quad f \in X_1.$$

Note that $C_c^1(0, +\infty)$ (the C^1-functions on $(0, +\infty)$ with compact supports) is a core for T_0. This follows from the fact that $C_c^1(0, +\infty)$ is included in $D(T_0)$ and $(U_0(t))_{t\geq 0}$ leaves invariant $C_c^1(0, +\infty)$ (see [97] Theorem 1.9). Since $C_c^1(0, +\infty) \subset D(T_0) \cap D(a)$ then $D(T_0) \cap D(a)$ is a core for T_0 and consequently (see Lemma 5.3.2) a is U_0-regular. In particular (see Remark 5.3.1)

$$\begin{cases} T_0^{(a)} = T_0 - a \\ D(T_0^{(a)}) = D(T_0) \cap D(a). \end{cases}$$

In addition to several preliminary results, we compute the spectral bound of the generator and show that it does not correspond to an eigenvalue (Theorem 11.5.1). We show also the existence of wave operators (Theorem 11.5.2).

11.5.1   A "Negative" Spectral Result

We start with

Lemma 11.5.1 *We assume (11.5.2). If*

$$\int_0^{+\infty} \frac{a(x)}{x}dx < +\infty \tag{11.5.3}$$

then the type of $\left(U_0^{(a)}(t)\right)_{t\geq 0}$ is equal to α.

Proof We have

$$\left\|U_0^{(a)}(t)f\right\|_{X_1} = \int_0^{+\infty} e^{-\int_{X(y,t)}^y \frac{a(p)}{r(p)}dp} f(X(y,t)) \frac{\partial X(y,t)}{\partial y} y\, dy.$$

The change of variable $x = X(y, t)$, i.e.,

$$\int_x^{y(x,t)} \frac{1}{r(\tau)} d\tau = t, \tag{11.5.4}$$

shows that $\left\| U_0^{(a)}(t) f \right\|_{X_1}$ is equal to

$$\int_0^{+\infty} e^{-\int_x^{y(x,t)} \frac{a(p)}{r(p)} dp} f(x) y(x,t) dx = \int_0^{+\infty} e^{-\int_x^{y(x,t)} \frac{a(p)}{r(p)} dp} \frac{y(x,t)}{x} f(x) x \, dx$$

and

$$\left\| U_0^{(a)}(t) \right\|_{\mathcal{L}(X_1)} = \sup_{x>0} \left(e^{-\int_x^{y(x,t)} \frac{a(p)}{r(p)} dp} \frac{y(x,t)}{x} \right).$$

Hence

$$\ln \left(\left\| U_0^{(a)}(t) \right\|_{\mathcal{L}(X_1)} \right)$$
$$= \ln \left(\sup_{x>0} e^{-\int_x^{y(x,t)} \frac{a(p)}{r(p)} dp} \frac{y(x,t)}{x} \right)$$
$$= \sup_{x>0} \ln \left(e^{-\int_x^{y(x,t)} \frac{a(p)}{r(p)} dp} \frac{y(x,t)}{x} \right)$$
$$= \sup_{x>0} \left[-\int_x^{y(x,t)} \frac{a(p)}{r(p)} dp + \ln \frac{y(x,t)}{x} \right].$$

Since $r(x) = \alpha x$ then (11.5.4) amounts to $\frac{y(x,t)}{x} = e^{\alpha t}$ and

$$t^{-1} \ln \left(\left\| U_0^{(a)}(t) \right\|_{\mathcal{L}(X_1)} \right)$$
$$= \sup_{x>0} \left[-t^{-1} \alpha^{-1} \int_x^{y(x,t)} \frac{a(p)}{p} dp \right] + \alpha.$$

Finally (11.5.3) implies

$$\lim_{t \to +\infty} t^{-1} \ln \left(\left\| U_0^{(a)}(t) \right\|_{\mathcal{L}(X_1)} \right) = \alpha$$

and ends the proof. $\square$

Let $T_0^{(a)}$ be the generator of $\left(U_0^{(a)}(t) \right)_{t \geq 0}$. Then, under (11.5.3), $s(T_0^{(a)}) = \alpha$ and

$$\left((\lambda - T_0^{(a)})^{-1} f \right)(y) = \frac{1}{r(y)} \int_0^y e^{-\int_x^y \frac{\lambda + a(\tau)}{r(\tau)} d\tau} f(x) dx \ (\lambda > \alpha);$$

moreover, we have the pointwise estimate

$$\left|(\lambda - T_0^{(a)})^{-1} f\right| \leq \frac{1}{yr(y)} \, \|f\|_{X_1} \, ; \tag{11.5.5}$$

(see [280] for the details). A second key preliminary result is

Lemma 11.5.2 *Let (11.5.3) be satisfied. Then* $B(\lambda - T_0^{(a)})^{-1}$ $(\lambda > \alpha)$ *is weakly compact on* X_1. *Moreover,* $B(\lambda - T_0^{(a)})^{-1}$ *extends to* $\lambda = \alpha$ *as a weakly compact strict contraction on* X_1.

Proof The pointwise estimate (11.5.5) shows that for $f \geq 0$

$$B(\lambda - T_0^{(a)})^{-1} f$$
$$\leq \int_x^{+\infty} a(y) b(x, y) \left((\lambda - T)^{-1} f\right)(y) dy$$
$$\leq \int_x^{+\infty} b(x, y) \frac{a(y)}{yr(y)} dy. \, \|f\|_{X_1}$$
$$= \psi(x) \|f\|_{X_1}$$

where $\psi \in X_1$ since

$$\int_0^{+\infty} \psi(x) x dx = \int_0^{+\infty} \left(\int_x^{+\infty} b(x, y) \frac{a(y)}{yr(y)} dy\right) x dx$$
$$= \int_0^{+\infty} \left(\int_0^y b(x, y) x dx\right) \frac{a(y)}{yr(y)} dy$$
$$= \int_0^{+\infty} \frac{a(y)}{r(y)} dy < +\infty.$$

Hence $B(\lambda - T_0^{(a)})^{-1}$ is dominated by the rank one operator

$$f \in X_1 \to \psi \int_0^{+\infty} f(y) y dy \in X_1$$

and consequently is weakly compact. Note that this estimate is uniform in $\lambda > \alpha$ so

$$f \in X_1 \to B(\alpha - T_0^{(a)})^{-1} f := \lim_{\lambda \to \alpha_+} B(\lambda - T_0^{(a)})^{-1} f$$

is not only a bounded operator on X_1 but also weakly compact on X_1. Note that

$$\left\|B(\alpha - T_0^{(a)})^{-1} f\right\|_{X_1} = \left\|a(\alpha - T_0^{(a)})^{-1} f\right\|_{X_1}$$

and

$$\left\|a(\alpha - T_0^{(a)})^{-1} f\right\|_{X_1}$$

$$
\begin{aligned}
&= \int_0^{+\infty} \frac{ya(y)}{r(y)}dy \int_0^y e^{-\int_x^y \frac{\alpha+a(\tau)}{r(\tau)}d\tau} f(x)dx \\
&= \int_0^{+\infty} \frac{ya(y)}{r(y)}dy \int_0^y e^{-\int_x^y \frac{1}{\tau}d\tau} e^{-\int_x^y \frac{a(\tau)}{\alpha\tau}d\tau} f(x)dx \\
&= \int_0^{+\infty} \frac{ya(y)}{r(y)}dy \int_0^y \frac{x}{y} e^{-\int_x^y \frac{a(\tau)}{\alpha\tau}d\tau} f(x)dx \\
&= \int_0^{+\infty} xf(x) \left(\int_x^{+\infty} \frac{a(y)}{\alpha y} e^{-\int_x^y \frac{a(\tau)}{\alpha\tau}d\tau} dy \right) dx.
\end{aligned}
$$

Finally

$$
\int_x^{+\infty} \frac{a(y)}{\alpha y} e^{-\int_x^y \frac{a(\tau)}{\alpha\tau}d\tau} dy = \int_x^{+\infty} -\frac{d}{dy} e^{-\int_x^y \frac{a(\tau)}{\alpha\tau}d\tau} dy
$$

$$
= 1 - e^{-\int_x^{+\infty} \frac{a(\tau)}{\alpha\tau}d\tau} \le 1 - e^{-\int_0^{+\infty} \frac{a(\tau)}{\alpha\tau}d\tau}
$$

shows that

$$
\left\| B(\alpha - T_0^{(a)})^{-1} \right\|_{\mathcal{L}(X_1)} \le 1 - e^{-\int_0^{+\infty} \frac{a(\tau)}{\alpha\tau}d\tau}. \tag{11.5.6}
$$

$\square$

We are ready to show

Theorem 11.5.1 *The spectral bound of $T_0^{(a)} + B$ is equal to α and is not an eigenvalue.*

Proof Note first that $s(T_0^{(a)} + B) \ge s(T_0^{(a)}) = \alpha$. The estimate (11.5.6) shows that

$$
r(B(\lambda - T_0^{(a)})^{-1}) < 1, \quad (\lambda \ge \alpha)
$$

so $s(T_0^{(a)} + B) \le \alpha$ and $s(T_0^{(a)} + B) = \alpha$. Finally, α is not an eigenvalue of $T_0^{(a)} + B$ because 1 is not eigenvalue of $B(\alpha - T_0^{(a)})^{-1}$. $\square$

11.5.2　Runaway Phenomenon

In fact, the above "negative" spectral result suggests that we are operating within the framework of scattering theory. We cannot directly derive this regime from the scattering theory for perturbed Frobenius–Perron semigroups developed in Chap. 8, as that framework is specifically designed for measure-preserving flows. It is more convenient to invoke the general scattering theory developed in Sect. 3.4 after appropriately rescaling the C_0-semigroups. We denote the stochastic C_0-group $\left(e^{-\alpha t}U_0(t)\right)_{t\in\mathbb{R}}$ by $\left(\widehat{U}_0(t)\right)_{t\in\mathbb{R}}$ and its generator by $\widehat{T}_0 := T_0 - \alpha$. Similarly, we denote the C_0-group

$\left(e^{-\alpha t}U_0^{(a)}(t)\right)_{t\in\mathbb{R}}$ by $\left(\widehat{U_0^{(a)}}(t)\right)_{t\in\mathbb{R}}$ and its generator by $\widehat{T_0^{(a)}} := \widehat{T_0} - a$. Note that the group $\left(\widehat{U_0^{(a)}}(t)\right)_{t\in\mathbb{R}}$ given by

$$\widehat{U_0^{(a)}}(t)f = e^{-\int_{X(y,t)}^{y}\frac{a(p)}{r(p)}dp}\,\widehat{U_0}(t)f$$

is contractive for $t \geq 0$, and uniformly bounded for $t \leq 0$ due to (11.5.3). Hence

$$\sigma(\widehat{T_0^{(a)}}) \subset i\mathbb{R}.$$

Finally, we denote the C_0-group $\left(e^{-\alpha t}V(t)\right)_{t\in\mathbb{R}}$ by $\left(\widehat{V}(t)\right)_{t\geq0}$ with generator $\widehat{T_0^{(a)}} + B$. Note that $\left(\widehat{V}(t)\right)_{t\geq0} = \left(e^{-\alpha t}V(t)\right)_{t\geq0}$ is contractive for $t \geq 0$. We are now ready to show

Theorem 11.5.2 *Let (11.5.3) be satisfied. Then the wave operators*

$$\begin{cases} \Omega^+ := s\lim_{t\to+\infty} \widehat{V}(t)\widehat{U_0^{(a)}}(-t) \\ \Omega^- : s\lim_{t\to+\infty} \widehat{U_0^{(a)}}(-t)\widehat{V}(t) \end{cases}$$

exist.

Proof We already know that $\sigma(\widehat{T_0^{(a)}}) \subset i\mathbb{R}$. We consider first Ω^+. According to Theorem 3.4.1, it suffices to check that the (left) strong limit

$$B\left(0_- - \widehat{T_0^{(a)}}\right)^{-1} := s\lim_{\lambda\to0_-} B\left(\lambda - \widehat{T_0^{(a)}}\right)^{-1} \text{ exists.}$$

The resolvent of $T_0^{(a)}$ is given by

$$\left(\lambda - \widehat{T_0^{(a)}}\right)^{-1}f = \int_0^{+\infty} e^{-\lambda t}e^{-\int_{ye^{-\alpha t}}^{y}\frac{a(p)}{\alpha p}dp}\,f(ye^{-\alpha t})e^{-\alpha t}dt \quad (\lambda > \alpha)$$

so

$$\left(\alpha_+ - \widehat{T_0^{(a)}}\right)^{-1}f = \int_0^{+\infty} e^{-\int_{ye^{-\alpha t}}^{y}\frac{a(p)}{\alpha p}dp}\,f(ye^{-\alpha t})e^{-2\alpha t}dt,$$

i.e.,

$$\left(0_+ - \widehat{T_0^{(a)}}\right)^{-1}f = \left(0_+ - (T_0^{(a)} - \alpha)\right)^{-1}f$$

$$= \int_0^{+\infty} e^{-\int_{ye^{-\alpha t}}^{y}\frac{a(p)}{\alpha p}dp}\,f(ye^{-\alpha t})e^{-2\alpha t}dt.$$

Similarly,

$$\left(0_- - \widehat{T_0^{(a)}}\right)^{-1} f = -\int_{-\infty}^0 e^{\int_y^{ye^{\alpha t}} \frac{a(p)}{\alpha p} dp} f(ye^{\alpha t}) e^{2\alpha t} dt.$$

We have

$$\left\| B\left(0_- - \widehat{T_0^{(a)}}\right)^{-1} \right\|_{X_1} = \left\| a\left(0_- - \widehat{T_0^{(a)}}\right)^{-1} f \right\|_{X_1}$$

and

$$\left\| a\left(0_- - \widehat{T_0^{(a)}}\right)^{-1} f \right\|_{X_1}$$

$$= \int_0^{+\infty} ya(y)dy \int_{-\infty}^0 e^{\int_y^{ye^{\alpha t}} \frac{a(p)}{\alpha p} dp} f(ye^{\alpha t}) e^{2\alpha t} dt$$

$$= \int_{-\infty}^0 e^{2\alpha t} dt \int_0^{+\infty} e^{\int_y^{ye^{\alpha t}} \frac{a(p)}{\alpha p} dp} f(ye^{\alpha t}) ya(y) dy$$

The change of variable $x = ye^{\alpha t}$ gives

$$\int_0^{+\infty} e^{\int_y^{ye^{\alpha t}} \frac{a(p)}{\alpha p} dp} f(ye^{\alpha t}) ya(y) dy$$

$$= \int_0^{+\infty} e^{\int_{xe^{-\alpha t}}^x \frac{a(p)}{\alpha p} dp} f(x) xe^{-\alpha t} a(xe^{-\alpha t}) e^{-\alpha t} dx$$

so $\left\| a\left(0_- - \widehat{T_0^{(a)}}\right)^{-1} f \right\|_{X_1}$ is equal to

$$\int_{-\infty}^0 e^{2\alpha t} dt \int_0^{+\infty} e^{\int_{xe^{-\alpha t}}^x \frac{a(p)}{\alpha p} dp} f(x) xe^{-\alpha t} a(xe^{-\alpha t}) e^{-\alpha t} dx$$

$$= \int_0^{+\infty} f(x) x dx \int_{-\infty}^0 a(xe^{-\alpha t}) e^{\int_{xe^{-\alpha t}}^x \frac{a(p)}{\alpha p} dp} dt.$$

On the other hand

$$\frac{d}{dt} e^{\int_{xe^{-\alpha t}}^x \frac{a(p)}{\alpha p} dp} = -\alpha xe^{-\alpha t} \left(\frac{a(xe^{-\alpha t})}{\alpha xe^{-\alpha t}} \right) e^{\int_{xe^{-\alpha t}}^x \frac{a(p)}{\alpha p} dp}$$

$$= -a(xe^{-\alpha t}) e^{\int_{xe^{-\alpha t}}^x \frac{a(p)}{\alpha p} dp}$$

so

$$\int_{-\infty}^{0} a(xe^{-\alpha t}) e^{\int_{xe^{-\alpha t}}^{x} \frac{a(p)}{\alpha p} dp} dt = \int_{-\infty}^{0} -\frac{d}{dt} e^{\int_{xe^{-\alpha t}}^{x} \frac{a(p)}{\alpha p} dp} dt$$

$$= e^{\int_{x}^{+\infty} \frac{a(p)}{\alpha p} dp} - 1$$

and

$$\left\| a \left(0_- - \widehat{T_0^{(a)}} \right)^{-1} f \right\|_{X_1} \leq \left(e^{\int_{0}^{+\infty} \frac{a(p)}{\alpha p} dp} - 1 \right) \| f \|_{X_1}.$$

This insures the existence of Ω^+. Finally, thanks to (11.5.3)

$$\left\| \widehat{U_0^{(a)}}(t) f \right\|_{X_1} \geq e^{-\alpha^{-1} \int_{0}^{+\infty} \frac{a(p)}{p} dp} \left\| \widehat{U_0}(t) f \right\|_{X_1}$$

$$= e^{-\alpha^{-1} \int_{0}^{+\infty} \frac{a(p)}{p} dp} \| f \|_{X_1}$$

and Theorem 3.4.2 insures the existence of Ω^-. $\qquad\qquad\square$

As in the particle swarm framework discussed in Sect. 10.13, we are now in a position to demonstrate a runaway effect, as a direct consequence of Theorem 3.4.3.

Theorem 11.5.3 *Let (11.5.3) be satisfied. Then*

$$e^{-\alpha t} \left\| V(t)\varphi - U_0^{(a)} \Omega^- \varphi \right\| \to 0 \ (t \to +\infty) \ (\varphi \in X_1),$$

i.e., (up to a scaling) any trajectory $(V(t)\varphi)_{t \geq 0}$ *of the full dynamics (with initial data* φ*) behaves asymptotically (as* $t \to +\infty$*) as a free trajectory* $\left(U_0^{(a)} \psi \right)_{t \geq 0}$ *(with initial data* $\psi = \Omega^- \varphi$*).*

11.6 Growth Equations with Non Local Boundary Condition

In this section, still under the condition (11.4.4), we deal with growth equations without fragmentation, but with a death rate $a(.)$

$$\frac{\partial}{\partial t} u(x, t) + \frac{\partial}{\partial x} (r(x)u(x, t)) + a(x)u(x, t) = 0 \qquad (11.6.1)$$

and nonlocal McKendrick-von Foerster boundary condition

$$\lim_{x \to 0_+} (r(x)u(x)) = \int_{0}^{+\infty} \beta(y)u(y)dy \qquad (11.6.2)$$

in the space

$$X_{0,\alpha} := L^1(\mathbb{R}_+;\ (1+x^\alpha)\,dx)$$

with

$$\beta_\alpha := \sup_{x>0} \frac{\beta(x)}{1+x^\alpha} < +\infty,$$

i.e., $\beta \in X_{0,\alpha}^*$. We denote by $\left(U_\beta^{(a)}(t)\right)_{t\geq 0}$ the growth C_0-semigroup with generator $T_\beta^{(a)}$ associated with (11.6.1)(11.6.2). We refer to [40] for the details. Our object here is to study the long-time asymptotic behavior of $\left(U_\beta^{(a)}(t)\right)_{t\geq 0}$ ($\beta \neq 0$). To the best of our knowledge, this problem has not yet been investigated, and the construction presented here is new. On the other hand, a simpler model in $L^1(\mathbb{R}_+;\ dx)$, known as the *age-dependent* population equation, corresponding to

$$r(x) = 1$$

is well documented and well understood, see e.g. [147] or ([120] Chap. VI, p. 216). We point out that the full growth-fragmentation equations were considered in [39] for the boundary condition

$$\lim_{x\to 0_+} (r(x)u(x)) = 0, \tag{11.6.3}$$

i.e., for $\beta = 0$ where the properties of the fragmentation operator play a key role in the analysis. We will show that, under the condition

$$\beta \neq 0,$$

a "similar" theory can be developed for growth equations without fragmentation operators. The construction is as technically involved as in the case with fragmentation; in fact, to some extent, the nonlocal boundary condition acts as a substitute for the fragmentation operator. Moreover, unlike [39], no threshold on α is needed here; in particular, we can cover the physical space

$$X_{0,1} := L^1(\mathbb{R}_+;\ (1+x)\,dx).$$

Actually, for the sake of simplicity, we restrict ourselves to the space $X_{0,1}$ i.e., to $\alpha = 1$ with

$$\beta_1 := \sup_{x>0} \frac{\beta(x)}{1+x} < +\infty.$$

We recall (see [40] Theorem 4.8) that in $X_{0,1}$

$$\left(\lambda - T_\beta^{(a)}\right)^{-1} f = \left(\lambda - T_0^{(a)}\right)^{-1} f + \frac{\langle \beta, \left(\lambda - T_0^{(a)}\right)^{-1} f \rangle_{X_{0,1}^*, X_{0,1}}}{1 - \langle \beta, e_\lambda \rangle} e_\lambda \qquad (11.6.4)$$

for

$$\lambda > \lambda^* := \beta_1 + 2C_r$$

where

$$e_\lambda(x) = e^{-\lambda \int_0^x r(s)^{-1} ds + \int_0^x a(s)r(s)^{-1} ds} \in X_{0,1}$$

and $\frac{1}{1-\langle \beta, e_\lambda \rangle} \leq \frac{1}{\lambda - \lambda^*}$ $(\lambda > \lambda^*)$.

We show that $\frac{e_{\lambda^*+\lambda}(x)}{1-\langle \beta, e_{\lambda^*+\lambda} \rangle}$ $(\lambda > 0)$ is the Laplace transform of a $X_{0,1}$-vector measure (see Lemma 11.6.1). This allows us to show that $U_\beta^{(a)}(t) - U_0^{(a)}(t)$ is weakly compact on $X_{0,1}$ (see Theorem 11.6.2). In the spirit of the analysis of growth-fragmentation equations with local boundary conditions, we show that $\left(U_\beta^{(a)}(t)\right)_{t\geq 0}$ has spectral gap when the sublevel sets of $a(.)$ are thin at infinity, e.g. when $\lim_{x\to+\infty} a(x) = +\infty$, (see Theorem 11.6.3). We deal also with spectral gaps when the sublevel sets of $a(.)$ are not thin at infinity (e.g. when $a(.)$ is bounded at infinity) by taking advantage of spectral gap results for absorption C_0-semigroups considered in Chap. 5, (see Theorem 11.6.4 for the generator $T_\beta^{(a)}$ and Theorem 11.6.5 for the semigroup $\left(U_\beta^{(a)}(t)\right)_{t\geq 0}$).

11.6.1 A Vector Measure

This subsection is devoted to a key technical lemma we need in the subsequent subsections. We recall first

Definition 11.6.1 A C^∞ function $f : (0, +\infty) \to \mathbb{R}$ is said to be completely monotone if $(-1)^k f^{(k)}(t) \geq 0$ for all integer $k \geq 0$.

Theorem 11.6.1 (Bernstein's theorem (see [170] Theorem 3.8.13, p. 164)) *A C^∞ function $f : (0, +\infty) \to \mathbb{R}$ is completely monotone if and only if there exists a (unique) positive measure μ on $[0, +\infty)$ such that*

$$f(\lambda) = \int_0^{+\infty} e^{-\lambda t} \mu(dt) \ (\lambda > 0).$$

The total mass of μ is given by $f(0_+) := \lim_{\lambda \to 0_+} f(\lambda)$.

We are ready to state

Lemma 11.6.1 *There exists a (unique) $X_{0,1}$-vector measure g on $[0, +\infty)$ such that*

$$\int_0^{+\infty} e^{-\lambda t} g(dt) = \frac{e_{\lambda^*+\lambda}(x)}{1 - \langle \beta, e_{\lambda^*+\lambda} \rangle} \quad (\lambda > 0).$$

Proof Note that

$$e_\lambda(x) = e^{-\lambda \int_0^x r(s)^{-1} ds + \int_0^x a(s) r(s)^{-1} ds} \in X_{0,1}$$

and

$$\frac{1}{1 - \langle \beta, e_\lambda \rangle} \le \frac{1}{\lambda - \lambda^*} \quad (\lambda > \lambda^*).$$

Step 1. If $g : (-R, R) \to \mathbb{R}$ is real analytic with nonnegative derivatives at 0 and if

$$f : (0, +\infty) \to (0, R)$$

is completely monotone then $g(f) : (0, +\infty) \to \mathbb{R}$ is completely monotone. Indeed, since

$$g(u) = g(0) + \sum_{k=1}^{\infty} \frac{g^{(k)}(0)}{k!} u^k \quad (|u| < R)$$

then

$$g(f(\lambda)) = g(0) + \sum_{k=1}^{\infty} \frac{g^{(k)}(0)}{k!} (f(\lambda))^k \quad (\lambda \in (0, +\infty)).$$

We note that f^k is the Laplace transform of $\mu * \dots * \mu$ (k times convolution of μ) so f^k is completely monotone. Since

$$g(f(\lambda)) = g(0) + \lim_{n \to +\infty} \sum_{k=1}^{n} \frac{g^{(k)}(0)}{k!} (f(\lambda))^k$$

and the set of completely monotone functions is a convex cone and is stable by pointwise limits (see [170] Proposition 3.8.16, p. 165) then $g(f)$ is completely monotone.

Step 2.

$$\lambda \in (0, +\infty) \to \langle \beta, e_{\lambda^*+\lambda} \rangle \text{ is completely monotone}$$

since

$$\frac{d^k}{d\lambda^k} \langle \beta, e_{\lambda^*+\lambda} \rangle = \frac{d^k}{d\lambda^k} \int_0^{+\infty} e^{-\lambda \int_0^x r(s)^{-1} ds} e^{-\lambda^* \int_0^x r(s)^{-1} ds + \int_0^x a(s) r(s)^{-1} ds} \beta(x)(1+x) dx$$

$$= (-1)^k \int_0^{+\infty} \left(\int_0^x r(s)^{-1} ds \right)^k e^{-\lambda \int_0^x r(s)^{-1} ds} e^{-\lambda^* \int_0^x r(s)^{-1} ds + \int_0^x a(s) r(s)^{-1} ds}$$

$$\times \beta(x)(1+x) dx.$$

Since $\langle \beta, e_{\lambda^*+\lambda} \rangle < 1$ $(\lambda > 0)$ and $\frac{1}{1-u} = \sum_{k=0}^{\infty} u^k$ $(|u| < 1)$ then, according to *Step1*,

$$\lambda \in (0, +\infty) \to \frac{1}{1 - \langle \beta, e_{\lambda^*+\lambda} \rangle} \text{ is completely monotone.}$$

Hence there exists a positive measure θ on $[0, +\infty)$ such that

$$\int_0^{+\infty} e^{-\lambda t} \theta(dt) = \frac{1}{1 - \langle \beta, e_{\lambda^*+\lambda} \rangle} \quad (\lambda > 0)$$

with total mass

$$\int_0^{+\infty} \theta(dt) = \frac{1}{1 - \langle \beta, e_{\lambda^*} \rangle}.$$

Similarly, for each $x > 0$

$$\lambda \in (0, +\infty) \to e_{\lambda^*+\lambda}(x) \text{ is completely monotone}$$

so there exists a positive measure ν_x on $[0, +\infty)$ such that

$$\int_0^{+\infty} e^{-\lambda t} \nu_x(dt) = e_{\lambda^*+\lambda}(x) \quad (\lambda > 0)$$

with total mass

$$\int_0^{+\infty} \nu_x(dt) = e_{\lambda^*}(x).$$

It follows that for each $x > 0$

$$\frac{e_{\lambda^*+\lambda}(x)}{1 - \langle \beta, e_{\lambda^*+\lambda} \rangle} \text{ is completely monotone}$$

and

$$\int_0^{+\infty} e^{-\lambda t} \gamma_x(dt) = \frac{e_{\lambda^*+\lambda}(x)}{1 - \langle \beta, e_{\lambda^*+\lambda} \rangle} \quad (\lambda > 0)$$

where $\gamma_x = \nu_x * \theta$ has total mass $\frac{e_{\lambda^*}(x)}{1-\langle \beta, e_{\lambda^*} \rangle}$. $\square$

11.6.2   *Stability of Essential Type Versus Boundary Perturbation*

We are now prepared to present a key spectral stability result concerning boundary perturbations of growth C_0-semigroups.

Theorem 11.6.2 *Under (11.4.4), $U_\beta^{(a)}(t) - U_0^{(a)}(t)$ is weakly compact on $X_{0,1}$.*

Proof We note that (11.6.4) shows that for $f \ge 0$ the Laplace transform of the mapping

$$t \to R_\beta(t)f := U_\beta^{(a)}(t)f - U_0^{(a)}(t)f$$

is given by

$$\frac{\langle \beta, \left(\lambda - T_0^{(a)}\right)^{-1} f\rangle}{1 - \langle \beta, e_\lambda\rangle} e_\lambda = \langle \beta, \int_0^{+\infty} e^{-\lambda t} U_0^{(a)}(t)f\, dt\rangle \frac{e_\lambda(x)}{1 - \langle \beta, e_\lambda\rangle}$$

$$= \int_0^{+\infty} e^{-\lambda t} \langle \beta, U_0^{(a)}(t)f\rangle dt . \frac{e_\lambda(x)}{1 - \langle \beta, e_\lambda\rangle}.$$

The change of variable

$$\lambda^* + \lambda \to \lambda$$

in Lemma 11.6.1 gives

$$\int_0^{+\infty} e^{-\lambda t} e^{\lambda^* t} \gamma_x(dt) = \frac{e_\lambda(x)}{1 - \langle \beta, e_\lambda\rangle} \quad (\lambda > \lambda^*)$$

so

$$\frac{\langle \beta, \left(\lambda - T_0^{(a)}\right)^{-1} f\rangle}{1 - \langle \beta, e_\lambda\rangle} e_\lambda = \int_0^{+\infty} e^{-\lambda t} \langle \beta, U_0^{(a)}(t)f\rangle dt . \int_0^{+\infty} e^{-\lambda t} e^{\lambda^* t} \gamma_x(dt)$$

$$= \int_0^{+\infty} e^{-\lambda t} \left[\int_0^t \langle \beta, U_0^{(a)}(t-s)f\rangle g(ds)\right]$$

where

$$g(ds) = e^{\lambda^* s} \gamma_x(ds).$$

By the injectivity of Laplace transform of measures

$$R_\beta(t)f = \int_0^t \langle \beta, U_0^{(a)}(t-s)f\rangle g(ds).$$

Note that

$$\sup_{s \le t} \langle \beta, U_0^{(a)}(t-s)f\rangle \le \|\beta\|_\infty \sup_{s \le t} \left\| U_0^{(a)}(t-s)f\right\|_1$$

$$= \|\beta\|_\infty \sup_{s \le t} \left\| U_0^{(a)}(s)\right\|_{\mathcal{L}(L^1)} \|f\|_1$$

so

$$\left| R_\beta(t) f \right| \leq \|\beta\|_\infty \sup_{s \leq t} \left\| U_0^{(a)}(s) \right\|_{\mathcal{L}(L^1)} \|f\|_1 \int_0^t g(ds)$$

where $\int_0^t g(ds) \in X_{0,1}$ because

$$\int_0^t g(ds) = \int_0^t e^{\lambda^* s} \gamma_x(ds) \leq e^{\lambda^* t} \int_0^{+\infty} \nu_x(dt) = e^{\lambda^* t} \frac{e_{\lambda^*}(x)}{1 - \langle \beta, e_{\lambda^*} \rangle} \in X_{0,1}.$$

Thus $R_\beta(t)$ is dominated by a rank one operator and consequently is weakly compact. $\qquad\square$

11.6.3 Spectral Gap Versus Thinness of Sublevel Sets

In this subsection, we investigate the spectral gap of $\left(U_\beta^{(a)}(t) \right)_{t \geq 0}$ by means of thinness at infinity of sublevel sets of death rate $a(.)$, in the spirit of the analysis of growth-fragmentation equations given in Sect. 11.4. We recall first that Lemma 11.4.3 gives

Lemma 11.6.2 *Let the sublevel sets Ω_c of $a(.)$ be thin at infinity in the sense (11.4.12). Then $T_0^{(a)}$ is resolvent compact.*

This implies

Lemma 11.6.3 *Let the sublevel sets Ω_c of $a(.)$ be thin at infinity in the sense (11.4.12). Then $T_\beta^{(a)}$ is resolvent compact. If additionally the support of $\beta(.)$ is not bounded then $s(T_\beta^{(a)}) > -\infty$.*

Proof Since

$$\left(\lambda - T_\beta^{(a)} \right)^{-1} f = \left(\lambda - T_0^{(a)} \right)^{-1} f + \frac{\langle \beta, \left(\lambda - T_0^{(a)} \right)^{-1} f \rangle_{X_{0,1}^*, X_{0,1}}}{1 - \langle \beta, e_\lambda \rangle} e_\lambda \quad (f \in X_{0,1})$$

then the compactness of $\left(\lambda - T_\beta^{(a)} \right)^{-1}$ is a direct consequence of Lemma 11.6.2 and the fact that

$$f \in X_{0,1} \to \frac{\langle \beta, \left(\lambda - T_0^{(a)} \right)^{-1} f \rangle_{X_{0,1}^*, X_{0,1}}}{1 - \langle \beta, e_\lambda \rangle} e_\lambda$$

is a rank-one operator. If the support of $\beta(.)$ is not bounded then $\left(U_\beta^{(a)}(t) \right)_{t \geq 0}$ is irreducible (see [40] Theorem 5.2) or equivalently $\left(\lambda - T_\beta^{(a)} \right)^{-1}$ is positivity improving and consequently irreducible. Finally, $\left(\lambda - T_\beta^{(a)} \right)^{-1}$ is compact and irreducible so

$$r\left(\left(\lambda - T_\beta^{(a)}\right)^{-1}\right) > 0$$

(see Proposition 2.4.2) and $s(T_\beta^{(a)}) > -\infty$ since $r\left(\left(\lambda - T_\beta^{(a)}\right)^{-1}\right) = (\lambda - s(T_\beta^{(a)}))^{-1}$. $\qquad\square$

We are ready to state the main result of this subsection.

Theorem 11.6.3 *Suppose that the sublevel sets Ω_c of $a(.)$ are thin at infinity in the sense (11.4.12) and the support of $\beta(.)$ is not bounded. Then the C_0-semigroup $\left(U_\beta^{(a)}(t)\right)_{t \geq 0}$ in $X_{0,1}$ is irreducible and has a spectral gap.*

Proof By Theorem 11.6.2, $\left(U_\beta^{(a)}(t)\right)_{t \geq 0}$, $\left(U_0^{(a)}(t)\right)_{t \geq 0}$ have the same essential spectrum and consequently the same essential type

$$\omega_{ess}(U_\beta^{(a)}) = \omega_{ess}(U_0^{(a)}).$$

Note that

$$\left(\lambda - T_{\frac{\beta}{2}}^{(a)}\right)^{-1} \leq \left(\lambda - T_\beta^{(a)}\right)^{-1} ; \; \left(\lambda - T_{\frac{\beta}{2}}^{(a)}\right)^{-1} \neq \left(\lambda - T_\beta^{(a)}\right)^{-1}$$

and both operators are compact and irreducible so, by Proposition 2.4.1,

$$r\left(\left(\lambda - T_{\frac{\beta}{2}}^{(a)}\right)^{-1}\right) < r\left(\left(\lambda - T_\beta^{(a)}\right)^{-1}\right)$$

or equivalently $s(T_{\frac{\beta}{2}}^{(a)}) < s(T_\beta^{(a)})$. Finally $\omega_{ess}(U_\beta^{(a)}) < \omega(U_\beta^{(a)})$ follows from

$$\omega_{ess}(U_\beta^{(a)}) = \omega_{ess}(U_0^{(a)}) = \omega_{ess}(U_{\frac{\beta}{2}}^{(a)}) \leq \omega(U_{\frac{\beta}{2}}^{(a)}) = s(T_{\frac{\beta}{2}}^{(a)}) < s(T_\beta^{(a)}) = \omega(U_\beta^{(a)})$$

and we are done. $\qquad\square$

11.6.4 Spectral Gap Versus Absorption Semigroups

In this subsection, we investigate the spectral gap of $\left(U_\beta^{(a)}(t)\right)_{t \geq 0}$ when the sublevel sets of death rate $a(.)$ are no longer thin at infinity; in particular $a(.)$ may be bounded at infinity. The construction relies on spectral gap theory of absorption semigroups developed in Chap. 5 for "locally" weakly compact resolvents. To this end, we explore first some spectral properties of $T_\beta^{(a)}$ in two preliminary lemmas.

Lemma 11.6.4 *We always have $s(T_\beta^{(a)}) > -\infty$ once $\beta(.)$ is not identically zero.*

Proof It follows from (11.6.4) that

$$\left(\lambda - T_\beta^{(a)}\right)^{-1} f \geq \frac{\langle \beta, \left(\lambda - T_0^{(a)}\right)^{-1} f \rangle_{X_{0,1}^*, X_{0,1}}}{1 - \langle \beta, e_\lambda \rangle} e_\lambda \ (f \geq 0)$$

for any $\lambda > \lambda^* := \beta_1 + 2C_r$. Thus $r(\left(\lambda - T_\beta^{(a)}\right)^{-1})$ is larger than or equal to the spectral radius of the one rank operator

$$f \in X_{0,1} \to \frac{\langle \beta, \left(\lambda - T_0^{(a)}\right)^{-1} f \rangle_{X_{0,1}^*, X_{0,1}}}{1 - \langle \beta, e_\lambda \rangle} e_\lambda$$

which admits the positive eigenvalue

$$\frac{\langle \beta, \left(\lambda - T_0^{(a)}\right)^{-1} e_\lambda \rangle_{X_{0,1}^*, X_{0,1}}}{1 - \langle \beta, e_\lambda \rangle}$$

associated with the eigenfunction

$$e_\lambda(x) = e^{-\lambda \int_0^x r(s)^{-1} ds + \int_0^x a(s) r(s)^{-1} ds}.$$

In particular

$$\frac{\langle \beta, \left(\lambda - T_0^{(a)}\right)^{-1} e_\lambda \rangle_{X_{0,1}^*, X_{0,1}}}{1 - \langle \beta, e_\lambda \rangle} > 0$$

once β is not identically zero. Finally

$$\frac{1}{\lambda - s(T_\beta^{(a)})} = r(\left(\lambda - T_\beta^{(a)}\right)^{-1}) \geq \frac{\langle \beta, \left(\lambda - T_0^{(a)}\right)^{-1} e_\lambda \rangle_{X_{0,1}^*, X_{0,1}}}{1 - \langle \beta, e_\lambda \rangle}$$

implies $s(T_\beta^{(a)}) > -\infty$. $\square$

We extract from the proof of Lemma 11.6.4 a more precise information.

Lemma 11.6.5 *If $\beta(.)$ is not identically zero then*

$$s(T_\beta^{(a)}) \geq \lambda - \frac{1 - \langle \beta, e_\lambda \rangle}{\int_0^{+\infty} e_\lambda(x) dx \int_x^{+\infty} e^{-\int_x^y \frac{\lambda + a(\tau)}{r(\tau)} d\tau} \frac{\beta(y)}{r(y)} (1 + y) dy} \tag{11.6.5}$$

for any $\lambda > \lambda^* := \beta_1 + 2C_r$. In particular

$$s(T_\beta^{(a)}) \geq \lambda^* - \frac{1 - \langle \beta, e_{\lambda^*} \rangle}{\int_0^{+\infty} e_{\lambda^*}(x)dx \int_x^{+\infty} e^{-\int_x^y \frac{\lambda^* + a(\tau)}{r(\tau)}d\tau} \frac{\beta(y)}{r(y)}(1+y)dy}. \tag{11.6.6}$$

Proof The last estimate given in Lemma 11.6.4 amount to

$$s(T_\beta^{(a)}) \geq \lambda - \frac{1 - \langle \beta, e_\lambda \rangle}{\langle \beta, \left(\lambda - T_0^{(a)}\right)^{-1} e_\lambda \rangle_{X_{0,1}^*, X_{0,1}}}.$$

Note that

$$(\lambda - T_0^{(a)})^{-1} f = \frac{1}{r(y)} \int_0^y e^{-\int_x^y \frac{\lambda + a(\tau)}{r(\tau)}d\tau} f(x)dx \quad (\lambda > s(T_0^{(a)}))$$

so $\langle \beta, \left(\lambda - T_0^{(a)}\right)^{-1} e_\lambda \rangle_{X_{0,1}^*, X_{0,1}}$ is equal to

$$\int_0^{+\infty} \frac{\beta(y)}{r(y)}(1+y)dy \int_0^y e^{-\int_x^y \frac{\lambda + a(\tau)}{r(\tau)}d\tau} e_\lambda(x)dx$$
$$= \int_0^{+\infty} e_\lambda(x)dx \int_x^{+\infty} e^{-\int_x^y \frac{\lambda + a(\tau)}{r(\tau)}d\tau} \frac{\beta(y)}{r(y)}(1+y)dy$$

and finally

$$s(T_\beta^{(a)}) \geq \lambda - \frac{1 - \langle \beta, e_\lambda \rangle}{\int_0^{+\infty} e_\lambda(x)dx \int_x^{+\infty} e^{-\int_x^y \frac{\lambda + a(\tau)}{r(\tau)}d\tau} \frac{\beta(y)}{r(y)}(1+y)dy}.$$

Passing to the limit $\lambda \to \lambda^*$ ends the proof. $\qquad\qquad\square$

We note that (11.6.6) is useful only if

$$\frac{1 - \langle \beta, e_{\lambda^*} \rangle}{\int_0^{+\infty} e_{\lambda^*}(x)dx \int_x^{+\infty} e^{-\int_x^y \frac{\lambda^* + a(\tau)}{r(\tau)}d\tau} \frac{\beta(y)}{r(y)}(1+y)dy} < +\infty.$$

We are now ready to give an estimate of the essential spectrum of $T_\beta^{(a)}$.

Theorem 11.6.4 *Let* $\beta(.) \neq 0$. *Suppose that* $\int_0^c \frac{a(s)}{r(s)}ds < +\infty$ *for any finite* $c > 0$ *and* $a(.)e_\lambda(.) \in X_{0,1}$. *Let*

$$a_\infty := \lim_{x \to \infty} \inf \, a(x);$$

(the case $a_\infty = +\infty$ *is allowed). Then*

$$\begin{cases} \sigma_{ess}(T_\beta^{(a)}) \subset \left\{ \mathrm{Re}\, \lambda \leq s(T_0^{(a)}) - a_\infty \right\} & \text{if } a_\infty < +\infty \\ \sigma_{ess}(T_\beta^{(a)}) = \emptyset & \text{if } a_\infty = +\infty. \end{cases}$$

In particular, $T_\beta^{(a)}$ has a "spectral gap" if

$$a_\infty > s(T_\beta^{(0)}) - s(T_\beta^{(a)}). \qquad (11.6.7)$$

Proof Note first that Lemma 11.6.4 insures that $s(T_\beta^{(0)})$ and $s(T_\beta^{(a)})$ are finite. According to Theorem 5.10.2 complemented by Remark 5.10.1, it suffices to show that

$$1_{[0,c]}a\left(\lambda - T_\beta^{(0)}\right)^{-1} \quad \text{is weakly compact}$$

in $X_{0,1} := L^1(\mathbb{R}_+; \ (1+x)\,dx)$. We note that

$$\left(\lambda - T_\beta^{(0)}\right)^{-1} f = \left(\lambda - T_0^{(0)}\right)^{-1} f + \frac{\langle \beta, \left(\lambda - T_0^{(0)}\right)^{-1} f \rangle_{X_{0,1}^*, X_{0,1}}}{1 - \langle \beta, e_\lambda \rangle} e_\lambda \quad (f \in X_{0,1})$$

and (see the pointwise estimate (11.4.7))

$$\left|(\lambda - T_0^{(0)})^{-1} f\right| \le \frac{1}{(1+y)r(y)} \|f\|_{X_{0,1}} \quad (\lambda > C_r)$$

so

$$1_{[0,c]}a(y)\left|(\lambda - T_0^{(0)})^{-1} f\right| \le \frac{1_{[0,c]}a(y)}{(1+y)r(y)} \|f\|_{X_{0,1}} \quad (\lambda > C_r).$$

Since

$$\frac{1_{[0,c]}a(y)}{(1+y)r(y)} \in L^1(\mathbb{R}_+; \ (1+x)\,dx)$$

then $1_{[0,c]}a\left(\lambda - T_\beta^{(0)}\right)^{-1}$ is weakly compact. Since

$$\frac{\langle \beta, \left(\lambda - T_0^{(0)}\right)^{-1} f \rangle_{X_{0,1}^*, X_{0,1}}}{1 - \langle \beta, e_\lambda \rangle} e_\lambda$$

is a rank one operator then $a(.)e_\lambda(.) \in X_{0,1}$ ends the proof. $\square$

A priori, both a_∞ and $s(T_\beta^{(a)})$ depend on $a(.)$ so it is more convenient to express (11.6.7) as

$$\lim_{x \to \infty} \inf a(x) + s(T_\beta^{(a)}) > s(T_\beta^{(0)}).$$

By Lemma 11.6.5, this condition is fulfilled if

$$a_\infty + \lambda^* - \frac{1 - \langle \beta, e_{\lambda^*} \rangle}{\int_0^{+\infty} e_{\lambda^*}(x)dx \int_x^{+\infty} e^{-\int_x^y \frac{\lambda^* + a(\tau)}{r(\tau)}d\tau} \frac{\beta(y)}{r(y)}(1+y)dy} > s(T_\beta^{(0)}).$$

This shows that the presence of a "spectral gap" of $T_\beta^{(a)}$ relies on both $\liminf_{x\to\infty} a(x)$ and on the non local quantity

$$\frac{1 - \langle \beta, e_{\lambda^*}\rangle}{\int_0^{+\infty} e_{\lambda^*}(x)dx \int_x^{+\infty} e^{-\int_x^y \frac{\lambda^* + a(\tau)}{r(\tau)}d\tau} \frac{\beta(y)}{r(y)}(1+y)dy}.$$

We are now ready to deal with the spectral gap of the semigroup $\left(U_\beta^{(a)}(t)\right)_{t\geq 0}$.

Theorem 11.6.5 *Suppose that the support of $\beta(.)$ is not bounded. Let $\int_0^c \frac{a(s)}{r(s)}ds < +\infty$ $(0 < c < +\infty)$ and let $a(.)e_\lambda(.) \in X_{0,1}$. If*

$$\lim_{x\to\infty} \inf a(x) > s(T_\beta^{(0)}) - s(T_\beta^{(a)})$$

then $\left(U_\beta^{(a)}(t)\right)_{t\geq 0}$ has a spectral gap in $X_{0,1}$.

Proof The general strategy of the proof is quite similar to that of Theorem 11.6.3 but the details are different because a priori the resolvent of $T_\beta^{(a)}$ and $T_0^{(a)}$ are not compact. It follows from Theorem 11.6.2 that $U_\beta^{(a)}(t)$ and $U_0^{(a)}(t)$ have the same essential spectrum and consequently the same essential type

$$\omega_{ess}(U_\beta^{(a)}) = \omega_{ess}(U_0^{(a)}).$$

Thus $\left(U_\beta^{(a)}(t)\right)_{t\geq 0}$ has a spectral gap if

$$\omega_{ess}(U_0^{(a)}) < \omega(U_\beta^{(a)}).$$

To this end, it suffices to have

$$\omega(U_0^{(a)}) < \omega(U_\beta^{(a)}),$$

i.e., $s(T_0^{(a)}) < s(T_\beta^{(a)})$ or equivalently

$$r\left(\left(\lambda - T_0^{(a)}\right)^{-1}\right) < r\left(\left(\lambda - T_\beta^{(a)}\right)^{-1}\right). \tag{11.6.8}$$

We point out that Theorem 11.6.4 expresses that $\left(\lambda - T_\beta^{(a)}\right)^{-1}$ has a spectral gap, i.e.,

$$r_{ess}\left(\left(\lambda - T_\beta^{(a)}\right)^{-1}\right) < r\left(\left(\lambda - T_\beta^{(a)}\right)^{-1}\right). \tag{11.6.9}$$

A priori two cases may occur.

Case 1: $\left(\lambda - T_0^{(a)}\right)^{-1}$ has a spectral gap

$$r_{ess}\left(\left(\lambda - T_0^{(a)}\right)^{-1}\right) < r\left(\left(\lambda - T_0^{(a)}\right)^{-1}\right).$$

In this case, $r\left(\left(\lambda - T_0^{(a)}\right)^{-1}\right)$ and $r\left(\left(\lambda - T_\beta^{(a)}\right)^{-1}\right)$ are isolated eigenvalues with finite algebraic multiplicities and (since $\left(\lambda - T_\beta^{(a)}\right)^{-1}$ is irreducible) all the conditions in Proposition 2.4.1 are fulfilled and consequently (11.6.8) is true.

Case 2: $\left(\lambda - T_0^{(a)}\right)^{-1}$ has not a spectral gap, i.e.,

$$r_{ess}\left(\left(\lambda - T_0^{(a)}\right)^{-1}\right) = r\left(\left(\lambda - T_0^{(a)}\right)^{-1}\right). \tag{11.6.10}$$

Since $\left(\lambda - T_0^{(a)}\right)^{-1} \leq \left(\lambda - T_\beta^{(a)}\right)^{-1}$ then the monotonicity of the *essential radius* in L^1 spaces (see Theorem 2.4.1) gives

$$r_{ess}\left(\left(\lambda - T_0^{(a)}\right)^{-1}\right) \leq r_{ess}\left(\left(\lambda - T_\beta^{(a)}\right)^{-1}\right).$$

By combining (11.6.10) and (11.6.9) we get again (11.6.8) and we are done. $\quad\square$

11.7 Comments

The literature on Population Dynamics is extensive and cannot be comprehensively summarized here. Numerous monographs are available, e.g. [15, 51, 218, 302, 320]. Among the classical spectral works on structured population models, we highlight e.g. [14, 109, 374]. Both Population Dynamics and Neutron Transport constitute major application areas for the spectral theory of positive C_0-semigroups.

Besides the work of [124], which motivated the study in Sect. 11.3, diffusion operators with respect to various physical or biological variables appear in other biological models as well; see e.g. [66] and the references therein.

The literature on growth-fragmentation equations is extensive, see e.g. [16, 32, 76, 121, 131]. For more recent developments, we refer to [54, 55, 126, 235, 280]. The probabilistic perspective is treated in [56]. Notably, [126] addresses a variety of related problems that rely fundamentally on positivity methods.

In connection with Theorem 11.3.2: We wonder whether the semigroup $(T(t))_{t\geq 0}$ is holomorphic? If so then both $(U(t))_{t\geq 0}$ and $(T(t))_{t\geq 0}$ are compact C_0-semigroups.

Around Theorem 11.4.3: The role of the mass loss assumption appears at two key places: In the proof that $T_0^{(a)} + B : D(T_0^{(a)}) \to X_{0,1}$ is a generator via Theorem

2.6.1 and (consequently) in the fact that the resolvent compactness of $T_0^{(a)}$ implies the resolvent compactness of $T_0^{(a)} + B$. A priori, the first challenge can, in principle, be overcome by suitably modifying honesty theory (see Sect. 2.11), without mass loss assumption (i.e., with $\eta(.) = 0$); in this case, $T_0^{(a)} + B$ need not be a generator but there exists a unique minimal extension $T_B \supset T_0^{(a)} + B$ which generates a positive C_0-semigroup $(V(t))_{t\geq 0}$ in $X_{0,1}$. Unfortunately, even in the honest case (i.e., $T_B = \overline{T_0^{(a)} + B}$), if $T_0^{(a)} + B$ is not closed, a priori we cannot conclude that T_B is resolvent compact when $T_0^{(a)}$ is. This constitutes the main obstacle to understanding the time asymptotics of the semigroup $(V(t))_{t\geq 0}$ in the natural functional spaces $X_{0,1}$ without assuming mass loss.

In the same spirit, we can also treat conservative models (10.14.4), that is, by removing the mass loss condition (11.4.3), although this requires working in higher moment spaces

$$X_{0,\alpha} := L^1(\mathbb{R}_+; \ (1 + x^\alpha)\,dx)\ (\alpha > \alpha_*)$$

for a suitable threshold $\alpha_* \geq 1$ [39].

We refer to [40] for a construction, analogous to that in [39], addressing the nonlocal (McKendrick-von Foerster) boundary condition (see 11.6.2).

As far as we know, the runaway phenomenon in Theorem 11.5.2 is presented here for the first time.

In [54] the authors consider growth fragmentation equations (in suitable moment spaces) and show that we cannot expect the existence of a spectral gap if the fragmentation rate $a(.)$ is bounded. This is coherent with the spectral gap theory (in moment spaces) given in [39] with (11.4.4) which relies on the condition that the sublevel sets of $a(.)$ are "thin at infinity". In the space X_1 under (11.4.13), the spectral gap theory relies on the condition that the sublevel sets of $a(.)$ are "thin" at infinity and at zero (typically $a(.)$ tends to infinity at zero and at infinity) [280]. We strongly suspect that no spectral gap exists when $a(.)$ is bounded near zero or at infinity. Theorem 11.5.1 further establishes that if $a(.)$ vanishes both at zero and at infinity, then not only does a spectral gap fail to exist, but the spectral bound of the generator is also not an eigenvalue. In connection with Theorem 11.6.5: A similar construction in the "finite agregates number" space $X_0 := L^1(\mathbb{R}_+, dx)$ in place of $X_{0,1}$ is possible.

It remains an open question whether the spectral analysis of growth equations with nonlocal McKendrick-von Foerster boundary conditions, as presented in Sect. 11.6, can also be addressed using recent abstract results on boundary perturbations of the essential spectral radius [67].

Chapter 12
Linearized Non Local Allen-Cahn Equations

12.1   Chapter Aims

In this chapter, we investigate the spectral gap properties of a class of C_0-semigroups, generated by convolution perturbations of the first-order differentiation operator in one spatial dimension. This analysis arises in the context of the nonlinear theory of nonlocal Allen-Cahn equations. We begin by developing a general framework in L^1-spaces, and subsequently extend the results to L^p spaces using an interpolation argument.

12.2   Chapter Overview

This short chapter is devoted to peripheral spectral (gap) analysis and related topics for a class of C_0-semigroups arising in the nonlinear theory of nonlocal Allen–Cahn equations, as studied in [44]. We complement the L^2 spectral analysis of the generator provided in [44] with a corresponding analysis of the semigroup itself. Indeed, due to the absence of a spectral mapping theorem in this context, the spectrum of the generator does not determine that of the semigroup, making a direct analysis of the latter necessary. In line with the spirit of this monograph, we first develop a general theory in L^1-spaces, which turn out to be particularly well-suited to this class of C_0-semigroups, more so than, for example, L^2-spaces. The extension to L^p spaces $(1 < p < \infty)$ then follows by interpolation arguments. This construction is new and appears here for the first time.

There exists a literature on traveling waves for non local Allen-Cahn equations

$$u_t = J * u - u + f(u)$$

where $J * u$ is the convolution of u and

© The Author(s), under exclusive license to Springer Nature Switzerland AG 2026
M. Mokhtar-Kharroubi, *Peripheral Spectra of Perturbed Positive Semigroups*, Lecture Notes in Mathematics 2388, https://doi.org/10.1007/978-3-032-11173-9_12

$$J \in C^1(\mathbb{R}) \cap L^1_+(\mathbb{R}), \quad \int_{\mathbb{R}} J(x)dx = 1; \tag{12.2.1}$$

see [44] and references therein. Actually, this is part of a very broad field [45] which is quite far from the main subjects of this monograph. However, the spectral analysis of its linearized version which will occupy us in this chapter is covered by the tools developed in this monograph.

The function f is smooth with three zeros ± 1 and $a \in (-1, 1)$ satisfying $f'(\pm 1) < 0$ and $f'(a) > 0$. A typical example is given by $f(u) = (u - a)(1 - u^2)$. There exist traveling wave solutions of the form $u(t, x) = \phi(x - ct)$ where

$$c = \frac{\int_{-1}^{1} f(u)du}{\int_{-\infty}^{\infty} (\phi'(x))^2 \, dx},$$

ϕ is continuous, nondecreasing and (if $c \neq 0$)

$$\begin{cases} c\phi' + J * \phi - \phi + f(\phi) = 0 \\ \phi(-\infty) = -1, \quad \phi(+\infty) = 1; \end{cases} \tag{12.2.2}$$

see [44]. The spectral analysis in $L^2(\mathbb{R})$ of the linearized operator of (12.2.2) about the traveling wave ϕ

$$L\psi = c\psi' + J * \psi - \psi + f'(\phi)\psi, \quad D(L) = H^1(\mathbb{R})$$

is given in [44]. The authors show that L has spectral gap, more precisely:

Theorem 12.2.1 ([44] Theorem 2.1) *If $c \neq 0$ then*
(i) $\sigma(L) \cap \{\mathrm{Re}\, \lambda \geq 0\} = \{0\}$
(ii) 0 is an algebraically simple eigenvalue with positive eigenfunction ϕ'
(iii) There exists $\gamma > 0$ such that $\sigma_{ess}(L) \subset \{\mathrm{Re}\, \lambda \leq -\gamma\}$.

In ([44] Corollary 2.2), the authors assert, as an immediate consequence of Theorem 12.2.1, that the semigroup $\left(e^{tL}\right)_{t \geq 0}$ generated by L possesses a spectral gap; the authors refer to the monograph [156] as a source for a proof of this fact but do not specify the relevant theorem, or even the chapter, within [156]. Although the claim is certainly correct, we have not been able to locate a proof in [156]. It is worth emphasizing that this statement is far from elementary. Indeed, since the semigroup $\left(e^{tL}\right)_{t \geq 0}$ is not immediately norm-continuous, the existence of a spectral gap for the generator L does not, a priori, imply a spectral gap for the semigroup itself, due to the absence, in general, of a spectral mapping theorem in this setting. This observation served as our initial motivation to revisit the spectral theory and to establish directly that $\left(e^{tL}\right)_{t \geq 0}$ indeed exhibits a spectral gap. In addition to several preliminary results of independent interest, we provide an L^1 version of Theorem 12.2.1, (see Theorem 12.5.1). We also establish a spectral gap result for the semigroup in L^1 spaces, (see Theorem 12.6.1). The results extend to L^p spaces, as discussed in Remark 12.6.1.

12.3 L^1 **Semigroups**

Let $\phi : \mathbb{R} \to \mathbb{R}$ be a C^1 nondecreasing function satisfying (12.2.2). Note that

$$h := \phi' \in L^1_+(\mathbb{R}) \text{ and } \int_{\mathbb{R}} h(x)dx = 2$$

and the differentiation of (12.2.2) gives

$$ch' + J * h - h + f'(\phi)h = 0$$

so $h' \in L^1(\mathbb{R})$, i.e., $h \in W^{1,1}(\mathbb{R})$. In this chapter, we are concerned with the linearized operator of (12.2.2) about the traveling wave ϕ

$$\varphi \to L\varphi = c\varphi' + J * \varphi - \varphi + f'(\phi)\varphi$$

as unbounded operator on $L^1(\mathbb{R})$. We introduce the positive C_0-group $(U(t))_{t \in \mathbb{R}}$ on $L^1(\mathbb{R})$ defined by

$$U(t)\varphi = e^{-t}\varphi(x + ct)$$

with generator

$$\begin{cases} T\varphi = c\varphi' - \varphi \\ D(T) = W^{1,1}(\mathbb{R}). \end{cases}$$

and note that

$$\begin{cases} Th + J * h + f'(\phi)h = 0 \\ \quad h \in D(T) \cap L^1_+(\mathbb{R}). \end{cases}$$

Note that

$$\mathcal{J} : \varphi \in L^1(\mathbb{R}) \to J * \varphi \in L^1(\mathbb{R})$$

is a stochastic operator and

$$\int_{\mathbb{R}} T\varphi + \mathcal{J}\varphi = 0, \quad \varphi \in D(T)$$

so

$$T + \mathcal{J} : D(T) \to L^1(\mathbb{R})$$

generates a stochastic C_0-semigroup $(Z(t))_{t \geq 0}$ on $L^1(\mathbb{R})$. Let

$$b(x) := f'(\phi(x))$$

and denote also by b the corresponding multiplication operator. Then

$$T + \mathcal{J} + b : D(T) \to L^1(\mathbb{R})$$

generates a positive C_0-semigroup $(V(t))_{t\geq 0}$ on $L^1(\mathbb{R})$. The main result in this section is an explicitation of $(Z(t))_{t\geq 0}$.

Theorem 12.3.1 *There exists $\widehat{J_t} \in L^1(\mathbb{R})$ ($t \in \mathbb{R}$) such that*

$$Z(t)\varphi = U(t)\varphi + \widehat{J_t} * (U(t)\varphi), \quad \varphi \in L^1(\mathbb{R}).$$

Proof We start with the Dyson-Phillips expansion of $(Z(t))_{t\geq 0}$

$$Z(t) = \sum_{n=0}^{\infty} U_n(t)$$

where

$$U_0(t) = U(t), \quad U_{n+1}(t) = \int_0^t U(t-s)\mathcal{J}U_n(s)ds.$$

A key observation is

$$\begin{aligned}
U(t)\,(J * \varphi) &= e^{-t}\,(J * \varphi)\,(x + ct) \\
&= e^{-t} \int_{\mathbb{R}} J(x + ct - y)\varphi(y)dy \\
&= e^{-t} J_{ct} * \varphi
\end{aligned}$$

where

$$J_{ct} : x \to J(x + ct)$$

so

$$U(t)\mathcal{J} = e^{-t}\mathcal{J}_{ct}$$

where

$$\mathcal{J}_{ct} : \varphi \in L^1(\mathbb{R}) \to J_{ct} * \varphi \in L^1(\mathbb{R}).$$

Similarly

$$\mathcal{J}U(t) = U(t)\mathcal{J} = e^{-t}\mathcal{J}_{ct}.$$

Hence

$$U_{n+1}(t) = \mathcal{J} \int_0^t U(t-s)U_n(s)ds.$$

In particular

$$U_1(t) = t\mathcal{J}U(t) = te^{-t}\mathcal{J}_{ct}$$

and

$$U_2(t) = \mathcal{J} \int_0^t U(t-s)U_1(s)ds$$

$$= \mathcal{J} \int_0^t U(t-s)s\,\mathcal{J}U(s)sds$$

$$= \frac{t^2}{2}\mathcal{J}^{(2)}U(t)$$

where

$$\mathcal{J}^{(2)} : \mathcal{J} : \varphi \in L^1(\mathbb{R}) \rightarrow J^{(2)} * \varphi \in L^1(\mathbb{R})$$

with $J^{(2)} := J * J$. Let us check by induction that

$$U_n(t) = \frac{t^n}{n!}\mathcal{J}^{(n)}U(t)$$

where

$$\mathcal{J}^{(n)} : \mathcal{J} : \varphi \in L^1(\mathbb{R}) \rightarrow J^{(n)} * \varphi \in L^1(\mathbb{R})$$

with

$$J^{(n)} := J * J ... * J \quad (n \text{ times}).$$

We have

$$U_{n+1}(t) = \mathcal{J} \int_0^t U(t-s)U_n(s)ds$$

$$= \mathcal{J} \int_0^t U(t-s)\mathcal{J}^{(n)}U(s)\frac{s^n}{n!}ds$$

$$= \frac{t^{n+1}}{(n+1)!}\mathcal{J}^{(n+1)}U(t)$$

and therefore

$$Z(t)\varphi = U(t)\varphi + \sum_{n=1}^{\infty} \frac{t^n}{n!}J^{(n)} * (U(t)\varphi).$$

Let

$$\widehat{J_t} := \sum_{n=1}^{\infty} \frac{t^n}{n!}J^{(n)}$$

with L^1 norm

$$\|\widehat{J_t}\| = \sum_{n=1}^{\infty} \frac{t^n}{n!}\|J^{(n)}\| = \sum_{n=1}^{\infty} \frac{t^n}{n!} = e^t - 1.$$

Finally

$$Z(t)\varphi = U(t)\varphi + \widehat{J_t} * (U(t)\varphi) \,. \qquad\qquad \square$$

We find again the fact that $(Z(t))_{t\geq 0}$ is stochastic. Indeed, for $\varphi \in L^1_+(\mathbb{R})$

$$\|Z(t)\varphi\| = \|U(t)\varphi\| + \left\|\widehat{J_t} * (U(t)\varphi)\right\|$$
$$= e^{-t}\|\varphi\| + \left(e^t - 1\right)e^{-t}\|\varphi\| = \|\varphi\| \,.$$

12.4 Irreducibility

We consider now the irreducibilityof $(Z(t))_{t\geq 0}$

Theorem 12.4.1 *In the case $c > 0$ we suppose that the support of J is not included in a set of the form $(-\infty, \alpha]$. In the case $c < 0$ we suppose that the support of J is not included in a set of the form $[\alpha, +\infty)$. Then $(Z(t))_{t\geq 0}$ is irreducible.*

Proof Since

$$(\lambda - T)^{-1}\varphi = \int_0^{+\infty} e^{-\lambda t} U(t)\varphi \, dt$$

then

$$\mathcal{J}(\lambda - T)^{-1}\varphi = J * \left(\int_0^{+\infty} e^{-\lambda t} U(t)\varphi \, dt\right)$$
$$= \int_0^{+\infty} (J * U(t)\varphi)\, e^{-\lambda t} dt$$
$$= \int_0^{+\infty} e^{-t}\, (J_{ct} * \varphi)\, e^{-\lambda t} dt$$
$$= \left(\int_0^{+\infty} J_{ct} e^{-(\lambda+1)t} dt\right) * \varphi$$

where

$$\int_0^{+\infty} J_{ct}(x) e^{-(\lambda+1)t} dt = \int_0^{+\infty} J(x + ct) e^{-(\lambda+1)t} dt$$

is equal to

$$\begin{cases} c^{-1}\int_x^{+\infty} J(y) e^{-c^{-1}(\lambda+1)(y-x)} dy & \text{if } c > 0 \\ -c^{-1}\int_{-\infty}^x J(y) e^{-c^{-1}(\lambda+1)(y-x)} dy & \text{if } c < 0. \end{cases}$$

This shows that $\int_0^{+\infty} J(x + ct) e^{-(\lambda+1)t} dt > 0$ everywhere and consequently $\mathcal{J}(\lambda - T)^{-1}$ is positivity improving, i.e., for any non trivial $\varphi \in L^1_+(\mathbb{R})$

$$\mathcal{J}(\lambda - T)^{-1}\varphi > 0 \text{ everywhere.}$$

Hence

$$(\lambda - T - \mathcal{J})^{-1} = (\lambda - T)^{-1} \sum_{j=0}^{\infty} \left(\mathcal{J}(\lambda - T)^{-1}\right)^j$$

$$\geq (\lambda - T)^{-1} \mathcal{J}(\lambda - T)^{-1}$$

is also positivity improving or equivalently $(Z(t))_{t \geq 0}$ is irreducible. $\qquad\square$

Since the irreducibility is stable by perturbing the generator by a bounded multiplication operator then we get

Corollary 12.4.1 *Let the conditions in Theorem 12.4.1 be satisfied. Then $(V(t))_{t \geq 0}$ is irreducible.*

12.5 Spectral Analysis of The Generator

We start with a key property

Lemma 12.5.1 *Let $(V(t))_{t \geq 0}$ be irreducible. Then $s(T + \mathcal{J} + b) = 0$.*

Proof We already know that

$$Th + \mathcal{J}h + bh = 0$$

so $s(T + \mathcal{J} + b) \geq 0$. We argue by contradiction by assuming that

$$s(T + \mathcal{J} + b) > 0.$$

We decompose b as

$$b(x) = b_1(x) + b_2(x) := b(x)1_{\{b \leq 0\}} + b(x)1_{\{b > 0\}}.$$

Since $b_1 \leq 0$ then $T + \mathcal{J} + b_1$ generates a contraction semigroup so

$$s(T + \mathcal{J} + b_1) \leq 0.$$

We note also that $b_2(x)$ has a bounded support so b_2 is T-compact because the imbedding of $D(T) = W^{1,1}(\mathbb{R})$ into $L^1_{loc}(\mathbb{R})$ is compact. It follows that b_2 is also $(T + \mathcal{J} + b_1)$-compact and

$$\sigma_{ess}(T + \mathcal{J} + b) = \sigma_{ess}(T + \mathcal{J} + b_1) \subset \{\mathrm{Re}\,\lambda \leq 0\}.$$

Since $s(T + \mathcal{J} + b) > 0$ then $s(T + \mathcal{J} + b)$ must be an isolated algebraically simple eigenvalue associated to a positive eigenfunction $\widehat{h}$. Let $\widehat{h}_*$ be the dual eigenfunction associated to the spectral bound of $(T + \mathcal{J} + b)^*$ in $L^\infty(\mathbb{R})$. Hence

$$
\begin{aligned}
0 &= \langle (T + \mathcal{J} + b)h, \widehat{h}_* \rangle \\
&= \langle h, (T + \mathcal{J} + b)^*\widehat{h}_* \rangle \\
&= s(T + \mathcal{J} + b)\langle h, \widehat{h}_* \rangle,
\end{aligned}
$$

i.e., $\langle h, \widehat{h}_* \rangle = 0$ which is not true. This ends the proof. $\qquad\square$

The main result of this section is the following L^1 version of Theorem 12.2.1.

Theorem 12.5.1 *Let $(Z(t))_{t \geq 0}$ be irreducible. Then*

$$
\sigma_{ess}(T + \mathcal{J} + b) \subset \left\{ \operatorname{Re} \lambda \leq \max \left\{ f'(-1), f'(1) \right\} \right\}.
$$

Proof Let $\alpha := \inf_{x \in \mathbb{R}} (b(x))$ and

$$
b(x) = \widehat{b}(x) + \alpha
$$

where $\widehat{b}(x) := b(x) - \alpha$. Let

$$
\left(\widehat{V}(t) \right)_{t \geq 0} := \left(e^{-\alpha t} V(t) \right)_{t \geq 0}
$$

with generator

$$
T + \mathcal{J} + b - \alpha = T + \mathcal{J} + \widehat{b}.
$$

Since $s(T + \mathcal{J} + b) = 0$ then

$$
s(T + \mathcal{J} + \widehat{b}) = -\alpha.
$$

One sees also that $\left(\widehat{V}(t) \right)_{t \geq 0}$ is nothing but the absorption semigroup obtained from $(Z(t))_{t \geq 0}$ and the potential $-\widehat{b}$, i.e.,

$$
\left(\widehat{V}(t) \right)_{t \geq 0} = \left(Z_{-\widehat{b}}(t) \right)_{t \geq 0}.
$$

The weak compactness assumption in Theorem 5.10.2 is met because of the (locally) compact imbedding of $W^{1,1}(\mathbb{R})$ into $L^1_{loc}(\mathbb{R})$. According to Theorem 5.10.2

$$
\sigma_{ess}(T + \mathcal{J} + \widehat{b}) \subset \left\{ \operatorname{Re} \lambda \leq s(T + \mathcal{J}) - \lim_{|x| \to \infty} \inf \left(-\widehat{b}(x) \right) \right\}.
$$

Since $s(T + \mathcal{J}) = 0$ then

$$\sigma_{ess}(T + \mathcal{J} + \widehat{b}) \subset \left\{ \mathrm{Re}\,\lambda \leq \lim_{|x|\to\infty}\sup \widehat{b}(x) \right\}$$

and a shift by α gives

$$\sigma_{ess}(T + \mathcal{J} + b) \subset \left\{ \mathrm{Re}\,\lambda \leq \lim_{|x|\to\infty}\sup b(x) \right\}.$$

This ends the proof since

$$\lim_{|x|\to\infty}\sup b(x) = \lim_{|x|\to\infty}\sup f'(\phi(x)) = \max\left\{ f'(-1), f'(1) \right\}. \qquad \square$$

12.6 Spectral Analysis of The Semigroup

In this section, we deal with spectral gap of the semigroup $(V(t))_{t\geq 0}$; this property is much stronger than that given in Theorem 12.5.1 and cannot be deduced from it because $(V(t))_{t\geq 0}$ is not norm continuous. Finally, since $(V(t))_{t\geq 0}$ is not "locally" weakly compact, unlike the resolvent of its generator, Theorem 5.10.1 does not apply. Instead, we develop a direct approach tailored to this specific problem.

Exceptionally, in this section only, for the simplicity of notations, we denote by $\left(Z_q(t)\right)_{t\geq 0}$ the positive C_0-semigroup with generator

$$T + \mathcal{J} + q : D(T) \subset L^1(\mathbb{R}) \to L^1(\mathbb{R}),$$

for $q \in L^\infty(\mathbb{R})$ (a correct notation, coherent with that for absorption semigroups, should be $\left(Z_{-q}(t)\right)_{t\geq 0}$). We start with a key preliminary result.

Lemma 12.6.1 *Let $q \in L^\infty(\mathbb{R})$. If $\lim_{|x|\to\infty} q(x) = 0$ then $\omega_{ess}(Z_q) \leq 0$.*

Proof Note that q need not be nonnegative. However, $\left(Z_q(t)\right)_{t\geq 0}$ is always positive. We have

$$Z_q(t) = Z(t) + \int_0^t Z(t-s)q\,Z(s)ds.$$

Since

$$Z(s)f = U(s)f + \widehat{J}_s * (U(s)f)$$

then $\int_0^t Z(t-s)q\,Z(s)f\,ds$ is equal to

$$\int_0^t Z(t-s)q\left[U(s)f + \widehat{J}_s * (U(s)f)\right]ds$$

$$= \int_0^t Z(t-s)q\,[U(s)f]\,ds$$
$$+ \int_0^t Z(t-s)q\left[\widehat{J}_s * (U(s)f)\right]ds.$$

Note first that q times a convolution operator is compact. Indeed, since $q \to 0$ at infinity, it suffices to check this property for q with bounded support which is true (see e.g. [73] Corollary 4.28, p. 114). It follows from Theorem 2.9.1 that

$$f \to \int_0^t Z(t-s)q\left[\widehat{J}_s * (U(s)f)\right]ds$$

is weakly compact (actually, it is compact, see Remark 2.9.1) and consequently

$$r_{ess}(Z_q(t)) = r_{ess}\left[Z(t) + \int_0^t Z(t-s)qU(s)ds\right]$$
$$\leq \left\|Z(t) + \int_0^t Z(t-s)qU(s)ds\right\|$$
$$\leq 1 + \|q\|_\infty (1 - e^{-t}).$$

Since $r_{ess}(Z_q(t)) = e^{\omega_{ess}(Z_q)t}$ then

$$\omega_{ess}(Z_q) \leq t^{-1}\ln\left[1 + \|q\|_\infty (1 - e^{-t})\right] \quad (t > 0)$$

so letting $t \to +\infty$ ends the proof. $\qquad\square$

We are ready to show that the semigroup $(V(t))_{t\geq 0}$ possesses a spectral gap.

Theorem 12.6.1 *We have* $\omega_{ess}(V) \leq \max\left\{f'(1),\,f'(-1))\right\} < 0 = \omega(V)$.

Proof We already know that $\omega(V) = s(T + \mathcal{J} + b) = 0$. Let

$$\alpha := \max\left\{f'(1),\,f'(-1))\right\}.$$

Let $\varepsilon > 0$ be fixed. Then

$$b(x) \leq \alpha_\varepsilon := \alpha + \varepsilon$$

for $|x|$ large enough, say $|x| \geq m$. Let

$$b_\varepsilon(x) = \begin{cases} b(x) \text{ if } |x| \geq m \\ \alpha_\varepsilon \text{ if } |x| < m. \end{cases}$$

Then

$$b(x) = b_\varepsilon(x) + \widehat{b}_\varepsilon(x)$$

where

$$\widehat{b_\varepsilon}(x) := b(x) - b_\varepsilon(x)$$

has its support in $[-m, m]$. Hence, by Lemma 12.6.1

$$\omega_{ess}(Z_{\widehat{b_\varepsilon}}) \le 0$$

where $\left(Z_{\widehat{b_\varepsilon}}(t)\right)_{t \ge 0}$ is generated by

$$T + J + \widehat{b_\varepsilon}.$$

Note that $b_\varepsilon(x) \le \alpha_\varepsilon$ so

$$b_\varepsilon(x) = (b_\varepsilon(x) - \alpha_\varepsilon) + \alpha_\varepsilon$$

and the fact that $b_\varepsilon(x) - \alpha_\varepsilon \le 0$ imply

$$Z_{\widehat{b_\varepsilon} + b_\varepsilon(x) - \alpha_\varepsilon}(t) \le Z_{\widehat{b_\varepsilon}}(t).$$

It follows from Theorem 2.4.1 that

$$r_{ess}(Z_{\widehat{b_\varepsilon} + b_\varepsilon(x) - \alpha_\varepsilon}(t)) \le r_{ess}(Z_{\widehat{b_\varepsilon}}(t)).$$

Finally, since α_ε is a constant then

$$V(t) = Z_b(t) = Z_{(\widehat{b_\varepsilon} + b_\varepsilon(x) - \alpha_\varepsilon) + \alpha_\varepsilon}(t) = e^{\alpha_\varepsilon t} Z_{\widehat{b_\varepsilon} + b_\varepsilon(x) - \alpha_\varepsilon}(t)$$

and

$$\begin{aligned} r_{ess}(V(t)) &= e^{\alpha_\varepsilon t} r_{ess}(Z_{\widehat{b_\varepsilon} + b_\varepsilon(x) - \alpha_\varepsilon}(t)) \\ &\le e^{\alpha_\varepsilon t} r_{ess}(Z_{\widehat{b_\varepsilon}}(t)), \end{aligned}$$

i.e.,

$$\omega_{ess}(V) \le \alpha_\varepsilon + \omega_{ess}(Z_{\widehat{b_\varepsilon}}) \le \alpha_\varepsilon = \alpha + \varepsilon.$$

This ends the proof since $\varepsilon > 0$ is arbitrary. $\qquad\square$

Remark 12.6.1 The spectral gap theory of this chapter extends to $L^p(\mathbb{R})$ ($1 \le p < +\infty$) by interpolation arguments. Indeed, since $(U(t))_{t \in \mathbb{R}}$ and J operate consistently in all $L^p(\mathbb{R})$ then so does $(V(t))_{t \ge 0}$ and it suffices to appeal to [160] Corollary 1.4 and Remark 1.5 (a).

Chapter 13
Spectra of Perturbed Convolution Semigroups

13.1 Chapter Aims

This chapter is devoted to the peripheral spectral analysis of absorption semigroups generated by formal operators of the form $M - V$, where M is the generator of a convolution C_0-semigroup $(\mathcal{M}(t))_{t \geq 0}$ on $L^p(\mathbb{R}^d)$ and V is a singular (i.e., unbounded) indefinite potential. We develop a general spectral gap theory and present various applications, including weighted Laplacians on $\mathbb{R}^d$, Poincaré inequalities for probability measures, and Witten Laplacians on 1-forms on $\mathbb{R}^d$, with further applications to the Helffer–Sjöstrand covariance formula. Notably, the analysis does not rely on any convexity assumptions for the weight functions.

13.2 Chapter Overview

In this chapter, we show how the spectral theory of absorption semigroups developed in Chap. 5 naturally applies to convolution C_0-semigroups $(\mathcal{M}(t))_{t \geq 0}$ on $L^p(\mathbb{R}^d)$ where the generator M is perturbed by indefinite potentials V. Specifically, we investigate whether the formal operators

$$\text{``} M - V \text{''}$$

define generators, and we explore their spectral properties, such as discreteness and (local) spectral gaps, as well as those of the associated C_0-semigroups. As useful illustrations, we present results on weighted Laplacians on $\mathbb{R}^d$, with applications to Poincaré inequalities for probability measures, and on Witten Laplacians on 1-forms on $\mathbb{R}^d$, with applications to Helffer-Sjöstrand's covariance formula that does not rely on any convexity assumptions for the weight functions.

M. Mokhtar-Kharroubi, *Peripheral Spectra of Perturbed Positive Semigroups*, Lecture Notes in Mathematics 2388, https://doi.org/10.1007/978-3-032-11173-9_13

Our main goal here is to revisit and extend, in several directions, a number of spectral results previously established in [271]. We restrict ourselves to symmetric convolution semigroups

$$\mathcal{M}^p(t) : f \in L^p(\mathbb{R}^d) \to \int_{\mathbb{R}^d} f(x-y)m_t(dy) \in L^p(\mathbb{R}^d)$$

where $\{m_t\}_{t \geq 0}$ is a family of symmetric (with respect to the origin) Borel sub-probability measures on $\mathbb{R}^d$ such that $m_0 = \delta_0$ (Dirac measure at zero), $m_t * m_s = m_{t+s}$ and $m_t \to m_0$ vaguely as $t \to 0_+$. Such convolution semigroups are closely related to Lévy processes and encompass many examples of practical interest, including Gaussian semigroups, α-stable semigroups, relativistic Schrödinger semigroups. Note that $(\mathcal{M}^p(t))_{t \geq 0}$ is a positive contraction C_0-semigroup on $L^p(\mathbb{R}^d)$ $(1 \leq p < +\infty)$ with generator

$$M^p : D(M^p) \subset L^p(\mathbb{R}^d) \to L^p(\mathbb{R}^d).$$

The sub-probability measures $\{m_t\}_{t \geq 0}$ are characterized via their Fourier transforms by

$$\widehat{m_t}(\zeta) := (2\pi)^{-\frac{d}{2}} \int e^{-i\zeta.x} m_t(dx) = (2\pi)^{-\frac{d}{2}} e^{-tF(\zeta)}, \ \zeta \in \mathbb{R}^d$$

where F is the so-called characteristic exponent, a continuous negative definite function (see [170] Definition 3.6.5, p. 122), and admits the representation

$$F(\zeta) = c + \zeta.C\zeta + \int_{\mathbb{R}^d \setminus \{0\}} [1 - \cos(x.\zeta)]\,\mu(dx)$$

for some constant $c \geqslant 0$, a symmetric positive semi-definite matrix C, and a Lévy measure μ on $\mathbb{R}^d \setminus \{0\}$, symmetric with respect to the origin, satisfying the condition

$$\int \min(1, |x|^2)\mu(dx) < +\infty.$$

Note that $\{m_t\}_{t \geq 0}$ forms a family of probability measures if and only if $F(0) = 0$, which corresponds to the condition $c = 0$. We also recall that the characteristic exponent F satisfies the bounds

$$F(\zeta) \geqslant 0, \ F(\zeta) \leq c_F(1 + |\zeta|^2) \ (\zeta \in \mathbb{R}^d)$$

and

$$M^2\varphi = -(2\pi)^{-\frac{d}{2}} \int_{\mathbb{R}^d} e^{i\zeta.x} F(\zeta)\widehat{\varphi}(\zeta)d\zeta$$

with domain

$$D(M^2) = \left\{ \varphi \in L^2(\mathbb{R}^d); \ F\widehat{\varphi} \in L^2(\mathbb{R}^d) \right\}.$$

It is known (see e.g. [263] Theorem 2) that $C_c^\infty(\mathbb{R}^d)$ is a core of $D(M^2)$ and consequently a core of $D(\sqrt{-M^2})$ as $D(M^2)$ is a core of $D(\sqrt{-M^2})$, (see e.g. [166] Theorem 4.1, p. 78). We point out that $-M^2$ is unitarily equivalent to the multiplication operator $\psi \to F\psi$ in $L^2(\mathbb{R}^d)$ defined on its maximal domain and consequently

$$\inf_{\left\{\varphi \in C_c^\infty(\mathbb{R}^d), \ \|\varphi\|_{L^2}=1\right\}} \left(-M^2\varphi, \varphi\right)_{L^2(\mathbb{R}^d)} = \inf_{\zeta \in \mathbb{R}^d} F(\zeta). \tag{13.2.1}$$

We refer the reader to [170] for a comprehensive treatment of convolution semigroups. Note first that the resolvent $(\lambda - M_1)^{-1}$ is a convolution operator of the form

$$(\lambda - M_1)^{-1} f = \int_{\mathbb{R}^d} f(x - y) m^\lambda(dy)$$

where $m^\lambda = \int_0^{+\infty} e^{-\lambda t} m_t dt \ \ (\lambda > 0)$ is a vaguely convergent integral, i.e.,

$$\int_{\mathbb{R}^d} f(x) m^\lambda(dx) := \int_0^{+\infty} e^{-\lambda t} \left[\int_{\mathbb{R}^d} f(x) m_t(dx) \right] dt, \ \ f \in C_0(\mathbb{R}^d).$$

Our standing assumption here is that m^λ is absolutely continuous with respect to Lebesgue measure, i.e.,

$$m^\lambda(dx) = E_\lambda(x) dx, \ \ E_\lambda(.) \in L^1(\mathbb{R}^d). \tag{13.2.2}$$

Thus, for $f \in L^1(\mathbb{R}^d)$, the resolvent operator can be expressed as

$$(\lambda - M^1)^{-1} f = \int_{\mathbb{R}^d} E_\lambda(x - y) f(y) dy$$

where $E_\lambda(.)$ is called the λ-potential kernel in the terminology in [78]. We recall that for the classical examples, the characteristic exponent F satisfies

$$\int_{\mathbb{R}^d} e^{-t F(\zeta)} d\zeta < +\infty \ \ (t > 0) \tag{13.2.3}$$

which ensures that the bounded measure $m_t \ (t > 0)$ is absolutely continuous with respect to Lebesgue measure, i.e.,

$$m_t(dx) = k_t(x) dx \ \ (t > 0)$$

where the density k_t belongs to $L^1_+(\mathbb{R}^d) \cap C_0(\mathbb{R}^d)$, is even and

$$E_\lambda(z) = \int_0^{+\infty} e^{-\lambda t} k_t(z)\, dt \quad (\lambda > 0).$$

A significant part of this chapter is devoted to the L^1-spectral properties of absorption semigroups $\left(\mathcal{M}_V^1(t)\right)_{t\geq 0}$ with indefinite potentials

$$V = V^{(+)} - V^{(-)} \tag{13.2.4}$$

where $V^{(+)}$ is regular for $(\mathcal{M}(t))_{t\geq 0}$,

$$V^{(+)} \text{ is bounded from below} \tag{13.2.5}$$

($V^{(+)}$ need not be nonnegative) and

$$V^{(-)} \in L^1_+(\mathbb{R}^d). \tag{13.2.6}$$

We point out that the above decomposition of V need not coincide with its canonical decomposition into positive and negative parts. This chapter addresses several key issues:

(1) Well posedness of the absorption semigroups $\left(\mathcal{M}_V^1(t)\right)_{t\geq 0}$ for indefinite potentials V.
(2) Compactness (or spectral gap) properties of the generators M_V^1 and the semigroups $\left(\mathcal{M}_V^1(t)\right)_{t\geq 0}$.
(3) Spectral analysis of weighted Laplacians with application to Poincaré inequality for probability measures.
(4) Spectral analysis of Witten Laplacians on 1-forms with application to Helffer-Sjöstrand's covariance formula.

As noted in the general Introduction, absorption semigroups arise in various fields [27, 78, 105, 149, 285, 338]. These issues were systematically studied from an L^1-perspective in [271], exploiting the key contraction property (see 5.2.2) in the L^1-theory of absorption semigroups. However, it turned out that the abstract spectral gap results obtained there, which rely heavily on kernel estimates, are not well suited for unbounded geometries. The aim of this chapter is to revisit these results in greater depth, guided by the insights developed in Chap. 5.

For the simplicity of notations, throughout the sequel we write $(\mathcal{M}(t))_{t\geq 0}$, $(\mathcal{M}_V(t))_{t\geq 0}$ in place of $\left(\mathcal{M}^1(t)\right)_{t\geq 0}$, $\left(\mathcal{M}_V^1(t)\right)_{t\geq 0}$. Similarly, we denote the generators by M and M_V instead of M^1 and M_V^1.

For potentials V bounded from below, we recall the known resolvent compactness of M_V on $L^1(\mathbb{R}^d)$ as well as the compactness of $\left(\mathcal{M}^1(t)\right)_{t\geq 0}$ on $L^1(\mathbb{R}^d)$ (provided that $(\mathcal{M}(t))_{t\geq 0}$ is holomorphic) under the condition that the sublevel sets $\{V \leq C\}$ of V are thin at infinity in the sense (13.3.1), (see Theorem 13.3.1). We also show that if

$$\lim_{|x|\to\infty} \inf V(x) > -s(M_V)$$

then M_V has locally a spectral gap in $L^1(\mathbb{R}^d)$, (see Theorem 13.3.2).

These results are useful, for example, in the study of weighted Laplacians (see Theorem 13.6.1), with applications to the Poincaré inequality for probability measures (see Corollary 13.6.1), as well as for Witten Laplacians on 1-forms (see Theorem 13.7.1), with applications to Helffer–Sjöstrand's covariance formula (see Theorem 13.7.2) without requiring any convexity assumption on the weight functions.

For indefinite potentials $V = V^{(+)} - V^{(-)}$, besides several preliminary results, we establish generation and spectral gap theorems relying on weak compactness arguments. These results cover, in particular, Kato class potentials $V^{(-)}$, (see Theorems 13.4.1 and 13.4.3). Some remarks, comments or open questions are relegated to the end of this chapter in the section "Comments".

13.3 Potentials Bounded from Below

We begin by exploring spectral properties of absorption semigroups $(\mathcal{M}_V(t))_{t\geq0}$ associated with potentials V bounded from below.We denote by $B(z, 1)$ the ball in $\mathbb{R}^d$ centered at z with radius 1 and, for any measurable set $\Xi \subset \mathbb{R}^d$, we denote by $|\Xi|$ its Lebesgue measure. The following first result concerning L^1 compactness is known.

Theorem 13.3.1 ([271] Theorem 30) *Let* $V = V^{(+)}$ *be bounded from below (i.e.,* $V \geq -c$ *for some constant* $c > 0$*). Suppose that the sublevel sets* $\{V \leq C\}$ *of* V *are thin at infinity in the sense*

$$|\{V \leq C\} \cap B(z, 1)| \to 0 \; (z \to \infty) \; (C > 0) \tag{13.3.1}$$

(e.g. $\lim_{|x|\to\infty} V(x) = +\infty$*). Then* M_V *is resolvent compact in* $L^1(\mathbb{R}^d)$*. If* $(\mathcal{M}(t))_{t\geq0}$ *is holomorphic then* $(\mathcal{M}_V(t))_{t\geq0}$ *is compact in* $L^1(\mathbb{R}^d)$*.*

The second result is concerned with *local* spectral gapsin $L^1(\mathbb{R}^d)$.

Theorem 13.3.2 *Let* $V = V^{(+)}$ *be bounded from below (i.e.,* $V \geq -c$ *for some constant* $c > 0$*). If*

$$V_\infty := \lim_{|x|\to\infty} \inf \, V(x) > -s(M_V) \tag{13.3.2}$$

then M_V *has locally a spectral gap.*

Proof Note first that any convolution operator in $L^1(\mathbb{R}^d)$ is "locally compact" (see e.g. [73] Corollary 4.28, p. 114). Let

$$\widetilde{V} := V + c. \tag{13.3.3}$$

Note that $\widetilde{V} \geq 0$. Hence, by Theorem 5.7.1, $M_{\widetilde{V}}$ has locally a spectral gap if

$$\tilde{V}_\infty = V_\infty + c > -s(M_{\tilde{V}}) = -s(M_V) + c \qquad (13.3.4)$$

(i.e., if $V_\infty > -s(M_V)$). By a shift, M_V has also locally a spectral gap. $\qquad\square$

Remark 13.3.1 Even though the semigroup $(\mathcal{M}_V(t))_{t\geq 0}$ interpolates to all $L^p(\mathbb{R}^d)$ spaces and is self-adjoint on $L^2(\mathbb{R}^d)$, so that the spectrum of its generator is real in $L^2(\mathbb{R}^d)$, it is, to the best of our knowledge, not known in general whether the spectrum of M_V in $L^1(\mathbb{R}^d)$ is real for arbitrary convolution C_0-semigroups. If this is the case, then M_V has a spectral gap. Moreover, if the underlying semigroup $(\mathcal{M}(t))_{t\geq 0}$ is holomorphic, then so is $(\mathcal{M}_V(t))_{t\geq 0}$ (see [169]), and in that case, $(\mathcal{M}_V(t))_{t\geq 0}$ does have a spectral gap.

13.4   Indefinite Potentials

In this section, we suppose that V decomposes as (13.2.4) and (13.2.5) (13.2.6) are satisfied. Our additional assumption is

$$V^{(-)} \text{ is } M\text{-bounded in } L^1(\mathbb{R}^d), \qquad (13.4.1)$$

i.e., $\widehat{G} := V^{(-)}(1 - M)^{-1} : L^1(\mathbb{R}^d) \to L^1(\mathbb{R}^d)$ is bounded.

13.4.1   Generation

Note that

$$G := V^{(-)}(1 - M_{V^{(+)}})^{-1} : L^1(\mathbb{R}^d) \to L^1(\mathbb{R}^d)$$

is dominated by $\widehat{G}$. Let $H(x - y)$ be the kernel of $(1 - M)^{-1}$. Then for $\varphi \geq 0$

$$\begin{aligned}
G\varphi &= V^{(-)}(1 - M_{V^{(+)}})^{-1}\varphi \\
&\leq V^{(-)}(1 - M)^{-1}\varphi \\
&= \widehat{G}\varphi = \int_{\mathbb{R}^d} V^{(-)}(x)H(x - y)\varphi(y)dy.
\end{aligned}$$

We decompose G as

$$1_{\{V^{(-)}(x)\leq j\}}G + 1_{\{V^{(-)}(x)>j\}}G = G_j^{(1)} + G_j^{(2)}$$

and decompose the kernel of $\widehat{G}$ as

$$V^{(-)}(x)H(x - y) = 1_{\{V^{(-)}(x)\leq j\}}V^{(-)}(x)H(x - y) + 1_{\{V^{(-)}(x)>j\}}V^{(-)}(x)H(x - y).$$

Therefore $G_j^{(1)} \le \widehat{G}_j^{(1)}$ and $G_j^{(2)} \le \widehat{G}_j^{(2)}$ where

$$\widehat{G}_j^{(1)} = \int_{\mathbb{R}^d} 1_{\{V^{(-)}(x)>j\}} V^{(-)}(x) H(x-y)\varphi(y)dy$$

has a norm equal to $\sup_y \int_{\{V^{(-)}(x)>j\}} V^{(-)}(x) H(x-y)dx$. We are ready to give a generation result.

Theorem 13.4.1 *Let (13.2.5) (13.2.6) (13.4.1) be satisfied and*

$$\lim_{j\to\infty} \sup_y \int_{\{V^{(-)}(x)>j\}} V^{(-)}(x) H_1(x-y)dx < 1. \tag{13.4.2}$$

Then G is a positive weakly compact perturbation of a strict contraction. In particular, G is quasi-compact in $L^1(\mathbb{R}^d)$ and

$$J_V := M_{V^{(+)}} + V^{(-)} : D(M_{V^{(+)}}) \subset L^1(\mathbb{R}^d) \to L^1(\mathbb{R}^d)$$

generates a positive C_0-semigroup $(\mathcal{J}_V(t))_{t\ge 0}$. Moreover $(\mathcal{J}_V(t))_{t\ge 0}$ is holomorphic if $(\mathcal{M}(t))_{t\ge 0}$ is.

Proof By assumption $\left\| G_j^{(1)} \right\| < 1$ for j large enough. Let us show that $G_j^{(2)}$ is weakly compact. Since $G_j^{(2)} \le \widehat{G}_j^{(2)}$, it suffices to show that $\widehat{G}_j^{(2)}$ is compact. We decompose

$$\widehat{G}_j^{(2)}\varphi = 1_{\{V^{(-)}(x)\le j\}} V^{(-)}(x) \int_{\mathbb{R}^d} H(x-y)\varphi(y)dy$$

into two parts

$$1_{\{|x<R|\}} 1_{\{V^{(-)}(x)\le j\}} V^{(-)}(x) \int_{\mathbb{R}^d} H(x-y)\varphi(y)dy$$
$$+1_{\{|x\ge R|,\ V^{(-)}(x)\le j\}} V^{(-)}(x) \int_{\mathbb{R}^d} H(x-y)\varphi(y)dy.$$

The second part itself is decomposed into two pieces

$$1_{\{|x|\ge R,\ \varepsilon\le V^{(-)}(x)\le j\}} V^{(-)}(x) \int_{\mathbb{R}^d} H(x-y)\varphi(y)dy$$
$$+1_{\{|x|\ge R,\ V^{(-)}(x)\le\varepsilon\}} V^{(-)}(x) \int_{\mathbb{R}^d} H(x-y)\varphi(y)dy;$$

the second piece, with norm $\le \varepsilon \|H\|_{L^1(\mathbb{R}^d)}$, is as small as we want provided that we choose ε small enough.

For a fixed $\varepsilon > 0$, let us show that the norm of the first piece tends to zero as $R \to +\infty$. It is easy to see that this piece depends *linearly* on H and its norm is $\leq j \, \|H\|_{L^1(\mathbb{R}^d)}$. Hence, by approximation, we may assume that $H \in L^1(\mathbb{R}^d) \cap L^\infty(\mathbb{R}^d)$. Let $\Xi_R = \{|x| \geq R \; V^{(-)}(x) \geq \varepsilon\}$. We have

$$1_{\{|x| \geq R, \; \varepsilon \leq V^{(-)}(x) \leq j\}} V^{(-)}(x) \int_{\mathbb{R}^d} H(x-y)\varphi(y)dy$$

$$\leq \left(1_{\Xi_R}(x)\right) j \int_{\mathbb{R}^d} H(x-y)\varphi(y)dy.$$

Since

$$\int_{\{|x \geq R|\}} V^{(-)}(x)dx \geq \int_{\{|x \geq R|, \; \varepsilon \leq V^{(-)}(x)\}} V^{(-)}(x)dx$$

$$\geq \varepsilon \left|\{|x \geq R|, \; \varepsilon \leq V^{(-)}(x)\}\right|$$

then

$$|\Xi_R| \leq \varepsilon^{-1} \int_{\{|x \geq R|\}} V^{(-)}(x)dx \to 0 \; (R \to +\infty)$$

because $V^{(-)} \in L^1_+(\mathbb{R}^d)$. Therefore

$$\int_{\mathbb{R}^d} 1_{\{|x| \geq R, \; \varepsilon \leq V^{(-)}(x) \leq j\}} V^{(-)}(x) \int_{\mathbb{R}^d} H(x-y)\varphi(y)dy$$

$$\leq j \int_{\Xi_R} dx \int_{\mathbb{R}^d} H(x-y)\varphi(y)dy$$

$$= j \int_{\mathbb{R}^d} \left(\int_{\Xi_R} H(x-y)dx\right) \varphi(y)dy$$

$$= j \int_{\mathbb{R}^d} \left(\int_{\Xi_R - y} H(z)dz\right) \varphi(y)dy$$

$$\leq j \sup_y \left(\int_{\Xi_R - y} H(z)dz\right) \|\varphi\|$$

$$\leq j \, \|H\|_{L^\infty(\mathbb{R}^d)} \sup_y |\Xi_R - y| \, \|\varphi\|$$

$$= j \, \|H\|_{L^\infty(\mathbb{R}^d)} |\Xi_R| \, \|\varphi\|$$

since Lebesgue measure is invariant by translation $|\Xi_R - y| = |\Xi_R|$ so, since

$$\leq j \, \|H\|_{L^\infty(\mathbb{R}^d)} |\Xi_R| \to 0 \; (R \to +\infty),$$

the norm of the first piece is as small as we want provided that R is large enough. Finally, the operator $L^1(\mathbb{R}^d) \to L^1(\mathbb{R}^d)$

$$\varphi \to 1_{\{|x < R|\}} 1_{\{V^{(-)}(x) \leq j\}} V^{(-)}(x) \int_{\mathbb{R}^d} H(x-y)\varphi(y)dy$$

is compact as a composition of the mulitplication operator by $1_{\{V^{(-)}(x)\leq j\}}V^{(-)}(x)$ and the operator

$$\varphi \in L^1(\mathbb{R}^d) \to 1_{\{|x<R|\}}\int_{\mathbb{R}^d}H(x-y)\varphi(y)dy \in L^1(\mathbb{R}^d) \qquad (13.4.3)$$

which is known (see e.g. [73] Corollary 4.28, p. 114) to be compact. Thus $\widehat{G}_j^{(2)}$ is compact since it is as close to a compact operator as we want. Then Theorems 2.6.4 and 2.6.5 end the proof. $\qquad\square$

Let us inspect the assumption (13.4.2). To this end, we recall first

Definition 13.4.1 We say that $V^{(-)}$ belongs to the Kato classof $(\mathcal{M}(t))_{t\geq 0}$ if

$$\lim_{\varepsilon\to 0}\sup_y \int_{\{|x-y|\leq\varepsilon\}}V^{(-)}(x)H_1(x-y)dx = 0,$$

or belongs to the generalized Kato class of $(\mathcal{M}(t))_{t\geq 0}$ if

$$\lim_{\varepsilon\to 0}\sup_y \int_{\{|x-y|\leq\varepsilon\}}V^{(-)}(x)H_1(x-y)dx < 1.$$

We refer e.g. to [78, 338, 370] for the interest of Kato class potentials in the theory of Schrödinger operators.

Proposition 13.4.1 *Let* $V^{(-)} \in L^1(\mathbb{R}^d)$. *Then (13.4.2) is satisfied if* $V^{(-)}$ *belongs to the generalized Kato class of* $(\mathcal{M}(t))_{t\geq 0}$.

Proof We have

$$\int_{\{V^{(-)}(x)>j\}}V^{(-)}(x)H_1(x-y)dx$$

$$=\int_{\{V^{(-)}(x)>j\}\cap\{|x-y|\leq\varepsilon\}}V^{(-)}(x)H_1(x-y)dx$$

$$+\int_{\{V^{(-)}(x)>j\}\cap\{|x-y|\geq\varepsilon\}}V^{(-)}(x)H_1(x-y)dx$$

$$\leq\int_{\{|x-y|\leq\varepsilon\}}V^{(-)}(x)H_1(x-y)dx + \left(\sup_{|z|\geq\varepsilon}H_1(z)\right)\int_{\{V^{(-)}(x)>j\}}V^{(-)}(x)dx.$$

By assumption

$$\lim_{\varepsilon\to 0}\sup_y \int_{\{|x-y|\leq\varepsilon\}}V^{(-)}(x)H_1(x-y)dx < 1.$$

For a fixed small $\varepsilon > 0$,

$$\int_{\{V^{(-)}(x)>j\}} V^{(-)}(x)dx \to 0 \ (j \to \infty)$$

and we are done. $\square$

13.4.2   Spectral Analysis

We begin with a compactness result in $L^1(\mathbb{R}^d)$.

Theorem 13.4.2 *Let (13.4.2) be satisfied. If the sublevel sets of $V^{(+)}$ are thin at infinity (e.g. if $\lim_{|x|\to\infty} V^{(+)}(x) = +\infty$) then J_V is resolvent compact in $L^1(\mathbb{R}^d)$. If moreover $(\mathcal{M}(t))_{t\geq 0}$ is holomorphic then $(\mathcal{J}_V(t))_{t\geq 0}$ is (holomorphic and) compact in $L^1(\mathbb{R}^d)$.*

Proof It is a direct consequence of Theorem 13.4.1 and ([271] Theorem 65). $\square$

The thinness of $V^{(+)}$ at infinity captures the idea that $V^{(+)}$ "tends to infinity" as $|x| \to \infty$. We now present a local spectral gap result under the weaker assumption that $V^{(+)}$ is merely sufficiently large at infinity.

Theorem 13.4.3 *Let (13.2.5) (13.2.6) (13.4.1) (13.4.2) be satisfied. Suppose that*

$$V_\infty^{(+)} := \lim_{|x|\to\infty} \inf V^{(+)}(x) > -s(M_{V^{(+)}}) \tag{13.4.4}$$

and

$$\lim_{j\to\infty} \sup_y \int_{\{V^{(-)}(x)>j\}} V^{(-)}(x)H_1(x-y)dx = 0 \tag{13.4.5}$$

(e.g. $V^{(-)}$ belongs to the Kato class of $(\mathcal{M}(t))_{t\geq 0}$). Then J_V has locally a spectral gap.

Proof We already know that $s(J_V) \geq s(M_{V^{(+)}})$. We note that (13.4.5) implies that G is weakly compact, i.e., $V^{(-)}$ is $M_{V^{(+)}}$-weakly compact so J_V and $M_{V^{(+)}}$ have the same essential spectrum. It follows from Theorem 13.3.2 that J_V has also locally a spectral gap. $\square$

Note that (13.4.4) is satisfied if $\lim\inf_{|x|\to\infty} V^{(+)}(x) > c_{V^{(+)}}$ where $c_{V^{(+)}}$ is given by

$$\inf_{\left\{\varphi\in C_c^\infty(\mathbb{R}^d),\ \|\varphi\|_{L^2(\mathbb{R}^d)}^2=1\right\}} \left(\left\|\sqrt{-M^2}\varphi\right\|_{L^2(\mathbb{R}^d)}^2 + \int_{\mathbb{R}^d} V^{(+)}(x)\,|\varphi(x)|^2\,dx \right).$$

As in Remark 13.3.1, we can make a similar observation on a possible spectral gap of $(\mathcal{J}_V(t))_{t\geq 0}$.

13.5 Locally Bounded Potentials

We now make use of the alternative approach to spectral gaps for absorption semi-groups developed in Sect. 5.10. Its translation to the setting of convolution semigroups yields the following consequence of Theorem 5.10.1.

Theorem 13.5.1 *Let $(\mathcal{M}(t))_{t\geq 0}$ be a symmetric convolution semigroup on $L^p(\mathbb{R}^d)$ with generator M. Suppose that (13.2.3) is satisfied. Let $V \in L^{\infty}_{loc}(\mathbb{R}^d)$[1] and let $(\mathcal{M}_V(t))_{t\geq 0}$ be the corresponding absorption semigroup with generator $M_V = M - V$. If*

$$\lim_{x\to\infty} \inf V(x) > s(M) - s(M_V) \tag{13.5.1}$$

then

$$\omega_{ess}(\mathcal{M}_V) \leq s(M) - \lim_{x\to\infty} \inf V(x);$$

in particular, $(\mathcal{M}_V(t))_{t\geq 0}$ has a spectral gap in $L^p(\mathbb{R}^d)$.

As noted in Remark 5.7.1, (13.5.1) is satisfied if there exists a bounded open set $O \subset \mathbb{R}^d$ and $\lim\inf_{|x|\to\infty} V(x) > c_O$ where c_O is given by

$$\inf_{\left\{\varphi\in C_c^{\infty}(O),\ \|\varphi\|^2_{L^2(\mathbb{R}^d)}=1\right\}} \left(\left\| \sqrt{-M^2}\varphi \right\|^2_{L^2(\mathbb{R}^d)} + \int_O V(x)\,|\varphi(x)|^2\,dx \right).$$

It may happen that $s(M_V) = s(M) = 0$; in this case, the semigroup $(\mathcal{M}_V(t))_{t\geq 0}$ exhibits a spectral gap as soon as $\lim\inf_{x\to\infty} V(x) > 0$. This situation arises, for instance, in the case of the weighted Laplacians studied in the next section.

13.6 Weighted Laplacians on $\mathbb{R}^d$

A spectral gap result for weighted Laplacians on general non compact riemannian manifolds is presented in Chap. 14 (see Sect. 14.4). We briefly recall it here in the specific context of Euclidean spaces. In particular, this result improves, in several directions, the earlier findings given in ([271] Chap. 7). Let Φ be a real $\mathcal{C}^2$ function on $\mathbb{R}^d$ such that

$$\int_{\mathbb{R}^d} e^{-\Phi(x)}\,dx = 1 \tag{13.6.1}$$

and let $L^2(\mu) = L^2(\mathbb{R}^d; \mu(dx))$ where

$$\mu(dx) = e^{-\Phi(x)}\,dx.$$

---

[1] See Remark 5.10.1 for a more general assumption.

We denote by $(.,.)_\mu$ and $\|\ \|_\mu$ the scalar product and the norm in $L^2(\mu)$. We define the weighted Laplacian on $L^2(\mu)$

$$\Delta_\mu = \Delta - \nabla\Phi.\nabla$$

which is (minus) the nonnegative self-adjoint operator $\Delta_{(\mu)}$ in $L^2(\mu)$ associated to the Dirichlet form

$$\int_{\mathbb{R}^d} |\nabla\varphi|^2 \, \mu(dx), \ \varphi \in H^1(\mu)$$

where

$$H^1(\mu) := \left\{\varphi \in L^2(\mu), \ \frac{\partial\varphi}{\partial x_i} \in L^2(\mu), \ 1 \le i \le N\right\}.$$

Then the weighted Laplacian Δ_μ in $L^2(\mu)$ is unitarily equivalent to a Schrödinger operator $\Delta - V$ on $L^2(\mathbb{R}^d; dx)$ via the unitary transformation

$$I : \varphi \in L^2(\mu) \to e^{-\frac{\Phi}{2}}\varphi \in L^2(\mathbb{R}^d; dx)$$

where

$$V = -\frac{1}{2}\Delta\Phi + \frac{1}{4}|\nabla\Phi|^2.$$

We note that $\sigma(\Delta_{(\mu)}) \subset [0, +\infty)$ and $0 \in \sigma(\Delta_{(\mu)})$ since $1 \in D(\Delta_{(\mu)})$ and $\Delta_{(\mu)}1 = 0$. The (minus) Schrödinger operator

$$\Delta_{(\mu)} = -\Delta + \frac{1}{4}|\nabla\Phi|^2 - \frac{1}{2}\Delta\Phi$$

in $L^2(\mathbb{R}^d; dx)$ is known as the Witten Laplacian (on 0-forms) and was studied in particular in [155] in connection with Fokker-Planck operators.

13.6.1   On Discrete Spectra

Since the spectral bound of the Laplacian on $L^2(\mathbb{R}^d)$ is zero then a straightforward consequence of Theorem 14.4.1 yields the following:

Theorem 13.6.1 *Let (13.6.1) be satisfied and let*

$$V_\infty := \lim_{|x|\to\infty} \inf \left(\frac{1}{4}|\nabla\Phi|^2 - \frac{1}{2}\Delta\Phi\right) > 0$$

($V_\infty = +\infty$ is allowed). Then

$$\begin{cases} \sigma_{ess}(\Delta_\mu) \subset \{\operatorname{Re}\lambda \le -V_\infty\} & \text{if } V_\infty < +\infty \\ \sigma_{ess}(\Delta_\mu) = \varnothing & \text{if } V_\infty = +\infty. \end{cases}$$

The fact that $\sigma_{ess}(\Delta_{(\mu)}) \subset [V_\infty, +\infty)$ is known (see [173] Proposition 6.1) where the proof relies on Person formula [301].

13.6.2 *Poincaré Inequality for Probability Measures*

Similarly, Corollary 14.4.1 gives a Poincaré inequality for probability measures $e^{-\Phi(x)}dx$ on $\mathbb{R}^d$.

Corollary 13.6.1 *Let (13.6.1) be satisfied.If*

$$V_\infty := \lim_{|x|\to\infty} \inf \left(\frac{1}{4}|\nabla\Phi(x)|^2 - \frac{1}{2}\Delta\Phi(x) \right) > 0$$

($V_\infty = +\infty$ is allowed) then there exists $c > 0$ such that

$$var_\mu(f) := \|f - \langle f\rangle\|^2_{L^2(\mathbb{R}^d,\mu)} \le c \int_{\mathbb{R}^d} |\nabla f|^2\, d\mu, \quad f \in H^1(\mathbb{R}^d, \mu)$$

where $\langle f\rangle := \int_{\mathbb{R}^d} f(x)\mu(dx)$.

13.7 Witten Laplacians on 1-Forms on $\mathbb{R}^d$

The aim of this section is to extend and refine, in various directions, several spectral results on Witten Laplacians on 1-forms in $\mathbb{R}^d$ previously established in ([271] Chap. 8). Let Φ be a real C^2 function on $\mathbb{R}^m$ and let $\mu(dx) = e^{-\Phi(x)}dx$ be a probability measure. We point out that we do not assume that Φ is convex. Let

$$L^2(\mu) := L^2(\mathbb{R}^m,\ \mu(dx))$$

with scalar product $(.,.)_\mu$ and norm $\|\ \|_\mu$. The d-Complex in weighted L^2 spaces is given by

$$\Omega^0 \xrightarrow{d^{(0)}} \Omega^1 \xrightarrow{d^{(1)}} \Omega^2 \to \cdots \Omega^N \to 0$$

where $\Omega^p := \Omega^p(\mathbb{R}^m)$ $(p \le m)$ denotes the space of $L^2(\mu)$ p-forms (i.e., p-forms with coefficients in $L^2(\mu)$) equipped with its

$$L^2\left(\mathbb{R}^m,\ \mu;\ \wedge^p\mathbb{R}^m\right)$$

structure (Ω^0 is identified to $L^2(\mu)$). For the sake of simplicity, we still keep in Ω^p the notations $(.,.)_\mu$ and $\|\,\|_\mu$. Here

$$d^{(p)} : \Omega^p \to \Omega^{p+1}$$

is the restriction to Ω^p of the exterior differential d and is considered as an unbounded operator

$$L^2\left(\mathbb{R}^m,\ \mu;\ \wedge^p\mathbb{R}^m\right) \to L^2\left(\mathbb{R}^m,\ \mu;\ \wedge^{p+1}\mathbb{R}^m\right)$$

with domain

$$\left\{\omega \in \Omega^p;\ d\omega \in \Omega^{p+1}\right\}$$

where $d\omega$ is computed in the distributional sense. We denote by

$$d^{*(p)} : \Omega^{p+1} \to \Omega^p$$

the adjoint of $d^{(p)}$. The Laplacian $\triangle^{(p)}$ on Ω^p ($p \geqslant 1$) is then defined by

$$\triangle^{(p)} = d^{*(p)} \circ d^{(p)} + d^{(p-1)} \circ d^{*(p-1)} \tag{13.7.1}$$

and

$$\triangle^{(0)} = d^{*(0)} \circ d^{(0)}.$$

Actually, the unbounded operator $\triangle^{(p)}$ is defined by means of its quadratic form

$$\left\|d^{(p)}\omega\right\|_\mu^2 + \left\|d^{*(p-1)}\omega\right\|_\mu^2,\quad \omega \in \Omega^p,$$

we refer to [153, 154, 173, 339] for the details. It turns out that the Laplacian operator on weighted 0-forms

$$\triangle^{(0)} : L^2(\mu) \to L^2(\mu)$$

is unitarily equivalent to the following one

$$\triangle^{(0)}_{(\mu)} := \triangle_{(\mu)} = -\triangle + \frac{1}{4}|\nabla\Phi|^2 - \frac{1}{2}\triangle\Phi$$

on $L^2(\mathbb{R}^m,\ dx)$ considered in the previous section while the Laplacian on weighted 1-forms

$$\triangle^{(1)} = d^{*(1)} \circ d^{(1)} + d^{(0)} \circ d^{*(0)}$$

on $L^2\left(\mathbb{R}^m,\ \mu;\ \wedge^1\mathbb{R}^m\right)$ is unitarily equivalent to the following one

$$\triangle^{(1)}_{(\mu)} = \triangle^{(0)}_{(\mu)} \otimes Id + Hess\,\Phi$$

on the unweighted space

$$L^2(\mathbb{R}^m,\ dx;\ \wedge^1 \mathbb{R}^m)$$

where $Hess\,\Phi$ is the hessian of Φ; see [154, 173].

To avoid any potential confusion between the d-Complex and the dimension d of $\mathbb{R}^d$ used in the previous sections, we now denote the ambient space dimension by m. We identify any 1-form to its coefficients and therefore the spaces

$$L^2(\mathbb{R}^m,\ dx;\ \wedge^1 \mathbb{R}^m) = \left(L^2(\mathbb{R}^m,\ dx)\right)^m.$$

By construction, $\triangle^{(0)}_{(\mu)}$ and $\triangle^{(1)}_{(\mu)}$ are nonnegative. Our goal now is to establish a spectral connection between $\triangle^{(0)}_{\Phi}$ and $\triangle^{(1)}_{\Phi}$, (see e.g. [173] Theorem 1.3 for other spectral links). To this end, we first recall a basic functional analytic result related to Glazman's Lemma.

Lemma 13.7.1 ([296] Proposition 6.1.4, Corollaries 6.1.1 and 6.1.2, p. 72). *Let A and B be two self-adjoint operators in a Hilbert space $\mathcal{H}$ such that*

$$(Au, u) \leq (Bu, u), \ u \in \mathcal{D}$$

where $\mathcal{D} \subset \mathcal{H}$ is a core for both A and B. Then:

(i) For any real λ, if $\sigma(A) \cap (-\infty, \lambda)$ is discrete (i.e., consists of isolated eigenvalues with finite multiplicities) then so si $\sigma(B) \cap (-\infty, \lambda)$.

(ii) If we denote by $\lambda^A_1 \leq \lambda^A_2 \leq \cdots \leq \lambda^A_k \leq \cdots$ and $\lambda^B_1 \leq \lambda^B_2 \leq \cdots \leq \lambda^B_k \leq \cdots$ their eigenvalues in $(-\infty, \lambda)$, numbered according to their multiplicities, then $\lambda^A_k \leq \lambda^B_k$.

We are now ready to present the main result of this section, where the function Φ is not assumed to be convex.

Theorem 13.7.1 *Let Φ be a C^2 function such that $\int_{\mathbb{R}^m} e^{-\Phi(x)} dx = 1$. We denote by $\lambda_\Phi(x)$ the lowest eigenvalue of $Hess\,\Phi(x)$. Let*

$$W := \frac{1}{4}|\nabla\Phi|^2 - \frac{1}{2}\triangle\Phi + \lambda_\Phi$$

be such that

$$W_\infty := \lim_{|x|\to\infty} \inf\ W(x) > 0$$

($W_\infty = +\infty$ is allowed). Then, in $L^2\left(\mathbb{R}^m,\ \mu;\ \wedge^1 \mathbb{R}^m\right)$,

$$\sigma(\triangle^{(1)}) \cap [0, V_\infty)$$

consists (at most) of isolated eigenvalues with finite algebraic multiplicities.

Proof We deal with $\triangle^{(1)}_{(\mu)}$ since $\triangle^{(1)}$ is unitarily equivalent to $\triangle^{(1)}_{(\mu)}$. We know that

$$\triangle^{(1)}_{(\mu)} = \triangle^{(0)}_{(\mu)} \otimes Id + Hess\,\Phi \geq \left(\triangle^{(0)}_{(\mu)} + \lambda_\Phi\right) \otimes Id \qquad (13.7.2)$$

in the sense of forms. Notice that

$$\triangle^{(0)}_{(\mu)} + \lambda_\Phi = -\triangle + \frac{1}{4}|\nabla\Phi|^2 - \frac{1}{2}\triangle\Phi + \lambda_\Phi = -\triangle + W.$$

Since $s(\triangle) = 0$ in $L^2(\mathbb{R}^m,\, dx)$ then Theorem 5.10.2 shows that

$$\begin{cases} \sigma_{ess}(\triangle - W) \subset \{\mathrm{Re}\,\lambda \leq -W_\infty\} \text{ if } W_\infty < +\infty \\ \qquad \sigma_{ess}(\triangle - W) = \varnothing \text{ if } W_\infty = +\infty \end{cases}$$

or equivalently $\sigma(\triangle^{(0)}_{(\mu)} + \lambda_\Phi) \cap [0,\, W_\infty)$ consists of isolated eigenvalues with finite algebraic multiplicities. This is also the case for $\left(\triangle^{(0)}_{(\mu)} + \lambda_\Phi\right) \otimes Id$ which is nothing but m copies of $\triangle^{(0)}_{(\mu)} + \lambda_\Phi$. Finally Lemma 13.7.1 and (13.7.2) end the proof. $\qquad\square$

Unlike the set $\sigma(\triangle^{(0)}) \cap [0,\, W_\infty)$, which contains at least the botom $\lambda^{(0)} := 0$ of $\sigma(\triangle^{(0)})$, it is generally not known a priori whether $\sigma(\triangle^{(1)}) \cap [0,\, V_\infty)$ is empty or not.

Suppose that $\Phi(.)$ is convex and the lowest eigenvalue $\lambda_\Phi(.)$ of $Hess\,\Phi(.)$ is not identically zero. If $\sigma(\triangle^{(1)}) \cap [0,\, V_\infty)$ is not empty then the botom eigenvalue $\lambda^{(1)}$ of $\sigma(\triangle^{(1)})$ (which is isolated by Theorem 13.7.1) is positive

$$\lambda^{(1)} > 0 = \lambda^{(0)};$$

see the proof of ([271] Theorem 61).

There exist subtle and intricate connections between $\sigma(\triangle^{(0)})$ and $\sigma(\triangle^{(1)})$, see [173] and references therein for further details. We extract the following results from [173]:

Theorem 13.7.2 ([173]). *Suppose that* 0 *is an isolated eigenvalue of* $\triangle^{(0)}$. *Then*
(i) The exterior differentiation $d^{(0)} : H^1(\mu) \subset L^2(\mu) \to L^2\left(\mathbb{R}^m,\, \mu;\, \wedge^1\mathbb{R}^m\right)$ *has a closed range* X_1.
(ii) X_1 *is invariant under* $\triangle^{(1)}$ *and* $\triangle^{(1)}_{|X_1}$ *(the restriction of* $\triangle^{(1)}$ *to* X_1) *is unitarily equivalent to* $\triangle^{(0)}_{|R_{(0)}}$ *(the restriction of* $\triangle^{(0)}$ *to its range* $R_{(0)}$). *In particular,* $\triangle^{(1)}_{|X_1}$ *and* $\triangle^{(0)}_{|R_{(0)}}$ *share the same spectrum and the same essential spectrum.*
(iii) (Helffer-Sjöstrand's covariance formula)For all $f, g \in H^1(\mathbb{R}^d;\, \mu)$

$$(f - \langle f\rangle,\, g - \langle g\rangle)_\mu = \left((\triangle^{(1)})^{-1}df,\, (\triangle^{(1)})^{-1}dg\right)_\mu \qquad (13.7.3)$$

where df *is the differential of* $f \in H^1(\mathbb{R}^d;\, \mu)$.

Note that the botom of $\sigma(\triangle_{|R_{(0)}}^{(0)})$ (or equivalently the botom of $\sigma(\triangle_{|X_1}^{(1)})$) is strictly positive, since 0 is an isolated eigenvalue of $\triangle^{(0)}$; thus the formula (13.7.3) is meaningful. If $\sigma(\triangle^{(0)}) \cap [0, V_\infty)$ contains more than one eigenvalue, then the botom of $\sigma(\triangle_{|R_{(0)}}^{(0)})$ coincides with the second eigenvalue of $\triangle^{(0)}$. Otherwise, this botom is larger than or equal to V_∞. We also note that, even though $\sigma(\triangle_{|R_{(0)}}^{(0)}) = \sigma(\triangle_{|X_1}^{(1)})$, we cannot exclude a priori that the bottom of $\sigma(\triangle^{(1)})$ is strictly less than the second eigenvalue of $\triangle^{(0)}$.

13.8 Comments

The importance of Schrödinger semigroups hardly needs to be emphasized. A fundamental reference is the classical review paper by Simon [338], which highlights the role of heat semigroups on $L^1(\mathbb{R}^d)$ perturbed by Kato class potentials, particularly due to their significance in probabilistic contexts. In particular, if V is a Kato class potential, then its form-bound with respect to the Laplacian is zero, and the form sum $(-\triangle) \dotplus V$ is well-defined. More systematic results on absorption semigroups were developed by Voigt [370]. While the extensive literature on Schrödinger semigroups is beyond the scope of this work, we wish to emphasize once more the richness of the L^1-point of view. Naturally, these ideas are not confined to the Laplacian and can be extended to a broader class of generators, such as those arising from convolution semigroups. See, for instance, [78] for relativistic Schrödinger operators, or [157] for even more general cases. The idea of systematically applying Desch's perturbation theorem to general Schrödinger-type semigroups with generators of the form "$T - V$", where V is T-bounded in $L^1(\mathbb{R}^d)$, originates in [260, 263]; in particular the (L^2) form-bound of V respective to $-T$ is less than or equal to the limit

$$\delta := \lim_{\lambda \to +\infty} r\left(V(\lambda - T)^{-1}\right)$$

where $r\left(V(\lambda - T)^{-1}\right)$ denotes the spectral radius in $L^1(\mathbb{R}^d)$. It remains an open question whether this form-bound is actually *equal* to δ. Notably, the form-bound is zero if V is T-weakly compact in $L^1(\mathbb{R}^d)$. These ideas can be extended to the form-perturbation theory of higher-order elliptic operators and systems with singular potentials; see [276]. In such contexts, the lack of positivity can be compensated by domination arguments and the availability of Gaussian heat kernel estimates [162].

The first systematic approach to L^1 spectral analysis, such as discreteness and spectral gap results, for abstract absorption semigroups, with applications to convolution semigroups, dates back to [271]. This framework crucially relied on the key contraction property (5.2.2) in the L^1-theory of absorption semigroups. However, the spectral gap results in [271], which were based on rather abstract kernel estimates, proved to be of limited practical use. This limitation motivated the development of the more refined methods presented in Chap. 5 of this monograph.

Numerous results concerning the resolvent compactness in $L^1(\mathbb{R}^d)$ of operators of the form $\Delta - V$, for various classes of potentials V, are presented in [271] Chap. 7 with applications to weighted Laplacians.

In connection with Sect. 13.6.2: Many examples found in the literature on the Poincaré inequality focus on the case

$$\lim_{|x|\to\infty} \left(\frac{1}{4}|\nabla\Phi|^2 - \frac{1}{2}\Delta\Phi\right) = +\infty$$

under which the Schrödinger operator $\Delta - V$ is resolvent compact in $L^1(\mathbb{R}^d)$. This condition, for example, holds if Φ is uniformly strictly convex. As an illustrative example of Corollary 13.6.1, consider the function $\Phi_c(x) = c + \sqrt{1 + |x|^2}$ where $c > 0$ is chosen appropriately to ensure the normalization condition $\int_{\mathbb{R}^d} e^{-\Phi_c(x)}dx = 1$. Simple calculations show that

$$\frac{1}{4}|\nabla\Phi|^2 - \frac{1}{2}\Delta\Phi \to \frac{1}{4}\ (|x| \to \infty). \tag{13.8.1}$$

Consequently, the probability measure $\mu(dx) := e^{-\left(c+\sqrt{1+|x|^2}\right)}dx$ satisfies the Poincaré inequality. If we replace $\Phi_c(x)$ by

$$\Phi_c^{(\alpha)}(x) = c + \left(1 + |x|^2\right)^{\frac{\alpha}{2}}\ (0 < \alpha \le 2),$$

then simple calculations show that the limit in (13.8.1) is $+\infty$ for $\alpha > 1$ and 0 for $\alpha < 1$. Consequently, the Poincaré inequality holds true for $\alpha \ge 1$ while

$$\sigma(\Delta^\mu) = \sigma_{ess}(\Delta^\mu) = \sigma_{ess}(-\Delta) = [0, +\infty)$$

for $\alpha < 1$ and the Poincaré inequality is not satisfied. For more examples, we refer the reader to [155]. Further results on fractional Poincaré inequalities and related topics can be found in [284] and references therein.

In connection with Theorem 13.7.2, if Φ is uniformly strictly convex, we recover the famous Brascamp-Lieb's inequality [70]

$$(f - \langle f\rangle, g - \langle g\rangle)_\mu \le \left((Hess\,\Phi)^{-1}df, dg\right);$$

see [173] for more details and further developments.

The Poincaré inequality stated in Corollary 13.6.1 for the weighted Laplacian on $\mathbb{R}^d$ is, in fact, a particular case of a more general result concerning weighted Laplacians on non compact Riemannian manifolds without boundary, see Corollary 14.4.1.

Chapter 14
Miscellaneous

14.1 Chapter Aims

This chapter collects several unrelated results and remarks concerning various topics discussed throughout the monograph. These include: a self-contained proof of Desch's perturbation theorem; a suitable bounded variation (BV) condition underlying Miyadera perturbations; a spectral gap property for general weighted Laplacians on non-compact Riemannian manifolds; a predual approach to L^p spectral properties of neutron transport; and a short discussion of symmetrizable operators arising in Transport Theory.

14.2 Proof of Desch's Theorem in AL Spaces

A proof of Desch's theorem (see Theorem 14.2.1 below), based on Miyadera perturbations [362], is presented in [371]. Here, we provide a (slightly) different proof that offers additional advantages exploited throughout this monograph. To this end, we first establish a useful technical result.

Lemma 14.2.1 ([245] Lemma 8.3) *Let $C \in \mathcal{L}_+(L^1(\nu))$ be such that $r(C) < 1$. Then there exists a new norm on $L^1(\nu)$, equivalent to the usual one $\|.\|_{L^1(\nu)}$, additive on the positive cone $L^1_+(\nu)$ and such that C is a strict contraction with respect to the new norm.*

Proof For reader's convenience, we recall the proof. We have

$$\left(\left\| C^j \right\|_{\mathcal{L}(L^1(\nu))} \right)^{\frac{1}{j}} \to r(C) < 1 \quad (j \to +\infty).$$

Let $r(C) < c < c' < 1$ and j_0 such that

M. Mokhtar-Kharroubi, *Peripheral Spectra of Perturbed Positive Semigroups*, Lecture Notes in Mathematics 2388, https://doi.org/10.1007/978-3-032-11173-9_14

$$\left\|C^j\right\|_{\mathcal{L}(L^1(\nu))} \le c^j < \left(c'\right)^j \quad (j \ge j_0).$$

We set

$$\|x\|_1 := \sum_{j\ge 0} \frac{\left\|C^j x\right\|_{L^1(\nu)}}{(c')^j}$$

where the series converges since $\left\|C^j x\right\|_{L^1(\nu)} \le c^j \|x\|_{L^1(\nu)}$ for $j \ge j_0$. Of course $\|x\|_1 \ge \|x\|_{L^1(\nu)}$. One easily checks that there exists $c'' > 0$ such that

$$\|x\|_1 \le c'' \|x\|_{L^1(\nu)}.$$

Moreover, if $x, y \in L^1_+(\nu)$ then

$$\begin{aligned}
\|x + y\|_1 &= \sum_{j\ge 0} \left\|C^j(x + y)\right\|_{L^1(\nu)} \\
&= \sum_{j\ge 0} \left\|C^j x + C^j y\right\|_{L^1(\nu)} \\
&= \sum_{j\ge 0} \left(\left\|C^j x\right\|_{L^1(\nu)} + \left\|C^j y\right\|_{L^1(\nu)}\right) \\
&= \sum_{j\ge 0} \left\|C^j x\right\|_{L^1(\nu)} + \sum_{j\ge 0} \left\|C^j y\right\|_{L^1(\nu)} \\
&= \|x\|_1 + \|y\|_1.
\end{aligned}$$

Finally,

$$\begin{aligned}
\|Cx\|_1 &= \sum_{j\ge 0} \frac{\left\|C^{j+1} x\right\|_{L^1(\nu)}}{(c')^j} \\
&= c' \sum_{j\ge 0} \frac{\left\|C^{j+1} x\right\|_{L^1(\nu)}}{(c')^{j+1}} \\
&= c' \sum_{j\ge 1} \frac{\left\|C^j x\right\|_{L^1(\nu)}}{(c')^j} \le c' \|x\|_1
\end{aligned}$$

shows that the norm of C relative to $\|.\|_1$ (we denote by $\|C\|_1$ for the sake of simplicity) is $\le c'$. This ends the proof. $\qquad\square$

We state again Desch's theorem.

Theorem 14.2.1 *Let $(U(t))_{t\ge 0}$ be a positive C_0-semigroup on $L^1(\nu)$ with generator T and let $B : D(T) \subset L^1(\nu) \to L^1(\nu)$ be positive. Then*

$$T + B : D(T) \subset L^1(\nu) \to L^1(\nu)$$

generates a positive semigroup $(V(t))_{t\geq 0}$ *if and only if*

$$\lim_{\lambda\to+\infty} r\left(B\left(\lambda-T\right)^{-1}\right) < 1. \tag{14.2.1}$$

In addition, $(V(t))_{t\geq 0}$ *is given by a Dyson-Phillips series*

$$V(t)x = \sum_{j=0}^{\infty} U_j(t)x \ (x \in D(T))$$

defined by $U_0(t) = U(t)$ *and*

$$U_{j+1}(t)x = \int_0^t U_j(t-s)BU(s)xds \ (x \in D(T))$$

where these operators extend uniquely to the whole space as bounded operators and the Dyson Plillips series converges in operator norm uniformly in $t \in [0, c]$ $(c > 0)$. *Finally, we have Duhamel identities*

$$\begin{cases} V(t)x = U(t)x + \int_0^t V(t-s)BU(s)xds, & (x \in D(T)) \\ V(t)x = U(t)x + \int_0^t U(t-s)BV(s)xds, & (x \in D(T)). \end{cases}$$

Proof Note first that if $T + B$ is a generator then it is resolvent positive and consequently (14.2.1) holds according to Proposition 2.5.1. Conversely, we fix λ large enough so that $r\left(B(\lambda-T)^{-1}\right) < 1$. The Duhamel equation

$$V(t)x = U(t)x + \int_0^t V(t-s)BU(s)xds \tag{14.2.2}$$

is quivalent to

$$e^{-\lambda t}V(t)x = \left(e^{-\lambda t}U(t)\right)x + \int_0^t e^{-\lambda(t-s)}V(t-s)B\left(e^{-\lambda s}U(s)\right)xds.$$

To solve it, let

$$Z(t) = e^{-\lambda t}V(t)$$

and $W(t) = e^{-\lambda t}U(t)$. We are going to solve

$$Z(t)x = W(t)x + \int_0^t Z(t-s)BW(s)xds \tag{14.2.3}$$

in the space $\mathcal{E}$ of strongly continuous mappings

$$Z : [0, +\infty) \to \mathcal{L}(L^1(\nu))$$

such that

$$\sup_{t\geq0} \|Z(t)x\|_{L^1(\nu)} < +\infty \quad (x \in L^1(\nu)).$$

By Banach-Steinhaus theorem

$$\sup_{t\geq0} \|Z(t)\|_{\mathcal{L}(L^1(\nu))} < +\infty, \quad (Z \in \mathcal{E}).$$

We endow $\mathcal{E}$ with the norm

$$\|Z\|_{\mathcal{E}} = \sup_{t\geq0} \|Z(t)\|_{\mathcal{L}(L^1(\nu))}$$

and we easily check that $\mathcal{E}$ is a Banach space. Thus, solving (14.2.3) amounts to finding a fixed point of the affine mapping

$$\mathcal{E} \ni Z \to W + \mathcal{L}Z$$

where

$$\mathcal{L}Z(t) = \int_0^t Z(.-s)BW(s)ds \quad \text{(stong integral)}.$$

Let

$$C := B(\lambda - T)^{-1}.$$

We are going to use the new (equivalent) norm $\|.\|_1$ on $L^1(\nu)$. If $x \in D_+(T)$ then

$$\left\| \int_0^t Z(t-s)BW(s)xds \right\|_1 = \int_0^t \|Z(t-s)BW(s)x\|_1\, ds$$

$$\leq \|Z\|_{\mathcal{E}} \int_0^t \|BW(s)x\|_1\, ds$$

$$= \|Z\|_{\mathcal{E}} \left\| \int_0^t BW(s)xds \right\|_1$$

$$\leq \|Z\|_{\mathcal{E}} \left\| \int_0^{+\infty} Be^{-\lambda s}U(s)xds \right\|_1$$

$$= \|Z\|_{\mathcal{E}} \left\| B\int_0^{+\infty} e^{-\lambda s}U(s)xds \right\|_1$$

$$= \|Z\|_{\mathcal{E}} \left\| B(\lambda - T)^{-1}x \right\|_1$$

$$\leq \|Z\|_{\mathcal{E}} \left\| B(\lambda - T)^{-1} \right\|_1 \|x\|_1.$$

Note that

$$y = \lim_{j \to \infty} j \int_0^{j^{-1}} W(s) y \, ds = \lim_{j \to \infty} j \int_0^{j^{-1}} W(s) y_+ \, ds - \lim_{j \to \infty} j \int_0^{j^{-1}} W(s) y_- \, ds$$

where the convergence holds in $L^1(\nu)$ for all y in $L^1(\nu)$ and in $D(T)$ (with graph norm) if $y \in D(T)$. For $x \in D(T)$, let

$$x_j^+ := j \int_0^{j^{-1}} W(s) x_+ \, ds, \quad x_j^- := j \int_0^{j^{-1}} W(s) x_- \, ds.$$

Then

$$\left\| \int_0^t Z(t-s) B W(s) \left(x_j^+ - x_j^- \right) ds \right\|_1 \leq \|Z\| \left\| B \left(\lambda - T \right)^{-1} \right\|_1 \left(\left\| x_j^+ \right\|_1 + \left\| x_j^+ \right\|_1 \right).$$

Since $x_j^+ - x_j^- \to x$ in $D(T)$ then

$$\left\| \int_0^t Z(t-s) B W(s) x \, ds \right\|_1 \leq \|Z\|_{\mathcal{E}} \left\| B \left(\lambda - T \right)^{-1} \right\|_1 \left(\|x_+\|_1 + \|x_-\|_1 \right)$$

$$= \|Z\|_{\mathcal{E}} \left\| B \left(\lambda - T \right)^{-1} \right\|_1 \|x\|_1 .$$

Thus

$$x \in D(T) \to \int_0^t Z(t-s) B W(s) x \, ds$$

extends uniquely to $L^1(\nu)$ and

$$\| (\mathcal{L}Z)(t) \| \leq \|Z\|_{\mathcal{E}} \left\| B \left(\lambda - T \right)^{-1} \right\|_1 .$$

Let us check that for all $x \in L^1(\nu)$

$$[0, +\infty) \ni t \to (\mathcal{L}Z)(t) x \in \mathcal{L}(L^1(\nu))$$

is continuous. This is true if $x \in D(T)$. Let $x \in L^1(\nu)$ and $(x_j)_j \subset D(T)$ such that $x_j \to x$ $(j \to \infty)$. Then the estimate

$$\left\| \int_0^t Z(t-s) B W(s)(x_n - x_m) ds \right\| \leq \|Z\|_{\mathcal{E}} \left\| B \left(\lambda - T \right)^{-1} \right\|_1 \|x_n - x_m\|_1$$

shows that

$$\left(\int_0^t Z(t-s) B W(s) x_j \, ds \right)_j$$

is a Cauchy sequence in the space of continuous and bounded (on $[0, +\infty)$) $L^1(\nu)$-valued mappings endowed with supremum norm. This shows the strong continuity of $t \to (\mathcal{L}Z)(t) \in \mathcal{L}(L^1(\nu))$. Finally $\mathcal{L} \in \mathcal{L}(\mathcal{E})$ and

$$\|\mathcal{L}Z\|_{\mathcal{E}} := \sup_{t \geq 0} \|(\mathcal{L}Z)(t)\| \leq \|Z\|_{\mathcal{E}} \left\| B(\lambda - T)^{-1} \right\|_1$$

so

$$\|\mathcal{L}\|_{\mathcal{L}(\mathcal{E})} \leq \left\| B(\lambda - T)^{-1} \right\|_1 < 1.$$

Thus (14.2.3) gives

$$(Z(t))_{t \geq 0} = \left(e^{-\lambda t} V(t) \right)_{t \geq 0} \tag{14.2.4}$$

by iterations

$$Z(t) = \sum_{j=0}^{\infty} \mathcal{L}^j W$$

(where the series converges in operator norm uniformly in $t \in \mathbb{R}$) or equivalently

$$Z(t) = \sum_{j=0}^{\infty} \widetilde{U}_j(t)$$

where $\widetilde{U}_0(t) = e^{-\lambda t} U(t)$ and

$$\widetilde{U}_{j+1}(t)x = \int_0^t \widetilde{U}_j(t-s) B \widetilde{U}_0(s)x\,ds.$$

This amounts to

$$V(t) = \sum_{j=0}^{\infty} U_j(t) \tag{14.2.5}$$

(where the series converges in operator norm uniformly in $t \in [0, c]$ $c > 0$) with $U_0(t) = U(t)$ and

$$U_{j+1}(t) = \int_0^t U_j(t-s) B U(s)\,ds \ (j \geq 0).$$

One can check (see e.g. [245] Chap. 7) that $(V(t))_{t \geq 0}$ is a semigroup. The proof above shows that this semigroup is strongly continuous. Let

$$G : D(G) \subset L^1(\nu) \to L^1(\nu)$$

be the generator of $(V(t))_{t \geq 0}$. Since

$$s \geq 0 \to BU(s)x \in L^1(\nu), \quad (x \in D(T))$$

is continuous then

$$\frac{1}{t}\int_0^t V(t-s)BU(s)x\,ds \to Bx \quad (t \to 0_+).$$

It follows from (14.2.2) that for $x \in D(T)$

$$\frac{V(t)x-x}{t} = \frac{U(t)x-x}{t} + \frac{1}{t}\int_0^t V(t-s)BU(s)x\,ds \to Tx + Bx$$

so $T + B \subset G$. The fact that for λ large enough

$$\lambda - (T+B) : D(G) \to L^1(\nu)$$

and

$$\lambda - G : D(T) \to L^1(\nu)$$

are bijective shows that $D(G) = D(T)$ and consequently $G = T + B$. It follows from

$$(\lambda - T - B)^{-1} = (\lambda - T)^{-1}\sum_{j=0}^{\infty}\left(B(\lambda - T)^{-1}\right)^j$$

that

$$(\lambda - T - B)^{-1} - (\lambda - T)^{-1} = (\lambda - T)^{-1}\sum_{j=1}^{\infty}\left(B(\lambda - T)^{-1}\right)^j$$

and

$$(\lambda - T - B)^{-1}B(\lambda - T)^{-1} = (\lambda - T)^{-1}\sum_{j=1}^{\infty}\left(B(\lambda - T)^{-1}\right)^j$$
$$= (\lambda - T - B)^{-1} - (\lambda - T)^{-1}.$$

We have also

$$B(\lambda - T - B)^{-1} = B(\lambda - T)^{-1}\sum_{j=0}^{\infty}\left(B(\lambda - T)^{-1}\right)^j$$
$$= \sum_{j=1}^{\infty}\left(B(\lambda - T)^{-1}\right)^j$$

and

$$(\lambda - T)^{-1} B (\lambda - T - B)^{-1} = (\lambda - T)^{-1} \sum_{j=1}^{\infty} \left(B(\lambda - T)^{-1} \right)^j$$
$$= (\lambda - T - B)^{-1} - (\lambda - T)^{-1}.$$

Finally, for $x \in D(T)$,

$$(\lambda - T)^{-1} B (\lambda - T - B)^{-1} x = (\lambda - T - B)^{-1} B (\lambda - T)^{-1} x$$

are nothing but the Laplace transform of

$$\int_0^t U(s) B V(t - s) x \, ds \text{ and } \int_0^t V(s) B U(t - s) x \, ds;$$

this ends the proof of the Duhamel identities by injectivity of the Laplace transform. $\qquad\square$

Remark 14.2.1 (*Proof of Theorem* 2.6.2) If $(U(t))_{t \geq 0}$ is immediately norm continuous then $U_j(.)$ ($j \geq 1$) are also immediately norm continuous and then so is the perturbed semigroup $(V(t))_{t \geq 0}$. Indeed, this result holds true for Miyadera perturbations [226] and any Desch perturbation in AL space is a Miyadera perturbation, see Remark 14.3.1 below.

14.3 Miyadera Perturbations Versus BV Vector Valued Functions

The objective of this section is to elucidate the connection between Miyadera's perturbation theorem and a suitable bounded variation (BV) property. This perspective clarifies why the theorem is especially well-suited for ordered spaces whose positive cone is equipped with an additive norm. This result was originally announced in [244] but has not been published until now. We begin with

Definition 14.3.1 Let X be a Banach space and let $(U(t))_{t \geq 0}$ be C_0-semigroup on X with generator T. Let $B : D(T) \to X$ be continuous where $D(T)$ is endowed with the graph norm. We say that B is a Miyadera perturbationof T if there exist $\alpha > 0$ and $\gamma \in [0, 1)$ such that

$$\int_0^{\alpha} \| B U(t) x \| \, dt \leq \gamma \| x \| \quad \forall x \in D(T). \tag{14.3.1}$$

We recall Miyadera perturbation theorem on Banach spaces.

Theorem 14.3.1 (see [314, 362]) *Let X be a Banach space and let $(U(t))_{t \geq 0}$ be a C_0-semigroup on X with generator T. Let B be a Miyadera perturbation of T.*

Then $T + B : D(T) \to X$ generates a C_0-semigroup $(V(t))_{t \geq 0}$ on X given by a Dyson-Phillips expansion $V(t) = \sum\limits_{j=0}^{\infty} U_j(t)$ (converging in operator norm) where $U_j(t) \in \mathcal{L}(X)$ $(j \geq 0)$ are such that

$$U_0(t) = U(t), \; U_{j+1}(t)x = \int_0^t U_j(t-s)BU(s)x\,ds \; (x \in D(T))$$

and

$$V(t)x = U(t)x + \int_0^t V(t-s)BU(s)x\,ds \; (x \in D(T)).$$

Remark 14.3.1 We note that Desch's theorem can be deduced from Theorem 14.3.1. Indeed, by renorming if necessary (see Lemma 14.2.1), we can assume that $\left\| B(\lambda - T)^{-1} \right\| < 1$ for some real λ. In this case, for $x \in D(T) \cap X_+$

$$
\begin{aligned}
\int_0^{+\infty} \left\| Be^{-\lambda t}U(t)x \right\| dt &= \left\| \int_0^{+\infty} Be^{-\lambda t}U(t)x\,dt \right\| \\
&= \left\| B \int_0^{+\infty} e^{-\lambda t}U(t)x\,dt \right\| \\
&= \left\| B(\lambda - T)^{-1}x \right\| \leq \left\| B(\lambda - T)^{-1} \right\| \|x\|
\end{aligned}
$$

and, arguing as previously, we end up with

$$\int_0^{+\infty} \left\| Be^{-\lambda t}U(t)x \right\| dt \leq \left\| B(\lambda - T)^{-1} \right\| \|x\| \; (x \in D(T))$$

so B is a Miyadera perturbation of $T - \lambda$ and we are done.

For a given $\alpha > 0$, our goal here is to characterize a portion of the Miyadera property, specifically

$$\exists \gamma > 0; \; \int_0^{\alpha} \|BS(t)x\| \, dt \leq \gamma \|x\| \; \forall x \in D(T), \tag{14.3.2}$$

without worrying about the smallness condition $\gamma < 1$. We first note that

$$\int_0^t S(s)x\,ds \in D(T) \text{ and } T\int_0^t S(s)x\,ds = S(t)x - x \; (x \in X)$$

so

$$x \in X \to B\int_0^t S(s)x\,ds \in X \text{ is continuous } (t \geq 0)$$

and, for all $x \in X$,

$$L_x : \left[0, +\infty\right[\ni t \to B \int_0^t S(s)x\,ds \in X \text{ is continuous.}$$

Here is the main result.

Theorem 14.3.2 *Let X be a Banach space and let $(U(t))_{t\geq 0}$ be a C_0-semigroup on X with generator T. Let $B : D(T) \to X$ be continuous where $D(T)$ is endowed with the graph norm. Then (14.3.2) is satisfied if and only if for all $x \in X$*

$$[0, \alpha] \ni t \to L_x(t) \in X \text{ has a bounded variation.}$$

Proof Note first that for $x \in D(T)$

$$h^{-1} \left(\int_0^{t+h} S(s)x\,ds - \int_0^t S(s)x\,ds \right) \to S(t)x$$

in $D(T)$ endowed with the graph norm so that

$$\left[0, +\infty\right[\ni t \to B \int_0^t S(s)x\,ds \in X \tag{14.3.3}$$

is of class C^1 with derivative

$$\left[0, +\infty\right[\ni t \to BS(t)x$$

and (14.3.3) has always a bounded variation $V(x)$ on $[0, \alpha]$ equal to

$$\int_0^\alpha \|BS(s)x\|\,ds \quad (x \in D(T)).$$

Let (14.3.2) be satisfied. Then

$$V(x) \leq \gamma \|x\| \quad \forall x \in D(T). \tag{14.3.4}$$

Let $x \in X$ be arbitrary and let $(x_n)_n \subset D(T)$ such that $x_n \to x$. We have

$$V(x_n) \leq \gamma \|x_n\| \quad \forall n. \tag{14.3.5}$$

Let

$$0 = t_0 < t_1 < \cdots < t_m := \alpha$$

be an arbitrary subdivision $\mathcal{S}_m$ of $[0, \alpha]$. The variation of $L_x(.)$ on $\mathcal{S}_m$ is given by

$$\sum_{j=0}^{m-1} \|L_x(t_{j+1}) - L_x(t_j)\|.$$

In particular

$$\sum_{j=0}^{m-1} \left\| L_{x_n}(t_{j+1}) - L_{x_n}(t_j) \right\| \leq \gamma \|x_n\| \quad \forall n. \tag{14.3.6}$$

Since

$$L_{x_n}(t_{j+1}) - L_{x_n}(t_j) \to L_x(t_{j+1}) - L_x(t_j) \ (n \to \infty)$$

then, passing to the limit in (14.3.6),

$$\sum_{j=0}^{m-1} \left\| L_x(t_{j+1}) - L_x(t_j) \right\| \leq \gamma \|x\|$$

and therefore $L_x(.)$ has a bounded variation

$$V(x) \leq \gamma \|x\| \quad \forall x \in X.$$

Conversely, assume that for all $x \in X$, $L_x(.)$ has a bounded variation on $[0, \alpha]$, i.e.,

$$V(x) := \sup_{\{\mathcal{S}_m\}} \sum_{j=0}^{m-1} \left\| L_x(t_{j+1}) - L_x(t_j) \right\| < +\infty, \quad (x \in X)$$

where $\{\mathcal{S}_m\}$ denotes the set of all possible subdivisions of $[0, \alpha]$. Let

$$X_k := \{x \in X; \ V(x) \leq k\}.$$

Thus $X = \cup_k X_k$ and it is easy to see that X_k is closed in X. It follows from Baire's lemma that some X_k (say $X_{\bar{k}}$) has an interior point $y \in X$ so, for some $\varepsilon > 0$,

$$V(y + \varepsilon z) \leq \bar{k}, \quad (\|z\| \leq 1).$$

Since $L_x(t)$ is linear in x then

$$\begin{aligned}
\varepsilon V(z) =V(\varepsilon z) &= V(y + \varepsilon z - y) \\
&\leq V(y + \varepsilon z) + V(-y) \\
&= V(y + \varepsilon z) + V(y) \leq 2\bar{k}
\end{aligned}$$

and $V(z) \leq \frac{2\bar{k}}{\varepsilon}$ $(\|z\| \leq 1)$ or equivalently

$$V(x) \leq \frac{2\bar{k}}{\varepsilon} \|x\| \quad \forall x \in X.$$

Since

$$V(x) = \int_0^\alpha \|BS(s)x\| \, ds \quad \forall x \in D(T)$$

then

$$\int_0^\alpha \|BS(s)x\| \, ds \leq \frac{2\overline{k}}{\varepsilon} \|x\| \quad \forall x \in D(T)$$

and (14.3.2) is satisfied. $\qquad\square$

If $(U(t))_{t\geq 0}$ is a positive C_0-semigroup on an AL space X with generator T and if $B : D(T) \to X$ is positive then, due to the additivity of the norm on the positive cone, for each $x \in X_+$ the non decreasing map

$$[0, \alpha] \ni t \to L_x(t) \in X$$

has a bounded variation. For a general element $x = x_+ - x_-$

$$[0, \alpha] \ni t \to L_x(t) = L_{x_+}(t) - L_{x_-}(t)$$

is BV as a difference of two BV functions. In general, the argument fails outside the realm of ordered spaces X with additive norm on X_+. However, the above argument extends to general ordered spaces provided that the pertubation B has finite rank:

$$Bx = \sum_{i \in I} x_i^*(x) x_i$$

where I is finite and each x_i^* is a positive linear form on $D(T)$. Indeed, in this case,

$$B \int_0^t S(s)x \, ds = \sum_{i \in I} x_i^* \left(\int_0^t S(s)x \, ds \right) x_i$$

and the mappings

$$[0, \alpha] \ni t \to x_i^* \left(\int_0^t S(s)x \, ds \right) \in \mathbb{R}$$

are real-valued functions of bounded variation. (The arguments extend to suitable nuclear perturbations B). This allowed a partial extension of the L^1 scattering theory given in [243] to Banach lattices for finite rank perturbations [84].

14.4 Spectra of Weighted Laplacians on Riemannian Manifolds

The objective of this section is to present a very general spectral gap result for weighted Laplacians on non-compact Riemannian manifolds without boundary. For further motivation on weighted Laplacians, we refer to [149, 285]. This result follows as a straightforward consequence of the abstract spectral theory developed in Chap. 5 (see Theorem 5.10.2). This purely functional-analytic approach complements the existing literature that often relies on *geometric* methods to study the essential spectrum of non-compact Riemannian manifolds; see e.g. [317, 336], and references therein.

Let M be a (connected) d-dimensional non compact riemannian manifold without boundary with a riemannian metric g. Let

$$\Delta = \operatorname{div} \circ \nabla$$

be the Laplace operator on M and let ν be the riemannian volume on M. Let

$$H^1(\nu) := \left\{ u \in L^2(M; \nu); \ \nabla u \in L^2(M; \nu) \right\}$$

(∇u is the gradient in a distributional sense) be the Hilbert space endowed with the scalar product

$$(u, v)_{H^1(\nu)} = \int_M u\bar{v}d\nu + \int_M \langle \nabla u, \nabla v \rangle d\nu$$

where

$$\langle \nabla u, \nabla v \rangle := g(\nabla u, \nabla v).$$

We define the Laplacian operator Δ_ν on $L^2(\nu) := L^2(M; \nu)$ as the self-adjoint operator associated to the Dirichlet form

$$\int_M \langle \nabla u, \nabla u \rangle d\nu, \ \ u \in H^1(\nu).$$

The domain of Δ_ν is

$$D(\Delta_\nu) = \left\{ u \in H^1(\nu); \ \Delta u \in L^2(\nu) \right\}$$

(Δu is computed in a distributional sense) and $\Delta_\nu u = \Delta u$ for $u \in D(\Delta_\nu)$; in particular

$$-\int_M (\Delta_\nu u)\bar{v}d\nu = \int_M \langle \nabla u, \nabla v \rangle d\nu, \ \ u \in D(\Delta_\nu), \ v \in H^1(\nu)$$

and $\Delta_\nu \leq 0$. The spectral properties of Laplace-Beltrami operators Δ_ν in non compact riemannian manifolds constitute a vast area with deep geometric content (see e.g. [100]), which lies beyond the scope of this monograph. We content ourselves with the elementary information on its spectral bound

$$s(\Delta_\nu) \leq 0.$$

We note that the symmetric C_0-semigroup $\left(e^{t\Delta_\nu}\right)_{t \geq 0}$ is a Markov semigroup, i.e., extends consistently to all $L^p(\nu)$ $(1 \leq p \leq +\infty)$ as a contraction semigroup and is strongly continuous if $p < +\infty$. For details, we refer to ([99] Chaps. 1 and 5).

Let h be a smooth function on M such that $h(x) > 0 \; \forall x \in M$. The two cases

$$\int_M h^2(x)\nu(dx) < +\infty \tag{14.4.1}$$

and

$$\int_M h^2(x)\nu(dx) = +\infty \tag{14.4.2}$$

will be considered simultaneously. We introduce the measure

$$\mu(dx) = h^2(x)\nu(dx)$$

and the associated Hilbert space $L^2(\mu)$. Similarly, let

$$H^1(\mu) := \left\{u \in L^2(\mu); \; \nabla u \in L^2(\mu)\right\}$$

and introduce a weighted Dirichlet form

$$\int_M \langle \nabla u, \nabla u \rangle d\mu, \; u \in H^1(\mu)$$

and the corresponding weighted Laplacian Δ_μ on $L^2(\mu)$ such that

$$-\int_M (\Delta_\mu u)\,\overline{v}d\mu = \int_M \langle \nabla u, \nabla v \rangle d\mu, \; u \in D(\Delta_\mu), \; v \in H^1(\mu).$$

This new Laplacian is nothing but a drifted Laplacian in $L^2(\nu)$

$$\Delta_\mu u = \Delta_\nu u + 2\frac{\langle \nabla h, \nabla u \rangle}{h}.$$

If we introduce $V := \frac{\Delta h}{h}$ then

$$\Delta_\mu u = \frac{1}{h}\left[h\Delta_\nu u + 2\langle \nabla h, \nabla u \rangle + u\Delta h - Vuh\right]$$

$$= \frac{1}{h}\left[\Delta_\nu uh - Vuh\right]$$

shows that

$$\Delta_\mu = \frac{1}{h}\circ(\Delta_\nu - V)\circ h$$

so the weighted Laplacian Δ_μ on $L^2(\mu)$ is unitarily equivalent to the self-adjoint operator $\Delta_\nu - V$ on $L^2(\nu)$ via the unitary transformation

$$I : u \in L^2(\mu) \rightarrow hu \in L^2(\nu).$$

Note that $\Delta_\nu - V$ is to be interpreted as an absorption Laplacian $(\Delta_\nu)_V$ with potential (or absorption)

$$V = \frac{\Delta h}{h}$$

in the sense of the theory of absorption semigroups (see Chap. 5). In particular, the spectral properties of Δ_μ on $L^2(\mu)$ are nothing but the spectral properties of $(\Delta_\nu)_V$ on $L^2(\nu)$. In general, h is given in the form $h(x) := e^{-\frac{\Phi}{2}(x)}$ where Φ is a smooth function on M so

$$\Delta_\mu u = \Delta_\nu u - \langle \nabla\Phi, \nabla u\rangle$$

and a simple calculation shows that

$$\frac{\Delta h}{h} = -\frac{1}{2}\Delta\Phi + \frac{1}{4}|\nabla\Phi|^2.$$

In this case $\mu(dx) = e^{-\Phi(x)}\nu(dx)$. Note that in the case (14.4.1), by replacing $\Phi(x)$ by $\Phi(x)+ \ln C$ if necessary with a suitable constant $C > 0$, we may assume that $\mu(dx)$ is a probability measure $\int_M e^{-\Phi(x)}\nu(dx) = 1$. For details, we refer to [149, 285]. We note that in the integrable weight case (14.4.1), the spectral bound $s(\Delta_\mu)$ of Δ_μ, given by

$$\sup_{u\in H^1(\mu),\ |u|_{L^2(\mu)}=1}\left(-\int_M |\nabla u|^2\,d\mu\right) = -\inf_{u\in H^1(\mu),\ |u|_{L^2(\mu)}=1}\int_M |\nabla u|^2\,d\mu,$$

is equal to zero since $1 \in H^1(\mu)$. On the other hand, in the non-integrable weight case (14.4.2), it is possible that $s(\Delta_\mu) < 0$. Note, however, that we always have $s(\Delta_\mu) > -\infty$ since the spectrum of a self-adjoint operator is never empty.

Since Δ_μ is unitarily equivalent to the self-adjoint operator $(\Delta_\nu)_V$ on $L^2(\nu)$, then we can state the main result of this section as a straightforward consequence of Theorem 5.10.2.

Theorem 14.4.1 *Suppose that*

$$\alpha := \lim_{x \to \infty} \inf \frac{\Delta h}{h} > -\infty.$$

If $\alpha > s(\Delta_\nu) - s(\Delta_\mu)$ then

$$\begin{cases} \sigma_{ess}(\Delta_\mu) \subset \{\mathrm{Re}\,\lambda \leq s(\Delta_\nu) - \alpha\} \ \textit{if } \alpha < +\infty \\ \qquad \sigma_{ess}(\Delta_\mu) = \varnothing \ \textit{if } \alpha = +\infty. \end{cases}$$

In the case (14.4.1), it is a classical result that the presence of a spectral gap implies the so-called Poincaré (or variance) inequality. For the reader's convenience, we provide a proof.

Corollary 14.4.1 *Let (14.4.1) be satisfied. Let $s(\Delta_\nu)$ be the spectral bound of Δ_ν. If $\alpha : \lim\inf_{x \to \infty} \frac{\Delta h}{h} > s(\Delta_\nu)$ then there exists a positive constant $\delta > 0$ such that*

$$var_\mu(f) := \|f - \langle f \rangle\|^2_{L^2(\mu)} \leq \delta^{-1} \int_M |\nabla f|^2 \, d\mu, \quad f \in H^1(\mu)$$

where $\langle f \rangle := \int_M f(x)\mu(dx)$.

Proof Let

$$\Delta_{(\mu)} := -\Delta_\mu.$$

According to Theorem 14.4.1, $\sigma(\Delta_{(\mu)}) \cap [0, \alpha)$ consists of isolated eigenvalues. We know that $0 \in \sigma(\Delta_{(\mu)})$ is associated to the eigenvector 1. If $\Delta_{(\mu)}$ has non zero eigenvalues in $[0, \alpha)$ then we choose δ as the smallest non zero eigenvalue of $\Delta_{(\mu)}$. Otherwise, we choose $\delta = \alpha$. Since the spectrum of the restriction of $\Delta_{(\mu)}$ to the orthogonal space to Vect(1) is included in $[\delta, +\infty)$ then for any $h \in H^1(\mu)$ such that $\int_M h d\mu = 0$ we have

$$\left\| \sqrt{\Delta_{(\mu)}} h \right\|^2_{L^2(\mu)} \geq \delta \|h\|^2_{L^2(\mu)}.$$

It suffices to take $h = f - \langle f \rangle$ where $f \in H^1(\mu)$. $\square$

14.5 Predual Analysis for Neutron Transport

In Chap. 10, we investigated numerous topics for Neutron Transport in the physical L^1-framework. From a mathematical standpoint, however, it is worthwhile to study their behavior in L^p-spaces ($1 < p < +\infty$). Here, we show how to capture some results in L^p-spaces via a suitable predual construction, similar to that presented in

Sect. 4.6. We focus exclusively on the topic of Sect. 10.8, outlining the main steps of the construction without revisiting all the details.

We consider the general assumptions in Sect. 10.8; in particular, C is assumed to be a Σ-regular collision operator on $L^1(D \times \mathbb{R}^d)$. Let

$$\Sigma_*(x, \zeta) := \int_{\mathbb{R}^d} c(x, \zeta, \zeta')\nu(d\zeta').$$

If we assume

$$\frac{\Sigma_*(., .)}{\Sigma(., .)} \in L^\infty(D \times \mathbb{R}^d) \tag{14.5.1}$$

then (the "predual" collision operator)

$$C_* : D(C_*) \subset L^1(D \times \mathbb{R}^d) \to L^1(D \times \mathbb{R}^d)$$

(with maximal domain) defined by

$$C_*\varphi = \int_{\mathbb{R}^d} c(x, \zeta', \zeta)\varphi(x, \zeta')\nu(d\zeta')$$

is Σ-bounded. Instead of T_Σ we consider the unbounded operator on $L^1(D \times \mathbb{R}^d)$

$$T_\Sigma^{(*)}\varphi = \zeta.\frac{\partial\varphi}{\partial x} + \Sigma\varphi$$

with domain

$$\left\{\varphi \in L^1; \zeta.\frac{\partial\varphi}{\partial x} \in L^1, \ \varphi_{|\Gamma_+} = 0\right\}$$

whose resolvent is nothing but

$$\left(\lambda - T_\Sigma^{(*)}\right)^{-1}\varphi = \int_0^{\tau(x,-\zeta)} e^{-\lambda t} e^{-\int_0^t \Sigma(x+s\zeta,\zeta)ds}\varphi(x + t\zeta, \zeta)dt.$$

According to Remark 10.15.2, the spectral bound of $T_\Sigma^{(*)}$ is equal to that of T_Σ

$$s(T_\Sigma^{(*)}) = s(T_\Sigma).$$

As T_Σ, the operator $T_\Sigma^{(*)}$ is a generator of a substochastic semigroup $\left(U_\Sigma^{(*)}(t)\right)_{t\geq 0}$ on $L^1(D \times \mathbb{R}^d)$ which coincides with the dual semigroup of $(U_\Sigma(t))_{t\geq 0}$ on $L^1 \cap L^\infty$.

By assuming that C_* is also a Σ-regular collision operator (this is the case e.g. if $c(x, \zeta', \zeta) = c(x, \zeta, \zeta')$), we can build again the whole theory of Sect. 10.8 in $L^1(D \times \mathbb{R}^d)$ with $T_\Sigma^{(*)}$ and C_* in place of T_Σ and C. In particular, $C_*\left(\lambda - T_\Sigma^{(*)}\right)^{-1}$ is power compact in $L^1(D \times \mathbb{R}^d)$ and

$$T_\Sigma^{(*)} + C_* : D(T_\Sigma^{(*)}) \subset L^1(D \times \mathbb{R}^d) \to L^1(D \times \mathbb{R}^d)$$

generates a positive (not necessarily substochastic) C_0-semigroup $\left(V^{(*)}(t)\right)_{t \geq 0}$ on $L^1(D \times \mathbb{R}^d)$. This semigroup coincides with the dual semigroup of $(V(t))_{t \geq 0}$ on $L^1 \cap L^\infty$. It follows from Riesz-Thorin interpolation theorem that $(V(t))_{t \geq 0}$ extends to all $L^p(D \times \mathbb{R}^d)$ as a positive C_0-semigroup $\left(V^{(p)}(t)\right)_{t \geq 0}$ with generator $A^{(p)}$ with p-independent "asymptotic spectra", (see more details in Sect. 4.6). Finally, we end up with

Theorem 14.5.1 *Let (14.5.1) be satisfied. Suppose that both C and C_* are Σ-regular collision operators on $L^1(D \times \mathbb{R}^d)$. Suppose that D has a finite volume and the velocity measure $\nu(d\zeta)$ satisfies (10.8.10). Then $(V(t))_{t \geq 0}$ interpolates to all $L^p(D \times \mathbb{R}^d)$ $(1 \leq p < +\infty)$ as a positive C_0-semigroup $\left(V^{(p)}(t)\right)_{t \geq 0}$ with generator $A^{(p)}$. Moreover, $\omega_{ess}(V) \leq \omega(U_\Sigma)$ and the "asymptotic spectra" of both $A^{(p)}$ and $\left(V^{(p)}(t)\right)_{t \geq 0}$ are p-independent; In particular $\omega_{ess}(V^{(p)}) < \omega(V^{(p)})$ if $\omega_{ess}(V) < \omega(V)$.*

A priori, C need not be T_Σ-bounded in $L^p(D \times \mathbb{R}^d)$ $(1 < p < +\infty)$ and the domain of $A^{(p)}$ is not precised in Theorem 14.5.1. Note that $(U_\Sigma(t))_{t \geq 0}$ is defined in all L^p spaces and we use the same notation T_Σ for its generator in L^p. We complement this theorem by

Proposition 14.5.1 *Besides the conditions in Theorem 14.5.1, we suppose that*

$$C \left(\lambda - T_\Sigma\right)^{-1} 1 \in L^\infty(D \times \mathbb{R}^d). \tag{14.5.2}$$

Then C is T_Σ-bounded in all $L^p(D \times \mathbb{R}^d)$ and

$$A^{(p)} = T_\Sigma + C : D(T_\Sigma) \subset L^p(D \times \mathbb{R}^d) \to L^p(D \times \mathbb{R}^d).$$

Proof By Riesz-Thorin interpolation theorem

$$C \left(\lambda - T_\Sigma\right)^{-1} \in \mathcal{L}(L^p(D \times \mathbb{R}^d)), \quad p \in [1, +\infty]$$

and C is T_Σ-bounded in all $L^p(D \times \mathbb{R}^d)$. Because of

$$C \left(\lambda - T_\Sigma\right)^{-1} 1 \leq \int_{\mathbb{R}^d} \frac{c(x, \zeta, \zeta')}{\lambda + \underline{\Sigma}(\zeta')} \nu(d\zeta')$$

with $\underline{\Sigma}(\zeta) := \inf_{x \in D} \Sigma(x, \zeta)$, (14.5.2) is satisfied e.g. if

$$\sup_\zeta \int_{\mathbb{R}^d} \frac{c(x, \zeta, \zeta')}{\lambda + \underline{\Sigma}(\zeta')} \nu(d\zeta') < +\infty. \tag{14.5.3}$$

According to Theorem 10.8.3, if the velocity measure $\nu(d\zeta)$ satisfies (10.8.10) then $C(\lambda - T_\Sigma)^{-1}$ is power compact in $L^1(D \times \mathbb{R}^d)$ and consequently, by interpolation, $C(\lambda - T_\Sigma)^{-1}$ is power compact in $L^p(D \times \mathbb{R}^d)$ $(1 < p < +\infty)$. This implies that

$$r\left[C\left(\lambda - T_\Sigma\right)^{-1}\right] \to 0 \ (\lambda \to +\infty)$$

in all $L^p(D \times \mathbb{R}^d)$ and consequently

$$T_\Sigma + C : D(T_\Sigma) \subset L^p(D \times \mathbb{R}^d) \to L^p(D \times \mathbb{R}^d) \qquad (14.5.4)$$

is resolvent positive. Finally $A^{(p)} = T_\Sigma + C$. $\qquad\qquad\qquad\qquad\square$

14.6 Symmetrizable Operators in Transport Theory

We began this monograph by emphasizing that neutron transport operators are inherently highly non-self-adjoint, which is why, in general, one cannot expect to recover more than their peripheral spectral properties. Nevertheless, for a particular class of neutron transport operators, significantly deeper spectral information can be obtained, owing to subtle and often unexpected links with self-adjoint operators. It is with a few brief observations in this direction that we now close this monograph.

We know for (say bounded) regular collision operator C on L^p space ($p > 1$) that $(\lambda - T_\Sigma)^{-1} C$ is compact in $L^p(D \times \mathbb{R}^d)$ and the so called asymptotic spectrum

$$\sigma_{as}(T_\Sigma + C) := \sigma(T_\Sigma + C) \cap \{\mathrm{Re}\,\lambda > s(T_\Sigma)\}$$

consists (at most) of isolated eigenvalues with finite algebraic multiplicity. However, since the seminal work of Lehner and Wing [204] on the simplest transport model in slab geometry, significantly more refined Hilbertian spectral results have been developed throughout the 1950s and 1960s for suitable isotropic models [2, 177, 233, 291, 307, 353, 354]. Indeed, for such models

$$\sigma_{as}(T_\Sigma + C) \cap \{\mathrm{Re}\,\lambda > -\inf \Sigma\} \qquad (14.6.1)$$

turns out to be *real*; but complex eigenvalues may exist in the region

$$\{s(T_\Sigma) < \mathrm{Re}\,\lambda < -\inf \Sigma\}$$

(see [282]) and the set (14.6.1) can be finite or infinite according to the type of moderators [354]. Outside this class of isotropic models, the reality of this portion of the spectrum remains a difficult and unresolved problem to this day. However, it is possible to "capture" the point spectrum located on the *real axis* for more general models [239, 241] (see also [245] Chap. 6). Indeed, $\lambda \in \mathbb{C}$ (such that $\mathrm{Re}\,\lambda > s(T_\Sigma)$) is an eigenvalue of $T_\Sigma + C$ if and only if 1 is an eigenvalue of $(\lambda - T_\Sigma)^{-1} C$. Let C be form-positive in $L^2(D \times \mathbb{R}^d)$, i.e.,

$$(C\varphi, \varphi) \geq 0$$

(and $c(x, \zeta, \zeta') = c(x, -\zeta, \zeta')$) and let $\lambda \in \mathbb{R}$ be such that

$$\lambda > s(T_\Sigma).$$

Then 1 is an eigenvalue of $(\lambda - T_\Sigma)^{-1} C$ if and only if 1 is an eigenvalue of

$$H_\lambda := \sqrt{C}\,(\lambda - T_\Sigma)^{-1}\,\sqrt{C} \qquad\qquad (14.6.2)$$

where H_λ is a *self-adjoint* (compact) operator in $L^2(D \times \mathbb{R}^d)$. Therefore, if we introduce the (positive) curve eigenvalues of H_λ

$$\rho_1^+(\lambda) \ge \rho_2^+(\lambda) \cdots \ge \rho_n^+(\lambda) \ge \cdots$$

then, a priori, we can "capture" the (asymptotic) *real* point spectrum of $T_\Sigma + C$ as the λ-solutions to

$$\rho_n^+(\lambda) = 1 \quad (n = 1, 2, \ldots). \qquad\qquad (14.6.3)$$

On the other hand, rather than appealing directly to the positive eigenvalues $\rho_n^+(\lambda)$ of the self-adjoint compact operator H_λ, one can adopt the perspective of *symmetrizable* operators [200, 312]. Specifically, the compact operator $(\lambda - T_\Sigma)^{-1} C$ is symmetrizable by the form-positive (though a priori not injective) self-adjoint operator C, thereby enabling the use of variational characterizations of eigenvalues for symmetrizable compact operators [257].

We point out that if a symmetrizable operator is not compact but possesses a spectral gap, then similar variational characterizations remain available for those eigenvalues located outside its essential spectral disc [236]. This phenomenon arises in the case of partly elastic neutron transport operators (see Sect. 10.14) where $(\lambda - T_\Sigma)^{-1} C$ is not compact due to the presence of the elastic collision operator C_e. Nevertheless, it remains essentially compact and symmetrizable by the form-positive operator $C = C_e + C_i$. A priori, we can "capture" the real spectrum of $T_\Sigma + C$ in the half line $(s(T_\Sigma + C_e), +\infty)$ by "solving" (14.6.3) under the condition $\rho_n^+(\lambda) > r_{ess}((\lambda - T_\Sigma)^{-1} C)$ [237].

Correction to: Introduction

**Correction to:
Chapter 1 in: M. Mokhtar-Kharroubi, *Peripheral Spectra
of Perturbed Positive Semigroups*, Lecture Notes
in Mathematics 2388,
https://doi.org/10.1007/978-3-032-11173-9_1**

The original version of the book was updated with the belated corrections in the Chapter 1. The reference 281 citation has been included. The book and the correction chapter have been updated with the changes.

The updated version of this chapter can be found at
https://doi.org/10.1007/978-3-032-11173-9_1

Bibliography

1. Abramovich, Y., Aliprantis, C., Burkinshaw, O.: On the spectral radius of positive operators. Math. Z **211**, 593–607 (1992)
2. Albertoni, S., Montagnini, B.: On the spectrum of neutron transport equations in finite bodies. J. Math. Anal. Appl. **13**, 19–48 (1966)
3. Aliprantis, C.D., Burkinshaw, O.: Positive Operators. Academic Press, New York (1985)
4. Aliprantis, D., Burkinshaw, O.: Positive compact operators on Banach lattices. Math. Z **174**, 289–298 (1980)
5. Ammar-Khodja, F., Mokhtar-Kharroubi, M.: On the stability of essential spectra of semigroups generated by one dimensional hyperbolic systems. In: Discrete and Continuous Dynamical Systems Series S, vol. 17 (5–6), pp. 1792–1820 (2024)
6. Applebaum., D.: Semigroups of Linear Operators with Applications to Analysis, Probability and Physics. London Mathematical Society Student Texts Series, vol. 93. Cambridge University Press (2019)
7. Arendt, W.: Resolvent positive operators. Proc. Lond. Math. Soc. **54**, 321–349 (1987)
8. Arendt, W., Rhandi, A.: Perturbation of positive semigroups. Arch. Math. **56**, 107–119 (1991)
9. Arendt, W., Batty, C.J.K.: Absorption semigroups and Dirichlet boundary conditions. Math. Ann. **295**, 427–448 (1993)
10. Arendt, W., ter Elst, A.F.M.: Gaussian estimates for second order elliptic operators with boundary conditions. J. Oper. Theory **38**(1), 87–130 (1997)
11. Arendt, W.: Positive semigroups of Kernel operators. Positivity **12**, 25–44 (2008)
12. Arendt, W., Batty, C.J.K., Hieber, M., Neubrander, F.: Vector-Valued Laplace Transforms and Cauchy Problems. Birkhaüser (2011)
13. Arendt, W., Glück, J.: Positive irreducible semigroups and their long-time behaviour. Philos. Trans. Roy Soc. A **378**, 20190611 (2020)
14. Arino, O.: Some spectral properties for the asymptotic behavior of semigroups connected to population dynamics. Siam. Rev. **34**(4), 445–476 (1992)
15. Arino, O., Axelrod, D., Kimmel, M. (eds.): Advances in Mathematical Population Dynamics-Molecules, Cells and Man. Series in Mathematical Biology et Medicine, vol. 6. World Scientific (1998)
16. Arino, O., Rudnicki, R.: Stability of phytoplankton dynamics. C. R. Biol. **327**, 961–969 (2004)
17. Arlotti, L.: The Cauchy problem for the linear Maxwell-Bolzmann equation. J. Diff. Eqn. **69**, 166–184 (1987)
18. Arlotti, L., Frosali, G.: Long time behaviour of particle swarms in runaway regime. In: Operator Theory: Advances and Applications, vol. 51, pp. 131–143. Birkhäuser Verlag Basel (1991)

M. Mokhtar-Kharroubi, *Peripheral Spectra of Perturbed Positive Semigroups*, Lecture Notes in Mathematics 2388, https://doi.org/10.1007/978-3-032-11173-9

19. Arlotti, L.: On the asymptotic behaviour of electrons in an ionized gas. In: Nelson, P., et al. (eds.) Transport Theory, Invariant Imbedding and Integral Equations. Lecture Notes in Pure and Applications, vol. 115, pp. 81–96. M. Dekker, New York (1989)

20. Arlotti, L.: A perturbation theorem for positive contraction semigroups on L^1 spaces with applications to transport equations and Kolmogorov's differential equations. Acta. Appl. Math. **23**, 129–144 (1991)

21. Arlotti, L., Lods, B.: Substochastic semigroups for transport equations with conservative boundary conditions. J. Evol. Eqn. **5**, 485–508 (2005)

22. Arlotti, L., Banasiak, J., Ciake-Ciake, F.L.: On well-posedness of linear Boltzmann equation of semiconductor theory. Math. Models Methods Appl. Sci. **16**(9), 1441–1468 (2006)

23. Arlotti, L., Banasiak, J., Lods, B.: A new approach to transport equations associated to a regular field: trace results and well-posedness. Mediterr. J. Math. **6**, 367–402 (2009)

24. Arlotti, L., Banasiak, J., Lods, B.: On general transport equations with abstract boundary conditions. The case of divergence free force field. Mediterr. J. Math. **8**, 1–35 (2011)

25. Arlotti, L., Lods, B., Mokhtar-Kharroubi, M.: On perturbed substochastic semigroups in abstract state spaces. Z. Anal. Anwend **30**(4), 457–495 (2011)

26. Arlotti, L., Lods, B., Mokhtar-Kharroubi, M.: Non-autonomous honesty theory in abstract state spaces with applications to linear Kinetic equations. Comm. Pure Appl. Anal. **13**(2), 729–771 (2014)

27. Bakry, D., Gentil, I., Ledoux, M.: Analysis and Geometry of Markov Diffusion Operators. Springer (2014)

28. Banasiak, J.: Some spectral properties of the linear Boltzmann equation of semiconductor theory with application to its asymptotic analysis. In: Proceedings of the Prague Mathematical Conference, Praha, Czech Republic, July 8–12, 1996

29. Banasiak, J.: Spectral theorems of Voigt type for linear Boltzmann equation with external field. Transp. Theory Stat. Phys. **27**(3–4), 241–255 (1998)

30. Banasiak, J.: On an extension of Kato-Voigt perturbation theorem for substochastic semigroups and its applications. Taiwanese J. Math. **5**, 169–191 (2001)

31. Banasiak, J., Mokhtar-Kharroubi, M.: Universality of dishonesty of substochastic semigroups: shattering fragmentation and explosive birth-and-death processes. Discrete Con. Dyn. Syst., Ser. B **5**, 529–542 (2005)

32. Banasiak, J.: On conservativity and shattering for an equation of phytoplankton dynamics. C. R. Biol.s **327**, 1025–1036 (2004)

33. Banasiak, J., Arlotti, L.: Perturbations of Positive Semigroups with Applications. Springer Monographs in Mathematics. Springer (2006)

34. Banasiak, J., Lachowicz, M.: Around Kato generation theorem for semigroups. Studia. Math. **179**(3), 217–238 (2007)

35. Banasiak, J., Pichór, K., Rudnicki, R.: Asynchronous exponential growth of a general structured population model. Acta. Appl. Math. **119**, 149–166 (2012)

36. Banasiak, J., Falkiewicz, A., Namayanja, P.: Semigroup approach to diffusion and transport problems on networks. Semigroup Forum **93**, 427–443 (2016)

37. Banasiak, J., Lamb, W., Laurençot, P.: Analytic Methods for Coagulation-Fragmentation Models, vol. I. CRC Press (2019)

38. Banasiak, J., Lamb, W.: Growth-fragmentation-coagulation equations with unbounded coagulation kernels. To appear in Philosophical Transactions of the Royal Society A (2020)

39. Banasiak, J., Mokhtar-Kharroubi, M.: On spectral gaps of growth-fragmentation semigroups in higher moment spaces. Kinetic. Related. Models **15**(2), 147–185 (2022)

40. Banasiak, J., Poka, D.W., Shindin, S.: Growth-fragmentation equations with McKendrick-von Foerster boundary condition. Discrete Con. Dyn. Syst.—Ser. S **17**(5–6), 2030–2057 (2024)

41. Bardos, C.: Problèmes aux limites pour les équations aux dérivées partielles du premier ordre à coefficients réels; théorèmes d'approximation; application à l'équation de transport. Ann. Sci. Ecole Norm. Sup **3**, 185–233 (1970)

42. Bardos, C., Santos, R., Sentis, R.: Diffusion approximation and computation of the critical size. Trans. Am. Math. Soc. **284**(2), 617–649 (1984)

43. Bardos, C., Bernard, E., Golse, F., Sentis, R.: The diffusion approximation for the linear Boltzmann equation with vanishing scattering coeffcient. Commun. Math. Sci. **13**(3), 641–671 (2015)
44. Bates, P.W., Chen, F.: Spectral analysis and multidimensional stability of traveling waves for nonlocal Allen-Cahn equation. J. Math. Anal. Appl. **273**, 45–57 (2002)
45. Bates, P.W.: On some nonlocal evolution equations arising in materials science. Fields Inst. Commun. **48**, 1–40 (2006)
46. Bátkai, A., Fijavž, M.K., Rhandi, A.: Positive Operator Semigroups. From Finite to Infinite Dimensions. Operator Theory: Advances and Applications, vol. 257. Birkhäuser, Berlin (2017)
47. Batty, C.J., Robinson, D.W.: Positive one-parameter semigroups on ordered Banach spaces. Acta Appl. Math. **1**, 221–296 (1984)
48. Bellini-Morante, A.: On Chapman-Kolmogorov equations for neutron population in a multiplying assembly. J. Math. Phys **16**(3), 585–589 (1975)
49. Bellini-Morante, A.: A rigorous derivation of the reactor Kinetics equation from the Chapman-Kolmogorov system. Riv. Math. Univ. Parma **4**, 239–250 (1976)
50. Bell, G.I.: Stochastic theory of Neutron Transport. J. Nucl. Sc. Eng **21**, 390–401 (1965)
51. Bellomo, N., Bellouquid, A., Gibelli, L., Outada, N.: A Quest Towards a Mathematical Theory of Living Systems. Birkhauser (2017)
52. Berestycki, H., Niremberg, L., Varadhan, S.R.S.: The principal eigenvalue and maximum principle for second-order elliptic operators in general domains. Commun. Pure Appl. Math., vol. XLVII, 47–92 (1994)
53. Bernard, E., Salvarani, F.: On the exponential decay to equilibrium of the degenerate linear Boltzmann equation. J. Funct. Anal. **265**, 1934–1954 (2013)
54. Bernard, E., Gabriel, P.: Asymptotic behavior of the growth-fragmentation equation with bounded fragmentation rate. J. Funct. Anal. **272**, 3455–3485 (2017)
55. Bernard, E., Gabriel, P.: Asynchronous exponential growth of the growth-fragmentation equation with unbounded fragmentation rate. J. Evol. Eqn. **20**, 375–401 (2020)
56. Bertoin, J., Watson, A.R.: A probabilistic approach to spectral analysis of growth-fragmentation equations. J. Funct. Anal. **274**, 2163–2204 (2018)
57. Biedrzycka, W., Tyran-Kaminska, M.: Existence of invariant densities for semiflow with jumps. J. Math. Anal. Appl. **435**, 61–84 (2016)
58. Birkhoff, G.: Lattice Theory, vol. American Mathematical Society Colloquium Publications, XXV (1948)
59. Birkhoff, G., Varga, R.S.: Reactor criticality and nonnegative matrices. J. Soc. Indus. Appl. Math. **6**, 354–377 (1958)
60. Birkhoff, G.: Reactor criticality in transport theory. Proc. Nat. Acad. Sci. **45**, 567–569 (1959)
61. Birkhoff, G.: Positivity and criticality. In: Birkhoff, G., Wigner, E.P. (eds.) Proceedings of Symposia in Applied Mathematics. Nuclear Reactor Theory, vol. XI. American Mathematical Society, Providence (1961)
62. Birkhoff, G.: Reactor criticality in neutron transport theory. Ren. Math. **22**, 102–126 (1963)
63. Bobrowski, A.: Functional Analysis for Probability and Stochastic Processes: An Introduction. Cambridge University Press (2005)
64. Bobrowski, A.: Generators of Markov Chains: From a Walk in the Interior to a Dance on the Boundary. Cambridge Studies in Advanced Mathematics, vol. 190 (2020)
65. Borysiewicz, M., Mika, J.: Time behaviour of thermal neutrons in moderating media. J. Math. Anal. Appl. **26**, 461–478 (1969)
66. Boujijane, S., Boulite, S., Halloumi, M., Maniar, L., Rhandi, A.: Well-posedness and asynchronous exponential growth of an age weighted structured fish population model with diffusion in L^1. J. Evol. Eqn. **24**, 14–31 (2024)
67. Boujijane, S., Boulite, S., Halloumi, M., Maniar, L., Rhandi, A.: Boundary perturbation theorem for the essential spectral radius with application to a spatial tumor invasion with cell age. In: Evolution Equations and Control Theory. https://doi.org/10.3934/eect.2025021.

68. Boulanouar, M.: A Mathematical study for a Rotenberg model. J. Math. Anal. Appl. **265**, 371–394 (2002)

69. Boulouz, A., Bounit, H., Driouich, A., Hadd, S.: On norm continuity, differentiability and compactness of perturbed semigroups. Semigroup Forum **101**, 547–570 (2020)

70. Brascamp, H.J., Lieb, E.H.: On extensions of the Brunn-Minkowski and Prékopa-Leindler theorems including inequalities for log concave functions, and with applications to the diffusion equation. J. Funct. Anal. **22**, 366–389 (1976)

71. Brendle, S., Nagel, R., Poland, J.: On the spectral mapping theorem for perturbed strongly continuous semigroups. Archiv. Math. **74**, 365–378 (2000)

72. Brendle, S.: On the asymptotic behaviour of perturbed strongly continuous semigroups. Math. Nachr. **226**, 35–47 (2001)

73. Brezis, H.: Functional Analysis. Springer, Sobolev Spaces and Partial Differential Equations (2011)

74. Brunel, A., Revuz, D.: Quelques applications probabilistes de la quasi-compacité. Annales de l'I. H. P, Section B, **10**(3), 301–337 (1974)

75. Busoni, G.: Asymptotic behaviour and wave operators in charge exchange. J. Math. Anal. Appl. **212**, 190–208 (1997)

76. Calvez, V., Lenuzza, N., Doumic, M., Deslys, J.P., Mouthon, F., Perthame, B.: Prion dynamics with size dependency-strain phenomena. J. Biol Dyn. **4**(1), 28–42 (2010)

77. Cañizo, J.A., Gabriel, P., Yoldasz, H.: Spectral gap for the growth-fragmentation equation via Harris's Theorem. SIAM. J. Math. Anal. **53**(5), 5185–5214 (2021)

78. Carmona, R., Masters, W.C., Simon, B.: Relativistic Schrödinger operators: asymptotic behaviour of the eigenfunction. J. Funct. Anal. **91**, 117–142 (1990)

79. Case, K.M., Zweifel, P.F.: Linear Transport Theory. Addison-Wesley (1967)

80. Cercignani, C.: The Boltzmann Equation and its Applications. Applied Mathematical Sciences, vol. 67. Springer (1988)

81. Cercignani, C.: Ludwig Boltzmann: The Man Who Trusted Atoms. Oxford University Press (1998)

82. Cessenat, M.: Théorèmes de trace L^p pour des espaces de fonctions de la neutronique. C. R. Acad. Sci, Paris, Ser. I, **299**, 831–834 (1984)

83. Cessenat, M.: Théorèmes de trace pour les espaces de fonctions de la neutronique. C. R. Acad. Sci, Paris, Ser. I, **300**, 89–92 (1985)

84. Chabi, M., Mokhtar-Kharroubi, M.: On perturbations of positive C_0-(semi)groups on Banach lattices and applications. J. Math. Anal. Appl. **202**, 843–861 (1996)

85. Chabi, M., Mokhtar-Kharroubi, M., Stefanov, P.: Scattering theory with two L^1 spaces: application to transport equations with obstacles. Annales de la faculté des sciences de Toulouse **6**(3), 511–523 (1997)

86. Chacon, R.V., Krengel, U.: Linear modulus of a linear operator. Proc. Am. Math. Soc. **15**(4), 553–559 (1964)

87. Cheng, G., Wang, S., Yuan, W.: Spectral analysis of transport operator in Lebowitz-Rubinow model. WSEAS. Trans. Math. **13**, 324–334 (2014)

88. Chill, R., Seifert, D.: Quantified versions of Ingham's theorem. Bull. Lond. Math. Soc. **48**(3), 519–532 (2016)

89. Choquet, G.: Cours d'Analyse:Topologie. Masson, Tome II (1964)

90. Clement, P., Heijmans, H., Angenent, S., van Duijn, C.J., de Pagter, B.: One-Parameter Semigroups, vol. 5. North-Holland Publishing Co., Amsterdam (1987)

91. Comets, F., Popov, S., Schütz, G.M., Vachkovskaia, M.: Billiards in a general domain with random reflections. Arch. Ration. Mech. Anal. **191**, 497–537 (2009)

92. Corngold, N., Kuscer, I.: Discrete relaxation times in neutron thermalization. Phys. Rev. **139**(3A), 981–990 (1965)

93. Coville, J., Li, F., Wang, X.: On eigenvalue problems arising from nonlocal diffusion models. Discrete Cont. Dyn. Syst. **37**(2), 879–903 (2017)

94. Dautray, R., Lions, J.L. (eds.): Analyse mathématique et calcul numérique pour les sciences et les techniques. Tome 1. Masson, Paris (1985)

95. Dautray, R. (ed.): Méthodes probabilistes pour les équations de la physique. Collection du CEA, Eyrolles (1989)
96. Davies, E.B.: Quantum dynamical semigroups and the neutron diffusion equation. Rep. Math. Phys. **11**, 169–188 (1977)
97. Davies, E.B.: One-Parameter Semigroups. Academic Press (1980)
98. Davies, E.B.: L^1 Properties of second order elliptic operators. Bull. Lond. Math. Soc. **17**(5), 417–436 (1985)
99. Davies, E.B.: Heat Kernels and Spectral Theory. Cambridge Tracts in Mathematics, vol. 92. Cambridge University Press (1989)
100. Davies, E.B., Safarov, Y. (eds.): Spectral Theory and Geometry. ICMS Instructional Conference, Edinburgh 1998. London Mathematical Society, Lecture Notes Series, vol. 273. Cambridge University Press (1999)
101. Davies, E.B.: Linear Operators and their Spectra. Cambridge Studies in Advanced Mathematics, vol. 106. Cambridge University Press (2007)
102. Degond, P., Mas-Gallic, S.: Existence of solutions and diffusion approximation for a model Fokker-Planck equation. Transp. Theory Stat. Phys. **16**, 589–636 (1987)
103. Degond, P., Lemou, M.: Dispersion relations for the linearized Fokker-Planck equation. Arch. Rational. Mech. Anal. **138**, 137–167 (1997)
104. Demeru, M.L., Montagnini, B.: Complete continuity of the free gas scattering operator in neutron thermalization theory. J. Math. Anal. Appl. **12**, 49–57 (1965)
105. Demuth, M., van Casteren, J.: Stochastic Spectral Theory for Self-Adjoint Feller Operators—A Functional Integration Approach, Probability and Its Applications. Birkhäuser Verlag, Basel (2000)
106. Desch, W.: Perturbations of positive semigroups in AL-spaces. Unpublished manuscript (1988)
107. Desvillettes, L.: On an asymptotic of the Boltzmann equation when the collisions become grazing. Transp. Theory. Stat. Phys. **21**, 259–276 (1992)
108. DiBenedetto, E.: Partial Differential Equations. Birkhäuser (1995)
109. Diekmann, O., Heijmans, H.J.A.M., Thieme, H.R.: On the stability of the cell size distribution. J. Math. Biol **19**(2), 227–248 (1984)
110. Ding, J., Zhou, A.: Nonnegatives Matrices, Positive Operators and Applications. World Scientific (2009)
111. Ding, J., Zhou, A.: Statistical Properties of Deterministic Systems. Tsinghua University Texts, Springer (2009)
112. Dolbeault, J.: An introduction to Kinetic equations: the Vlasov-Poisson system and the Boltzmann equation. Discr. Cont. Dyn. Syst. **8**(2), 361–380 (2002)
113. Dolbeault, J., Mouhot, C., Schmeiser, C.: Hypocoercivity for linear Kinetic equations conserving mass. Trans. Am. Math. Soc. **367**(6), 3807–3828 (2015)
114. Duderstadt, J.J., Martin, W.R.: Transport Theory. Wiley (1979)
115. Dunford, N., Schwartz, J.: Linear Operators. Part I, Wiley Classics Library (1988)
116. Edmunds, D.E., Evans, W.D.: Spectral Theory and Differential Operators. Clarendon Press, Oxford (1989)
117. Eisner, T., Farkas, B., Haase, M., Nagel, R.: Operator Theoretic Aspects of Ergodic Theory. Graduate Texts in Mathematics. Springer (2015)
118. Emamirad, H.: On the Lax and Phillips scattering theory for transport equation. J. Funct. Anal. **62**, 276–303 (1985)
119. Emamirad, H.: Scattering Theory for Transport Phenomena. Mathematical Physics Studies. Springer (2021)
120. Engel, K.J., Nagel, R.: A Short Course on Operator Semigroups. Universitext, Springer (2006)
121. Engler, H., Prüss, J., Webb, G.F.: Analysis of a model for the dynamics of prions. II. J. Math. Anal. Appl. **324**(1), 98–117 (2006)
122. Eshima, N., Tabata, M.: The spectrum of the transport operator with a potential term under the spatial periodicity condition. Rend. Sem. Mat. Univ. Padova **97**, 211–233 (1997)

123. Eshima, N., Tabata, M.: The spectrum of the linear transport operator with a force in a torus. Acta. Math. Hungar **74**(1–2), 63–81 (1997)

124. Farkas, J.Z., Hinow, P.: Physiologically structured populations with diffusion and dynamic boundary conditions. Math. Biosci. Eng. **8**, 503–513 (2011)

125. Foguel, S.R.: The Ergodic Theory of Markov Processes. Van Nostrand Reinhold Comp, New York (1969)

126. Fonte Sanchez, C., Gabriel, P., Mischler, S.: On the Krein-Rutman theorem and beyond (2023). hal-04093201

127. Frosali, G., van der Mee, C.V.M., Paveri-Fontana, S.L.: Conditions for runaway phenomena in the Kinetic theory of particle swarms. J. Math. Phys. **30**(5), 1177–1186 (1989)

128. Frosali, G., van der Mee, C.V.M.: Scattering theory relevant to the linear of particle swarms. J. Stat. Phys. **56**(1/2), 139–148 (1989)

129. Frosali., G.: Functional-analytic techniques in the study of time dependent electron swarms in weakly ionized gases. In: Proceedings of III International Workshop on Mathematical Aspects of Fluid and Plasma Dynamics (Salice Terme, Italy, 1988). Lecture Notes in Mathematics, vol. 1460, pp. 107–139 (1991)

130. Frosali, G., van der Mee, C.V.M., Mugelli, F.: A characterization theorem for the evolution semigroup generated by the sum of two unbounded operators. Math. Methods Appl. Sci. **27**, 669–685 (2004)

131. Gabriel, P., Doumic Jauffret, M.: Eigenelements of a general agregation-fragmentation model. Math. Models Methods Appl. Sci. **20**(5), 757–783 (2010)

132. Gallagher, I., Saint-Raymond, L., Texier, B.: From Newton to Boltzmann: hard spheres and short-range potentials. In: Zurich Lectures in Advanced Mathematics. European Mathematical Society, Zurich (2014)

133. Gao, N.: Extensions of Perron-Frobenius theory. Positivity **17**, 965–977 (2013)

134. Gerlach, M.: On the peripheral point spectrum and the asymptotic behavior of irreducible semigroups of Harris operators. Positivity **17**(3), 875–898 (2013)

135. Gerlach, M., Glück, J.: Convergence of positive operator semigroups. Trans. Am. Math. Soc. **372**(9), 6603–6627 (2019)

136. Glassey, R.T., Strauss, W.A.: Perturbation of essential spectra of evolution operators and the Vlasov-Poisson-Boltzmann system. Discrete Cont. Dyn. Syst. Ser. **5**(3), 457–472 (1999)

137. Glück, J.: On the peripheral spectrum of positive operators. Positivity **20**, 307–336 (2016)

138. Glück, J.: Spectral gaps for hyperbounded operators. Adv. Math. **362**, 106958 (2020)

139. Glück, J.: The essential spectrum on L^1 via weakly compact operators; Private Communication, February 7, 2023

140. Goldstein, J. A.: Semigroups of Linear Operators and Applications. Oxford Mathematical Monographs (1985)

141. Goldstein, J.A., Nagel, R.: The evolution of operator semigroups. In: Semigroups of Linear Operators-Theory and Applications, vol. 113, Bedlewo, Poland, pp. 89–103. Springer (2015)

142. Golse, F., Lions, P.L., Perthame, B., Sentis, R.: Regularity of the moments of the solution of a transport equation. J. Funct. Anal. **76**, 110–125 (1988)

143. Gorban, A.N.: Hilbert's sixth problem: the endless road to rigour. Philos. Trans. R. Soc. A **376**, 20170238 (2018)

144. Greenberg, W., Van der Mee, C., Protopopescu, V.: Boundary Value Problems in Abstract Kinetic Theory. Birkhäuser Verlag (1987)

145. Greiner, G.: Spektrum und Asymptotik stark stetiger Halbgruppen positiver Operatoren. Sitzungsber. Heidelbg. Akad. Wiss, Math.-Natur. Kl, pp. 55–80 (1982)

146. Greiner, G.: Spectral properties and asymptotic behavior of the linear transport equation. Math. Z **185**, 167–177 (1984)

147. Greiner, G.: A typical Perron-Frobenius theorem with applications to age-dependent population equation. In: Kappel, F., Schappacher, W. (eds.) Infinite Dimensional Systems. Lecture Notes in Mathematics, vol. 1076, pp. 86–100. Springer (1984)

148. Greiner, G.: Perturbing the boundary conditions of a generator. Houston J. Math. **13**, 213–229 (1987)

149. Grigor'yan, A.: Heat kernels on weighted manifolds and applications. Contemp. Math. **398**, 93–191 (2006)
150. Habetler, G.J., Martino, A.: Existence theorems and spectral theory for the multigroup diffusion model. In: Birkhoff, G., Wigner, E.P. (eds.) Proceedings of Symposia in Applied Mathematics, Nuclear Reactor Theory, vol. XI. American Mathematical Society, Providence (1961)
151. Hangelbroek, R.J.: A functional analytic approach to the linear transport equation. Ph.D. Thesis, University of Groningen, Netherlands (1973) (published also in TTSP, 5 (1976), 1–85)
152. Han-Kwan, D., Léautaud, M.: Geometric analysis of the linear Boltzmann equation I. Trend to equilibrium. Ann. PDE **1**, 3 (2015). https://doi.org/10.1007/s40818-015-0003-z
153. Helffer, B.: Remarks on decay of correlations and Witten Laplacians Brascamp-Lieb inequalities and semiclassical limit. J. Funct. Anal. **155**, 571–586 (1998)
154. B. Helffer. Semiclassical Analysis, Witten Laplacians and Statistical Mechanics. Series on Partial Differential Equation Applications, vol. 1. World Scientific (2002)
155. Helffer, B., Nier, F.: Hypoelliptic Estimates and Spectral Theory for Fokker-Planck Operators and Witten Laplacians. Lecture Notes in Mathematics, vol. 1862. Springer (2005)
156. Henry, D.: Geometric Theory of Semilinear Parabolic Equations. Lecture Notes in Mathematics, vol. 840. Springer, New York (1981)
157. Herbst, I., Sloan, A.D.: Perturbation of translation invariant positivity preserving semigroups on $L^2(\mathbb{R}^N)$. Trans. Am. Math. Soc. **236**, 325–360 (1978)
158. Hejtmanek, J.: Scattering theory of the linear Boltzmann operator. Comm. Math. Phys. **43**, 109–120 (1975)
159. Hejtmanek, J.: Dynamics and spectrum of the linear multiple scattering operator in the Banach lattice $L^1(\mathbb{R}^3 \times \mathbb{R}^3)$. Transp. Theory. Stat. Phys. **8**, 29–44 (1979)
160. Hemple, R., Voigt, J.: On L^p spectrum of Schrödinger operators. J. Math. Anal. Appl. **121**, 138–159 (1987)
161. Hernandez, F.L., Semenov, E.M., Tradacete, P.: Strictly singular operators on L^p spaces and interpolation. Proc. Am. Math. Soc. **138**(2), 675–686 (2010)
162. Hieber, M.: Gaussian estimates and holomorphy of semigroups in L^p spaces. J. Lond. Math. Soc. **54**, 148–160 (1996). https://doi.org/10.1112/jlms/54.1.148
163. Horton, E., Kyprianou, A.E.: Stochastic Neutron Transport and Non-local Branching Markov Processes. Probability and Its Applications, Birkhäuser (2023)
164. Huang, J., Edwards, B.F., Levine, A.D.: General solutions and scaling violation for fragmentation with mass loss. J. Phys., A, **24**, 3967–3977
165. Huber, A.: Spectral properties of the linear multiple scattering operator in L^1 Banach lattice. Int. Eqn. Oper. Theory **6**, 357–371 (1983)
166. Huet, D.: Décomposition spectrale et opérateurs. Presses universitaires de France (1976)
167. Hutson, V., Martinez, S., Mischaikow, K., Vickers, G.T.: The evolution of dispersal. J. Math. Biol. **47**, 483–517 (2003)
168. Ishikawa, M.: Admissible and regular potentials for positive C0-semigroups and application to heat semigroups. Semigroup Forum **48**, 96–106 (1994)
169. Ishikawa, M.: Analyticity of absorption semigroups. Semigroup Forum **50**, 307–315 (1995)
170. Jacob, N.: Pseudo-Differential Operators & Markov Processes. Fourier Analysis and Semigroups, vol. 1. Imperial College Press (2001)
171. Jarmouni-Idrissi, K., Mokhtar-Kharroubi, M.: A class of non-linear problems arising in the stochastic theory of Neutron Transport. Nonlinear Anal. **31**(3–4), 265–293 (1998)
172. Jarmouni, K., Thevenot, L.: Homogenization of a nonlinear neutron transport equation. Transp. Theory Stat. Phys **31**(2), 93–123 (2002)
173. Johnsen, J.: On the spectral properties of Witten-Laplacians, their ranges projections and Brascamp-Lieb s inequality. Int. Eqn. Oper. Theory **36**, 288–324 (2000)
174. Jorgens, K.: An asymptotic expansion in the theory of Neutron Transport. Commun. Pure. Appl. Math. **11**, 219–242 (1958)
175. Jüngle, A.: Transport Equations for Semiconductors. Lecture Notes in Physics, vol. 773. Springer, Berlin (2009)

176. Jüngel, A.: Entropy Methods for Diffusive Partial Differential Equations. BCAM Springer-Briefs (2016)
177. Kaper, H.G.: The initial value problem for monoenergetic neutrons in an infinite slab with delayed neutron production. J. Math. Anal. Appl. **19**, 207–230 (1967)
178. Kaper, H.G., Lekkerkerker, C.G., Hejtmanek, J.: Spectral Methods in Linear Transport Theory. Birkhäuser Verlag (1982)
179. Karlin, S.: Positive operators. J. Math. Mech. **8**, 907–937 (1959)
180. Kato, T.: On the semigroups generated by Kolmogoroff's differential equations. J. Math. Soc. Jpn. **6**, 1–15 (1954)
181. Kato, T.: Perturbation theory for nullity, deficiency and other quantities of linear operators. J. Anal. Math. **6**, 261–322 (1958)
182. Kato, T.: Superconvexity of the spectral radius and convexity of the spectral bound and the type. Math. Z **180**, 265–273 (1982)
183. Kato, T.: Perturbation Theory of Linear Operators. Springer (1984)
184. Keller, M., Lenz, D.: Unbounded laplacians on graphs: basic spectral properties and the heat equation. Math. Model. Nat. Phenom **5**(2), 198–224 (2010)
185. Keller, M., Lenz, D.: Dirichlet forms and stochastic completeness of graphs and subgraphs. J. Reine Angew. Math. **666**, 189–223 (2012)
186. Kingman, J.F.C.: A convexity property of positive matrices. Quart. J. Math. Oxford **12**(2), 283–284 (1961)
187. Kipnis, C.: Majoration des semigroupes de contraction de L^1 et applications. Ann. Inst. Henri Poincaré, Section B, **10**(4), 369–384 (1974)
188. Kulczycki, T., Siudeja, B.: Intrinsic ultracontractivity of the Feynman-Kac semigroup for relativistic stable processes. Trans. Am. Math. Soc. **358**(11), 5025–5057 (2006)
189. Kunze, M., Schluchtermann, G.: Strongly generated Banach spaces and measures of non-compactness. Math. Nachr. **191**, 197–214 (1998)
190. I. Kuscer. A survey of neutron transport theory. Acta. Physica. Austriaca. Suppl X, 491–528 (1973)
191. Larsen, E.W., Habetler, G.J.: A functional-analytic derivation of case's full and half range formulas. Commun. Pure. Appl. Math. **26**, 525–537 (1973)
192. Larsen, E.W., Zweifel, P.F.: On the spectrum of the linear transport operator. J. Math. Phys. **15**(11), 1987–1997 (1974)
193. Larsen, E., Keller, J.B.: Asymptotic solutions of neutron transport problems. J. Math. Phys **15**, 75–81 (1974)
194. Lasota, A., Mackey, M.C.: Chaos, Fractals and Noise. Stochastic Aspects of Dynamics. Springer (1995)
195. Latrach, K., Mokhtar-Kharroubi, M.: On an unbounded linear operator arising in the theory of growing cell population. J. Math. Anal. Appl. **211**, 273–294 (1997)
196. Latrach, K.: Essential spectra on spaces with the Dunford-Pettis property. J. Math. Anal. Appl. **233**, 607–622 (1999)
197. Latrach, K., Lods, B.: Spectral analysis of transport equations with bounce-back boundary conditions. Math. Meth. Appl. Sci. **32**, 1325–1344 (2009)
198. Latrach, K., Megdiche, H., Taoudi, M.A.: A compactness result for perturbed semigroups and application to a transport model. J. Math. Anal. Appl. **359**, 88–94 (2009)
199. Laurençot, P., Perthame, B.: Exponential decay for the growth-fragmentation/cell-division equation. Commun. Math. Sci. **7**(2), 503–510 (2009)
200. Lax, P.D.: Symmetrizable linear transformations. Comm. Pure Appl. Math. **7**, 633–647 (1954)
201. Lax, P.D., Phillips, R.: Scattering theory for transport phenomena. In: Proceedings Conference on Functional Analysis, University of California, Irvine, Calif, pp. 119–130 (1967)
202. Lax, P.D.: Mathematics and physics. Bull. Am. Math. Soc. **45**(1), 135–152 (2008)
203. Lederman, W., Reuter, G.E.H.: On differential equations for the transition probabilities of Markov processes with denumerably many states. Proc. Cambridge Philos. Soc. **49**(2), 247–262 (1953)

204. Lehner, J., Wing, G.M.: On the spectrum of an unsymmetric operator arising in the transport theory of neutrons. Commun. Pure. Appl. Math. **8**, 217–234 (1955)
205. Lin, S.C.: Wave operators and similarity for generators of semigroups in Banach spaces. Trans. Am. Math. Soc. **139**, 469–494 (1969)
206. Littmann, W.: Fourier transforms of surface-carried measures and differentiability of surface averages. Bull. Am. Math. Soc. **69**, 766–770 (1963)
207. Lods, B.: On linear Kinetic equations involving unbounded cross-sections. Math. Meth. Appl. Sci. **27**, 1049–1075 (2004)
208. Lods, B., Sbihi, M.: Stability of the essential spectrum for 2D-transport models with Maxwell boundary conditions. Math. Meth. Appl. Sci. **29**, 499–523 (2006)
209. Lods, B.: Variational characterizations of the effective multiplication factor of a nuclear reactor core. Kinetic Relat. Models **2**, 307–331 (2009)
210. Lods, B., Mokhtar-Kharroubi, M., Sbihi, M.: Spectral properties of general advection operators and weighted translation semigroups. Commun. Pure. Appl. Anal. **8**(5), 1–24 (2009)
211. Lods, B., Mokhtar-Kharroubi, M.: Convergence to equilibrium for linear spatially homogeneous Boltzmann equation with hard and soft potentials: a semigroup approach in L^1 spaces. Math. Meth. Appl. Sci., 1–29 (2017)
212. Lods, B., Mokhtar-Kharroubi, M., Rudnicki, R.: Invariant density and time asymptotics for collisionless Kinetic equations with partly diffuse boundary operators. Ann. I. H. Poincaré, AN **37**, 877–923 (2020)
213. Lods, B., Mokhtar-Kharroubi, M.: Convergence rate to equilibrium for collisionless transport equations with diffuse boundary operators: a new Tauberian approach. J. Funct. Anal. **283**(10) (2022)
214. Lods, B., Mokhtar-Kharroubi, M.: On eventual compactness of collisionless Kinetic semigroups with velocities bounded away from zero. J. Evol. Eqn. **22**, 25 (2022)
215. Lods, B., Mokhtar-Kharroubi, M.: Convergence rate to equilibrium for conservative scattering models on the torus. a new Tauberian approach. Trans. Am. Math. Soc. **377**(4), 2741–2820 (2024)
216. Loskot, K., Rudnicki, R.: Relative entropy and stability of stochastic semigroups. Annales Polonici Mathematici, LIII. **2**, 139–145 (1991)
217. Lyons, R.: Seventy years of Rajchman measures. In: Proceedings of the Conference in Honor of Jean-Pierre Kahane (Orsay, 1993). J. Fourier Anal. Appl. (Special Issue), 363–377 (1995)
218. Magel, P., Ruan, S. (eds.): Structured Population Models in Biology and Epidemiology. Lecture Notes in Mathematics. Mathematical Biosciences Subseries, vol. 1936 (2008)
219. Majorana, A.: Trend to equilibrium of electron gas in a semiconductor according to the Boltzmann equation. Transp. Theory. Stat. Phys. **27**(5–7), 547–571 (1998)
220. Majorana, A., Milazo, C.: Space homogeneous solutions of the linear semiconductor Boltzmann equation. J. Math. Anal. Appl. **259**, 609–629 (2001)
221. Manavi, A., Vogt, H., Voigt, J.: A note on absorption semigroups and regularity. Archiv der Mathematik **106**, 485–488 (2016)
222. Marek, I.: Frobenius theory of positive operators: comparison theorems and applications. SIAM J. Appl. Math. **19**, 607–628 (1970)
223. Marek, I.: Fundamental decay mode and asymptotic behaviour of positive semigroups. Czechoslovak Math. J. **30**(4), 579–590 (1980)
224. Martcheva, M., Thieme, H.R., Dhirasakdanon, T.: Kolmogorov's differential equations and positive semigroups on first moment sequence spaces. J. Math. Biol. **53**(4), 642–671 (2006)
225. Martinez, J., Mazon, J.M.: Quasi-compactness of dominated positive operators and C_0-semigroups. Math. Z **207**, 109–120 (1991)
226. Mátrai, T.: On perturbations preserving the immediate norm continuity of semigroups. J. Math. Anal. Appl. **341**, 961–974 (2008)
227. Maz'ya, V., Shubin, M.: Discreteness of spectrum and positivity criteria for Schrödinger operators. Ann. Math. **162**(2), 919–942 (2005)
228. Mellet, A., Mischler, S., Mouhot, C.: Fractional diffusion limit for collisional Kinetic equations. Arch. Rat. Mech. Anal. **199**, 493–525 (2011)

229. Metafune, G., Pallara, D.: On the location of the essential spectrum of Schrödinger operators. Proc. Am. Math. Soc. **130**(6), 1779–1786 (2001)
230. Metz, J.A.J., Diekmann, O. (eds.): The Dynamics of Physiologically Structured Populations. Springer Lecture Notes in Biomathematics, vol. 68. Springer, New York (1986)
231. Michel, P., Mischler, S., Perthame, B.: General relative entropy inequality: an illustration on growth models. J. Math. Pures Appl. **84**(9), 1235–1260 (2005)
232. Miclo, L.: On hyperboundedness and spectrum of Markov operators. Invent. Math. **200**, 311–343 (2015)
233. Mika, J.: Time dependent Neutron Transport in plane geometry. Nucleonik, **9** Bd Helft 4, 200–205 (1967)
234. Mika, J.: Fundamental eigenvalues of the linear transport equation. J. Quant. Spectr. Radiat. Transfert **11**, 879–891 (1971)
235. Mischler, S., Scher, J.: Spectral analysis of semigroups and growth-fragmentation equations. Ann. Inst. H. Poincaré Anal. Non Linéaire, **33**(3), 849–898 (2016)
236. Mohamed, Y., Mokhtar-Kharroubi, M.: Spectral analysis of non-compact symmetrizable operators on Hilbert spaces. Math. Methods Appl. Sci. **38**(11), 2316–2335 (2015)
237. Mohamed, Y., Mokhtar-Kharroubi, M.: Fine spectral analysis of isotropic partly elastic neutron transport operators. Acta Appl. Math. **156**, 33–78 (2018)
238. Mokhtar-Kharroubi, M.: La compacité dans la théorie du transport des neutrons. C. R. Acad. Sci., Paris, Ser I, **303**, 617–619 (1986)
239. Mokhtar-Kharroubi, M.: Spectral theory of the neutron transport operator in bounded geometries. Transp. Theory. Stat. Phys. **16**, 467–502 (1987)
240. Mokhtar-Kharroubi, M.: Les équations de la neutronique: Positivité, Compacité, Théorie spectrale, Comportement asymptotique en temps. Thèse d'Etat (former french habilitation), Paris (1987)
241. Mokhtar-Kharroubi, M.: Spectral theory of the multigroup neutron transport operator. Eur J. Mech. B/Fluids **9**(2), 197–222 (1990)
242. Mokhtar-Kharroubi, M.: Compactness properties for positive semigroups on Banach lattices and applications. Houston J. Math. **117**(1), 25–38 (1991)
243. Mokhtar-Kharroubi, M.: Limiting absorption principles and wave operators on $L^1(\mu)$ spaces. Application to transport theory. J. Funct. Anal. **115**, 119–145 (1993)
244. Mokhtar-Kharroubi, M.: BV characterization of Miyadera perturbations of C_0-semigroups on Banach spaces. In: Workshop "Evolution Equations, Control Theory, and Biomathematics", Luminy, France, 8–12 March 1993
245. Mokhtar-Kharroubi, M.: Mathematical Topics in Neutron Transport Theory. New aspects, vol. 46. World Scientific (1997)
246. Mokhtar-Kharroubi, M., Thevenot, L.: On the diffusion theory of Neutron Transport on the torus. Asymp. Anal. **30**(3–4), 273–300 (2002)
247. Mokhtar-Kharroubi, M.: On the essential spectrum of transport operators in L^1 spaces. Prépublication du Laboratoire de Mathématiques de Besançon, vol. 27 (2002)
248. Mokhtar-Kharroubi, M.: On the leading eigenvalue of neutron transport models. Prépublication du Laboratoire de Mathématiques de Besançon, vol. 42 (2003)
249. Mokhtar-Kharroubi, M.: On the convex compactness property for the strong operator topology and related topics. Math. Methods Appl. Sci. **27**(6), 687–701 (2004)
250. Mokhtar-Kharroubi, M.: Homogenization of boundary value problems and spectral problems for Neutron Transport in locally periodic media. Math. Models Methods Appl. Sci. **14**(01), 47–78 (2004)
251. Mokhtar-Kharroubi, M.: On L^1-spectral theory of Neutron Transport. Diff. Int. Eqn. **18**(11), 1221–1242 (2005)
252. Mokhtar-Kharroubi, M.: Optimal spectral theory of the linear Boltzmann equation. J. Funct. Anal. **226**, 21–47 (2005)
253. Mokhtar-Kharroubi, M., Sbihi, M.: Critical spectrum and spectral mapping theorems in transport theory. Semigroup Forum **70**, 406–435 (2005)

254. Mokhtar-Kharroubi, M., Sbihi, M.: Spectral mapping theorems for Neutron Transport, L^1-theory. Semigroup Forum **72**, 249–282 (2006)
255. Mokhtar-Kharroubi, M.: Spectral properties of a class of positive semigroups on Banach lattices and streaming operators. Positivity **10**(2), 231–249 (2006)
256. Mokhtar-Kharroubi, M.: On the leading eigenvalue of neutron transport models. J. Math. Anal. Appl. **315**, 263–275 (2006)
257. Mokhtar-Kharroubi, H., Mokhtar-Kharroubi, M.: On symmetrizable operators on Hilbert spaces. Acta. Appl. Math. **102**(1), 1–24 (2008)
258. Mokhtar-Kharroubi, M.: On perturbed positive semigroups on the Banach space of trace class operators. In: Infinite Dimensional Analysis, Quantum Probability and Related Topics, vol. 11, pp. 405–425 (2008)
259. Mokhtar-Kharroubi, M.: On collisionless transport semigroups with boundary operators of norm one. J. Evol. Eqn. **8**, 327–362 (2008)
260. Mokhtar-Kharroubi, M.: On Schrödinger semigroups and related topics. J. Funct. Anal. **256**, 1998–2025 (2009)
261. Mokhtar-Kharroubi, M., Voigt, J.: On honesty of perturbed substochastic C_0-semigroups in L^1-spaces. J. Oper. Theory **64**(1), 131–147 (2010)
262. Mokhtar-Kharroubi, M.: On permanent regimes for non-autonomous linear evolution equations in Banach spaces with applications to transport theory. Kinetic. Relat. Models **3**(3), 473–499 (2010)
263. Mokhtar-Kharroubi, M.: Perturbation theory for convolution semigroups. J. Funct. Anal. **259**, 780–816 (2010)
264. Mokhtar-Kharroubi, M.: New generation theorems in transport theory. Afrika Matematika **22**, 153–176 (2011)
265. Mokhtar-Kharroubi, M., Salvarani, F.: Convergence rates to equilibrium for neutron chain fissions. Acta. Appl. Math. **113**, 145–165 (2011)
266. Mokhtar-Kharroubi, M.: On L^1 exponential trend to equilibrium for conservative linear Kinetic equations on the torus. J. Funct. Anal. **266**(11), 6418–6455 (2014)
267. Mokhtar-Kharroubi, M.: On some measure convolution operators in neutron transport theory. Acta. Appl. Math. **134**(1), 1–20 (2014)
268. Mokhtar-Kharroubi, M.: Spectral theory for Neutron Transport. In: Banasiak, J., Mokhtar-Kharroubi, M. (eds.) Evolutionary Equations with Applications in Natural Sciences. Lectures Notes in Mathematics N^0, vol. 2126, pp. 319–386. Springer (2015)
269. Mokhtar-Kharroubi, M.: On strong convergence to ergodic projection for perturbed substochastic semigroups. In: Semigroups of Linear Operators—Theory and Applications, vol. 113, Bedlewo, Poland, 2013, pp. 89–103. Springer (2015)
270. Mokhtar-Kharroubi, M.: On L^1 asymptotics of conservative partly elastic neutron transport semigroups on the torus, talk given in the Conference "Modern Applications of Operator Theory" held in Bedlewo, Poland, April 11–16, 2016
271. Mokhtar-Kharroubi, M.: Compactness properties of perturbed sub-stochastic C_0-semigroups on $L^1(\mu)$ with applications to discreteness and spectral gaps. Mém. Soc. Math. France (148) (2016)
272. Mokhtar-Kharroubi, M.: Contractivity results in ordered spaces. Applications to relative operator bounds and projections with norm one. Math. Nachr. **290**, 1732–1752 (2017)
273. Mokhtar-Kharroubi, M., Rudnicki, R.: On asymptotic stability and sweeping of collisionless Kinetic equations. Acta. Appl. Math. **147**, 19–38 (2017)
274. Mokhtar-Kharroubi, M., Seifert, D.: Rates of convergence to equilibrium for collisionless Kinetic equations in slab geometry. J. Funct. Anal. **275**, 2404–2452 (2018)
275. Mokhtar-Kharroubi, M., Richard, Q.: Time asymptotics of structured populations with diffusion and dynamic boundary conditions. Discrete Cont. Dyn. Syst., Ser. B **23**(10), 4087–4116 (2018)
276. Mokhtar-Kharroubi, M.: Form-perturbation theory for higher-order elliptic operators and systems by singular potentials. Philos. Trans. R. Soc., A **378**, 20190621 (2020)

277. Mokhtar-Kharroubi, M., Richard, Q.: Spectral theory and time asymptotics of size-structured two-phase population models. Discrete Cont. Dyn. Syst., Ser. B **25**(8), 2969–3004 (2020)

278. Mokhtar-Kharroubi, M.: Spectra of structured diffusive population equations with generalized Wentzell-Robin boundary conditions and related topics. Discrete Cont. Dyn. Syst., Ser. S **13**(12), 3551–3563 (2020)

279. Mokhtar-Kharroubi, M.: Existence of invariant densities and time asymptotics of conservative linear Kinetic equations on the torus without spectral gaps. Acta. Appl. Math. **175**, 8 (2021)

280. Mokhtar-Kharroubi, M.: On spectral gaps of growth-fragmentation semigroups with mass loss or death. Commun. Pure. Appl. Anal. **21**(4), 1293–1327 (2022)

281. Mokhtar-Kharroubi, M.: Spectral gap and irreducibility for collisional transport semigroups with general vector fields. (2025). hal-05390754

282. Montagnini, B.: Existence of complex eigenvalues for the mono energetic neutron transport operator. Transp. Theory Stat. Phys. **5**, 127–167 (1976)

283. Montagnini, B.: The eigenvalue spectrum of the linear Boltzmann operator in $L^1(R^N)$ and $L^2(R^N)$. Meccanica **14**, 134–144 (1979)

284. Mouhot, C., Russ, E., Sire, Y.: Fractional Poincaré inequalities for general measures. J. Math. Pures Appl. **95**, 72–84 (2011)

285. Munteanua, O., Wang, J.: Geometry of manifolds with densities. Adv. Math. **259**, 269–305 (2014)

286. Nagel, R. (ed.): One-Parameter Semigroups of Positive Operators. Lecture Notes in Mathematics, vol. 1184 (1986)

287. Nagel, R., Poland, J.: The critical spectrum of a strongly continuous semigroup. Adv. Math. **152**, 120–133 (2000)

288. Nagel, R., Rhandi, A. (ed.): Semigroup and applications everywhere. Philos. Trans. Roy Soc. A. Math., Phys. Eng. Sci. **378**(2185) (2020)

289. Nelkin, M.S.: Neutron thermalization. In: Birkhoff, G., Wigner, E. (eds.) Nuclear Reactor Theory. Proceedings of the Symposium in Applied Mathematics, vol. XI. American Mathematical Society, Providence (1961)

290. Nickel, G.: A new look at boundary perturbations of generators. Electron. J. Diff. Eqn. **2004**(95), 1–14 (2004)

291. Norton, R.V.: On the real spectrum of monoenergetic neutron transport operator. Commun. Pure. Appl. Math. **15**, 149–158 (1962)

292. Oinarov, R.: On the separability of the Schrödinger operator in the space of summability functions. Dokl Akad Nauk SSSR **285**, 1062–1064 (1985)

293. Pal, I.: Statistical theory of neutron chain reactions. Acta. Phys. Hung **21**, 390 (1962)

294. Palczewski, A.: Spectral properties of the space nonhomogeneous linearized Boltzmann operator. Transp. Theory. Stat. Phys. **13**(3–4), 409–430 (1984)

295. Palczewski, A.: Evolution operators generated by the space and time nonhomogeneous linearized Boltzmann operator. Transp. Theory. Stat. Phys **14**(1), 1–33 (1985)

296. Pankov, A.: Lecture Notes on Schrödinger Equations. Contemporary Mathematical Studies. Nova Science Publishers Inc. (2007)

297. Papanicolaou, G.C.: Asymptotic analysis of transport processes. Bull. Am. Math. Soc. **81**, 330–392 (1975)

298. Pazy, A., Rabinowitz, P.: A nonlinear integral equation with applications to neutron transport theory. Arch. Rat. Mech. Anal. **32**, 226–246 (1969)

299. Pazy, A., Rabinowitz, P.: On a branching process in neutron transport theory. Arch. Rat. Mech. Anal. **51**, 153–164 (1973)

300. Pelczynski, A.: On strictly singular and strictly cosingular operators. II. Strictly singular and strictly cosingular operators in $L^1(\nu)$-spaces. Bull. Acad. Polon. Sci. Ser. Sci. Math. Astronom. Phys. **13**, 7–41 (1965)

301. Persson, A.: Bounds for the discrete part of the spectrum of a semi-bounded Schrödinger operator. Mathematica Scand. **8**, 143–153 (1960)

302. Perthame, B.: Transport Equations in Biology. Frontiers in Mathematics. Birkhäuser Verlag, Basel (2007)

303. Perthame, B.: Mathematical tools for Kinetic equations. Bull. (New Series) of AMS, **41**(2), 205–244 (2004)
304. Pettersson, R.: On weak and strong convergence to equilibrium for solution to the linear Boltzmann equation. J. Stat. Phys. **72**(1/2), 355–380 (1993)
305. Pichor, K., Rudnicki, R.: Continuous Markov semigroups and stability of transport equations. J. Math. Anal. Appl. **249**, 668–685 (2000)
306. Pichor, K., Rudnicki, R.: Asymptotic decomposition of substochatic operators and semigroups. J. Math. Anal. Appl. **436**, 305–321 (2016)
307. Pimbley, G.H.: Solution of the initial value problem for the multivelocity neutron transport equation with a slab geometry. J. Math. Mech. **8** (1958)
308. Poupaud, F.: Runaway phenomena and fluid approximation under high fields in semiconductor Kinetic theory. Z. Angew. Math. Mech. **72**(8), 359–372 (1992)
309. Protopopescu, V.: On the scattering matrix for the linear Boltzmann equation. Rev. Roum. Phys. **21**, 991–994 (1976)
310. Ragozin, D.: Rotation invariant measure algebras on Euclidean space. Indiana. Univ. Math. J. **23**, 1139–1154 (1973/74)
311. Reed, M., Simon, B.: Methods of Modern Mathematical Physics: Fourier Analysis. Self-adjointness. II. Academic Press, New York (1975)
312. Reid, W.T.: Symmetrizable completely continuous transformations in Hilbert spaces. Duke. Math. J. **18**, 41–56 (1951)
313. Reuter, G.E.H.: Denumerable Markov processes and the associated contraction semigroup. Acta. Math. **97**, 1–46 (1957)
314. Rhandi, A.: Dyson-Phillips expansions and unbounded perturbations of linear C_0-semigroups. J. Comp. Appl. Math. **44**, 339–349 (1992)
315. Ribaric, M., Vidav, I.: Analytic properties of the inverse $A(z)^{-1}$ of an analytic linear operator valued function $A(z)$. Arch. Rat. Mech. Anal. **32**, 298–310 (1969)
316. Ricard, E.: An inequality in noncommutative Lp-spaces. J. Math. Anal. Appl. **438** (2016)
317. Rocha, A.: Essential spectrum of the weighted Laplacian on noncompact manifolds and applications. Geom Dedicata **186**, 197–219 (2017)
318. Rudnicki, R.: On asymptotic stability and sweeping for Markov operators. Bull. Pol. Acad. Sci., Math. **43**, 245–262 (1995)
319. Rudnicki, R.: Stochastic semigroups and their applications in physics and biology. In: Lecture given in CIMPA School "Evolutionary Equations with Applications in Natural Sciences", South Africa, Muizenberg, July 22–August 2, 2013
320. Rudnicki, R., Tyran-Kaminska, M.: Piecewise Deterministic Processes in Biological Models. Springer Briefs in Applied Sciences and Technology. Mathematical Methods. Springer (2017)
321. Saint-Raymond, L.: Hydrodynamic Limits of the Boltzmann Equation. Lecture Notes in Mathematics, vol. 1971 (2009)
322. Sanchez, R.: The criticality eigenvalue problem for the transport with general boundary conditions. Transp. Theory. Stat. Phys. **35**, 159–185 (2006)
323. Sawashima, I.: On spectral properties of some positive operators. Natural Science Report, Ochanomizu University, vol. 15, no. 2, pp. 53–64 (1964)
324. Sbihi, M.: Spectral theory of neutron transport semigroups with partly elastic collision operators. J. Math. Phys. **47** (2006)
325. Sbihi, M.: A resolvent approach to the stability of essential and critical spectra of perturbed C_0-semigroups on Hilbert spaces with applications to transport theory. J. Evol. Eqn. **7**(1), 35–58 (2007)
326. Schaefer, H.: Some spectral properties of positive operators. Pac. J. Math. **10**, 1009–1019 (1960)
327. Schaefer, H.: Banach Lattices and Positive Operators. Springer (1974)
328. Schaefer, H.: Existence of spectral values for irreducible C_0-semigroups. J. Funct. Anal. **74**, 139–145 (1987)
329. Schappacher, W.: Scattering theory for the linear Boltzmann equation. Ber. Math. Stat. Sekt. Forschungszentrum, **69** (1976)

330. Schluchtermann, G.: On weakly compact operators. Math. Ann. **292**, 263–266 (1992)
331. Schluchtermann, G.: Perturbation of linear semigroups. In: Recent Progress in Operator Theory (Regensburg 1995). Oper. Theory Adv. Appl. **103**, 263–277, Birkhauser, Basel (1998)
332. Schwartz, J.: The pernicious influence of mathematics on science. In: Hersh, R. (ed.) 18 Unconventional Essays on the Nature of Mathematics. Springer Science & Business Media Inc., New York (2006)
333. Shizuta, Y.: On fundamental equations of spatially independent problems in neutron thermalization theory. Progr. Theor. Phys. **32**(4), 489–511 (1964)
334. Shizuta, Y.: On the classical solutions of the Boltzmann equation. Commun. Pure. Appl. Math. **36**, 705–754 (1983)
335. Shore, S.N.: Blue sky and hot piles: the evolution of radiative transfer theory from atmospheres to nuclear reactors. Historia Math. **29**, 463–489 (2002)
336. Silvares, L.: On the essential spectrum of the Laplacian and the drifted Laplacian. J. Funct. Anal. **266**, 3906–3936 (2014)
337. Simon, B.: Existence of the scattering matrix for the linearized Boltzmann equation. Commun. Math. Phys. **41**, 99–108 (1975)
338. Simon, B.: Schrödinger semigroups. Bull. Am. Math. Soc. **7**, 447–526 (1982)
339. Sjöstrand, J.: Correlation asymptotics and Witten Laplacians. St Petersburg. Math. J. **8**, 123–148 (1997)
340. Song, D., Greenberg, W.: Spectral properties of transport equations for slab geometry in L^1 with reentry boundary conditions. Transp. Theory. Stat. Phys. **30**(4–6), 325–355 (2001)
341. Stefanov, P.: Spectral and scattering theory for the linear Boltzmann equation in exterior domain. Math. Nachr. **137**, 63–77 (1988)
342. Stewart, H.B.: Spectral theory of heterogeneous diffusion systems. J. Math. Anal. Appl. **54**, 59–78 (1976)
343. Suhadolc, A.: Linearized Boltzmann equation in L^1 spaces. J. Math. Anal. Appl. **35**, 1–13 (1971)
344. Takak, P.: A spectral mapping theorem for the exponential function in linear transport theory. Transp. Theory Stat. Phys. **14**(5), 655–667 (1985)
345. Teschl, G.: Mathematical Methods in Quantum Mechanics with Applications to Schrodinger Operators. Graduate Studies in Mathematics, vol. 99. American Mathematical Society (2009)
346. Thieme, H.R.: Positive perturbation of operator semigroups: growth bounds, essential compactness and asynchronous exponential growth. Discrete Cont. Dyn. Syst. **4**(4), 735–764 (1998)
347. Thieme, H.R., Voigt, J.: Stochastic semigroups: their construction by perturbation and approximation. In: Proceedings Positivity IV—Theory and Applications, Dresden, pp. 135–146 (2006)
348. Thieme, H.R., Voigt, J.: Relatively bounded extensions of generator perturbations. Rocky. Mountain. J. Math. **39**(3), 947–969 (2009)
349. Tyran-Kaminska, M.: Ergodic theorems and perturbations of contraction semigroups. Studia. Math. **195**(2), 147–155 (2009)
350. Tyran-Kaminska, M.: Substochastic semigroups and densities of piecewise deterministic Markov processes. J. Math. Anal. Appl. **357**, 385–402 (2009)
351. Uchiyama, J.: On the spectra of integral operators connected with Boltzmann and Schrödinger operators. Publ. Rims. Kyoto Univ. Ser. A **3**, 101–127 (1967)
352. Ukaï, S.: On the spectrum of the space-independent Boltzmann operator. J. Nucl. Energy, Part A/B, **19**, 833–848 (1965)
353. Ukaï, S.: Real eigenvalues of the monoenergetic transport operator for a homogeneous medium. J. Nucl. Sci. Technol. **3**, 263–266 (1966)
354. Ukaï, S.: Eigenvalues of the neutron transport operator for a homogeneous finite moderator. J. Math. Anal. Appl. **30**, 297–314 (1967)
355. Ukaï, S., Yang, T.: Mathematical Theory of Boltzmann Equation. Lecture Notes Series, vol. 8, Liu Bie Ju Center of Mathematical Sciences, City University of Hongkong, Hongkong (2006)

356. Verwaerde, D.: Une approche non deterministe de la Neutronique: Modelisation. Note CEA, vol. 2731 (1993)
357. Verwaerde, D.: Une approche non deterministe de la Neutronique. Existence d'une solution aux equations de probabilité de présence. Note CEA, vol. 2791 (1995)
358. Vidav, I.: Existence and uniqueness of nonnegative eigenfunctions of the Boltzmann operator. J. Math. Anal. Appl. **22**, 144–155 (1968)
359. Vidav, I.: Spectra of perturbed semigroups with application to transport theory. J. Math. Anal. Appl. **30**, 264–279 (1970)
360. Villani, C.: A review of mathematical topics in collisional Kinetic theory. In: Handbook of Mathematical Fluid Dynamics, vol. I, 71-305, North-Holland, Amsterdam (2002)
361. Villani, C.: H-Theorem and beyond: Boltzmann's entropy in today's mathematics. In: Conference Proceedings "Boltzmann's Legacy", Erwin-Schrödinger Institute, Vienna, July 2007
362. Voigt, J.: On the perturbation theory for strongly continuous semigroups. Math. Ann. **229**, 163–171 (1977)
363. Voigt, J.: On the existence of the scattering operator for the linear Boltzmann equation. J. Math. Anal. Appl. **58**, 541–558 (1977)
364. Voigt, J.: A perturbation theorem for the essential spectral radius of strongly continuous semigroups. Monatsh. Math. **90**, 153–161 (1980)
365. Voigt, J.: Functional analytic treatment of the initial boundary value problem for collisionless gases. Habilitationsschrift, München (1980)
366. Voigt, J.: Stochastic operators, information and entropy. Commun. Math. Phys. **81**, 31–38 (1981)
367. Voigt, J.: Positivity in time dependent linear transport theory. Acta. Appl. Math. **2**, 311–331 (1984)
368. Voigt, J.: Spectral properties of the neutron transport equation. J. Math. Anal. Appl. **106**, 140–153 (1985)
369. Voigt, J.: On substochastic C_0-semigroups and their generators. Transp. Theory Stat. Phys. **16**, 453–466 (1987)
370. Voigt, J.: Absorption semigroups, their generators and Schrödinger semigroups. J. Funct. Anal. **67**, 167–205 (1986)
371. Voigt, J.: On resolvent positive operators and positive C_0-semigroups in AL-spaces. Semigroup Forum **38**, 263–266 (1989)
372. Voigt, J.: On the convex compactness property for the strong operator topology. Note di Math. XI **I**, 259–269 (1992)
373. Voigt, J.: Stability of essential type of strongly continuous semigroups. Proc. Steklov Inst. Math. **3**, 383–389 (1994)
374. Webb, G.F.: An operator-theoretic formulation of asynchronous exponential growth. Trans. Am. Math. Soc. **303**, 751–763 (1987)
375. Weis, L.: A generalization of the Vidav-Jorgens perturbation theorem for semigroups and its application to transport theory. J. Math. Anal. Appl. **129**, 6–23 (1988)
376. Weis, L.: A short proof for the stability theorem for positive semigroups on $L^p(\mu)$. Proc. Am. Math. Soc. **126**, 3253–3256 (1998)
377. Wigner, E.P.: Mathematical problems of nuclear reactor theory. In: Birkhoff, G., Wigner, E. (eds.) Nuclear Reactor Theory. Proceedings of Symposia in Applied Mathematics, vol. XI. American Mathematical Society, Providence (1961)
378. Wing, G.M.: An Introduction to Transport Theory. Wiley, Inc (1962)
379. Wong, C.P.: New approaches to honesty theory and applications in quantum dynamical semigroups. J. Oper. Theory **75**(2), 443–474 (2016)
380. Wong, C.P.: Stochastic completeness and honesty. J. Evol. Eqn. **15**, 961–978 (2015)
381. Xianwen, Z.: The spectrum of the linear Boltzmann operator with an external field. Transp. Theory Stat. Phys. **29**(6), 699–710 (2000)
382. Yang, T., Yu, H.: Spectrum analysis of some Kinetic equations. Arch. Rat. Mech. Anal. **222**, 731–768 (2016)
383. Zerner, M.: Quelques propriétés spectrales des opérateurs positifs. J. Funct. Anal. **72**, 381–417 (1987)

Index

LECTURE NOTES IN MATHEMATICS   Springer

Editors in Chief: J.-M. Morel, B. Teissier

Editorial Policy

1. Lecture Notes aim to report new developments in all areas of mathematics and their applications
   – quickly, informally and at a high level. Mathematical texts analysing new developments in
   modelling and numerical simulation are welcome. See Lecture Notes in Mathematics for the
   existing online volumes of LNM.

   Manuscripts should be reasonably self-contained and rounded off. Thus, they may, and often
   will, present not only results of the author but also related work by other people. They may
   be based on specialised lecture courses. Furthermore, the manuscripts should provide sufficient
   motivation, examples and applications. This clearly distinguishes Lecture Notes from journal
   articles or technical reports which normally are very concise. Articles intended for a journal
   but too long to be accepted by most journals, usually do not have this "lecture notes" character.
   For similar reasons it is unusual for doctoral theses to be accepted for the Lecture Notes series,
   though habilitation theses may be appropriate.

2. Besides monographs, multi-author manuscripts resulting from SUMMER SCHOOLS or similar
   INTENSIVE COURSES are welcome, provided their objective was held to present an active
   mathematical topic to an audience at the beginning or intermediate graduate level (a list of
   participants should be provided).

   The resulting manuscript should not be just a collection of course notes but should require
   advance planning and coordination among the main lecturers. The subject matter should dictate
   the structure of the book. This structure should be motivated and explained in a scientific introduction, and the notation, references, index and formulation of results should be, if possible,
   unified by the editors. Each contribution should have an abstract and an introduction referring
   to the other contributions. In other words, more preparatory work must go into a multi-authored
   volume than simply assembling a disparate collection of papers, communicated at the event.

3. Manuscripts should be submitted online either at www.editorialmanager.com/lnm to Springer's
   mathematics editorial in Heidelberg, or electronically to one of the series editors. Authors should
   be aware that incomplete or insufficiently close-to-final manuscripts almost always result in
   longer refereeing times and nevertheless unclear referees' recommendations, making further
   refereeing of a final draft necessary. The strict minimum amount of material that will be considered should include a detailed outline describing the planned contents of each chapter, a
   bibliography and several sample chapters. Parallel submission of a manuscript to another pub-
   lisher while under consideration for LNM is not acceptable and can lead to rejection.

4. The type of material considered for publication includes:

   - Research monographs
   - Lectures on a new field or presentations of a new angle in a classical field
   - Summer schools and intensive courses on topics of current research

In general, **monographs** will be sent out to at least 2 external referees for evaluation.

A final decision to publish can be made only on the basis of the complete manuscript, however a refereeing process leading to a preliminary decision can be based on a pre-final or incomplete manuscript.

Volume Editors of **multi-author works** are expected to arrange for the refereeing, to the usual scientific standards, of the individual contributions. If the resulting reports can be forwarded to the LNM Editorial Board, this is very helpful. If no reports are forwarded or if other questions remain unclear in respect of homogeneity etc., the series editors may wish to consult external referees for an overall evaluation of the volume.

5. Manuscripts should in general be submitted in English. Final manuscripts should contain at least 100 pages of mathematical text and should always include:

   - A table of contents
   - An informative introduction, with adequate motivation and perhaps some historical remarks: it should be accessible to a reader not intimately familiar with the topic treated
   - A subject index: as a rule, this is genuinely helpful for the reader
   - For evaluation purposes, manuscripts should be submitted as pdf files.

6. Careful preparation of the manuscripts will help keep production time short besides ensuring satisfactory appearance of the finished book in print and online. After acceptance of the manuscript authors will be asked to prepare the final LaTeX source files plus the corresponding pdf- or zipped ps-file (Preparing your Manuscript). You can also download our LaTeX package for monographs. The LaTeX source files are essential for producing the full-text online version of the book. For more detailed information, please visit the Book Manuscript Guidelines and Key Style Points. The technical production of a Lecture Notes volume takes approximately 12 weeks. Additional instructions, if necessary, are available on request from lnm@springer.com.

7. Authors receive a total of 30 free copies of their volume and free access to their book on SpringerLink, but no royalties. They are entitled to a discount of 40 % on the price of Springer books purchased for their personal use, if ordering directly from Springer.

8. Commitment to publish is made by a *Publishing Agreement*; contributing authors of multi-author books are requested to sign a *Consent to Publish form*. Springer-Verlag registers the copyright for each volume. Authors are free to reuse material contained in their LNM volumes in later publications: a brief written (or e-mail) request for formal permission is sufficient.

Addresses:

Professor Jean-Michel Morel, City University Hongkong, Kowloon Tong, Hong Kong
E-mail: moreljlmichel@gmail.com

Professor Bernard Teissier, Equipe Géométrie et Dynamique,
Institut de Mathématiques de Jussieu – Paris Rive Gauche, Paris, France
E-mail: bernard.teissier@imj-prg.fr

Springer: Ute McCrory, Editor, Mathematics, Heidelberg, Germany.
E-mail: lnm@springer.com